* Should the group appoint a coordinator for the project?
- Do you have collaborative software? How will you share the document as it develops?
- Where in your organizational plan can you begin confidently?

Revision

- Have you stated clearly and specifically the purpose of the report?
- Have you put into the report everything required? Do you have sufficient supporting evidence? Have you stated the implications of your information clearly?
- Are all your facts and numbers accurate?
- Have you answered the questions your readers are likely to have?
- Does the report contain anything that you would do well to cut out?
- Does your organization suit the needs of your content and your audience?
- Are your paragraphs clear, well organized, and of reasonable length? Are there suitable transitions from one point to the next?
- Is your prose style clear and readable?
- Is your tone appropriate to your audience?
- Have you satisfied the needs of an international audience?
- Are all your statements ethical? For example, have you avoided making ambiguous statements or statements that deliberately lead the reader to faulty inferences?
- Are your graphs and tables clear and accurate? Are they well placed? Do they present your information honestly?
- Is your document readable, accessible, and visually effective?
- Are there people you should share your draft with—for example, members of the target audience—before going on to a final draft?

Editing

- Have you checked thoroughly for misspellings and other mechanical errors?
- Have you included all the formal elements that your report needs?
- Are design elements such as headings, margins, spacing, typefaces, and documentation consistent throughout the draft?
- Are your headings and titles clear, properly worded, and parallel? Do your headings in the text match those in the table of contents?
- Is your documentation system the one required? Have you documented wherever appropriate? Do the numbers in the text match those in the notes?
- Have you keyed the tables and figures into your text, and have you sufficiently discussed these visual items?
- Are all parts and pages of the manuscript in the correct order?
- Will the format of the typed or printed report be functional, clear, and attractive?
- Does your manuscript satisfy stylebook specifications governing it?
- Have you included required notices, distribution lists, and identifying code numbers?
- Do you have written permission to reproduce extended quotations or other matter under copyright? (Permission is necessary only when your work is to be published or copyrighted.)
- While you were composing the manuscript, did you have any doubts or misgivings that you should now address?
- Have you edited your manuscript for matters both large and small?
- What remains to be done (such as proofreading the final copy)?

How to Use *Reporting Technical Information*

One of the leading texts in technical writing, *Reporting Technical Information* introduces you to all aspects of technical communication, including letters, proposals, progress reports, recommendation reports, research reports, instructions, and oral reports. Continuing the esteemed tradition of its predecessors, the eleventh edition provides a solid foundation in technical communication and offers material on the most recent developments in the field.

Integrated throughout the text are icons that direct you to the companion Web site for *Reporting Technical Information* (www.oup.com/us/houp). The Web site offers additional sample documents that both supplement and extend the text's examples, including other report types, such as procedure manuals, full-length feasibility studies, and environmental impact statements. It also features tutorials, document design templates, practice quizzes, checklists, interactive exercises, and annotated links for each chapter, providing effective and engaging tools to help you become a better communicator.

The following pages provide an overview of the additional resources offered by the Web site and show how they are integrated with the textbook using the margin icons, offering a fully interactive experience.

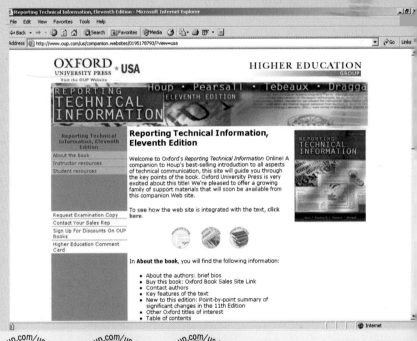

>>Follow the icons  When you see these icons in the text margin, go to the companion site and you will find further resources relevant to that section.

ICON	BOOK FEATURE	WEB RESOURCE
	>> Within each chapter the authors have provided an array of valuable document examples to illustrate their points. These documents include those published in traditional formats as well as those that reflect changes in technology and an increasing use of Web-based platforms.	>> This icon indicates where on the Web site you will find sample documents relevant to each chapter. 

>>

ICON	BOOK FEATURE	WEB RESOURCE

>> Following each chapter is a list of exercises to help you test your knowledge of the subject.

>> This icon indicates where on the Web site you will find a variety of items that will engage you in the chapter content and help you study. This includes multiple-choice self-tests, interactive scenarios, and key terms and concepts available in flashcard format.

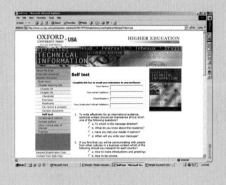

>> Throughout the book, the authors point out how important it is to consider your audience as you create a document.

>> Interactive scenarios provide the opportunity to create documents geared to specific audiences and let you see at a glance how a specific document would change for different audiences.

ICON	BOOK FEATURE	WEB RESOURCE
www.oup.com/us/houp	>> Within each chapter, key terms and concepts have been highlighted to further your learning of the chapter subject.	>> Flashcards let you test your knowledge of chapter terms and concepts.
www.oup.com/us/houp CHECKLIST	>> Throughout the text, the authors ask useful questions to help you think about your audience before you begin to write.	>> This icon indicates where on the Web site you will find downloadable versions of checklists highlighting important points in each chapter that will guide you as you create a document.

ICON	BOOK FEATURE	WEB RESOURCE
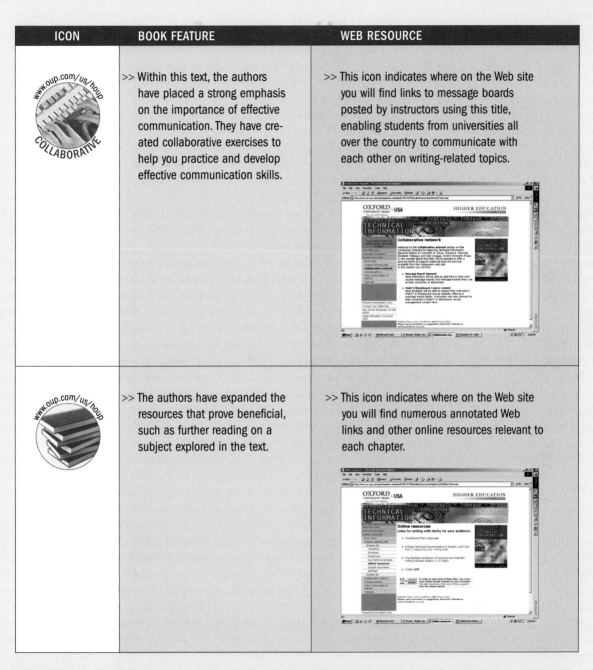	>> Within this text, the authors have placed a strong emphasis on the importance of effective communication. They have created collaborative exercises to help you practice and develop effective communication skills.	>> This icon indicates where on the Web site you will find links to message boards posted by instructors using this title, enabling students from universities all over the country to communicate with each other on writing-related topics.
	>> The authors have expanded the resources that prove beneficial, such as further reading on a subject explored in the text.	>> This icon indicates where on the Web site you will find numerous annotated Web links and other online resources relevant to each chapter.

The companion Web site also contains numerous resources and materials for instructors. In particular, instructors using *Reporting Technical Information* have submitted examples of their most interesting and successful writing assignments for posting to our companion Web site, where they can be shared with colleagues. You can find this valuable resource within the Better Writing—Success at Work section of the Instructor Resources.

If you are an instructor and would like more information about what is available, please visit www.oup.com/us/houp and click on Instructor Resources.

Reporting
Technical Information

ELEVENTH EDITION

Reporting
Technical Information

Kenneth W. Houp Late, The Pennsylvania State University

Thomas E. Pearsall Emeritus, University of Minnesota

Elizabeth Tebeaux Texas A&M University

Sam Dragga Texas Tech University

New York ■ Oxford
OXFORD UNIVERSITY PRESS
2006

Oxford University Press, Inc., publishes works that further Oxford University's
objective of excellence in research, scholarship, and education.

Oxford   New York
Auckland   Cape Town   Dar es Salaam   Hong Kong   Karachi
Kuala Lumpur   Madrid   Melbourne   Mexico City   Nairobi
New Delhi   Shanghai   Taipei   Toronto

With offices in
Argentina   Austria   Brazil   Chile   Czech Republic   France   Greece
Guatemala   Hungary   Italy   Japan   Poland   Portugal   Singapore
South Korea   Switzerland   Thailand   Turkey   Ukraine   Vietnam

Published by Oxford University Press, Inc.
198 Madison Avenue, New York, New York 10016
http://www.oup.com

Library of Congress Cataloging-in-Publication Data

Reporting technical information / Kenneth W. Houp ... [et al.].—11th ed.
    p. cm.
    ISBN 13: 978-0-19-517879-1

    1. Technical writing. I. Houp, Kenneth W., 1931-

T11.R426 2005
808'.0666—dc22

                           2005045083

Excerpts regarding the Cisco 800 Series Routers Installations instructions from Cisco Web site. Reproduced with the permission of
Cisco Systems, Inc. Copyright © 2004 Cisco Systems, Inc. All rights reserved. Excerpts from *DNA: The Secret of Life* by James Watson
and Andrew Berry. Copyright © 2003. Used by permission of Alfred A. Knopf, a division of Random House, Inc. Excerpts from
Final Environmental Impact Statement courtesy of the Virginia Department of Transportation. Excerpts from Hampton Roads
Crossing Study courtesy of the Virginia Department of Transportation. Excerpts from I-95/I-395 HOV Restriction Study courtesy of
the Virginia Department of Transportation. Excerpts from Northern Virginia Park and Ride Feasibility Study courtesy of the
Virginia Department of Transportation. Excerpts from "Reducing the Risk of Groundwater Contamination by Improving Pesticide
Storage and Handling" by Harris, Hoffman, and Mazec, courtesy of The Texas Cooperative Extension, Texas A&M University.
"Resin Shell Moulding" from *Precision Casting Processes* by A. J. Clegg. Reprinted by permission of Butterworth Heinmann, 1991.

Cover art: (left) from the U.S. Department of Energy, Office of Health and Environmental Research, *To Know Ourselves,* prepared
by the Lawrence Berkeley National Laboratory; (center) from the Texas State Soil and Water Conservation Board, *Fort Phantom Hill
Reservoir Watershed—Brush Control Assessment and Feasability Study,* prepared for by Brazos River Authority, TX; (right) from
NASA, *Exercise Countermeasures Demonstration Project during the Lunar-Mars Life Support Test Project Phase IIA,* January 1998
(NASA/TP-98-206537)

Printing number: 9 8 7 6 5 4 3 2

Printed in the United States of America
on acid-free paper

BRIEF **CONTENTS**

CONTENTS

PREFACE

A NEW DIRECTION

In December 2004 the *New York Times* reported some conclusions of a survey of American corporations: one third of employees in the nation's blue-chip companies write poorly, and businesses are spending as much as $3.1 billion annually on remedial training. With this and similar findings in mind, we have extensively rewritten and updated this edition of *Reporting Technical Information*, offering readers help in developing communication skills in terms of online, hardcopy, and oral presentations. By helping students to achieve a practical grasp of the rhetorical skills, we hope to make them both attractive as employees and effective as professionals.

Thus the eleventh edition of *Reporting Technical Information* takes a new direction—specifically targeting students in a wide variety of science, health, business, engineering, and technical majors to prepare them to develop documents of the kinds they will most likely need to write after they leave school and begin their careers. While students preparing for careers as professional communicators will find this book relevant and useful, we especially address the needs of college students who must be prepared to communicate effectively in a work environment, regardless of their profession.

OUR APPROACH AND ORGANIZATION

A writer in today's heavily media-oriented environment either writes documents that are inviting, clear, and readable or risks being ignored. Readers bombarded with information, much of it from sources that hardly existed a generation ago, must choose carefully which documents they will read and how much they will read of each one. Today effectiveness as a writer and a reader will be a critical factor—possibly *the* critical factor—in a person's professional success. Helping to develop that effective writer and reader is the purpose of this book.

Reporting Technical Information, Eleventh Edition, consists of twenty chapters in three parts: we begin with the foundation concepts (Part I), move to techniques (Part II), and then embed these concepts and techniques in a variety of document applications (Part III).

- The first chapter provides an overview of technical writing. In Part I, Foundations, we discuss the composing process (Chapter 2), writing for one's readers (Chapter 3), achieving a readable style (Chapter 4), ethical writing (Chapter 5), and writing for international readers (Chapter 6).
- Part II, Techniques, consists of five chapters. Chapter 9, focusing on creating and managing text, is new in this edition. Two chapters cover designing and formatting documents and developing the main elements of reports: Chapters 8 and 10, respectively, have been rewritten to reflect the growing importance of these areas. Two fully updated chapters look at gathering, evaluating, and documenting information (Chapter 7) and creating tables and figures (Chapter 11).
- Part III, Applications, puts into practice the skills learned in Parts I and II. We have included new material, Chapter 13, covering the skills required to create reports for any occasion. Two chapters have been thoroughly refined with an abundance of new examples: one on writing effective proposals and progress reports (Chapter 16) and one on understanding the strategies and communications of the job search (Chapter 20). The remaining six chapters in Part III, which have been fully updated for this edition, cover planning correspondence and e-mail (Chapter 12), developing analytical reports—recommendation reports and feasibility studies (Chapter 14), developing empirical research reports (Chapter 15), formulating instructions, procedures, and policies (Chapter 17), writing collaboratively (Chapter 18), and the basics of preparing oral reports (Chapter 19).
- An appendix serves as an easy-to-use reference handbook, which guides the student in the use of language and highlights important errors that should be avoided.

WHAT'S NEW IN THE ELEVENTH EDITION

In this new edition, we have continued our strong focus on the rhetorical nature of writing. Our goal is to help writers in three ways: to prompt them to identify both their readers and the context in which their documents will be read and used, to assist them in defining their purpose in writing, and to teach them to design documents that are properly targeted and appropriately styled. Ultimately, we focus on the **development of effective documents**—documents that will achieve the writer's purpose because they can be read, read correctly, and read quickly.

We have made **additional practical resources available on the fully integrated companion Web site,** such as sample documents, interactive exercises, and online tutorials. **Icons** appearing throughout the book, a feature introduced in this edition, direct students to these resources at www.oup.com/us/houp.

Further example documents have been added throughout. More than half the sample documents we have used in this text are new examples, and all are real-world cases, showing the growing influence of online delivery of documents. These additions fall throughout the book but are found primarily in the revised Chapter 16, Writing Proposals and Progress Reports, and the updated Chapter 20, Understanding Strategies and Communications of the Job Search. New Web reports and other related documents are included. Also, we offer a new Example Student Technical Report in full as an additional online appendix to the book, along with many other sample documents, on the book's companion Web site.

Because tables and figures are so critical to technical reports, we focus on the types of illustration that students and employees should be able to create with basic word processing software. To help students visualize these illustrations, **many more illustrative figures have been added** to the text for this edition. We have included these new figures throughout, but primarily in our new Chapter 8, Designing and Formatting Documents, and our fully revised Chapter 11, Creating Tables and Figures. Also, these figures represent a new approach, focusing on a broader range of disciplines.

Opening scenarios for each chapter, most researched and developed for this edition, demonstrate the impact of technical communication in the real world.

Additional end-of-chapter exercises include some projects specifically intended to encourage collaboration, often linked to an online component on the companion Web site.

Chapter 5

■ We offer an **increased coverage of ethics** via discussions of ethical principles throughout the text and have thoroughly updated Chapter 5, Writing Ethically.

Chapter 6

■ Our revised and updated Chapter 6, Writing for International Readers, features **extended coverage of international and global workplace issues.** Issues of cultural and social style differences are discussed in other chapters as appropriate. **The reading list on protocols for many countries has been updated.**

Chapters 7–11

■ Our examples show how to create reports that respond to various business and technical situations. In every application, **we emphasize the growth of online documents:** e-mail, correspondence delivered as e-mail attachments, reports prepared for either hardcopy distribution or online reading, procedures, instructions, and policies, proposals and related progress/status reports, and oral presentations. **Updated and increased current technology coverage** has been an important addition to our new edition. Because the work environment has become inextricable from the Internet and intranets, or LANS (local area networks), we describe the development of written and oral communications in terms of **both online and hardcopy presentation.** Chapters in this part on document design and usability, in particular, have been heavily revised to reflect changes in technology and an increasing use of Web-based documents and platforms. Many links to useful Web sites appear throughout.

Chapter 8

■ **Our new Chapter 8, Designing and Formatting Documents, focuses on achieving clarity for documents through design and format.** Also, we have used the design elements discussed in the chapter throughout the book itself. In this way, our scenarios and resulting examples show how the basics of design can and should be adjusted to meet the rhetorical needs of the communication context. We want students to understand that the choices they make in designing every workplace document are in fact rhetorical and should always reflect the needs of audience(s), purpose(s), and context.

Chapter 9

■ **Our important new Chapter 9, Creating and Managing Text, covers the topic of content management** with the aim of helping the writer develop the best strategy for developing and organizing content.

Chapter 13

■ The eleventh edition emphasizes **report development** and stresses repeatedly that reports should be designed for the occasion for which they will be used. Reports have common elements, which must be developed both to respond to the needs of readers and to accommodate the context. **Chapter 13, Creating Reports for Any Occasion, new in this edition, prepares students to be versatile and creative** in addressing many different workplace report situations—including online and oral presentations—that may arise. Our examples illustrate an important point: that technical communication often embodies both visual and textual rhetoric.

Chapter 19

■ Today, most presentations use PowerPoint slides. In this chapter we show **how to develop PowerPoint presentations** that will allow readers to follow content easily, whether they are in the audience or receiving the slides for reading only. We also discuss how to write papers that are to be read aloud.

ANCILLARIES

Instructor's Manual

Additional teaching resources for instructors, prepared by Elizabeth Robinson, Texas A&M University, are available as a CD-ROM and online. The CD includes quiz items and answer keys that are not found on the book's companion Web site.

Companion Web Site

A fully integrated companion Web site (www.oup.com/us/houp) offers additional practical resources for instructors and students. (See "**How to Use** *Reporting Technical Information*" at the beginning of the text for details on the use of the margin icons in the text and their relation to the companion Web site.)

Instructor Resources

■ *Instructor's Manual.* Organized by chapter, this extensive manual offers chapter objectives, teaching strategies, workshop activities for both traditional and computer classrooms, writing projects (both traditional and distance learning assignments), links, worksheets with sample problems and critical thinking questions, and overhead/PowerPoint slides.

■ *Better Writing—Success at Work (Sample Writing Assignments).* Instructors are encouraged to submit examples of their most interesting and most successful writing assignments to post to our companion Web site and share with their colleagues.

■ *Collaborative Network: Message Board Network.* Here instructors will be able to add links to their own course message boards and those they use at their university or elsewhere.

■ *WebCT/Blackboard Course Content.* To accompany this new edition, we now offer WebCT and Blackboard content packages for their interactive online course management systems. These easily adaptable programs enable instructors to customize content and functionality to meet their individual course needs.

Student Resources

■ *Chapter Learning Aids.* This section offers the following student resources:
  ■ *Overviews.* Chapter outlines identify key concepts and give a comprehensive overview for each chapter.
  ■ *Sample Documents.* The Web site offers a range of additional examples, scenarios, and interactive writing samples, as well as exercises where the student may construct and edit documents described in the text. Included are a large number of the examples and figures used within the eleventh edition text, as well as an example of a student technical report in full. The Web site also presents other report types, such as the following:
    ■ Visual Presentations as video files (discussed in Chapter 13)
    ■ Full-length Feasibility Studies and Environmental Impact Statements (introduced and discussed in Chapter 14)
    ■ Empirical Research Reports (discussed in Chapter 15)
    ■ Procedure Manuals (discussed in Chapter 17)
    ■ Oral Reports as audio files (introduced in Chapter 19)
  ■ *Checklists.* Downloadable versions of important question checklists from the book, available as pdf and Word files, can be accessed electronically or printed.
  ■ *Flashcards/Key Terms and Concepts.* Important terms and ideas are defined and explained as a useful resource, and also summarized in flashcard format to help students remember what they have learned.
  ■ *Online Resources: Current Topic Annotated Links and Additional Resources.* Students can also access links to relevant, contemporary sites with explanatory text on the various chapter topics.
  ■ *Self-Tests/Exercises.* Multiple-choice tests give students the chance to practice what they have learned and receive immediate feedback. Answers can be evaluated and a score given online. Writing exercises for each chapter are also provided. Students' self-test scores and completed exercises can be sent to the instructor's e-mail address.
■ *Book Icons.* Additional resources that correspond to the icons in the margin of the textbook can be accessed from the Web site. (See "**How to Use Reporting Technical Information**" at the beginning of the text for details on the use of these icons in the text and their relation to the companion Web site.)
■ *Tutorials.* These features give the students the opportunity to understand and put to use such on-screen programs as Microsoft PowerPoint for

creating slide presentations, Microsoft Word for document editing, and Adobe Photoshop for creating or editing artwork.
- ▧ *Collaborative Network.* This new addition will offer links to message boards posted by instructors, where students can work with their class-mates online on collaborative exercises assigned from the book and the Web site, and also perhaps with students from other schools across the country.
- ▧ *Author Contact Information.*

The vast majority of these valuable ancillary materials for students was prepared by Suzanne Karberg, Purdue University.

A FINAL NOTE

In all chapters, we focus on the application of basic rhetorical principles. Students are encouraged to study these principles, along with our illustrative examples. By using classroom assignments that allow them to experiment with the rhetoric of technical reports, we believe that students will be well prepared to function as writers in the high-tech world of work where technology has transformed both communication and work.

ACKNOWLEDGMENTS

We thank the many colleagues who took time to review our work and make useful suggestions: Mary Been, Oregon Institute of Technology; Anne Bliss, University of Colorado; John Brocato, Mississippi State University; Anthony Flinn, Eastern Washington University; Alexander Friedlander, Drexel University; Andrea J. Heugatter, Washington University in St. Louis; D'Wayne Hodgin, University of Idaho; Suzanne Karberg, Purdue University; Richard Keenan, University of Maryland Eastern Shore; Karen M. Kuralt, University of Arkansas at Little Rock; Steve Lazenby, University of North Carolina at Charlotte; Bob LeRoy, Chemeketa Community College; Dejun Liu, Prairie View A&M University; Zita McShane, Frostburg State University; Chester Mills, Southern University at New Orleans; Elizabeth Robinson, Texas A&M University; Scott Romine, University of North Carolina at Greensboro; and Joyce Swofford, Clayton College and State University.

We are altogether grateful to the dedicated book publishers (and book lovers) of Oxford University Press for their conscientious efforts to make this edition of *Reporting Technical Information* both elegant and eloquent: Janet M. Beatty, executive editor; Christine D'Antonio, production editor; Emily Pillars, development editor; Neil Abell, editorial assistant; and Annika Sarin, designer.

We offer special thanks to Elizabeth Robinson, Texas A&M University, for a splendid Instructor's Manual and to Suzanne Karberg, Purdue University, for exceptional assistance with the Student Resources on the book's companion Web site.

Finally we express our love and gratitude to our spouses, Jene and Linda, for their loving and loyal support.

PLEASE CONTACT US

We look forward to your comments on this edition of *Reporting Technical Information*. You may reach us by e-mail at e-tebeaux@tamu.edu and sam.dragga@ttu.edu.

Elizabeth Tebeaux
Sam Dragga

Reporting
Technical Information

Scenario

You are an assistant department manager with an organization considering purchase of a new information system. A committee of nine employees from departments across the organization has been asked to determine whether a new information system is needed and if so, the kind of system that should be purchased. What should a new system be able to do that the current one can't do?

The committee, which meets every two weeks for five months, surveys departments throughout the organization, interviews and tapes responses from key individuals from all areas, e-mails surveys to employees, and brings lists of needs discussed in other company meetings. Once the data have been collected, the committee chair, Sam Wilson, asks you to compile the committee's report for submission to the organization's vice president for operations. Creating the report will require analysis and summary of all the notes, memoranda, e-mails, and testimonies of individuals in the organization, as well as minutes of the committee meetings.

Your task, to compile a report that accurately reflects the results of the six months of study, clearly state the recommendations of the committee, and the different parts of the organization. You will need to write a report that can be easily read by everyone in the organization as well as by the board of directors, which will have to approve funding for a new system. But the report must persuade the board and the executive officers, of the need for a new system.

You realize that you have to resolve a number of issues: how to design the report—what elements should be included—how these elements should be arranged and presented to ensure that the report is read correctly and fully by those who will make the decision on whether to purchase a new system, how to organize the content so that even casual readers will know what the committee has recommended and what information it wants included, and how to persuade readers to accept the conclusions of the committee.

"This should be an easy report," remarks Kevin Gaither, a member of the committee. "Over 90% of the people we have talked to believe that we have to have a new system."

"Perhaps," you reply. "But knowing that we need a system and persuading the board that the money will be well spent is another matter. Unlike us, board members have not been involved in the daily discussions. They don't have the same perspective we have; they don't have to use this information system. In addition, the reasons we give for needing a new system are far from uniform: a number of people giving the same reasons express them in different ways. Summarizing a study that accurately reflects what we've discovered the past six months is not going to be easy. In addition, the board will want to be sure that the cost is justified."

"You can do it," Kevin remarks. "You are the best writer in the group. And, the rest of us will help you once you get a draft. My problem is that I never know how to get started."

Lisa and Darren, two other committee members, nod in agreement. Sam then asks if you can have a working draft in two weeks. "Send us all a copy by e-mail. I want everybody to read the draft and come ready to make helpful comments. To give everyone time to do an adequate review, we will meet in three weeks. This is an important report, but we need to get it done."

CHAPTER ONE

An Overview of Technical Writing

What is "technical" writing? Why does your academic program or business want you to study "technical" writing? This first chapter describes "technical" writing to provide a background for understanding content presented in the remaining chapters.

THE MATTER OF DEFINITION

As you work your way through this book, you will see that "technical writing" refers to writing done in a work environment. Technical writing may also be called professional, business, or workplace writing, since it occurs in a business or work setting. In contrast, academic writing is prepared for your course work and focuses on school learning activities. While many of the characteristics of academic writing apply to work-related writing, "technical" (or workplace) writing and academic writing have different goals. These differences define the respective types of writing.

WRITING AT WORK VERSUS WRITING AT SCHOOL

Throughout your educational experience, teachers have given you writing assignments that require you to show that you have mastered concepts and facts. Teachers use writing assignments to determine whether you have learned what they wish you to learn. In school, your teachers are a captive audience: they are paid to read what you write (no matter how good or bad, how clear or muddled it is), to assess the accuracy of the content, and to determine a grade for the assignment. The grade and a teacher's comments help you know whether you have met the desired performance objectives.

You write for teachers knowledgeable about the subject you are addressing in your assignment. They expect a specific response and will read what you write to determine the correctness of the content. Even when you do not express yourself clearly, professors, because of their expertise in the subject and their role as evaluators, can usually figure out what you are trying to say. And, in most cases, teachers work diligently to understand your intent.

Academic writing follows a predictable sequence. In a typical class, you submit assignments, which are graded and returned. Your teachers assign work, which they expect to receive by a specific date. These written assignments have relevance for a specific course of study for a specific time period (a semester or term). Even if you work as a part of a project team to develop a report, you will be working with students who have an academic background similar to yours. Everyone on the team knows that the effectiveness of the assignment will affect the grade.

Eight Basic Differences

Workplace, business, or technical writing differs significantly from academic writing because writing at work and writing in school have eight different purposes.

Writing at Work Achieves Job Goals   Without effective communication, a business organization cannot function. Communication among individuals and departments within the organization, and with individuals in other organizations, becomes the means by which the organization operates. Communication will be written, virtual, electronic, oral, or some combination. For example, you may talk with other employees and then follow up what you have said with written documents—such as a letter or a report.

Effective communication also enables the individual employee to complete job tasks. While correctness and accuracy in the use of English remain chief concerns when you write, your purpose is not to convince anyone that you know a subject for the sake of receiving a grade. You write to persuade, to instruct others on how to perform specific tasks, or to inform people about issues important to them. But your main purpose is to help you perform your job within the organization.

Writing at Work Addresses a Variety of Readers   The audience for whom you write will no longer be a single reader, a professor, a specialist in the subject area. Employee communication surveys indicate that employees write to many readers who have varied educational and technical backgrounds, but most often to people who are less knowledgeable about a topic than the writers. Your supervisor, for example, may have majored in a field of study very different from yours, or your supervisor's responsibilities may have channeled his or her technical knowledge into other areas. For example, you may report directly to a person whose educational background has been in physical chemistry or electrical engineering but whose responsibilities may now be in personnel management, database administration, quality control, or financial analysis. You may have been hired because you bring a specific kind of expertise to the organization, an expertise that your immediate supervisor does not have. In such a case, one of your tasks is to share your expertise with your supervisor and others who will need to be informed about your work as they do their own jobs.

You will often need to communicate with employees from other departments in the organization. They will read what you write based on their responsibilities, cultural and work backgrounds, educational profiles, and technical expertise. Early in your career, you may find yourself supervising employees who have a level of vocational expertise you have not had the time or opportunity to acquire. Your fellow workers' view of various job-related issues may differ markedly from yours owing to differences in age, education, and cultural background. You may also find yourself writing to clients or customers outside the organization. Your ability to communicate with them will depend heavily on how well you perceive the unique background each brings to the job and the way others' backgrounds differ from yours.

Given the international dimension of business, you may need to communicate with customers, clients, and employees in other countries. Your organization may have offices throughout the world or clients in several countries. You may want to sell your company's products to the European Union or to corporations located on the Pacific Rim or in Canada, South America, or Central America.

Writing at Work Addresses Readers with Different Perspectives These readers, all of whom have their own job tasks, will feel no commitment to read what you write unless your messages are useful to them in their work. As you write for different readers both inside and outside the organization, remember what interests each reader is how your message affects his or her job goals.

International readers will approach communications with a perspective different from yours. Communicating with readers in China, Mexico, and England will require different approaches for dealing with each group and culture.

Writing at Work Creates Excessive Paperwork and E-Mail Given how rapidly the quantity of information grows today, few documents—paper or virtual—are read completely. Most are skimmed. Your readers will read as little as possible, and they will read only the parts of a document that can help them. As they pick up your report or see your e-mail subject line, they will immediately ask:

- What is this?
- Why should I read it?
- How does it affect me?
- What am I going to have to do?

Readers will want to find the main points and ideas quickly, and they will become impatient if they are unable to find them by glancing at the page. People in business today will seldom read any document completely or respond to it unless the message at the beginning indicates that a response is mandatory. How they react to the first few sentences of the message you have written will often determine how much more of it they read.

In short, on the job, your readers do not have to read what you write. If you want your writing to be read, you must make your message clear, easy to read, and as interesting and relevant as possible. Because your readers will read selectively, conciseness and clarity are basic ingredients of effective business communication. Mechanical correctness remains a desirable quality, but correct writing that cannot be read easily and quickly will not be read at all.

Writing at Work May Be Read by Readers Unknown to the Writer Because you will write to a variety of readers, these readers fall into two main categories:

1. The primary reader(s), to whom the writing is addressed
2. The secondary reader(s), who receive copies of what you write because the information pertains to their job responsibilities

Because of the prevalence of e-mail, you really have no idea who will read what you write. Messages and reports sent as attachments can be easily forwarded. Documents posted online on an organization's Web site are far from secure from prying eyes and hackers.

Whoever your known readers will be, you should always anticipate unknown readers who may receive copies of your reports or e-mail. In any event, paper documents can be easily photocopied and distributed.

You cannot underestimate the problem that unknown readers present. Copies of your reports and letters will be placed in files accessible by people who may not know anything about you or the situation you are writing about. Unknown readers, for example, may use your report to gain understanding of a work situation they have inherited with a new job assignment. Often, your documents are used to assess your performance and to determine your promotion potential. How well you write suggests how well you have done your job.

Writing at Work Has an Indefinite Life Span Nearly everything you write for an organization will remain in the organization's database indefinitely. Because known and unknown readers may use the documents for an indefinite period of time, the problem of trying to determine who will likely read what you write is further complicated by the shelf life of each document—the length of time that it is accessible and usable. While academic writing is done in response to assignments, applicable only for a specific semester, course, and professor, workplace communications have an indefinite life span.

Writing at Work Creates Legal Liability for the Writer and the Organization Unlike academic writing, prepared and used to evaluate a student's knowledge of a given academic subject, writing at work can be used against you in lawsuits. Your signature on a report or letter makes you responsible for the content. When you write, therefore, bear in mind the indefinite life span of workplace documents and the difficulty of knowing just who will read what you write. Remember, as well, that people may use your writing for reasons you never considered, perhaps taking sentences and even paragraphs out of context for use in situations unrelated to the issue you addressed in the original document. They can then try to use what you say to support claims against you and the organization you represent. In today's litigious society, designing documents that will be difficult or impossible to misuse should be one of your primary goals.

Writing at Work Uses a Variety of Written Documents Most of the academic writing you have done comprises essays, essay examinations, research papers, and laboratory reports. However, surveys show that at work, employees can expect to be asked to create a variety of other documents: letters, e-mail, informational and procedural memoranda, proposals, progress reports, project reports, feasibility studies, economic justification reports, policy statements, travel reports, news releases, speeches, training procedures, budget forecasts, employee evaluations, user documentation, and perhaps articles for publication. What you write will change with your responsibilities, the kind of job you have, and your position in the organization. How you write each document will depend on:

- ▪ The topic being discussed
- ▪ The situation leading to the document
- ▪ Your readers' needs and perspectives
- ▪ Your purpose in writing

While writing in school has taught you important concepts such as paragraph development and correct, effective sentence development, your task in the workplace will be to apply these techniques to the design of whatever you need to write to perform your job.

WRITING AND COMMUNICATING AT WORK

You are probably wondering how you can develop such a wide range of documents and deal with the variety of communication demands you are likely to face. The answer involves your learning, understanding, and then practicing a method of writing, one that you can apply to any document. Chapter 2 explains this method. The remaining chapters help you learn to apply this method to develop a variety of typical technical documents.

Before you leave this chapter, however, you would do well to reflect on the role that communications plays in an organization.

Organizations produce technical writing for internal and external use. Internally, documents such as memoranda and e-mail are common types of communication. These documents enable members of an organization to share information and ideas and show that work has been done or is being done. Many such documents are written collaboratively (see Chapter 18).

The amount and variety of communication done in an organization has grown rapidly during the past three decades. Studies consistently indicate that college-educated employees spend 20% or more of their time at work writing. In fact, most college-educated workers identify the ability to write well as critically important to their job performance. For that reason, most job ads cite "excellent communication skills" as a requirement for positions offered to college graduates.

The manufacture of information has become a major industry in its own right. Much of that information is research related. Many government agencies, scientific laboratories, and commercial companies make research their principal business. They may undertake this research to satisfy their internal needs or those of related organizations. The people who conduct the research may include social scientists, computer scientists, chemists, physicists, mathematicians, psychologists—the whole array of professional specialists. They record and transmit much of this research via reports. The clients for such research may be government agencies or other institutions that are not equipped to do their own research. Reports may, in fact, be the only products of some companies and laboratories (see Chapters 13, 14, and 15).

Much technical writing goes on at universities and colleges. Research professors publicize results in books, journal articles, and papers for professional societies. Many faculty also write grant proposals to secure funding for their research. Effective grant writing requires skills in proposal development, progress reports, and summary reports (see Chapter 16). In short, knowing how to develop documents for any occasion will assist you in pursuing your career and achieving career goals.

Many organizations prepare reports for public use. For example, a state department of natural resources is entrusted not only with conserving woodlands, wetlands, and wildlife but also with making the public aware of these resources. State and federally supported agricultural extension services have as a major responsibility the preparation and dissemination of agricultural information for interested users. Profit-earning companies must create and improve a public image and also attract customers and employees. Airlines, railroads, and distributors of goods and services, all have to keep in the public view through pamphlets, posted notices, Web sites, and radio and television announcements.

A broad and sound foundation in the essential elements of workplace writing is a tremendous asset for writers of technical documents, for it gives them versatility both on and off the job. They can write a good letter, prepare an effective brochure, or compose an informative report. In this comprehensive sense, they are simply *writers*. The same writing skills that are important in a college classroom are important on the job. Surveys show that workers rank writing skills in this order of importance:

1. Clarity
2. Conciseness
3. Organization
4. Correctness—use of standard English

As the preceding examples of technical writing make clear, audience awareness is the foundation of successful technical writing. What is appropriate for your professional colleagues may be inappropriate for the general public. For example, terms are not normally defined if the audience is expected to know them. But the indispensable corollary to that proposition is that terms *must be defined* when your audience cannot be expected to know them. Fact-filled sentences are appropriate for an audience that is highly professional and highly motivated. However, when your readers do not share your motivation, professional knowledge, and enthusiasm, you should slow your pace and make your prose less dense. In technical writing, you have to know your audience as well as your objectives and adapt your style and material to both.

■ THE FOUNDATIONS OF EFFECTIVE TECHNICAL WRITING

Five imperatives underlie every chapter in this book:

1. Know your reader.
2. Know your objective.
3. Be simple, direct, and concise.
4. Know the context in which your communication will be received and used.
5. Design your communication with imperatives 1–4 as guideposts.

■ THE QUALITIES OF GOOD TECHNICAL WRITING

Because the qualities of good technical writing vary, depending on audience and objective, we cannot present a list that applies equally to everything you will write. In general, however, good technical writing does the following:

- Exemplifies effective design: makes a good impression when it is picked up, handled, and flipped through, or read online
- Accommodates selective reading—for instance, some readers will look only at the summary; others will skip from the introduction and to the conclusions; some, however, will read the entire report
- Presents a rational and readily discernible plan, revealed by the table of contents and a series of headings throughout the report
- Reads coherently and cumulatively from beginning to end, but can also be read selectively
- Anticipates readers' questions and answers them in timely fashion
- Has the necessary preliminary or front matter to characterize the report and disclose its purpose and scope
- Provides essential information that is written clearly, without jargon or padding
- Uses tables and graphs, when appropriate, to present and clarify its content
- Has, when needed, a summary or set of conclusions to reveal the results obtained
- Conveys an overall impression of authority, thoroughness, soundness, and honest work
- Can stand alone and be understood by readers who were not part of the initial readership
- Makes a positive statement about the writer and the organization

Beyond all these basic characteristics, good technical writing is free from typographical errors, grammatical slips, and misspelled words. Even minor flaws distract attention from the document's main points and call into question the writer's credibility.

■ ■ ■ ■ ■ ■ ■ ■   Exercises   ■ ■ ■ ■ ■ ■ ■

Further exercises related to the content of this chapter and all documents employed within the following exercises can be found on the book's companion Web site.

1. What kinds of writing do you anticipate having to do in the early years of your career? What do professional publications in your field suggest about communication in your career field?

2. Bring one or more examples of technical writing from your discipline. Be ready to explain how these examples reflect the characteristics of technical writing. How do you know that the examples you have selected are technical writing?

3. How is technical writing used in your disciplinary area? Ask a faculty member in your discipline about the kind of writing he or she does and the kind of writing that people working in related industries do.

4. Professional associations will often publish articles in a magazine or on the association's official Web site that describe a day in the life of a prominent individual in the field. Locate such a news article for a professional in your field. How much of that person's typical day includes speaking and writing? How much e-mail does he or she have to answer? What kinds of letters, memos, and reports does he or she have to write? Summarize your findings in a one-page memo to your instructor.

5. Examine a document you have received from your university. Is the document easy to read? If not, how would you improve it? Does this document exemplify the characteristics of good technical writing? How and why?

6. Examine the reports on the book's companion Web site. Be prepared in class to discuss the reports. Which one is the most effective from your perspective as a new reader? Why? Which report makes the best first impression? Why?

7. Examine the following document, which was submitted for a broad audience. Do you think it exemplifies the qualities of good technical writing? Why or why not? How would you improve it?

Implications of Recent Clinical Trials for the National Cholesterol Education Program Adult Treatment Panel III Guidelines

I. Cleeman; C. Noel Bairey Merz; H. Bryan Brewer, Jr; Luther T. Clark; Donald B. Hunninghake; Richard C. Pasternak; Sidney C. Smith, Jr; Neil J. Stone, for the Coordinating Committee of the National Cholesterol Education Program. Endorsed by the National Heart, Lung, and Blood Institute, American College of Cardiology Foundation, and American Heart Association.

Abstract

The Adult Treatment Panel III (ATP III) of the National Cholesterol Education Program issued an evidence-based set of guidelines on cholesterol management in 2001. Since the publication of ATP III, 5 major clinical trials of statin therapy with clinical end points have been published. These trials addressed issues that were not examined in previous clinical trials of cholesterol-lowering therapy. The present document reviews the results of these recent trials and assesses their implications for cholesterol management. Therapeutic lifestyle changes (TLC) remain an essential modality in clinical management. The trials confirm the benefit of cholesterol-lowering therapy in high-risk patients and support the ATP III treatment goal of low-density lipoprotein cholesterol (LDL-C) <100 mg/dL. They support the inclusion of patients with diabetes in the high-risk category and confirm the benefits of LDL-lowering therapy in these patients. They further confirm that older persons benefit from therapeutic lowering of LDL-C. The major recommendations for modifications to footnote the ATP III treatment algorithm are the following. In high-risk persons, the recommended LDL-C goal is <100 mg/dL, but when risk is very high, an LDL-C goal of <70 mg/dL is a therapeutic option, i.e., a reasonable clinical strategy, on the basis of available clinical trial evidence. This therapeutic option extends also to patients at very high risk who have a baseline LDL-C <100 mg/dL. Moreover, when a high-risk patient has high triglycerides or low high-density lipoprotein cholesterol (HDL-C), consideration can be given to combining a fibrate or nicotinic acid with an LDL-lowering drug. For moderately high-risk persons (2+ risk factors and 10-year risk 10% to 20%), the recommended LDL-C goal is <130 mg/dL, but an LDL-C goal <100 mg/dL is a therapeutic option on the basis of recent trial evidence. The latter option extends also to moderately high-risk persons with a baseline LDL-C of 100 to 129 mg/dL. When LDL-lowering drug therapy is employed in high-risk or moderately high-risk persons, it is advised that intensity of therapy be sufficient to achieve at least a 30% to 40% reduction in LDL-C levels. Moreover, any person at high risk or moderately high risk who has lifestyle-related risk factors (e.g., obesity, physical inactivity, elevated triglycerides, low HDL-C, or metabolic syndrome) is a candidate for TLC to modify these risk factors regardless of LDL-C level. Finally, for people in lower-risk categories, recent clinical trials do not modify the goals and cutpoints of therapy.

Source: http://circ.ahajournals.org/cgi/content/abstract/110/2/227

PART ONE

Foundations

Part I emphasizes the composing process, introducing the book's central focus. Chapter 2 discusses how to analyze a writing situation and how to discover, arrange, revise, and edit technical information. Chapter 3 shows how to adapt content for various audiences. Chapter 4 emphasizes elements of style at the paragraph, sentence, and language level. Chapter 5 explains what it means to be ethical and how to be ethical in your writing. Chapter 6 deals with the reality of globalization and offers advice on how to write for non-American readers.

■ ■ ■

Scenario

With your new engineering degree, you went to work for Southwest Coal Power (SCP), a company that builds coal-burning power plants. In the six months you've been with SCP, you've learned how new technologies are making the burning of coal cleaner and safer for the environment. SCP has contracted to build a new power plant in Roll, Arizona. Before SCP can build the plant, however, it must file an Environmental Impact Statement that details any environmental problems the plant may cause and the company's plans to mitigate such conditions. You have been assigned to the team writing the statement. The proposed plant will use a fluidized-bed boiler, and the head of the team has assigned you the task of explaining this equipment.

In your first draft, you wrote a highly technical textbook description of the fluidized-bed boiler, and the head of the team didn't like it at all. "Think of your readers," he said. "They're not engineers, they're bureaucrats and politicians and concerned citizens. They won't understand most of this, and when people don't understand something, they get suspicious and hostile. Give me something a nonengineer would understand."

So now you are putting on paper some thoughts about your readers. They won't understand engineering terminology. They need some sort of analogy, maybe. You've seen the fluidized-bed boiler likened to a giant pressure cooker. That might work. And you have some good drawings that show the boiler in action, making it pretty easy to visualize.

What's the reader's point of view? Probably that coal is a dirty fuel and that the new power plant will endanger the environment. You can show people how the new kind of boiler "fluidizes" more than 90 percent of the sulfur and nitrogen pollutants out of coal. You begin to realize that thinking about your readers can be a good way to discover the content you need and how to organize it.

This chapter explains how audience analysis fits into the composing process and identifies the other steps of the process.

Composing

For many writers, the hardest part of having to write is deciding what to say, or "getting started." You may feel that this is half the battle. So, what's an effective way to "get started"? Because technical writing differs from personal essays, you may find that "getting started" is easier and more interesting than trying to begin an essay, such as you wrote in freshman composition, or developing essay exam responses in courses such as history, philosophy, or psychology.

THE BASIC PARTS OF THE COMPOSING PROCESS

Writing effectively begins with an understanding that writing well requires a *process*. If you understand the process and why it's important, then you can see the value of using it. The more frequently you use it, the more efficient you will become in the process of composition. The result: a better percentage of the documents you write will achieve their intended goal.

This composing process has several parts:

- **Analyzing** the situation that requires a written response
- **Choosing** and arranging content
- **Drafting** and **revising**
- **Editing** the finished draft

Writing requires each activity, but most of the time the activities are recursive: that is, you begin with an analysis of a situation that requires you to write; but as you choose and then arrange content, you are probably drafting and doing some revising. When you are satisfied with your content and your arrangement, you may decide to focus on revision to improve the presentation of your content. Editing, the final stage, ensures that what you have written is correct in content, usage, punctuation, spelling, and sentence structure. During editing, you look at your document as a whole, to perform a "quality check."

Writing becomes extremely difficult if you try to do all the parts at once! It is equally difficult to begin by preparing an outline about the subject or topic first, then trying to follow the outline as you collect information. Neither method will produce an effective document. Research has shown that good writers usually follow a basic process.

1. Analyzing the Writing Situation: Audience and Purpose

The first step in composing is the most critical to the success of what you write. In this step, you need to know *why* you are writing, what you are attempting to achieve with your document, what situation or problem has led to need to create this particular document. Then, you must consider all your readers—those who will or may see your document.

Every technical or workplace document is written to respond to a specific situation. Each document has a targeted audience. Writing responds to both— the situation and the readers in that situation. Writing is NOT simply compiling information about a subject.

2. Choosing/Discovering Content

Sources for content may include reports already prepared by the organization, research material from databases, indexes, and the Web, and interviews, surveys, statistics, technical periodicals, and books. What you need to include should always be guided by three questions:

■ *Why* are you writing?
■ *What* does your reader need?
■ *How* does your reader perceive the subject?

As you search for information, think of your purpose: what you want your reader to know and do with what you write. In the workplace, writing solves problems and enables the organization to operate.

After you have considered your purpose and begun to research your topic, begin to list ideas that you can use to develop your topic. If you wish to do this activity on your computer, simply list ideas. Then, move, insert, and delete ideas. Then, on the basis of the ideas listed, ask yourself what additional information you will need to locate. Computers are ideal for making lists and turning them into larger units. Don't like what you wrote? Delete it. You may want to begin your document by writing your purpose at the beginning—to help you stay on track.

3. Arranging Content

As you collect and begin summarizing information and data, you will begin to consider how to arrange the material. In what order should you present your content? Most reports begin with an introduction, followed by a summary of the report. The abstract or summary may precede the introduction. Or, the introduction may be combined with a summary of the report. Next comes the discussion section, in which you present the supporting information. Most reports follow some version of this plan. You may be able to choose your arrangement, or you may be told how to organize your document. Proposals, for example, often have specific, required sections. Many business organizations have rules on how reports distributed to clients outside the firm should be written. Part III gives various ways to arrange and present material in technical documents of different types. We emphasize, however, that these are suggested approaches and guidelines: they can and should be modified depending on the needs of the topic, readers, and the purpose of the document. Effective writing addresses all three sets of needs.

A useful way to arrange content is to sort material in groupings that can be used as a resource when you begin writing. If you know what arrangement you will use, sort material so that you can easily find it when you begin drafting specific segments of your document, or so that another person can pick up your notes and continue with the document, if necessary. You can also sort material electronically: create folders of information on each segment of your report. Then, arrange material within each folder before you begin drafting. This method allows you to track material you use and insert appropriate citations when you use material from a specific source.

If you use electronic articles from your library's database, you can insert these articles into files that can be accessed later, when you begin to draft your document.

4. Drafting and Revising

Drafting is a highly individual activity. Few writers do it exactly the same way, but many now use their computers for the job. When you begin your draft, open a file and save it with the name of the report. Then, begin keyboarding ideas or sections, pasting in material already listed, arranged, and developed as necessary. To help keep track of the ideas you are developing, you may wish to type the names of your main segments, boldface those, and insert information points in the appropriate segment. Later, in Figure 2-1, you will see that some of the ideas in the list of ideas become headings. Some are combined with other ideas. You can arrange, delete, and add ideas as the need arises. Most writers work on a document in a start/stop fashion.

To show you how the composing process works when it is applied to a routine business document, read the following situation and track the development of this memorandum.

Situation Bob Johnson, an engineering project manager with a local civil engineering firm, has been named local arrangement chair of the forthcoming construction engineering conference. Two months before the conference, Bob sends a memo to everyone in his group to let them know what they will have to do. He has informed his group via e-mail that they will have certain responsibilities throughout the conference. The memo will tell them specifically what they will be doing. He makes notes to himself in the draft by inserting comments in brackets, [ ]. He highlights bracketed material to help himself remember major tasks he will need to do before he sends the memo. As he plans his memo, Bob keyboards the following list of topics:

Location, date, time info of conference
Specific duties of the SE Group
General instructions
Conference schedule
Other information SE Group needs to know

As he develops the memo, he inserts information beneath each heading and then revises several. His first draft looks like this:

Conference Location, Date, Time

CE conference—October 28–29—Lancaster Center. Our group's responsibility—serve as greeters, help prevent glitches. Over 150 engineers have already registered, and the cutoff date is still three weeks away. We need to be sure we are organized. Conference may be larger than last year. We want to do our part to ensure the success of the meeting. Help all attendees have a good conference. Be proactive in anticipating problems with people getting where they need to be.

Please be at the Lancaster Center at the following times:

Oct. 28: noon–end of the day
Oct. 29: 7:00–end of the conference. Last session begins at 3:30.

Our Responsibilities

Helping visitors locate the section meetings, answer any questions, deal with any hotel reservation glitches, transportation problems, questions about restaurants.

Remain available until after the dinner on Oct. 28.
Oct. 29: On site throughout the day.

General Information

- Number expected to register and attend: 200+.
- Visitors will arrive at the hotel by mid-morning on the 28th. Some will come the evening before. [Contact Ralph to see if we need to be at the hotel on Oct. 27 after 5:00. Check information folder for sponsor letters.]
- Be available no later than noon on 28th. If possible, arrive at the Lancaster Center earlier than that. Dress is business casual.
- If those flying in arrive late, contact Jim or Joanna via their cell phones to ensure that registrations are not canceled. Jim: 228-3459; Joanna: 322-1875.

Conference Schedule

Oct. 28

Light lunch: noon–1:15
Opening session: 1:30–3:00
Second session: 3:30–5:00 [Check sponsors for all sessions. These have to be correct!!!! Check with central planning group.]
Dinner: 6:30—Holcomb Room, 2nd floor of the LC

Oct. 29

Breakfast: Room 104 of the LC, 7:00–8:00 [sponsor?]
Third session: 8:30–10:00 [session sponsor?]
Closing Session: 10:30–noon [sponsor?]
Lunch: noon

Displays

Nine vendors will display software all day the 29th in Room 106. Consultants will be on hand to discuss compatibility issues. Be available to help vendors set up.

Conference Materials

Will be available at the check-in desk at the front door.
Each folder will contain brochures about new products and a schedule. Add a list of restaurants downtown?

Other

Breakout rooms will be available for the second part of the sessions.
Phone and faxes are available in Room 110 from 8:00–5:00.
Refreshments will be available during break periods. Water in all rooms.

As you continue to draft, you will revise. But during the drafting stage you should revise only for meaning. Try to avoid worrying about sentences that don't sound quite right. If the sentence you write captures what you want to say, even clumsily, don't stop to revise. You can clean up these sentences later. Word processing programs, like Word, will alert you to problems in sentence structure and spelling. Don't attempt to correct these mechanical problems unless you can do so without slowing progress. Focus on presenting your material to the readers: when you believe you have your basic ideas on the screen or page, you can begin a formal revision process.

5. Revision

During the formal revision process, you will want to revise from different perspectives.

Logic Does your presentation make sense? Try reading paragraphs aloud that seem to be "scrambled." Hearing what you have written often identifies and locates any problems in logic. Is the order in which your material is written appropriate for the purpose and for your readers?

Completeness Is your presentation complete in terms of the purpose of your document and your readers' needs and requirements? Is your information correct? Does your document contain all requested information? Have you noted the source of ideas and information you include?

Style   Examine each paragraph and each sentence. Are your paragraphs really paragraphs? Do they have topic sentences? Do all the sentences in the paragraph pertain to the meaning you are building in the paragraph? Try to open each paragraph with a topic sentence. Eliminate or recast sentences that provide little support for the topic sentence. Sentences should be concise. As we will discuss in Chapter 4, Achieving a Readable Style, clear, concise, precise sentences encourage readers to follow your ideas. Complicated, wordy sentences can be hard to understand. Avoid long paragraphs—they discourage readers and tend to become incoherent.

Visuals   Do you need visuals—photos, graphs, drawings, other illustrations—to help your reader "see" and remember key ideas? Chapter 11, Creating Tables and Figures, will provide guidelines on developing visuals that will be effective. Visuals combined with text often are the best means of communicating with readers.

6. Document Design

When you began drafting, if you used headings or names of report segments, you have already begun to design your document. Document design refers to the way information is arranged and displayed on the page. With word processing, you have many choices of font, typeface, and even color. Your choices in these areas, and then your placement of visuals, can encourage or discourage people from becoming interested enough in what you are saying to begin to digest it. If you want what you have written to be read, you will need to design pages so that the information is inviting and accessible. Examine Figure 2-1: note how the message looks after Bob's initial revision of his draft.

7. Editing

Editing is a critical writing requirement. In complex reports, you will want to perform several "edits," beginning with mechanics—spelling, usage, punctuation, sentence structure, and so on. When your word processing program, usually by means of a green line under a sentence or phrase, indicates a problem with a sentence, stop and check it carefully. We have included as an appendix a handbook that covers many common errors.

Another edit focuses on citing sources. Check your documentation to be sure that you have given credit or sources for all information you have used. Be sure that when you use graphics and ideas not your own, you give credit to the source.

A third edit focuses on the document as a whole: How does it look? How does it sound? Is the important information easy to locate? Is the document complete?

Don't try to check for every error at once, in one reading. Editing requires care, objectivity, and diligence.

TO: SE Group DATE: October 1, 2004

FROM: Bob Johnson

SUBJECT: Preparations for the Construction Engineering Conference

Conference Location, Date, Time

The construction engineering conference is scheduled October 28–29 at the Lancaster Center. Our group will serve as greeters. Over 150 engineers have already registered, and the cutoff date is still three weeks away. We need to be sure we are organized to help visitors as they arrive.

Please be at the Lancaster Center at the following times:

Oct. 28 noon–end of the day

Oct. 29 7:00–end of the conference. Last session begins at 3:30.

SE Group—Specific Duties

We will be responsible for helping visitors locate the section meetings, answer any questions, and deal with any hotel reservation and transportation glitches. We must remain available until after the dinner on Oct. 28. Oct. 29—We will be on site throughout the day and help guests who need to leave promptly at the close of the morning session.

General Information

- Number expected to register and attend: 200+

- Visitors will arrive at the hotel by mid morning on the 28th. Some will come the evening before.

- Be available no later than noon. If possible, arrive at the Lancaster Center earlier than that.

- If those flying in arrive late, contact Jim or Joanna via their cell phones to ensure that registrations are not canceled. Jim: 228-3459; Joanna: 322-1875.

FIGURE 2-1 · MEMO—SECOND DRAFT

Preparations for the Construction Engineering Conference—2

Conference Schedule

Oct. 28

Noon–1:15	Light lunch—Mellon Room (Sponsor: KLM Ltd.)
1:30–3:30	Room 105, opening session
3:30–5:00	Room 105, second session (Sponsor: Bickle and Lauren)
6:30	Buffet in Holcomb Room, 2nd floor of the LC

Oct. 29

7:00–8:00	Breakfast: Room 104 of the LC
8:30–10:00	Room 105, Third session (Sponsor: MERK Inc.)
10:30–noon	Room 105, closing session (Sponsor: Malcolm, Fisher, & Peabody)
Noon	Lunch: Mellon Room

Software Displays

Nine vendors will display software all day the 29th in Room 106. Consultants will be on hand to discuss compatibility issues. Be available to help vendors with setup.

Conference Materials

Available at the check-in desk at the front door.

Registration folder with name tags: will contain brochures about new products and a schedule of activities. List of restaurants in town for those who are staying for the weekend.

Other Information

- Phone and fax machines are available in Room 110 from 8:00 to 5:00.
- Refreshments will be available during break periods. Bottled water in all rooms.
- Call me on my cell phone 229–2905 if anything comes up that isn't covered here.

FIGURE 2-1 · *CONTINUED*

☑ Planning and Revision Checklist

As you pursue the major topics discussed in this book, you will find more detailed checklists for planning and revision. However, the following general checklist can be used in developing any document.

Analyzing the Situation

1. What is your subject or topic?
2. What is the purpose of the document you will write?
3. Who are your readers?
4. Why are you writing? Why is this document required? What is the situation that led to the need for this document?

Selecting Content

5. What topics must or should be covered? What do your readers need to know? What do you want your readers to do after they have read your document?
6. What structure do you plan to use? If you have required report sections, what are they?
7. What information resources do you have available? What resources do you need to locate?
8. What types of visuals (graphs, photos, diagrams, etc.) are you considering using? How will these improve your content?

Arrangement

9. Based on the information you are collecting and your purpose in writing, in what order should the information be placed? What do readers need to know first?
10. Have you sorted your material into groups?
11. Are you beginning to see a plan for headings that announce the content to the hurried reader?
12. Are all the information groupings (with their headings) relevant to your purpose?

Drafting

13. Have you begun to insert information under the headings noted in step 11?
14. Have you recorded the source of all information you will use so that you can insert correct citations after you have completed your draft?
15. Have you noted where you will use graphics? Have you noted the source of each graphic that will have to be credited?

Revising

16. Have you stated clearly the purpose of your report?
17. Does all your content support your purpose?
18. Will readers be able to follow your logic?
19. Have you included all required items—report sections and required information?
20. Have you checked all facts and numbers?
21. Could any material be deleted?

22. Is your document easy to read? Have you made all major points accessible and visible? Are your paragraphs well organized and reasonably short? Does each page have a clear focus?

23. Can you find someone who can read your draft and suggest improvements?

Editing

24. Have you checked for misspellings and for punctuation errors, such as misplaced commas, semicolons, colons, and quotation marks?

25. Have you checked all points of the completed draft at which your word processing program indicates errors in sentence structure, mechanics, or spelling?

26. Have you included all the necessary formal elements?

27. If you are following a style sheet (see Chapter 4), is your system of documentation complete and accurate?

28. Are your pages numbered?

29. Are all graphics placed in the appropriate locations within the text?

30. Is the format (font selected, size, placement, and content of headings) consistent? Are headings persuasive, informative, specific?

USING THE COMPOSING PROCESS IN A WORKPLACE ENVIRONMENT

As an employee, you may be working on several documents at once: status reports, project final reports, a proposal, routine memoranda, agendas for several meetings, a trip report, and so on. Each document will likely be in a different development phase. You may have an hour (or less), spread over a work day, to plan and write a routine memorandum, to be sent as an e-mail attachment. For collaborative projects, you may be sending/receiving, revising, re-revising, and editing content over several weeks or even months. For a proposal, you may be required to develop one part, which you draft and then send to the project director. This may entail researching the topic, visiting with the project director to learn about the primary readers of the proposal, and attending one or more planning meetings with the rest of the proposal team. Or, you may have to draft a report about a subject your supervisor will complete and send to parties you do not know. In short, the composing process for many employees develops at a different pace for every document. And, unlike academic writing, you may not have as long as you want or need to complete a writing project. Particularly when you are working with other employees, progress will depend on priorities: each individual will have his or her own work to do.

Most professionals will tell you that planning is the most important part of the writing process: knowing your readers (if that is possible) and knowing the purpose of the document—what you want to happen as a result of what you write. You can assume that documents that do not achieve the intended result were either ignored, read and ignored, or read and rejected. These situations are frustrating. While you cannot control what happens to what you write, you *can* plan your content and presentation to help you as you decide "what to write." Thus planning allows you to anticipate how your writing will be received.

Word processing can help with your writing: you can open or name a report file, write notes to yourself about who you think will read your document and what you want to accomplish, and use "free writing" to begin the document. Save your work whenever you need to move to another task. You may wish to note what you need to do the next time you open the file. Then, when you reopen the file, you can see exactly what you have done and know what you were thinking when you stopped working on this document. Much writing in the workplace is done in short time blocks, separated by interruptions to answer e-mail or field telephone calls.

After you finish your draft, it is preferable to save it and then return to that file later for revising and editing. Short documents can be finalized in one operation, but if you try to simultaneously revise and edit longer documents, you risk making major errors. A break between writing and revising allows the material to "cool": you can then return to what you have written with a fresh perspective. When revising, look at your ideas with some degree of objectivity, trying to read the content as someone in your intended audience will look at it. Editing is critical to eliminate errors in usage, punctuation, grammar, and sentence structure that make you and your organization look incompetent.

UNDERSTANDING THE COMPOSING PROCESS: WHY BOTHER?

You may be asking: if the writing process is not likely practiced in its pristine form, why bother? But the writing process—even in a truncated, condensed, inadequate format—can still occur and will help you develop effective documents that will be a credit to you professionally. Many poorly written documents come from writers who do not know the writing process or are unfamiliar with the demands of a busy workplace. As you complete this course, you should clearly understand that good writing is more than correct spelling and correct usage.

Let us use an analogy to explain why the importance of understanding and practicing the composing process. Few recreational golfers can follow all the guidelines on driving and putting, but their understanding of the principles behind the correct grip, stance, and swing still guide their efforts. Thus, they "play golf" better than they would if they knew nothing about how to choose the correct club, hold it, approach the ball, and "play the hole." Writing is like that: the more you know about it, and use the correct process, the better writer you will become. Bad habits produce badly written documents, and bad golf!

Clearly, some people have a gift for golf, and some for writing. But even excellent writers know the importance of the writing process. Understanding the composing process, which will underlie every topic we discuss, will help you develop your assignments and other writing tasks you undertake, both in college and in the workplace.

■ ■ ■ ■ ■ ■ ■ ■   Exercises   ■ ■ ■ ■ ■ ■ ■ ■

Further exercises related to the content of this chapter and all documents employed within the following exercises can be found on the book's companion Web site.

1. Your university has established a special committee to deal with parking problems. One recommendation requires increasing parking fees to help to pay for a new parking facility, which could be planned and built in less than three years. The Parking Office has asked students for their input. Since the parking fee will increase $75 per semester under this plan, you decide to write the director of parking, whom you know will share your response with the University Parking Council.

Think about your response. Focus on the questions 1–7 in the foregoing checklist.

· Who is your audience? Who would receive or review your comments?
· What is your purpose?
· What should you say and not say?
· What is the risk?
· How should you sound?
· Can you support your claims with facts, history, surveys, and so on?

2. Watch one of the weekly television news channels. Plan an e-mail response to a news item or feature story related to your profession or industry. Is the story correct? Fair? Complete? On what information do you base your conclusion? Assume that your response may be shared with the viewing audience. Plan your response by using the four questions above.

3. Plan a response to an item regarding your profession or industry in the business section of one of the local newspapers or a weekly news magazine. Was the reporting correct? Fair? Complete? Work independently, and then meet as a group with other majors in your field. Compare your response with those of other team members. Work together to plan a single response that will be sent via letter or e-mail to the appropriate editor at the newspaper.

4. What parts of the composing process do you enjoy or find easy? Which ones cause you difficulty? In a group of four or five students, share your self-evaluation of your strengths and weaknesses as a writer. Exchange ideas and strategies for managing the composing process. Compose a list of suggestions to share with the class in a brief and informal presentation.

5. How long does it take you to write a typical e-mail message? Memo? Letter? Report? What specific strategies could you adopt to cut that time by at least 20%. Compile a list to share with the class.

Scenario

Justin Arrington works as a systems analyst in charge of the help desk at a large insurance company. Justin's group responds to hardware and software problems at the company. When questions arise from various offices, Justin sends a technical person to investigate and then reports to the person who asked for help with the problem.

Justin receives a call from the customer service department, which has been trying to use a new customer complaint tracking program. Justin quickly learns that the customer service staff had not installed the program as Justin's team had instructed. Now the program has shut down. Justin types this response to Foster Davey, who had made the call for help:

> TO: f-davey@ign-mail
>
> If your employees had read the procedures before installing the program, AS WE SPECIFICALLY TOLD THEM TO DO, particularly the installation warnings, they would have seen that the program takes 40 full minutes to install. Not allowing the full 40 minutes, which includes a short break before the final 3 minutes, caused the entire program to shut down. This is a simple program to install, IF all instructions are followed. Tell your people to read the manual with specific attention to p. 1, which warns that the program is not fully loaded after 37 minutes. After a 40-second interval, the program will automatically begin the final installation segment. A "program installed" note will appear at the end.

Justin winces, and starts over:

> Foster, we identified the problem with your new CS tracking program. Please note, on page 1 of the installation manual, that the program takes a full 40 minutes to install. The program may appear to be ready to go after 37 minutes. However, a 40-second interval, during which the screen is blank, except for the blinking cursor, precedes the final 3 minutes of program installation. After this final stage, a "program installed" note will appear. As your group experienced today, if the program is not allowed to load correctly, it will shut down. Praba Raghavan will help your group reload the program before tomorrow morning. Everything else appears normal, so the program should operate correctly. Let us know if other problems arise.

Justin feels better about the second version and sends the message.

CHAPTER THREE

Writing for Your Readers

As we discussed in Chapter 2, developing effective documents requires a process involving at least five stages: planning the document, discovering content, arranging ideas, drafting and revising, and editing. While the stages can be undertaken separately, when you write you will be moving back and forth from one activity to the other. Following this process will help ensure that content is appropriate as well as correctly and effectively presented.

Planning is the most important of the five stages, but it is much more than just collecting information and then arranging it in some kind of order. Planning requires that you:

1. Understand as precisely as possible who will be reading and using what you write
2. Determine your purpose in sharing the content with your readers
3. Know the context in which your writing will be read and used and why your document is needed

Too often, writers become absorbed in the ideas and information they either want or need to write and forget that the person or group who will read the document may have a very different view of the content. As a writer, you must never forget that readers cannot climb into your mind and know exactly what you are thinking. Written documents, designed after careful analysis of the characteristics of those who will read them, become your way of helping readers understand what's in your mind.

GOALS OF COMMUNICATION

In developing any communication, you have three main goals that show the relationship between reader, purpose, and context. Keep these goals in mind as you study the planning process:

1. You want readers to understand your meaning exactly the way you intend.
2. You want your writing to achieve its goal with the designated readers.
3. You want to keep the goodwill of those with whom you communicate.

THE PLANNING PROCESS

To achieve the three goals just listed, you must pursue the following five tasks, both before you begin to write and while you are actually composing your document. Determine:

1. Who will read what you write
2. The goals you want your writing to achieve

www.oup.com/us/houp

CHECKLIST

3. Your role in the organization and how this should be reflected in what you write
4. The content by considering your readers' frame of reference and your purpose in writing
5. The business context in which you are communicating

Determining Your Readers

Academic versus Nonacademic Readers In understanding your readers' point of view, it's helpful to examine how your writing in school differs from the writing that you will do as an employee in an organization.

Writing in School Your writing as a student has been directed toward teachers and professors. Your goal has been to convince them that you understand and have mastered concepts and facts presented in a course. In school, your teachers are paid to read what you write (no matter how good, bad, clear or muddled it is), to assess the accuracy of the content, and to determine your grade. Your teachers are knowledgeable about the subject you are writing about. They expect a specific response, and part of their job is to determine what you are trying to say.

Academic writing follows a predictable sequence, and even when you do team projects, you usually work with the students whose backgrounds are similar to yours. In a work context, however, the writing situation is very different.

Writing at Work: Different Readers, Different Purposes On the job, employees write not for a single expert reader but for many readers, who have varied educational and technical backgrounds. The person for whom you work, for example, may have an educational background very different from yours; your supervisor's responsibilities may have channeled his or her technical knowledge into other areas. For example, you may report directly to a person with an educational background in physical chemistry or electrical engineering and have job responsibilities in personnel management, database administration, quality control, or financial analysis. You may have been hired because you bring a specific kind of expertise to the organization, an expertise that your immediate supervisor does not have. One of your tasks may be to share your knowledge with your supervisor and other employees who are not well informed about the work you have been hired to do. Sometimes you will write to people who are interested in the financial aspects of an issue. At other times, you will write to people who are interested only in the technical, the personnel, or perhaps the liability aspects of the same issue.

You will often need to communicate with employees from other departments within your organization. They will read what you write based on their own jobs, backgrounds, educational profiles, and technical expertise. You may also find yourself writing to customers outside the organization. Unlike students with whom you took classes in college, many of those with whom you work and communicate will be older than you and will have different educational and technical backgrounds. Your ability to communicate with these customers and coworkers will depend on your ability to perceive the unique background each brings to the job and the way individual backgrounds differ from yours.

In short, you may have a single reader or a variety of readers. Your reader(s) can and likely will change with every document you have to write. You can count on one important fact: these readers, all of whom have their own job tasks and come from a variety of age, cultural, educational, and disciplinary backgrounds, will feel no commitment to read what you write unless your message is useful to them as they do their own jobs.

Information Overload and Indifferent Readers Few of the very many documents created today are read completely. Most are skimmed. People who are doing their jobs are not selecting your documents for leisure reading; accordingly, they focus only on the parts that will be helpful to them. As they pick up your report, for example, they will immediately be asking:

- *What is this?*
- *Why should I read it?*
- *How does it affect me?*
- *What am I going to have to do?*

They will want to find the main points quickly, and they will become impatient if they are unable to do this by glancing at the page. They will not usually read any document completely or bother to respond to it unless the message at the beginning indicates that they should do so. How they respond to the first few sentences of your writing will often determine how much more of it they will read.

The Challenge of the Indifferent Reader As you make the transition from student writer to employee writer, remember that unlike your teachers, your supervisors and coworkers do not have to read your material. Because of this need to read selectively, you must make your message clear and easy to read, and as interesting and relevant to your readers as possible. Mechanical correctness remains a desirable quality, but correct writing that cannot be read easily and quickly will not be considered effective. Few readers will be either patient or impressed with verbose, disorganized writing, even if each sentence is mechanically correct. Without conciseness and clarity—topics that are discussed in Chapter 4—few people will read what you write, much less respond to it. Ask yourself:

- *Who will read what I write?*
- *Who will act on what I write?*

The Potential for Unknown Readers In a work context, you will never know for sure who will read what you write, thanks to the prevalence of copy machines and the ease with which e-mail messages can be forwarded. But in

TO:	Computing Support Staff	Primary readers
DATE:	November 21, 2001	
SUBJECT:	Customer Complaints, Jan–Oct 2001	
FROM:	Darren Herscowitz	
DIST:	Melanie Stuart, Operations	Secondary readers—distribution list
	Mack Schropshire, Sales	
	Aston Conolee,	Potential unknown readers who may have access to the report from this file
	Gif Small,	
	File 2301.5	

FIGURE 3-1 · INTERNAL REPORT HEADING

most organizations, you will be expected to send copies of your documents to the individuals on what is called a distribution list. However, you should anticipate that what you write also will be disseminated to others outside the distribution list. Thus, you will have a primary reader (the person to whom the report is addressed), secondary readers (those who receive copies—the distribution list, often labeled DIST or cc, for copy), and unknown readers (those who receive copies from any of your expected readers). The memorandum heading (Figure 3-1) shows how complex the problem of determining readers can be. And, copies of anything you write will be filed, either in paper or electronic form.

Among those receiving copies, only one or a few will act on what you write; you need to attempt to identify the person or persons. In many cases, your primary reader will transmit your document to someone else for action. Perhaps this individual is one of your secondary readers. Or, the person who will be responsible for acting on what you write may be unknown to you. Thus, determining WHO will act on what you write is critical because your assessment of the perspective of this small population will guide your composition of the message to be conveyed.

Asking Questions to Analyze Your Readers

You will want to determine as much as possible about your readers.

- *How much do they know about what you are writing?*
- *Are they experts in the area about which you are writing?*

Readers who are not technical experts have different needs from readers who are well informed about your subject area.

- *Do your readers know anything at all about your topic?*
- *What is their educational level?*
- *What is their cultural background?*

Before addressing readers from a culture other than your own, you will want to review material covered in Chapter 6, Writing for International Readers. Culture affects communication style. For example, in the United States, readers value direct, concise letters. Almost everywhere else, this approach is considered rude. In the United States, business documents focus on business only. In many other countries, business and personal relationships are intertwined. Therefore, business communication prepared for such readers will often include "personal," non-business-related elements. If you fail to adapt your content and your style to the perspectives of readers in the culture in which you wish to do business, you can easily ruin any possibility of a successful business relationship.

- *Will your readers be interested in what you write? If not, how could you present your message to make it appealing?*
- *What kind of relationship do you have with a given reader?*
- *What is the reader's attitude toward you, the subject matter you need to communicate, the job you have, and your department of the organization?*
- *Do you have credibility with this reader?*

The foregoing questions will help you to predict your readers' perceptions as determined by education, family, geographical and cultural background, job responsibilities, rank in the organization, age, and life experiences—just to name a few demographic categories that define how people see the world. How much a reader knows about your topic is critical because it determines what you say and the technical level of your presentation. The following situations illustrate this point.

Situation 1 Physicians, caregivers, and patients share an interest in cancer research but from different perspectives. As a result, the Web offers a major means of delivering information to all three levels of readers. Note the differences in information about the drug Avastin for patients and for physicians: compare Figure 3-2 with Figure 3-3. Patient information uses extensive definition, common terminology and descriptions of how the patient receives the drug. Note that the patient information provides visuals to supplement text and minimal medical nomenclature. In contrast, physician information, describing the biochemistry of the drug, highlights safety, uses medical nomenclature but no visuals, and includes references. Readability for the busy physician is enhanced by bulleted descriptors.

Situation 2 Similarly, discussions of laparoscopic colectomy—for patients and then for physicians, show even more differences. The article for patients (Figure 3-4) focuses on long-term survival, and the information for physicians (Figure 3-5) notes the same point. However, the article targeted at physicians focuses on safety of the procedures and the need for specific training. Note, too, that the article for physicians carefully compares surgical outcomes, while information for patients mentions cost, convenience, and pain levels. Compare Figure 3-4 and Figure 3-5.

About Avastin

How does Avastin work?
Information patients, families, and caregivers need to know

Avastin is not chemotherapy, but it is given in combination with chemotherapy. While chemotherapy attacks the tumor directly, Avastin attacks the blood vessels that surround the tumor.

In order to grow and spread, tumors need a constant supply of oxygen and other nutrients. Tumors get this supply by creating their own network of blood vessels. This process is called **angiogenesis** (an'-gee-o-jen'-i-sis). To start angiogenesis, a tumor sends out signals to nearby blood vessels. These signals cause new blood vessels to "sprout" toward the tumor. Once these new vessels reach the tumor, they provide the supply of blood that "feeds" the tumor.

Avastin: an anti-angiogenic agent

blood vessels

A tumor creates a network of blood vessels – a process called angiogenesis

tumor

blood vessels

Avastin, an anti-angiogenic agent, inhibits blood vessel formation, which starves the tumor

tumor

Avastin works by blocking angiogenesis. Because of this, it is often referred to as **anti-angiogenic** (an'-tee-an'-gee-o-jen'-ick) therapy. By preventing the growth of new blood vessels, Avastin helps "starve" the tumor. This makes it hard for the tumor to grow. On average, people taking Avastin in combination with chemotherapy were more likely to live longer, have a longer period of time before their tumors grew, and have their tumors become smaller in size. Avastin plus chemotherapy may work better than chemotherapy alone.

How is Avastin given?

Avastin is given by **infusion,** which means it is given through a needle placed in a vein in the arm, hand, or through a central line (such as a port). The first time Avastin is given, it will take about 90 minutes. After the first or second time, once the doctor makes sure you have no problems with the infusion, treatments with Avastin may require less time, usually about 60 or 30 minutes.

Avastin is given in combination with chemotherapy. The recommended dose of Avastin is 5 mg/kg given once every 14 days for as long as the doctor recommends therapy.

Genentech.   http://www.avastin.com/avastin/patBrochure2.m

FIGURE 3-2 · DRUG INFORMATION FOR A PATIENT OR CAREGIVER

Mechanism of Action

Anti-angiogenesis: inhibiting a tumor's blood supply

Angiogenesis is a critical process in tumor development[4–6]

- Occurs in response to environmental stimuli, such as hypoxia[4–6]
- Driven by the release of pro-angiogenic signals, such as vascular endothelial growth factor (VEGF), which binds to receptors on nearby vessel endothelial cells[4–6]

    Though other pro-angiogenic factors exist, VEGF has been identified as the most potent and predominant[4,5]

- Creates a growing vascular network, providing the nutrients needed to facilitate tumor development[4–6]

VEGF and angiogenesis play a critical role in the development of CRC[7,8]

- VEGF expression is directly correlated with the growth and proliferation of new blood vessels in colorectal tumors[7]
- VEGF may be activated in the early stages of CRC and is important in tumor progression, recurrence, and metastasis[7,8]
- Anti-angiogenic therapy has been shown to inhibit tumor growth in CRC[8]

Avastin targets VEGF to inhibit angiogenesis, as demonstrated in preclinical models

- Recombinant humanized monoclonal antibody to VEGF
- Binds to and inhibits the biologic activity of VEGF, which is overexpressed in tumors
- Prevents the interaction of VEGF with VEGFR-1 (Flt-1) and VEGFR-2 (KDR) on the surface of endothelial cells
- Reduces microvascular growth and inhibits progression of metastatic disease

Important safety information

In Genentech-sponsored clinical studies, the most serious adverse events associated with Avastin were **GI perforation, wound healing complication, hemorrhage,** hypertensive crisis, nephrotic syndrome, and congestive heart failure. The most common grade 3-4 adverse events were asthenia, pain, hypertension, diarrhea, and leukopenia.

Please see full Prescribing Information, including Boxed WARNINGS, and Safety Info Section for additional safety information.

References

4. Bergers G, Benjamin LE. Tumorigenesis and the angiogenic switch. *Nat Rev Cancer.* 2003;3:401–410.
5. McMahon G. VEGF receptor signaling in tumor angiogenesis. *The Oncologist.* 2000;5(suppl 1):3–10.
6. Rosen LS. Clinical experience with angiogenesis signaling inhibitors: focus on vascular endothelial growth factor (VEGF) blockers. *Cancer Control.* 2002;9(suppl 2):36–44.
7. Reinmuth N, Parikh AA, Ahmad SA, et al. Biology of angiogenesis in tumors of the gastrointestinal tract. *Microsc Res Tech.* 2003;60:199–207.
8. Ellis LM. A targeted approach for antiangiogenic therapy of metastatic human colon cancer. *Am Surg.* 2003;69:3–10.

Genentech. http://www.avastin.com/avastin/moaPro.m

FIGURE 3-3 · DRUG INFORMATION FOR A PHYSICIAN

Laparoscopic Surgery Does Not Affect Survival in Colorectal Cancer

According to an article recently published in *The Lancet*, laparoscopic surgery does not affect survival compared to conventional surgery for colorectal cancer.

Colorectal cancer remains the second leading cause of cancer deaths in the United States. The surgical removal of the cancer remains an integral part of the treatment strategy for patients with cancer that has not spread to distant and/or several sites in the body. The conventional surgical procedure involves the opening of the pelvis and/or abdomen to gain access to the large intestine. However, the use of laparoscopic surgery has been evaluated for the surgical treatment of colorectal cancer.

Laparoscopic surgery involves the placement of small probes into the area of surgery. The probes contain a camera, which displays images onto large television screens in the operating room. The surgeon can perform the surgery through the probes while watching his or her movements on the screen. This type of procedure prevents the need for large surgical incisions, and may reduce the risk of infection, healing complications, pain and/or blood loss. However, long-term effects of the use of laparoscopic surgery in colorectal cancer have remained unclear, and recommendations were to utilize this procedure only in the context of clinical trials.

Recently, researchers from China conducted a clinical study to evaluate laparoscopic surgery in the treatment of colorectal cancer. This trial included 403 patients with rectosigmoid cancer who underwent either the laparoscopic procedure to remove their cancer, or the conventional surgical procedure between 1993 and 2002. Approximately 5 years following surgery, survival rates were 76.1% for patients who underwent laparoscopic surgery, compared to 72.9% for patients who underwent conventional surgery. Cancer-free survival at 5 years was 75.3% for the laparoscopic group, compared to 78.3% for the conventional surgery group. Postoperative recovery was improved in the group undergoing laparoscopic surgery, including less pain, a shorter time to first bowel movement, and a shorter hospital stay. However, the laparoscopic surgical procedure took approximately 45 minutes longer to perform than conventional surgery, and resulted in approximately an additional $2,000.00 per patient.

The researchers concluded that laparoscopic-assisted surgery does not reduce survival compared to conventional surgery for the treatment of colorectal cancer while reducing pain and length of hospital stay. Patients with colorectal cancer who are to undergo surgery may wish to speak with their physician about the risks and benefits of laparoscopic surgery.

Reference: Leung K, Kwok S, Lam S, et al. Laparoscopic resection of rectosigmoid carcinoma: prospective randomized trial. *The Lancet.* 2004;363:1187–1192.

FIGURE 3-4 · SURGICAL INFORMATION FOR A PATIENT
Source: //patient.cancerconsultants.com/rectal_cancer_news.aspx?id=23596

Randomized Study Shows That Laparoscopic Colectomy for Colon Cancer Is Safe

Researchers affiliated with the Clinical Outcomes of Surgical Therapy Study Group of the Laparoscopic Colectomy Trial have reported that laparoscopic-assisted surgery is a safe and acceptable technique compared to open conventional colectomy. The details of this multi-center trial appeared in the May 13, 2004 issue of *The New England Journal of Medicine*.

Laparoscopic surgery has been available for over 10 years beginning with gall bladder surgery. However, there have been doubts about this technique for the treatment of colon cancer. These doubts concern the ability of the surgeon to perform an adequate colectomy and to determine accurately the correct stage of disease. There are also technical difficulties that require specific training to perform laparoscopically assisted colectomy, which many surgeons have not mastered.

The current study randomly allocated 872 patients with colon cancer to be treated with conventional surgery or laparoscopic surgery. Patients were excluded from this trial if they had advanced local or metastatic disease. Over 95% of patients on this trial had stage I–III colon cancer. The median age was 69–70 years. Results indicate that, with a median follow-up of 4.4 years, laparoscopic surgery does not result in an increase in cancer recurrences and does not compromise survival (see Table 1).

Table 1 ■ Data and Outcomes of Laparoscopic Surgery for Colon Cancer		
End Point	**Conventional Surgery**	**Laparoscopic Surgery**
Length of incision	18 cm	6 cm
Duration of surgery	95 min	150 min
Open surgery	100%	21%
Median oral pain administration	2 days	1 day
Median parenteral pain administration	4 days	3 days
Median hospitalization	6 days	5 days
Complications	20%	21%
30-day mortality	1%	<1%
Cumulative 5-year recurrence rate	20%	20%
3-year survival rate	85%	86%

Comments: Other than patient or physician preference there do not appear to be any significant advantages for laparoscopic over open surgery. Patients who elect to have laparoscopic surgery need to be informed that in 20% of cases, open surgery will be necessary anyway; mainly because of advanced disease or technical difficulties of visualization. The author of the accompanying editorial pointed out that there have been significant advances in laparoscopic techniques and equipment but that training has not kept pace with this progress.[2]

References
1. The Clinical Outcomes of Surgical Therapy Study Group. A Comparison of Laparoscopically Assisted and Open Colectomy for Colon Cancer. *New England Journal of Medicine* 2004;350:2050–2059.
2. Pappas TN, Jacobs DO. Laparoscopic Resection for Colon Cancer—The End of the Beginning. *New England Journal of Medicine* 2004;350:2091–2092.

FIGURE 3-5 · SURGICAL INFORMATION FOR A PHYSICIAN

Source: http://www.cancerconsultants.com/professional_new/education_news.php?article=colon

Situation 3 A reader's expertise about a topic will tell you how technical you can be, what level of language you can use. Consider a zoology graduate student who was asked to teach a short course on bird-watching for the university's continuing education program. The four paragraphs in Figure 3-6 were part of a longer explanation that he wrote for fellow class members, adults who were experienced bird-watchers.

The Function of Flocking in Long-Distance Soaring Migrants

Studies by ornithologists have shown that soaring birds migrate in flocks. Since most migratory broad-winged hawks are observed in flocks and form groups even in the early morning, flocking must have some specific advantages for these raptors.

Some researchers believe that flocking assists hawks in navigating and in orienting themselves in the proper direction. Other biologists conclude that flocking enables hawks to locate thermals, the rising currents of warm air that allow the birds to soar and thus gain altitude. A hawk that has reached the top of a thermal, can then glide down to the base of the next thermal, soar up, and glide down again, thermal-hopping until it reaches its destination.

Some researchers also suggest that thermal travel conserves energy and time for migrating raptors hawks' ability to find these thermals for soaring (wings spread for circular motion) and gliding (wings spread for forward motion). This method of flight is essential to conserve energy. In contrast, flapping flight uses over five times as much energy. Thermals also increase hawks' flight speed, since they use air currents both while soaring within the thermal and while gliding to the next one, rather than relying on their own powered flight.

These researchers also believe that flocking behavior enhances hawks' chances for encountering these life-saving thermals. A group of hawks moving together, as in (a) will more likely find thermals, which are produced randomly by the heating of the earth's surface, than will a bird traveling alone across the vast expanse of sky, as in (b). Interestingly, a computer simulation program has been designed to find optimum dimensions for encountering thermals produced by a geometric shape similar to that of hawk flocks.

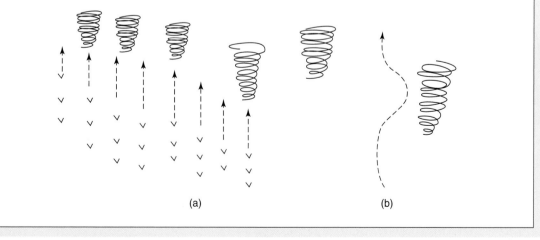

(a)   (b)

FIGURE 3-6 · SITUATION 3: RESPONSE FOR A KNOWLEDGEABLE AUDIENCE

The student was also asked to discuss migratory birds with members of a middle school science class who were also going to attend a continuing education session. He translated his material on "Flocking" to fit their knowledge level and what they may have observed about birds (Figure 3-7).

- *How well do you know your reader?*

Many times you will know your reader personally. However, if you also know the individual's level in the organization, the responsibilities associated with that level, and the kind of technical expertise your reader has, it will be easier to decide what to say and how to present the information. Knowing a

---

What Is Flocking?

Why do hawks fly together in a group? Haven't you noticed that you often see hawks flying together rather than just one hawk flying all by itself?

Soaring birds, such as hawks and vultures, migrate in flocks, groups of birds that fly close together. Scientists have studied why hawks like flying together in flocks. These scientists have concluded that hawks travel as a group to help each other fly in the right direction. Another possible reason is that a group of hawks traveling together can more easily find thermals than can one hawk flying all by itself.

What are thermals and why are they important to soaring birds? Thermals are bubbles of warm air that rise from the ground into the sky. Hawks get inside these thermals and circle high in the air. When they reach the top of the thermal, they glide down to the bottom of the next one, then up again and down until they arrive home. Their flight looks like this:

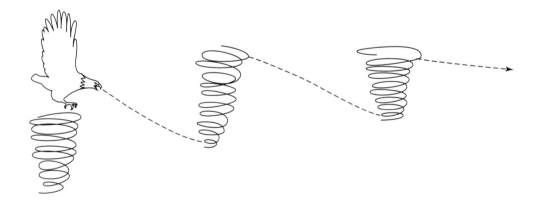

So hawks, by using thermals, which are natural warm air currents, can circle and glide rather than flap their wings. As a result, they save energy as well as time. That's why you see hawks flying together. Working together, they can find these thermals, rather than looking for them alone.

FIGURE 3-7 · SITUATION 3: RESPONSE FOR A NONTECHNICAL AUDIENCE

person's responsibilities in the organization can be particularly useful in helping you anticipate your reader's attitude—how interested he or she will be in
your subject. Because people tend to read only what is useful to them, try to
relate your message to your reader's job. Awareness of the readers' attitude
toward your message is another factor that will help you determine how to present your information.

- *Who else is likely to read what you write?*

Since the person to whom you are addressing your report may not be the
one to act on it, you need to know who else will read your document.

Most reports and letters have distribution lists: the names of those who
receive copies (Figure 3-1). The person who will ultimately act on what you
write may be someone on the distribution list. Thus, the needs and perceptions
of those who receive copies should be considered.

- *Why is each person on the distribution list receiving a copy?*
- *How much does each person on this list know about your topic?*

Sometimes your primary reader will know the situation you are discussing
and will use your report to share the information with others in the organization who also must be kept up to date.

- *What situation led to the need for this document?*

To be able to select content, level of language (technical or general), and the
amount of explanation needed in a business context, a writer must be careful to
determine the needs and perspective of each reader. Closely examining a situation requiring a written response may even help you determine what your
readers have to know and how you might best present your message.

Situation 4 Andrew Sadowski, a tax specialist for an accounting firm, needs
to write a client, ranch owner Michele Miller, to explain how she can deduct
from her income tax the costs of new ranch equipment. Michele, who studied
animal science in college, is not a tax expert. She has asked Andrew a specific
question, which Andrew must answer in language she can understand
(Figure 3-8).

Assume, however, that Andrew's supervisor, Paul Breaux, who has received
a paper copy of one letter to Michele, asks Andrew to provide a summary of
recent rules for deducting business expenses from an individual's annual
income tax obligation. Paul, a tax specialist, has the responsibility to stay current on the IRS rules governing business expenses as deductions. In a second
letter (Figure 3-9), Andrew uses the full range of specialized tax jargon to summarize the relevant rules.

You will find that your readers differ according to their knowledge of the
topic, what they need to know, and the ways in which they will use what you
write. In Situation 4, Michele wants a clear answer to a two-part question: Can
she deduct the cost of new equipment she has purchased for her ranch and, if

January 16, 2001

Ms. Michele Miller
St. Rt. 2, Box 483
Victoria, TX 75384

Dear Ms. Miller:

I have reviewed the financial information you have given me about your ranching operations during 1999–2000 and have determined that you are eligible to use a portion of the amount you have spent on ranch equipment to reduce your current taxable income. According to the information you have given me, you will be able to reduce your taxable income by $10,000, based on the $19,500 you spent on ranch equipment in the past year.

Qualifications for the Deduction

The IRS allows businesses to choose to deduct the costs of certain assets instead of recording those assets on the balance sheet and deducting only a small portion of their cost as depreciation each year. However, several qualifications must be met before the current deduction is allowed.

1. The purchased assets must be tangible assets that depreciate (such as barns or tractors, rather than land).

2. You must start using the asset in your business the same year that you choose to take the current deduction.

3. You must let the IRS know that you choose to take the current deduction by making an election— noting the amount your spent—on your tax return. You need to use Form 4562 to record your purchase and the deduction.

Limitations on the Deduction

The IRS also places limits on the amount of the deduction you can take each year for all equipment you purchased:

1. The maximum deduction allowed is $10,000, even if, as in your case, you spent more than you earned. If you had spent less than $10,000, the deduction would be limited to the amount you had spent.

2. The remaining deduction allowed cannot exceed the taxable income that your business earns for the year.

How to Take the Deduction

To be able to take the $10,000 deduction, complete Part I of Form 4562 and file that form when you file your 1040 tax return for the current year. Instructions for making the election are on the back of the form. A copy of Form 4562 can be obtained at your IRS office in Victoria.

If your business circumstances change, of if you have questions, please call me at 800 453-1859.

Sincerely,

Andrew Sadowski, C.P.A.

Pc: Paul Breaux

FIGURE 3-8 · SITUATION 4: RESPONSE TO A NONEXPERT READER

DATE: January 16, 2001

TO: Paul Breaux

FROM: Andrew Sadowski

SUBJECT: Section 179 Election to Expense Certain Depreciable Business Assets

Under Section 170 of the IRC, U.S. taxpayers are allowed to expense, at their election, the cost of certain tangible property, called Section 170 property, rather than capitalizing that property to an asset account. The deduction is allowed in the year that the property is placed in service.

Section 179 Property

The property that qualifies for expensing is called Section 179 property and is any property subject to depreciation that is

1. tangible personal property, other than air conditioning and heating units;

2. other tangible property, not including a building or structural component of a building;

3. elevators and escalators;

4. research facilities and facilities for the bulk storage of fundible commodities;

5. single-purpose agricultural or horticultural structure;

6. qualified rehabilitation expenditures;

7. qualified timber property; or

8. storage facilities used in connection with the distribution of petroleum or its primary products.

This list of properties and further explanation about them can be found in Section 48(a) of the IRC.

Limitations on Amount Expenses

Several limitations exist on the amount allowed to be expenses during one taxable year. These limitations are listed in Section 179(b) of the IRC:

1. The maximum amount that can be expensed in one year is $10,000 ($5,000 for married persons filing separately).

2. This $10,000 maximum is reduced dollar for dollar by the cost of Section 179 property placed in service during the taxable year that exceeds $200,000. Therefore, the deduction is completely phased out when the cost of Section 179 property in service during one year equals $210,000.

3. The remaining deduction (after limitations 1 and 2 have been applied) is further limited to the taxable income of the trade or business. In this instance, taxable income is computed without taking into account the Section 179 amounts in question. Any amounts disallowed under this limitation can be carried forward indefinitely.

Making the Election to Expense

The election to expense the cost of Section 179 property is made on Part I of Form 4562 and should be filed with the taxpayer's original tax return for the year the property is placed in service. This election must be made for each taxable year in which a section 179 expensing is claimed.

NOTE: Once this election has been made, it can be revoked only with the consent of the Secretary of the Treasury.

More detailed information on 179 expensing, including examples, can be found in Treasury Regulations, Sec. 1.179-1 through Sec. 1.179-5. If I can be of help, please contact me at extension 2905.

FIGURE 3-9 · SITUATION 4: RESPONSE TO AN EXPERT READER

so, how should the deduction be calculated and reported? In contrast, Paul wants to know the Internal Revenue Service rules for deducting purchases. How Andrew presents the information to these readers is determined by what each one knows and needs to know about the topic.

Determining Your Purpose

- *Why are you writing?*
- *What do you want to achieve with your document?*

Determining your purpose—why you are writing and what you want to achieve—is as important as determining your readers. And you may have more than one purpose. For example, you may be writing to

- Inform readers
- Provide information
- Recommend a course of action
- Prove that you have provided the information requested

Situation 5 Project director Karen Thorpe supervises nine engineers, who submit monthly project reports on their engineering work. Karen uses these monthly reports to compile reports to her supervisor and to keep track of how each project is progressing. For the past three months, Sharon Hall, one of the engineers, has submitted her project reports late. Although Karen has repeatedly asked Sharon to observe departmental deadlines, this past month the engineer did not submit her report until a week before the next one was due. Thus, Karen writes Sharon a memo that has two purposes: (1) to notify Sharon officially that she must submit all project reports on time from now on and (2) to provide documentation for Sharon's personnel file that she has been informed that her tardy reports are affecting the project group and asked to comply with team deadlines (Figure 3-10). In addition, the memo can emphasize that Sharon's contribution is important, while firmly but objectively warning her that a consistent pattern of lateness is unacceptable.

Understanding Your Role as a Writer

- *What is your position in the organization?*

As an employee in an organization, you will be communicating with employees above you, below you, and on your own level. In writing to individuals in any group, you will communicate, not as you would with a friend, but as the person responsible for the work associated with your position.

At work, if you are to have credibility as a writer, the image that you project should be appropriate to your position. What you say and how you say it should reflect your level of responsibility in the organization—the power relationship that exists between you and the reader. The image you project will differ depending on the reader. You will project the image of a subordinate when

DATE: February 6, 2001

TO: Sharon Hall

FROM: Karen Thorpe

SUBJECT: Timely Submission of Monthly Project Reports

As I discussed in our meeting of January 20, submission of status reports on each engineering project is a standard requirement for all team members in our group. These reports provide critical information for tracking progress and problems on each project. When one team member's reports are missing, the status of the group's effort is affected. Your input is important.

Since October 1999, your reports have been one to two weeks late. Last month, your January report was submitted three weeks late, one week before the February report was due. Late submission of any person's status report affects our ability to inform the division office of our progress.

Please see that your monthly reports are submitted by the first work day of each month.

FIGURE 3-10 · RESPONSE TO SITUATION 5

you write to those higher than you in the organization, but you will transmit the image of the supervisor to those who work directly under you. When you communicate with others on your own job level, you will convey the image of a colleague. Good writers have the ability to fit each message to its reader. Situation 6 provides examples that illustrate this technique.

Situation 6 Bill Ramirez, director of training, finds himself in a dilemma. He must write his supervisor (Marshall Collins), an associate (Kevin Wong), and his assistant director (Joyce Smith) about the same issue: unqualified people have been enrolled in a training course, and not enough seats are available for qualified employees who need the class. Kevin is frustrated that employees in his office are not being scheduled for the class. Joyce is responsible for scheduling the class. Marshall needs to know about how Bill is handling this dilemma Note how Bill's e-mails to Joyce (Figure 3-11) Kevin (Figure 3-12), and Marshall (Figure 3-13) differ.

Read the three e-mails and explain how Bill has shifted his message to reflect awareness of the different positions and needs of his readers. How does the perspective of each reader determine what Bill says in each e-mail?

TO: <Joyce Smith> jsmith@ibc.com

SUBJ: Unqualified Enrollees in DM Classes

Please remove everyone with less than six months experience from the June Database Management classes. I know you need help in determining the hiring dates of all the people who apply, but we have a three-month waiting list of people, not to mention district supervisors who are unhappy with our inability to meet their training needs. Please stop by before you leave at 5:00 so that we can figure out how to make the screening process easier. Please call Marc Jacz in personnel. He developed a selection procedure for our C++ courses last year and has agreed to work with us.

FIGURE 3-11

TO: <Kevin Wong> wong@ibc.com

SUBJ: Meeting Training Requirements in DM Classes

Thanks for your message. I appreciate your telling me about your problem.

We will see to it that your people who signed up for the June DM classes are taken care of. Joyce's group is understaffed, and they don't have enough people to check experience records of applicants. However, we are working on the problem. Your employees will receive priority. Kevin, I apologize for the situation, and I will appreciate your patience as we get this problem resolved. Please call me if you want to talk. We will schedule your people as quickly as possible.

FIGURE 3-12

TO: <Marshall Collins> mcollins@ibc.com

SUBJ: Update on Training Enrollments

You may hear that we have been having difficulties meeting the demand for our Database Management classes. That's true. We can enroll only those who have completed the basic six months with IBC. Determining who is qualified has been a problem because of the demand for these classes. We have to have priorities on who is admitted. My associate and I are working with personnel to develop an efficient plan for screening enrollees, but we need more people who are qualified to train new hires in database management. In the past six months we have been unprepared for the demand. I'll keep you informed.

FIGURE 3-13

Planning the Content

- *What ideas should be used to achieve the goals of the message?*
- *What ideas should be omitted?*
- *How should these ideas be arranged?*

Once you have analyzed your readership and your purpose, you can begin to decide what you want and need to say, then how you will phrase your ideas. Knowing what your audience needs to know will help you decide how to arrange your message.

- *What tone do you want to convey?*

Knowing how your message sounds will always be critical. Review the e-mails in Figures 3-11, 3-12, and 3-13. Note differences in the tone of each message. How a message is conveyed—whether its tone is respectful and commensurate with the writer's position—often is as important as the content.

Anticipating the Context in Which Your Writing Will Be Received

- *How will your writing be used?*

Knowing what is likely to happen to your writing when it is received also helps you plan your content. For example, a document that has reached its primary destination may be:

- Quickly skimmed and filed
- Skimmed and then routed to the person who will be responsible for acting on it
- Read, copied, and distributed to readers unknown to the writer
- Read and used as an agenda item for discussing a particular point
- Read carefully and later used as a reference

Being able to visualize the context in which a document will be read and used can often guide you in deciding what to say and also how to organize the information and arrange it on the page. Situation 7 explains how consideration of the context helped a writer plan a letter:

Situation 7 Katheryn Stone, who works for the American Farm Bureau Federation, tracks congressional legislation and reports to various community groups about laws that affect them. In the wake of an act passed by the 104th Congress, Katheryn wanted to make a report to the Dos Palos Ag Boosters.

Katheryn wrote the president of Dos Palos Ag Boosters, knowing that the letter would be copied and distributed widely in the community as well as in the organization. Thus, she designed a letter (Figure 3-14) that clearly summarized the main points of the legislation. She used a listing arrangement to enhance readability of the news. Because this letter would be read by a wide audience, she did not bring up the legal issues involved in transferring the property.

Katheryn extracted her letter from the House of Representatives bill shown in Figure 3-15.

January 4, 1997

Mr. Daniel Halpern
10046 Camelia Drive
Dos Palos, CA

Dear Mr. Halpern:

SUBJECT: News from Congress

Greetings from Washington! I am happy to inform you that Bill 4041 has been approved. This bill allows the Secretary of Agriculture to sell a parcel of government-owned land to the Dos Palos Ag Boosters to use as a farm school.

Contents of the Bill

In summary, Bill 4041 contains the following provision:

- Congress agreed to give 22 acres of land located at 18296 Elgin Avenue, Dos Palos, California, to the Dos Palos Ag Boosters to use as a farm school.
- The farm school will educate students and beginning farmers in principles of farming.
- The Dos Palos Ag Boosters will pay the Secretary of Agriculture for the land.
- The Secretary of Agriculture will survey the land to determine the exact acreage transferred to the Ag Boosters and the price of this land will be determined by the market price of the land.
- The Secretary of Agriculture may transfer the land to the Dos Palos School District if the Dos Palos Ag Boosters ask the Secretary to do so.
- The sale of this land is final.

Please contact our office or your state representative regarding this bill. Transfer of land must follow specific procedures and will take approximately one year to complete all legal requirements.

Sincerely,

Katheryn Stone

Katheryn Stone

FIGURE 3-14 · SITUATION 7: RESPONSE TO A BROAD NONEXPERT AUDIENCE

HR 4041 IH

104th CONGRESS

2d Session

To authorize the Secretary of Agriculture to convey a parcel of used agricultural land in Dos Palos, California, to the Dos Palos Ag Boosters for use as a farm school.

SECTION 1. LAND CONVEYANCE, UNUSED AGRICULTURAL LAND, DOS PALOS, CALIFORNIA.

1. CONVEYANCE AUTHORIZED—Notwithstanding any other provision of law, including section 335 (c) of the Consolidated Farm and Rural Development Act (7 U.S.C. 1985 (c)), the Secretary of Agriculture may convey to the Dos Palos Ag Boosters of Dos Palos, California, all right, title, and interest of the United States in and to a parcel of real property (including improvements thereon) held by the Secretary that consists of approximately 22 acres and is located at 18296 Elgin Avenue, Dos Palos, California, to be used as a farm school for the education and training of students and beginning farmers regarding farming. Any such conveyance shall be final with no future liability accruing to the Secretary of Agriculture.

2. CONSIDERATION—As consideration for the conveyance under subsection(a), the transferee shall pay to the Secretary an amount equal to the fair market value of the parcel conveyed under subsection (a).

3. ALTERNATIVE TRANSFEREE—At the request of the Dos Palos Ag Boosters, the Secretary may make the conveyance authorized by subsection (a) to the Dos Palos School District.

4. DETERMINATION OF FAIR MARKET VALUE AND PROPERTY DESCRIPTION—The secretary shall determine the fair market value of the parcel to be conveyed under subsection (a). The exact acreage and legal description of the parcels shall be determined by a survey satisfactory to the Secretary. The cost of any such survey shall be borne by the transferee.

5. ADDITIONAL TERMS AND CONDITIONS—The Secretary may require such additional terms and conditions in connection with the conveyance under this section as the Secretary considers appropriate to protect the interests of the United States.

FIGURE 3-15 · SITUATION 7: LEGAL NOTIFICATION

THINKING ABOUT YOUR READERS: A SUMMARY OF CONSIDERATIONS

As you think about those whom you want to read what you will write, keep in mind the following questions, which evolve from the preceding discussion. Throughout the writing process, you will need to consider each question. The process may seem complicated; but with practice, considering your readers will become a natural part of your planning, writing, and revision.

1. Who will read your message? Always try to assess your readers: Who will act on what you write? Who else may read the document? For example, your primary reader may forward an e-mail message to one or many other people; you can never be sure. Who will act on what you write?

2. How much does your reader know about your topic? Does your reader have a strong background in and knowledge about your topic? Or, does your reader know little about it? If your reader is not an expert on your topic, you will need to define terms, explain concepts, and provide background information.

3. Is your reader interested in your topic? Writing to an interested reader is much easier than writing to an uninterested or indifferent reader. Nevertheless, look for ways to make your writing interesting to your reader. Be concise: provide only what is necessary. Use graphics if possible to help people "see" what you are saying. Chapter 4 deals with ways to create clear sentences.

4. How technical should you be in explaining your position? The answer to this question depends on the education and level of knowledge of your readers. Sometimes, the technical level of a presentation depends on an individual's position in the organization.

5. What does your reader need to know? What is your purpose in writing? Based on your reader profile, you will need to determine how much information you must give, based on what you are trying to accomplish with your message and the readers' level of understanding and interest in the topic.

6. How will readers approach your document? Most readers will likely skim or scan a document and then read parts of it critically or analytically, depending on their own needs or interest level. **Note:** Do not get into the habit of including extra information in a document "just in case." Give readers what they need. Supply additional material in attachments. Remember that in this information age, people are presented with more to read than they can possibly handle. Designing documents that target the needs and backgrounds of readers helps ensure the effectiveness of written communications.

7. What is your business relationship to your reader? Knowing your role in the organization and the level in the organization of your readers will also help you decide what to say and how to say it. Writing to your supervisor will differ from writing to customers or to your peers in the organization. A person's job responsibilities will shape how the individual approaches any document. Try reading aloud what you have written. Does the message sound appropriate for your intended audience? In addition to being clear and concise, your writing should convey an attitude that is appropriate for the reader, the occasion, and the context.

In short, once you have identified your readers and your purpose in writing—these two aspects of communication are inextricable—you can make decisions about content, style, visual aids, and even the length of any document.

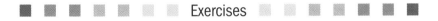 Exercises ■ ■ ■ ■ ■ ■ ■ ■

Further exercises related to the content of this chapter and all documents employed within the following exercises can be found on the book's companion Web site.

1. Select an article in a specialized trade journal or publication in your field of study. Write a summary of the article for an audience unfamiliar with the information. Describe your intended audience as precisely as possible. Identify the specific summary techniques that you used.

2. For a report exercise based on Chapters 13 and 14, use a database to locate three articles on the same topic. These articles should be selected from three different publications. Examine each article, and if possible the entire issue in which the article appeared. To what audience is each article directed? How do the articles differ? To what type of audience is each publication directed? How does each writer target a specific article to the audience of the publication?

3. You have just received a letter from the dean of your college explaining that a college-wide study is under way to evaluate the required courses in each department. As part of this evaluation, the dean wants to know the worst required course(s) you have taken. You also are asked to evaluate the best course(s) you have taken. In your analyses, you are to state specifically why you have so rated these courses. A committee composed of faculty from each department, from outside the department, and from the dean's office will be examining each curriculum.

Address the same issues (i.e., the best and worst courses in the curriculum) but this time as if writing to majors in your field. How will your letter differ, both in style and in the specific information you include?

4. Draft a letter to a faculty member from whom you have received a lower grade than you believe you should have received. Compare your letter with those of other students. Ask them to assess the effectiveness of your letter.

Scenario

As a new project engineer for the state highway department, you were made part of a team that was writing a new policy and procedure memorandum (PPM). The purpose of the PPM was to increase citizen participation in highway planning. The team spent an afternoon discussing the PPM and its purpose. The team leader, Chief Engineer Rosenberg, asked you to have a written purpose statement ready for the next meeting. "It will give us a good start for our next discussion," he said.

That night, you gave up watching your favorite TV shows to write the statement and have it ready for the next day. You wanted to be sure the statement covered everything and would be taken seriously by its readers. Of course, you also wanted it to impress Chief Engineer Rosenberg. By midnight you were satisfied with your statement:

> The purpose of this PPM (Policy and Procedure Memorandum) is to ensure, to the maximum extent practicable, that highway locations and designs reflect and are consistent with federal, state, and local goals and objectives. The rules, policies, and procedures established by this PPM are intended to afford full opportunity for effective public participation in the consideration of highway location and design proposals before submission to the federal Department of Transportation. The procedures provide a medium for free and open discussion and are designed to encourage early and amicable resolution of controversial issues that may arise.

At the next day's meeting the chief was impressed all right, but in the wrong way. He read your statement aloud, and said, "A bit on the pompous side, don't you think?" He then proceeded to rewrite the statement. You had to admit that his version was easier to read and understand.

This chapter discusses the principles the chief used to achieve clarity and avoid pomposity.

CHAPTER FOUR

Achieving a Readable Style

"Style" refers to the words and phrases you choose, the sentence structure you use, and the overall way you express your ideas in a document. Style can be discerned on the word/phrase/sentence level, and also on the paragraph level. An easily read paragraph is composed of easily read sentences that form a coherent idea. If readers must go over a paragraph several times to grasp its meaning, the writer's style is usually the culprit. You need to develop a style that will help readers to grasp the message the first time they approach any document you have written.

A good writer adjusts the style of any document to the audience and the purpose of the document. Let's begin by examining style from the paragraph level and then focus on sentences that work together to form the paragraph.

An Important Note: Because of the quantity of information—reports, e-mails, memos, and so on—that confront readers daily, the need for easily read documents has never been greater. You enhance the probability that people will read what you write and respond if the style is clear, from their perspective, and the information is easy to follow.

THE PARAGRAPH

The paragraph is a group of sentences that work together to produce a coherent idea. Paragraphs should be moderate in length—long paragraphs discourage readers—and should begin with a topic sentence (a central statement) of the content of the paragraph. The supporting sentences build on the idea stated in the topic sentences. The supporting sentences should occur in a logical order. Let's examine an example paragraph from a book justly famous for the clarity with which it explains a difficult subject (Watson, *DNA: The Story of Life*, 2003, 211):

> DNA analysis has already changed the face of microbiology [topic sentence]. Before DNA techniques were broadly applied, methods of identifying bacterial species were extremely limited in their powers of resolution: you could note the form of colonies growing in a Petri dish, view the shape of individual cells through the microscope, or use such relatively crude biochemical assays as the Gram test, by which species can be sorted as either "negative" or "positive" depending on features of their cell wall. With DNA sequencing, microbiologists suddenly had an identification factor that was discernibly, definitely different in every species. Even species, like those inhabiting the ocean depths, that cannot be cultured in a laboratory because of the difficulty of mimicking their natural growing conditions are amenable to DNA analysis, providing a sample can be collected from the deep.

In this paragraph, the sentences proceed in an order that builds the topic of the sentence—the importance of DNA in microbiology. Reordering the sentences would obscure the meaning of the paragraph.

If statements you make need to show an obvious connection to the first, or topic, sentence, you can enumerate points that you introduce. Examine this passage from another very well-written book (Levitt, *Take on the Street*, 2002, 19):

> Sadly, the brokerage industry still has numerous flaws [topic sentence]. That's not to say that all brokers are commission-hungry wolves on the prowl for naïve investors. Some are; others are just inept. Most are honest professionals. They are good people stuck in a bad system, whose problems remain fourfold. First, some brokers are not trained well enough for the enormous tasks they are expected to carry out. Second, the system in which brokers operate is still geared toward volume selling, not giving objective advice. Third, to increase sales, firms use contests to get brokers to sell securities that investors may not need. Most brokers rarely, if ever, disclose to their clients how they are paid or how their bonuses are structured, even though such disclosures would go a long way to resolving the conflict-of-interest problem. Fourth, branch-office managers and other supervisors, who are paid commissions just like their brokers, have an incentive to push everyone to sell more and to turn a blind eye to questionable practices.

Note that the paragraph has 10 sentences. The sentences link logically—they move in a tight order that builds the meaning of the paragraph—and the author uses a transition device to demarcate the "fourfold" points announced in sentence five.

Our next example paragraph is built around a list, which draws the reader's eyes to the central idea presented in the paragraph, the three phases of an explosion:

The exploding wire is a simple-to-perform yet very complex scientific phenomenon [topic sentence]. The course of any explosion depends not only on the materials and their shape but also on the electrical parameters of the circuit. An explosion consists primarily of three phases:

1. The current builds up and the wire explodes.
2. Current then flows during the dwell period.
3. "Postdwell conduction" begins with the reignition caused by impact ionization.

These phases may be run together by varying the circuit parameters.

Use of the list emphasizes the topic sentence: how explosions develop. In this case, listing provides a better method of topic sentence development for this particular idea than the same material presented in a linear paragraph:

The exploding wire is a simple-to-perform yet very complex scientific phenomenon. The course of any explosion depends not only on the material and shape of the wire but also on the electrical parameters of the circuit. In an explosion, the current builds up and the wife explodes, current flows during the dwell period, and "postdwell conduction" begins with the reignition caused by impact ionization. These phases may be run together by varying the circuit parameters.

You want to avoid excessive use of any writing technique, particularly indiscriminate use of boldface, bullets, multiple fonts, and lists. These devices should be used to draw the reader's attention to the important ideas. How you present the text on the page should be inviting and enable readers to find information easily. Combining format strategies (discussed in Chapter 8) with concise paragraphs developed with topic sentences and well-designed sentences produces a page that is easy to read. Compare Figure 4-1 and Figure 4-2. Why is Figure 4-2 easier and quicker to read?

FROM: Division Staff Manager—Customer Services, Birmingham

The Company's previous position has been not to place interduct direct buried for fiber optic cable. This directive reemphasizes this policy and explains why it is still in effect.

Recently, a company pursuing sales of interduct for this purpose has stated that their interduct will allow placement of fiber optic cable. Several demonstrations were held that showed the duct being buried and a similar-sized cable being pulled into the duct with some success.

A recent real-life trial of this direct buried interduct was very unsuccessful. It was found that after being placed and allowing the ground to settle for several days, the interduct conformed to the high and low spots in the trench. When these numerous small bends are introduced into the interduct, it becomes impossible to pull more than 400–600 feet of fiber optic cable into the interduct before the 600-pound pulling tension in exceeded.

Interduct itself offers little or no advantage as protection to a direct buried fiber optic cable. In fact, it has a negative advantage in that it will allow the cable to be pulled and fibers shattered or cracked for much greater distances.

As a result of this trial and previous recommendations, no interduct should be placed for direct buried.

Please direct further questions to H. I. Rogers at Ext. 6727

FIGURE 4-1 · EXAMPLE PARAGRAPHS

FROM: Division Manager—Customer Services, Birmingham

SUBJECT: **Company Policy: Interduct shall not be placed for buried
            lightweight cable**

**ACTION   Please ensure that this policy is conveyed to be understood by
REQUIRED: your construction managers.**

Contrary to the alleged claims of some overzealous vendors, the above policy
remains unchanged.

A field trial was conducted where the interduct was buried and cable placement
was attempted several days later. The negative observations were as follows:

1. Cable lengths are reduced between costly splices. The interduct conformed to the high and low spots in the trench. These numerous bends introduced
added physical resistance against the cable sheath during cable pulling. Even with
the application of cable lubricant, the average length pulled was 500 feet before
the 600-pound pulling tension was exceeded.

2. Maintenance liability is increased. Due to shorter cable lengths, the
number of splices increases. As the number of splices increases, maintenance
liability increases.

3. Added material costs are counterproductive. Interduct, while adding
16% to the material cost, offers little or no mechanical protection to buried fiber
optic cable. In fact, when occupied interduct was pulled at a 90 degree angle with
a backhoe bucket, the fibers were not broken only at the place of contact as they
would have been with a direct buried cable. Instead, due to the stress being
distributed along the interduct, the fibers shattered and cracked up to 100 feet
in each direction.

4. Increased labor costs are unnecessary. In about the same amount of time
required to place interduct, the fiber optic cable could be placed. The added 28%
of labor hours expended to pull cable after the interduct is placed cannot be
justified.

If you have questions, please call H. I. Rogers at Ext 6727.

FIGURE 4-2 · EXAMPLE PARAGRAPHS REFORMATTED

BASIC PRINCIPLES OF EFFECTIVE STYLE

Effective writers adjust their style in terms of:

1. Readers' knowledge of the subject
2. Readers' expectations about style based on the specific kind of writing
3. Readers' probable reading level based on the context in which the document will be read
4. Writer's relationship to readers—the professional roles of both parties

Determine Readers' Knowledge of the Subject

Reader familiarity with the subject will determine how many specialized terms you can use and where you will place them—in the text, in a glossary, in a sidebar, and so on. For readers thoroughly familiar with the subject, you can use acronyms, specialized nomenclature, and jargon that readers in a specific discipline are comfortable reading and using. Otherwise, limit the use of specialized vocabulary or perhaps define the terms. Another possibility: substitute phrases or words that will clearly express your meaning.

The following examples show how the same information can conveyed to individuals with different levels of knowledge about a subject. As you will see, you cannot separate analysis of your audience from the sentence structure and content used in a document.

An agricultural extension agent understands the technical description of a ruminant stomach:

The true (glandular) stomach in the ruminant is preceded by three divisions, or diverticula (lined with stratified squamous epithelium), where food is soaked and subjected to digestion by microorganisms before passing to the digestive tract.

The rumen, reticulum, and omasum of ruminants are collectively known as the fore-stomach. The cardia is located craniodorsally in the dome-shaped atrium ventriculi, which is common to both the rumen and the reticulum. The sulcus ruminoreticularis (esophageal groove), which extended from the cardia to the omasum, is formed by two heavy muscular folds or lips, which can close to direct material from the esophagus into the omasum directly, or open and permit the material to enter the rumen and reticulum.

But in explaining to 4-H members how the ruminant stomach works, the agricultural extension agent uses a different approach, relying on examples and analogies that will have meaning for children in primary school. He includes the technical terms for each part of the ruminant stomach, but he immediately links each term to descriptive words that would be familiar to his young readers.

The ruminant animals—such as sheep, goats, cattle, deer, antelope, elk, and camels—have a unique stomach system. The word *ruminant* comes from the Latin word ruminare, which means to chew over again and implies that ruminants are "cud-chewing" animals. Because of this need to chew their food over and over, their system differs from that of the human or other monogastric creature. Where the human stomach is one large tank, the ruminant's consists of four fermentation and storage tanks connected in series by an intricate network of flexible plumbing. The first three tanks make up the fore-stomach. The fourth tank is comparable to the human stomach and can be called the true stomach.

The rumen is the first tank, or stomach, and is quite large. It is responsible for about 75% of the digestive process. When it is full, the rumen holds up to 55 gallons of food, bacteria, and fluids. The main job of the rumen is to store food and keep it until the animal must chew it again. The rumen can be compared to the common blender. When food enters, the rumen begins mixing it with bacteria, which causes the food to start breaking down—or digesting.

The reticulum is the second stomach and is relatively small in comparison to the rumen. The reticulum occupies 5% of the total stomach. Its purpose, like that of the reticulum, is storage. The reticulum looks like a cheese grater. The common name for the reticulum is the honeycomb because it is lined with a mucous membrane that contains honeycomb-like compartments. When food enters this stomach, it passes through the honeycomb, which then breaks down the food and shreds it into small pieces. When the food stored in the reticulum has been broken down by the rumen and the reticulum, it moves to the omasum.

The omasum is the third stomach and completes the fore-stomach. It is small—occupying about 8% of the total stomach—but it is important to the process of digestion. The omasum's purpose is to make sure the food is broken down enough before it enters the true stomach. This stomach rips, shreds, and crushes the food into a liquid form so that it will not clog the pipe that connects the omasum to the abomasum.

The abomasum, which takes up about 7% of the digestive system, is the fourth stomach and is comparable to our own stomach. The abomasum digests what the rumen, reticulum, and omasum break down. At the end of the abomasum is the pipe that allows the food to enter the small intenstine. This pipe, called the pylorus, is similar to a strainer. Only properly digested food can enter the pipe.

Determine Whether a Particular Style Will Be Expected

In many organizations, "style" refers not only to the choice of words—such as salutations and closing phrases that are to be used in all the firm's correspondence—but also to items such as the heading used on all company reports, the format used for all letters, the information included on title pages of all reports, and the use of abstracts versus summaries. Large organizations often specify the stationery, the binders required for reports to clients and for intracompany reports, and the individuals to be included on distribution lists of major reports. Many organizations use document templates. When you begin working for an organization, be sure to ask whether there is a style manual or set of templates. If the company has such a document, make sure that your writing conforms to its requirements.

Many professional societies have "style sheets," which provide standards for all reports and articles published in that discipline. For example, computer engineering companies will often follow the style sheet of the IEEE Computer Society, which can be found at http://www.computer.org/author/style/index.htm. Similarly, petroleum engineering companies may continue to follow the style sheet of the Society of Petroleum Engineers in developing reports to clients and publications: http://www.spe.org/pdf/StyleGuide_March02.pdf. Companies often have their own internal style guide that explains how all reports, proposals, policies, procedures, and internal memoranda are to be prepared.

Anticipate Readers' Comprehension Level in a Given Context

The degree of difficulty in written material that readers can accommodate without misunderstanding the content will determine the style appropriate for addressing those particular readers. While readers' knowledge of the subject will determine this degree of difficulty, or level of comprehension, two other issues determine style: the context in which a reader reads, and the reader's interest in the subject.

People read analytically in some situations, whereas in others they will read rapidly or "skim" the document while looking for key words, phrases, and ideas. In an educational context, readers (like you) read evaluatively, particularly when they are trying to understand and then remember what they read. Readers will also shift from evaluative to surface reading when they are looking for information and when they need to know what to do in an emergency situation. In both situations, the technical level of language will remain, but the style will differ.

Note, the style in Figure 4-3, posted on an offshore oil platform. The instructions, using terms known by the workers on the platform, seem highly technical, but they are appropriate for the audience and the context. Note the use of lists to enhance readability. In determining your readers and assessing the context, consider what your readers may know and not know about the topic. Provide information at a level your readers will need to process your message.

Know Your Relationship to the Readers and How You Want to Sound

To relate properly to your reader, you need an idea of the kind of image you must project based on your position in the organization. In selecting the appropriate image, bear in mind that language expresses a range of human emotions, and the words you choose may sound formal, neutral, casual, or respectful—or effusive, rude, or dictatorial. Thus, in choosing a style, you must ensure its appropriateness to your position in the organization, to the context in which you are writing, and to your readers' positions and responsibilities in the organization.

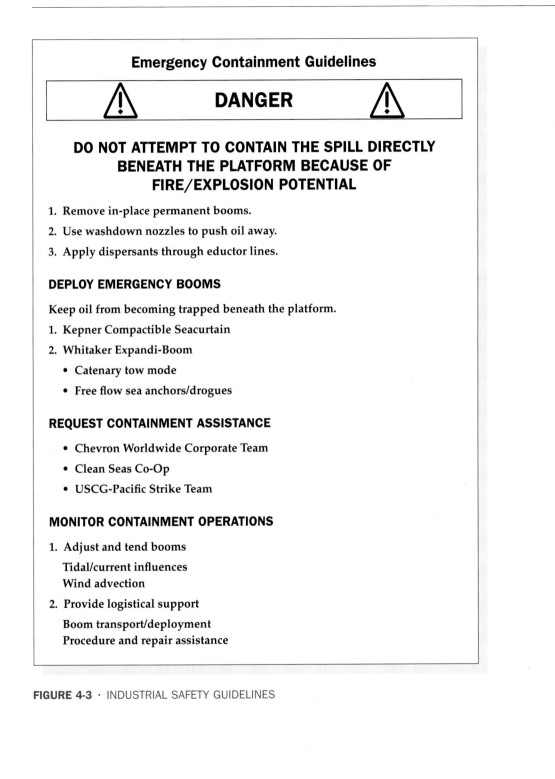

FIGURE 4-3 · INDUSTRIAL SAFETY GUIDELINES

Refer again to Figures 3-11, 3-12, and 3-13 in Chapter 3, which show samples of writing for different audiences, different purposes, and different contexts. Writing always conveys a tone or sound. Controlling how you sound ensures the effectiveness of your message. For example, the following note, written to a bank customer, exemplifies how highlighting techniques can change the tone of language, from concrete to abrupt and chastising.

Dear Mr. Green:

I am returning your mortgage papers so that you can add the mortgage amount on page 2. Since the amount line is <u>clearly</u> marked, you should be able to insert the amount and return it to me immediately.

I regret that this omission will cost you a penalty. You were <u>specifically</u> told during our meeting last week that you were to fill in <u>all</u> blanks and that the completed documents <u>had</u> to be submitted by the first of the month. We have *no* recourse except to charge you a late fee of $50 because of <u>your</u> error.

Your returning the completed form will be appreciated and will save you from further late charges, beginning on the 15th of this month.

Is the message clear? Absolutely. But does the tone convey a positive attitude toward the reader? Absolutely not. In short, concise sentences need to convey the appropriate attitude in addition to the appropriate message.

Adjust the Style to the Reader, the Purpose, and the Context

Business and technical writing should be as concise as possible because of the large quantity of information that most readers confront. Because of the increasing number of messages sent electronically, messages should focus on tightly constructed paragraphs and sentences. Even in complex, highly technical reports, readers value conciseness: the longer, the report, the less likely that anyone will read all of it.

- *To what extent does the table of contents exemplify "effective style"?*

We will discuss this point—the importance of report elements, such as the table of contents—in Chapter 10.

Select Your Level of Language; Adjust the Density of Information

Audiences familiar with your subject will expect a complex presentation. However, in routine reports and messages that you want readers to digest quickly, use writing that is less detailed and focuses on the main ideas. For reports such as those prepared for Congress by the Congressional Research Service, the broader the audience—in this case Congress and the public—the more carefully you must select easy-to-understand language. Refer again to the two descriptions of the ruminant stomach given earlier. The audience for each determines the density of the information and the level of nomenclature.

THE SENTENCE

Watch Sentence Length

Documents composed entirely of long sentences can be difficult to read. Sentence length should vary, but consider revising sentences that are longer than 20 to 25 words. Even legal documents can benefit from shorter sentences.

Before

This Appendix contains a brief discussion of certain economic and demographic characteristics of the Area in which the County is located and does not constitute a part of this Official Statement. Information in this Appendix has been obtained from the sources noted, where are believed to be reliable, although no investigation has been made to verify the accuracy of such information.

After

This Appendix contains a brief discussion of certain economic and demographic characteristics of the Area in which the County is located. The Appendix does not constitute part of this Official Statement. Information in the Appendix has been obtained from the sources noted, which are believed to be reliable. However, the accuracy of the information has not been verified.

Keep Subjects and Verbs Close Together

A recipe for sentence clarity: keep the subject of the sentence and the verb close together, and emphasize verbs. The more verbs in a sentence, the sharper and more direct the sentence. We call this verb/word ratio. For example, compare

    s   v   s   v

John *loves* Mary because she *inherited* money.   [verb/word ratio = 2/7]

   and

             s   v

Mary's inheritance of money *was* one of the reasons for John's interest in Mary.   [verb/word ratio = 1/12]

In this simple example, you can see the point: the more verbs, the sharper the sentence. If you examine the note to Mr. Green again, you will see that clarity is good. Only the inappropriate tone of the message impedes communication.

Let's take this method a step further: Lengthy sentences become less distracting to the reader if the writer structures them to enhance clarity and readability. To achieve clarity, build sentences with clauses, using as many verbs and verbals as possible. For example, the sentence,

When they plan investment portfolios, financial planners recommend a variety of investments because they resist rapid economic changes. (18 words)

develops about three clauses:

> *When they* plan *investment portfolios*
> *financial planners* recommend *a variety of investments*
> *because they* resist *economic change.*

Note that the sentence follows the three guidelines:

1. Interlocking clauses [3 in this sentence]
2. Specific action verbs: *plan, recommend,* and *resist*
3. Subject next to the verb in each clause:

> *they plan*
> *planners recommend*
> *they resist*

The verb/word ratio in this sentence is 3/17.
Assume that the writer did not follow the guidelines and avoided active verbs:

> In plans for investment portfolios, a variety of investments is recommended by financial planners because of their resistance to economic changes.

The verb/word ratio is 1/21. The sentence lacks directness and conciseness. Compare this sentence with the 18-word example that precedes it. Can you see the difference? Basically, the more verbs and verbals, the clearer the sentence:

Omit Verbiage; Use Concrete Verbs

Both e-mail and reports benefit from omission of verbiage, defined as the use of excessive words when one word will do. Choose direct verbs.

Instead of:	Use:
give approval to	approve
have a tendency to	tend to
have an influence on	influence
give notification of	notify
have a discussion about	discuss

Eliminate excessive or redundant words and phrases to eliminate sentence clutter:

Instead of:	Use:
at a later date	later
at the present time	now
by means of	by

due to the fact that	since/because
for the purpose of	to, for
for the reason that	since/because
in the event that	if
in the normal course of procedure	normally
on a daily basis	daily
we are not in a position to	we cannot
pursuant to our agreement	as we agreed
without further delay	now

Many writers and speakers develop a habit of using two or three words where one word will do. All the following expressions can be shortened without loss of clarity.

Absolutely free	Exactly identical
Background experience	Final end
Basic fundamentals	Past experience
Close proximity	Summarize briefly
Complete absence	Very complete
Consensus of opinion	Very unique

For most writing, use conversational language. Write to express, not to impress. Use specific, concrete language:

Instead of:

> There *is* now no effective existing mechanism for introducing into the beginning initiation and development stages requirements on how *to guide* employees on how *to minimize* errors in product development efforts. [verb/word ratio = 3/31]

Use:

> The company *has* no way *to guide* employees on *minimizing* product development errors during the early development stages. [verb/word ratio = 3/18]

Instead of the following sentence, which incorporates two clauses and two verbs,

> Our lack of pertinent data *prevented* determination of committee action effectiveness in funding targeting to areas that *needed* assistance the most. [verb/word ratio = 2/21]

Use a sentence with four interlocking clauses and four verbs:

> Because we *lacked* pertinent data, we *could not determine* whether the committee *had targeted* funds to areas that most *needed* assistance.
> [verb/word ratio = 4/21]

or:

> We *didn't have* enough data. Thus we *couldn't decide* whether the committee *had sent* funds to areas that most *needed* them.
> [verb/word ratio = 4/21]

When we break the sentence into two sentences, we still have four verbs. In addition, the short sentence, followed by the longer, explanatory sentence makes the meaning clear.

When clarity is your prime objective, consider using shorter, more familiar words.

Instead of:	Consider using:
accelerate	speed up
accumulate	gather
aggregate	total
anticipate	expect
cognizant	know
correspondence	e-mail, letters
endeavor	try
facilitate	help
inaugurate	begin
maintenance	upkeep
subsequent	later
terminated	ended

But always be aware of how direct/indirect words affect tone:

> (A) We encourage you to anticipate the amount of correspondence you will accumulate and suggest that you endeavor to respond in timely fashion.

> (B) Please expect large amounts of e-mails and try to answer these messages quickly.

Note that (B) is easier and quicker to read than (A). Now, go back to the opening scenario to this chapter (p. 52). Try rewriting the statement with the principles presented in this chapter.

Write "Clean" Prose

The following excerpt from Watson's *DNA: The Story of Life* (87–88), addressed to readers interested in science and possessing a basic understanding of genetics, shows how our guidelines work when they are applied to paragraphs and the paragraphs linked to form longer content units. Note the use of topic sentences, the structure of each sentence, and the development of each paragraph:

> The great size of DNA molecules posed a big problem in the early days of molecular biology. To come to grips with a particular gene—a particular stretch of DNA—we would have to devise some way of isolating it from all the rest of the DNA that sprawled around it in either direction. But it was not only a matter of isolating the gene; we also needed some way of "amplifying" it: obtaining a large enough sample of it to work with. In essence we needed a molecular editing system: a pair of molecular scissors that could cut the DNA text into manageable sections; a kind of molecular glue pot that would allow us to manipulate those pieces; and finally a molecular duplicating machine to amplify the pieces that we had cut out and isolated. We wanted to do the equivalent of what a word processor can now achieve: to cut, paste, and copy DNA.
>
> Developing the basic tools to perform these procedures seemed a tall order even after we cracked the genetic code. A number of discoveries made in the late sixties and early seventies, however, serendipitously came together in 1973 to give us so-called "recombinant DNA" technology—the capacity to edit DNA. This was no ordinary advance in lab techniques. Scientists were suddenly able to tailor DNA molecules, creating ones that had never before been seen in nature. We could "play God" with the molecular underpinning of all of life. This was an unsettling idea to many people. Jeremy Rifkin, an alarmist for whom every new genetic technology has about it the whiff of Dr. Frankenstein's monster, had it right when he remarked that recombinant DNA "rivaled the importance of the discovery of fire itself."

This excerpt uses a variety of sentences of moderate length, close subject–verb patterns, and familiar words; the description of recombinant DNA is given in terms easily understood by the nonexpert reader interested in science. In short, Watson has focused on simplicity in telling the story of DNA.

Avoid Ponderous Language

Simplicity is related to naturalness. Writers, however, often believe that they must sound learned and sophisticated to impress readers. The idea that direct writing is not sophisticated frequently derives from writing done in secondary school. Teachers encourage high school students to expand their vocabularies. Academic writing in college further enforces the importance of using jargon-laden language to convince the professor that the student knows the subject and the nomenclature of the discipline. Instructors may reward students including ponderous verbiage in research papers. On the job, however, verbose writing may be ignored or misread by readers who must quickly glean information relevant to their job needs.

Much business and technical writing contains words that have no specific meaning or words so overused that they are almost meaningless without further definition or descriptive adjectives:

concerns	logistical
dynamic	matrix
grid	program
hardware	strategic
infrastructure	synchronized
interface	systematized
integrated	

While writers do use words in this list because they are common to our reading and writing vocabularies, you should avoid overusing them.

Avoid Excessive Use of Is/Are Verb Forms

Choosing specific, concrete verbs for clarity means avoiding forms of the *be* verb, if possible. As the following sentences illustrate, excessive use of *be* verbs often obscures action verbs. Many times, a *be* verb is the best choice (as this sentence exemplifies). However, you can lessen the tendency to use *be* verbs by doing the following:

- Avoid beginning sentences with *There is* or *There are, There was* or *There were.*
- Avoid beginning sentences with phrases such as *It is clear that, It is evident that,* and *It should be noted that.*
- Choose a specific verb rather than the *be* forms *is, are, was,* and *were.*

Be verbs often create a longer, less direct sentence. Consider the following:

Delegation *is* a means of lessening the manager's workload.

vs

Managers who delegate *reduce* their workload.

Another example:

My decision *is based on the assumption* that his statement *is erroneous.*

vs

I *assumed* that his statement *is wrong.*

Two more examples: In the second part of the first one, notice that the two verbs are back to back.

Our office *has been provided* with the authority *to make* a determination about the selection of a computing system.

> vs

Our office *was authorized to select* a computing system.

Our hope *was to establish* new guidelines.

> vs

We *hoped to establish* new guidelines.

The clearest sentences focus on the agent and the action (the verb):

There *are* two systems presently available for testing job candidates.

> vs

Two available systems *can test* job candidates.

There *are* 10 steps that *must be performed* manually in the operation of the machine.

> vs

Operating the machine *requires* 10 manual steps.

Use Active Voice for Clarity

The structure of a sentence affects its clarity. In active voice, the agent that does the action occurs next to the verb. In active voice sentences, both the agent and the action appear in the sentence, and the agent becomes the subject of the sentence.

>   agent   verb
> The department teaches the course every spring term.

>   agent verb
> Our office submits all travel vouchers within 24 hours of their completion.

The result? Concise, direct sentences:

Before:

>   verb   agent   verb
> (A) Attempts were made by the division staff to assess the project.

After:

>   agent   verb
> (B) The division staff attempted to assess the project.

Sentence (A) uses passive voice. Sentence (B) uses active voice: the agent (staff) occurs as the subject and is located next to the verb (attempted).

Readability research indicates that active voice sentences may be more readable than passive sentences. It often helps readers to have the agent (the actor) placed near the action (the verb). Placing both the agent/actor and the action/ verb at the beginning alerts the reader to the basic meaning of the sentence, as contained in the subject and the verb. The following examples illustrate this concept.

> The door is to be locked at 6:00 P.M.

This sentence, which does not specify the agent, could mean either of the following:

> The guard (or some designated person) will lock the door at 6:00 P.M.
> Whoever is leaving the building at 6:00 P.M. must lock the door.

As these revisions illustrate, an understandable sentence clearly identifies both the agent and the action carried out by the agent. When you write, be sure your sentences indicate WHO or WHAT performs the ACTION.

Passive voice sentences often intentionally exclude the actor or agent doing the action. Such sentences are more verbose and less informative than an active voice version:

> Expansion of the disclosure investigation to discover all fraud has been initiated.
>
> vs
>
> The State Comptroller's office expanded the investigation to discover and disclose all fraud.

Define When Necessary

Much technical writing is generated to explain, and explanation is inevitably linked to definition and precise style. Note in the following example from Watson's book on DNA (165). The first paragraph opens with a topic sentence supported by an extended analogy.

¶1. The human body is bewilderingly complex. Traditionally, biologists have focused on one small part and tried to understand it in detail. This basic approach did not change with the advent of molecular biology. Scientists for the most part still specialize on one gene or on the genes involved in one biochemical pathway. But the parts of any machine do not operate independently. If I were to study the carburetor of my car engine, even in exquisite detail, I would still have no idea about the overall function of the engine, much less the entire car. To understand what an engine is for, and how it works, I'd need to study the whole thing—I'd need to place the carburetor in context, as one functioning part among many. The same is true of genes. To understand the genetic processes underpinning life, we need more than a detailed knowledge of particular genes or pathways; we need to place that knowledge in the context of the entire system—the genome.

Now read all four paragraphs (164–165) and note in the following that the topic sentences, combined, provide a summary of the material. Note, too, that the concise sentences follow a subject–verb–object pattern and vary in length. In each paragraph, the sentences are arranged in a way that allows the idea articulated by the topic sentence to develop linearly.

¶2. [The genome is the entire set of genetic instructions in the nucleus of every cell.] (In fact, each cell contains two genome, one derived from each parent: the two copies of each chromosome we inherit furnish us with two copies of each gene, and therefore two copies of the genome.) Genome sizes vary from species to species. From measurement of the amount of DNA in a single cell, we have been able to estimate that the human genome—half the DNA contents of a single nucleus—contains some 3.1 billion base pairs: 3,100,000,000 As, Ts, Gs, and Cs.

In this paragraph, how does the writer attempt to give readers a sense of the complexity of a single gene?

In paragraph 3, the [topic sentence] is supported by examples of human diseases and conditions that lead to death. In this paragraph, the concluding sentence, combined with the topic sentence, summarizes the main idea of the paragraph.

¶3. [Genes figure in our every success and woe, even the ultimate one:] they are implicated to some extent in all causes of mortality except accidents. In the most obvious cases, diseases like cystic fibrosis and Tay–Sachs are caused directly by mutations. But there are many other genes whose work is just as deadly, if more oblique, influencing our susceptibility to common killers like cancer and heart disease, both of which may run in families. Even our response to infectious diseases like measles and the common cold has a genetic component since the immune system is governed by our DNA. And aging is largely a genetic phenomenon as well: the effects we associate with getting older are to some extent a reflection of the lifelong accumulation of mutations in our genes. Thus, if we are to understand fully, and ultimately come to grips with, these life-or-death genetic factors, we must have a complete inventory of all the genetic players in the human body.

Paragraph 4 develops comparison with chimpanzees. The [topic sentence,] combined with the concluding sentence, summarizes the essence of the paragraph.

¶4. [Above all, the human genome contains the key to our humanity.] The freshly fertilized egg of a human and that of a chimpanzee are, superficially at least, indistinguishable, but one contains the human genome and the other the chimp genome. In each, it is the DNA that oversees the extraordinary transformation from a relatively simple single cell to the stunningly complex adult of the species, comprised, in the human instance, of 100 trillion cells. But only the chimp genome can make a chimp, and only the human genome a human. The human genome is the great set of assembly instructions that governs the development of every one of us. Human nature itself is inscribed in that book.

Writers must learn to use the terminology of their discipline, but they must know *when* to use it. One of the fundamental elements of clear style is definition—to clarify and educate. Effective writers know *when* to define words and concepts that may be confusing to the reader, and they know *how* to define or illustrate. Go back to the concise but clear definition of recombinant DNA on page 67. Note how many ways this passage defines ideas: by example, description, illustration, comparison and contrast, (analogy) and by cause–effect analysis. Let's examine another example from Watson (130):

> Since recombinant technologies allow us to harness cells to produce virtually any protein, the question has logically arisen: Why limit ourselves to pharmaceuticals? Consider the example of spider silk. So-called dragline silk, which forms the radiating spokes of a spider web, is an extraordinarily tough fiber. By weight, it is five times as strong as steel. Though there are ways spiders can be coaxed to spin more than their immediate needs require, unfortunately, attempts to create spider farms have foundered because the creatures are too territorial to be reared en masse. Now, however, the silk-protein-producing genes have been isolated and can be inserted into other organisms, which can thus serve as spider-silk factories. This very line of research is being funded by the Pentagon, which sees Spiderman in the U.S. Army's future: soldiers may one day be clad in protective suits of spider-silk body armor.

One definition, used in a health care brochure from the U.S. National Institutes of Health (NIH) shows, from another perspective, how definition devices relate to clear style (Figure 4-4). Note how definition, visual display, carefully structured sentences, and listing enhance the accessibility of the information.

Avoid Impersonal Language

Remember that writing exists for human beings, and few of us enjoy writing that is harder to read than it needs to be. What constitutes "difficult" writing depends on the reader, the topic, and the purpose of the document. But readers usually appreciate direct, concise writing that uses a conversational style. Using shorter, rather than longer sentences also helps readers follow your thoughts:

> Please give immediate attention to ensure that the pages of all documents prepared for distribution are numbered sequentially and in a place of optimum visibility. This is needed to facilitate our ability to refer to items during meetings.

> vs

> Please correctly number the pages of all documents. To help us locate material during meetings, place page numbers in the upper right-hand corner.

> or

> To help us locate material during meetings, please number all pages sequentially.

Diverticulosis and Diverticulitis

Many people have small pouches in their colons that bulge outward through weak spots, like an inner tube that pokes through weak places in a tire. Each pouch is called a diverticulum. Pouches (plural) are called diverticula. The condition of having diverticula is called diverticulosis. About 10 percent of Americans over the age of 40 have diverticulosis. The condition becomes more common as people age. About half of all people over the age of 60 have diverticulosis.

When the pouches become infected or inflamed, the condition is called diverticulitis. This happens in 10 to 25 percent of people with diverticulosis. Diverticulosis and diverticulitis are also called diverticular disease.

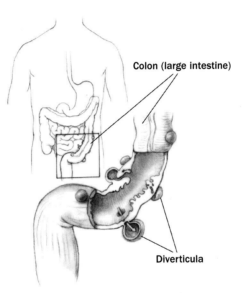

What Causes Diverticular Disease?

Although not proven, the dominant theory is that a low-fiber diet is the main cause of diverticular disease. The disease was first noticed in the United States in the early 1900s. At about the same time, processed foods were introduced into the American diet. Many processed foods contain refined, low-fiber flour. Unlike whole-wheat flour, refined flour has no wheat bran.

Diverticular disease is common in developed or industrialized countries—particularly the United States, England, and Australia—where low-fiber diets are common. The disease is rare in countries of Asia and Africa, where people eat high-fiber vegetable diets.

FIGURE 4-4 • GOVERNMENT HEALTH CARE BROCHURE
Source: http://digestive.niddk.nih.gov/ddiseases/pubs/diverticulosis/index.htm

Fiber is the part of fruits, vegetables, and grains that the body cannot digest. Some fiber dissolves easily in water (soluble fiber). It takes on a soft, jelly-like texture in the intestines. Some fiber passes almost unchanged through the intestines (insoluble fiber). Both kinds of fiber help make stools soft and easy to pass. Fiber also prevents constipation.

Constipation makes the muscles strain to move stool that is too hard. It is the main cause of increased pressure in the colon. This excess pressure might cause the weak spots in the colon to bulge out and become diverticula.

Diverticulitis occurs when diverticula become infected or inflamed. Doctors are not certain what causes the infection. It may begin when stool or bacteria are caught in the diverticula. An attack of diverticulitis can develop suddenly and without warning.

What Are the Symptoms?

Diverticulosis

Most people with diverticulosis do not have any discomfort or symptoms. However, symptoms may include mild cramps, bloating, and constipation. Other diseases such as irritable bowel syndrome (IBS) and stomach ulcers cause similar problems, so these symptoms do not always mean a person has diverticulosis. You should visit your doctor if you have these troubling symptoms.

Diverticulitis

The most common symptom of diverticulitis is abdominal pain. The most common sign is tenderness around the left side of the lower abdomen. If infection is the cause, fever, nausea, vomiting, chills, cramping, and constipation may occur as well. The severity of symptoms depends on the extent of the infection and complications.

What Is the Treatment for Diverticular Disease?

A high-fiber diet and, occasionally, mild pain medications will help relieve symptoms in most cases. Sometimes an attack of diverticulitis is serious enough to require a hospital stay and possibly surgery.

Diverticulosis

Increasing the amount of fiber in the diet may reduce symptoms of diverticulosis and prevent complications such as diverticulitis. Fiber keeps stool soft and lowers pressure inside the colon so that bowel contents can move through easily. The American Dietetic Association recommends 20 to 35 grams of fiber each day. The table below shows the amount of fiber in some foods that you can easily add to your diet.

Amount of Fiber in Some Foods

Fruits

Apple, raw, with skin	1 medium	=	4 grams
Peach, raw	1 medium	=	2 grams
Pear, raw	1 medium	=	4 grams
Tangerine, raw	1 medium	=	2 grams

FIGURE 4-4 · *CONTINUED*

Vegetables

Asparagus, fresh, cooked	4 spears	=	1 gram
Broccoli, fresh, cooked	1/2 cup	=	2.5 grams
Brussels sprouts, fresh, cooked	1/2 cup	=	2 grams
Cabbage, fresh, cooked	1/2 cup	=	1.5 grams
Carrot, fresh, cooked	1/2 cup	=	2.5 grams
Cauliflower, fresh, cooked	1/2 cup	=	1.5 grams
Romaine lettuce	1 cup	=	1 gram
Spinach, fresh, cooked	1/2 cup	=	2 grams
Summer squash, cooked	1 cup	=	3 grams
Tomato, raw	1	=	1 gram
Winter squash, cooked	1 cup	=	6 grams

Starchy Vegetables

Baked beans, canned, plain	1/2 cup	=	6.5 grams
Kidney beans, fresh, cooked	1/2 cup	=	8 grams
Lima beans, fresh, cooked	1/2 cup	=	6.5 grams
Potato, fresh, cooked	1	=	3 grams

Grains

Bread, whole-wheat	1 slice	=	2 grams
Brown rice, cooked	1 cup	=	2.5 grams
Cereal, bran flake	3/4 cup	=	5 grams
Oatmeal, plain, cooked	3/4 cup	=	3 grams
White rice, cooked	1 cup	=	1 gram

Source: United States Department of Agriculture (USDA). USDA Nutrient Database for Standard Reference Release 15. Available at www.nal.usda.gov/fnic/cgi-bin/nut_search.pl. Accessed March 20, 2003.

The doctor may also recommend taking a fiber product such as Citrucel or Metamucil once a day. These products are mixed with water and provide about 2 to 3.5 grams of fiber per tablespoon, mixed with 8 ounces of water.

Until recently, many doctors suggested avoiding foods with small seeds such as tomatoes or strawberries because they believed that particles could lodge in the diverticula and cause inflammation. However, it is now

FIGURE 4-4 · *CONTINUED*

generally accepted that only foods that may irritate or get caught in the diverticula cause problems. Foods such as nuts, popcorn hulls, and sunflower, pumpkin, caraway, and sesame seeds should be avoided. The seeds in tomatoes, zucchini, cucumbers, strawberries, and raspberries, as well as poppy seeds, are generally considered harmless. People differ in the amounts and types of foods they can eat. Decisions about diet should be made based on what works best for each person. Keeping a food diary may help identify individual items in one's diet.

If cramps, bloating, and constipation are problems, the doctor may prescribe a short course of pain medication. However, many medications affect emptying of the colon, an undesirable side effect for people with diverticulosis.

Diverticulitis

Treatment for diverticulitis focuses on clearing up the infection and inflammation, resting the colon, and preventing or minimizing complications. An attack of diverticulitis without complications may respond to antibiotics within a few days if treated early.

To help the colon rest, the doctor may recommend bed rest and a liquid diet, along with a pain reliever.

An acute attack with severe pain or severe infection may require a hospital stay. Most acute cases of diverticulitis are treated with antibiotics and a liquid diet. The antibiotics are given by injection into a vein. In some cases, however, surgery may be necessary.

Points to Remember

- Diverticulosis occurs when small pouches, called diverticula, bulge outward through weak spots in the colon (large intestine).
- The pouches form when pressure inside the colon builds, usually because of constipation.
- Most people with diverticulosis never have any discomfort or symptoms.
- The most likely cause of diverticulosis is a low-fiber diet because it increases constipation and pressure inside the colon.
- For most people with diverticulosis, eating a high-fiber diet is the only treatment needed.
- You can increase your fiber intake by eating these foods: whole-grain breads and cereals; fruit like apples and peaches; vegetables like broccoli, cabbage, spinach, carrots, asparagus, and squash; and starchy vegetables like kidney beans and lima beans.
- Diverticulitis occurs when the pouches become infected or inflamed and cause pain and tenderness around the left side of the lower abdomen.

NIH Publication No. 04-1163
April 2004

FIGURE 4-4 · *CONTINUED*

Three additional examples:

> It has recently been brought to my attention that only a small percentage of the employees in our division are contributors to the citizens' health research fund supported by this firm. This fund is a major source of money for the encouragement of significant discoveries and innovations made in behalf of research relevant to community health.

> vs

> Only a small percentage of employees in our division contribute the citizens' health research fund. Our firm supports this research because the results it produces improve community health.

> As a result of their expertise, the consulting team is provided with the opportunity to make a reasonable determination of the appropriate direction to proceed regarding their selection of information systems.

> vs

> The consulting team has the expertise to select the best information systems.

> It is our contention that the necessary modifications should be made to make the system operational because completely replacing it is economically prohibitive.

> vs

> We believe that the system should be modified to make it operational. Complete replacement would cost over $18,000.

To check for pompous writing, read your documents aloud. Read aloud the statement in the opening scenario on page 52. How does this comment sound?

Now read aloud a few sentences you have written. Is the final sentence easy to speak? How does it sound to you as you speak it? Speak aloud the preceding sentences and the revisions. Which ones are easier to say? Remember: Sentences that are hard to speak are also hard to read.

■ ■ ■ ■ ■ ■ ■   Exercises   ■ ■ ■ ■ ■ ■ ■

Further exercises related to the content of this chapter and all documents employed within the following exercises can be found on the book's companion Web site.

1. Examine for its style a letter, memo, or e-mail message that you have written. Which of the suggestions in this chapter for achieving a readable style did you seem to follow? Which of the guidelines do you seem to violate? In a group of four or five students, compare your answers. Share your original documents with members of the group. Five suggestions on how to improve each person's style.

2. Find the style sheet used by your discipline or professional association. Bring a copy to class. What suggestions about style does it offer?

3. Ben Smith, a graduate student, works as a summer intern for a landscape and environmental design firm. A senior designer has asked
Ben to submit a proposal for the summer project he would like to
undertake. Because the firm requires such proposals to be concise,
revise the document Ben wrote, incorporating the principles discussed in this chapter.

TO: Clyde Van Cleeve

FROM: Benjamin Smith

SUBJECT: Summer Project Proposal—Study of Jones Park

Project Objectives

The purpose of this project is to conduct an in-depth study of Jones Park this
summer. Four main objectives have been established. These are to collect site
evaluation for the park, to provide analysis of the data collected, to submit evaluations of past attempts and efforts toward a design for the park, and finally, to
make a recommendation of a design for the park based on the findings of this
investigation.

Background Information about Jones Park

Jones Park has a unique character that can only be captured through a detailed
study of the site. The park is located in southwest Felixville, bordered on the
north by Adams Street and on the east by Johnson Boulevard. The south
boundary is Charles Parkway, and the west boundary is 6th Street.

Proposed Data Gathering

A. Historical uses. In order to initiate the data-gathering aspect of the study, the
historical use of the park area will be determined, as well as how it has developed into the park as it is today. Land title records for the City of Felixville,
as well as the land title records for the county, will be the main sources of
information. At one time, the Jones Park area was owned by a corporation of
five residents living in the area. Consequently, a study of city ordinances,
abstracts and deed restrictions, and state laws will be necessary, as they facilitate understanding, and pertain to the city's ownership of the land.

Besides the historical legal use of Jones Park, a firsthand account of the historical
events influential to its development will be of vital importance. Personal
interviews will be conducted with the surviving members of the corporation
that once owned the park area, as well as with residents who have grown up in
the area. By including their views of the park as a part of the neighborhood, a
better understanding of the changes that have taken place can be developed.

B. Present uses. Present uses of the park emphasize the importance of the area. Easements within the park are located for drainage and utility. Another study of city ordinances will determine what the easement should be legally. Subsequent interviews with representatives of the utility company and Mayborough College will be conducted to determine if the requirements are being met.

Pedestrian uses of the park include recreation, transitional viewing, and nature study, and identify elements for the neighborhood. These uses will be studied by on-site and off-site observations. Detailed recordings will be maintained throughout an extended study period to determine the specific uses within the park. Survey questionnaires will also be distributed to users of the park, including residents adjacent to the area. Why do they use the park? What have they observed about the park as it exists today? What do they like or dislike about the area? What changes, if any, should be made for future use?

Proposed Analysis of Data

After the historical and present site use data have been gathered, they will be organized in a way that lends itself to a systematic analysis of the site. This analysis will involve the legal aspects of how much control organizations have over the park in relation to the city's legal powers. Also involved will be the deed restrictions and stipulations the one-time owners of the Jones Park area made to the city at the time of the transfer of ownership from the corporation to the city. Furthermore, the control that these restrictions have over the park in relation to the city's power today becomes pertinent. An analysis of the degree of power the city actually has over the Jones Park area is a critical point.

Furthermore, the existing site uses of the park, as well as the existing site conditions, must be recorded together in a site analysis. From the site analysis, relationships between park uses and site conditions will become apparent.

Proposed Evaluation

After the analysis of the site has been completed, an evaluation of various other park proposals that have been submitted to the Parks Department of Felixville will be undertaken. The park proposals that are available to evaluate have been submitted by other interested groups. These plans and park recommendations will be evaluated in relation to the findings of the analysis of the site uses and site conditions.

Final Recommendation

To aggregate all of the knowledge attained from the project, a site design for Jones Park will be recommended. The site design will be directed toward the ascertainable goal of providing an area that is most beneficial to the users of the park and to the neighborhood surrounding the park.

Introducing the Human Genome

The Recipe for Life

For all the diversity of the world's five and a half billion people, full of creativity and contradictions, the machinery of every human mind and body is built and run with fewer than 100,000 kinds of protein molecules. And for each of these proteins, we can imagine a single corresponding gene (though there is sometimes some redundancy) whose job it is to ensure an adequate and timely supply. In a material sense, then, all of the subtlety of our species, all of our art and science, is ultimately accounted for by a surprisingly small set of discrete genetic instructions. More surprising still, the differences between two unrelated individuals, between the man next door and Mozart, may reflect a mere handful of differences in their genomic recipes—perhaps one altered word in five hundred. We are far more alike than we are different. At the same time, there is room for near-infinite variety.

It is no overstatement to say that to decode our 30,000 genes in some fundamental way would be an epochal step toward unraveling the manifold mysteries of life.

Some Definitions

The **human genome** is the full complement of genetic material in a human cell. (Despite five and a half billion variations on a theme, the differences from one genome to the next are minute; hence, we hear about the human genome—as if there were only one.) The genome, in turn, is distributed among 23 sets of **chromosomes,** which, in each of us, have been replicated

Some DNA details

and re-replicated since the fusion of sperm and egg that marked our conception. The source of our personal uniqueness, our full genome, is therefore preserved in each of our body's several trillion cells. At a more basic level, the genome is DNA, deoxyribonucleic acid, a natural polymer built up of repeating **nucleotides,** each consisting of a simple sugar, a phosphate group, and one of four nitrogenous bases. The hierarchy of structure from chromosome to nucleotide is shown in *Some DNA details*. In the chromosomes, two DNA strands are twisted together into an entwined spiral—the famous double helix—held together by weak bonds between complementary bases, adenine (A) in one strand to thymine (T) in the other, and cytosine to guanine (C-G). In the language of molecular genetics, each of these linkages constitutes a **base pair.** All told, if we count only one of each pair of chromosomes, the human genome comprises about three billion base pairs.

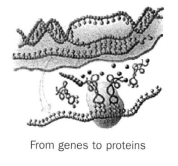

From genes to proteins

The specificity of these base-pair linkages underlies all that is wonderful about DNA. First, replication becomes straightforward. Unzipping the double helix provides unambiguous templates for the synthesis of daughter molecules: one helix begets two with near-perfect fidelity. Second, by a similar template-based process, depicted in *From genes to proteins*, a means is also available for producing a DNA-like messenger to the cell

FIGURE 4-5 • EXCERPT FROM THE INTRODUCTION TO THE HUMAN GENOME PROJECT
Source: http://www.ornl.gov/sci/techresources/Human_Genome/publicat/tko/03_introducing.html

cytoplasm. There, this **messenger RNA,** the faithful complement of a particular DNA segment, directs the synthesis of a particular protein. Many subtleties are entailed in the synthesis of proteins, but in a schematic sense, the process is elegantly simple.

Every **protein** is made up of one or more polypeptide chains, each a series of (typically) several hundred molecules known as **amino acids,** linked by so-called peptide bonds. Remarkably, only 20 different kinds of amino acids suffice as the building blocks for all human proteins. The synthesis of a protein chain, then, is simply a matter of specifying a particular sequence of amino acids. This is the role of the messenger RNA. (The same nitrogenous bases are at work in RNA as in DNA, except that uracil takes the place of the DNA base thymine.) Each linear sequence of three bases (both in RNA and in DNA) corresponds uniquely to a single amino acid. The RNA sequence AAU thus dictates that the amino acid asparagine should be added to a polypeptide chain, GCA specifies alanine—and so on. A segment of the chromosomal DNA that directs the synthesis of a single type of protein constitutes a single **gene.**

FIGURE 4-5 · *CONTINUED*

4. Examine the excerpt from the Human Genome Project Introduction in Figure 4-5. The complete report can be found on the Web site of the Oak Ridge National Laboratory: http://www.ornl.gov/sci/techresources/ Human_Genome/publicat/primer/prim1.html

   a. How does the writer use definition to develop the introduction?
   b. How many definition techniques are used?
   c. What strategies does the writer use to make the content as clear as possible?
   d. How does the writer develop the paragraphs?

5. Read the following excerpt from *Take on the Street,* by Arthur Levitt (2002, 153–154). Note the use of definition and topic sentences in each paragraph: the way each paragraph develops, begins, and ends. Also note the conversational structure of sentences, the length of sentences, and use of familiar language to explain "cash flow": how does the author use each of these stylistic features to produce text that is easy to follow?

> To really understand the quality of a company's earnings, you need to analyze the cash-flow statement, the last document in a company's financial reports. As its name implies, this statement shows the actual cash that came into the company and the actual cash that flowed out. Remember that the balance sheet reveals a company's assets, liabilities, and shareholders' equity at the close of the fiscal year or the most recent quarter. The income statement reflects the changes that have occurred in the balance sheet items, including promises of money that the company has made or received. The cash-flow

statement differs in that it reveals the changes in actual cash that the business has generated and the actual cash raised from creditors and investors. It also shows how the company invested that cash between the start of one fiscal year to the end of another. In short, cash flows shows where the money is coming from and how the money is being spent.

The cash-flow statement has three parts. First you'll see "cash from operating activities," which is the money that comes in from sales of the company's products and services and the money going out to produce those sales. It also includes interest and tax payments. Recall that accounting rules allow companies to list revenue on the income statement before they receive actual payment. Not so on the cash-flow statement, which lists only revenues actually collected.

You may see negative cash flow from operations. While on its face that may seem like a bad sign, it isn't always a signal to sell. Fast-flowing start-ups, which consume more cash than they can generate in the first few years of the business, will especially show negative cash flow. They cover the shortfall by borrowing money or issuing stock. But sometimes negative operating cash flow indicates that a company is in serious trouble, especially if the company is disposing of assets, as by selling off pieces of the company, because it can't persuade investors to buy its stock or bankers to lend it money.

The second item on the cash-flow statement is "cash from investing activities." This is where the company reveals how it's using its excess cash, either by investing in other companies or by expanding its own business. If the company, like thousands of others, invested in the stock market in the go-go '90s, this is where you'll see how much cash it received from stock sales, or how much it spent to buy shares. Any gains or losses on such stock, however, are reported in the operating activities section of the cash flow statement. The investing part of the cash flow statement also reveals any investments in long-term assets, such as an acquisition (or a sale) of a manufacturing plant or equipment, or the opening of new retail stores. Finally, if the company lent money to its executives to allow them to buy stock, that loan will be listed here.

The third part, "cash from financing activities," includes money that comes in or goes out when a company sells or buys shares of its own stock. Issuing new debt or bank borrowings also will show up here an increase in cash. Paying dividends or paying down debt will show up as a decrease in cash. A start-up company with little or no sales is more likely than a mature company to show lots of financing activity, since the cash to run the business has to come from somewhere.

So there's the cash-flow statement in a nutshell. If you look back at the balance sheet from the amount of cash the company has—listed under current assets—that figure comes from the final cash balance on the cash-flow statement.

REFERENCES

Levitt, Arthur (with Paula Dwyer). *Take on the Street*. New York: Pantheon Books, 2002.

Watson, James (with Andrew Berry). *DNA: The Secret of Life*. New York: Alfred A. Knopf, 2003.

Scenario

You are a supervisor at Bright Engineering, a civil engineering and construction company. It's been a tough year at your company: you typically do a lot of projects for cities, such as fire stations and jails, but business has been sluggish.

The Chief Financial Officer, Maria Trevino, has notified you that no raises will be given this year, but you have been authorized to give a single merit bonus of $1,000 to the leading engineer in your division. This stipulation creates a dilemma for you as two candidates come immediately to mind.

Karissa Hopper has been at Bright Engineering for only two years, but she is always cordial and conscientious and has been doing a great job of keeping projects within their budgets. She is also a single mother raising two young daughters. She's been having a tough time with medical bills for the younger child and could really use the extra money.

Lew Vecchio is also a good candidate for the bonus. He has been at Bright for five years and has never missed a deadline on a project. He is gregarious, industrious, and a favorite of the clients. He is also your neighbor and friend. His wife is your mother's doctor. You know he doesn't really need the money, but would appreciate the recognition that comes with the merit bonus. If you choose Lew, however, individuals might question your integrity and think you made the selection because of your friendship.

You decide to give the bonus to Karissa and write the following e-mail message to the CFO:

> Maria:
>
> I recommend Karissa Hopper as the candidate for the merit bonus. She is scrupulous on budgets and conscientious on meeting deadlines (see attached list of projects and clients). The job she did this year on the Lubbock jail was exceptional and likely to lead to additional business with the city. I can't think of a better contributor to the productivity of my division.

Is your e-mail message ethical?

Would you write the same message if you thought it would be made public?

As you'll see in this chapter, your communications on the job are often riddled with ethical implications. You'll also find in this chapter ways of identifying and managing unethical communication practices.

The key advice of this chapter, however, is this: always assume that sooner or later, your communications may be made public. If you believe that you can't hide your actions, you're more likely to act ethically.

CHAPTER FIVE

Writing Ethically

ETHICAL PERSPECTIVES

Philosophers of ethics espouse two basic perspectives. The first focuses on character, encouraging us to develop virtues such as wisdom, courage, justice, prudence, truthfulness, and generosity. According to this perspective, if you cultivate such virtues, you will develop ethical habits and will live a moral life. This perspective asks the question "Who will I be?" Will I be a manager recognized for my fairness and honesty? Will I be a writer appreciated for being caring and candid? Will I be a colleague who is trusted and admired?

The second perspective is analytical, focusing on a logical determination of right and wrong behavior, especially how to decide ethical dilemmas. Ethics, according to this perspective, asks the question "What will I do?" Answers to this question emphasize obedience to obligations or consequences of behavior or a negotiation of the two. For example, in the case of the merit bonus, you would determine your obligations:

- Are you obliged to consider only work-related factors in making your decision?
- Are you permitted to consider the candidate's financial situation?
- Are you obliged to report all the factors that influenced your decision or do you have the right to omit the factors that might damage your case?

Or do you consider only the consequences of your decision and try to determine the solution that offers the greatest benefit or the least cost? In business typically, a decision is good only if it leads to good results (and the sooner the better). What will be the impact on Karissa if she doesn't receive the bonus? What will be the impact on Lew if she does? What will happen to your reputation? Will your boss think you made the right decision? Will your colleagues? If Lew ever sees your e-mail message, would that damage your friendship? If there's little chance he'll see it, does that make it okay to write it?

Or do you weigh all such questions in making your decision?

On the job you typically have limited time for the rational analysis of ethical dilemmas and thus you won't always have the opportunity to consider all the issues or answer all the questions relating to your dilemma. You might have to act or make a decision quickly—in minutes or seconds. In such situations, you will need to operate on intuitions of conscience and virtuous habits of mind. Nevertheless, you'll need to know the nature and scope of your obligations and practice anticipating the conseqences of your decisions. Such knowledge and practice will assist you in making rational decisions if time permits and in making good intuitive judgments if it doesn't.

YOUR PROFESSIONAL OBLIGATIONS

None of us are isolated individuals, operating entirely separate from the traffic of human society. Each of your ethical obligations has several aspects, often intersecting, and from time to time competing. Consider, for example, your duties to the following:

■ **To yourself** You will have to make decisions and take actions that allow you to support yourself financially while establishing (and maintaining) your reputation in your field. You can't quit (or lose) your job every time you object to a supervisor's policy.

■ **To your discipline and profession** As a member of your profession, you have a responsibility to advance the knowledge and reputation of your field. You must share with colleagues information that will improve the practices of your profession, clarify understanding, offer new insights, and promote better training of students of your discipline. You must communicate in a manner that both brings credit to your profession and inspires the next generation to join it.

■ **To your academic institution** You have a moral obligation to the institution that trained you for your profession. Your successes or failures will be indicative of the merits of that institution and its faculty. If you disgrace yourself by illegal or unethical actions, for example, investigating officials and the public might ask why you weren't taught better behavior.

■ **To your employer** Your responsibility as an employee is to serve the interests of your organization, to help it make money, to promote its products and services, and to shield confidential information and intellectual property, especially if maintaining confidentiality offers a competitive advantage.

■ **To your colleagues** You have a duty to your colleagues on the job to do your fair share of the work assigned and to do it with integrity, accuracy, and efficiency. You also have responsibility to use no more than your fair share of the resources allotted and to take no more than your fair share of the credit (or blame) given.

■ **To the public** Your obligation to society is to promote the public good through greater safety, fuller liberty, and a better quality of life. Your decisions and actions on the job could create opportunities for communities to thrive or could cause public aspirations to fall victim to private greed.

In communicating on the job, you will have to juggle your various obligations and determine which has priority. Because you also have important responsibilities to yourself, your profession, your schools and teachers, your colleagues, and the public, you can't always simply do whatever the boss requests. You will need to make every effort to avoid being either submissive or self-righteous.

CODES OF CONDUCT

Both your professional association and your employing organization are likely to have codes of conduct or ethical guidelines that specify their expectations regarding appropriate behavior. For example, the Society for Technical Communication, the international association for technical writers and graphic artists, expresses its guidelines as a list of six principles (see Figure 5-1).

www.oup.com/us/houp

STC Ethical Principles for Technical Communication

As technical communicators, we observe the following ethical principles in our professional activities.

Legality

We observe the laws and regulations governing our profession. We meet the terms of contracts we undertake. We ensure that all terms are consistent with laws and regulations locally and globally, as applicable, and with STC ethical principles.

Honesty

We seek to promote the public good in our activities. To the best of our ability, we provide truthful and accurate communications. We also dedicate ourselves to conciseness, clarity, coherence, and creativity, striving to meet the needs of those who use our products and services. We alert our clients and employers when we believe that material is ambiguous. Before using another person's work, we obtain permission. We attribute authorship of material and ideas only to those who make an original and substantive contribution. We do not perform work outside our job scope during hours compensated by clients or employers, except with their permission; nor do we use their facilities, equipment, or supplies without their approval. When we advertise our services, we do so truthfully.

Confidentiality

We respect the confidentiality of our clients, employers, and professional organizations. We disclose business-sensitive information only with their consent or when legally required to do so. We obtain releases from clients and employers before including any business sensitive materials in our portfolios or commercial demonstrations or before using such materials for another client or employer.

Quality

We endeavor to produce excellence in our communication products. We negotiate realistic agreements with clients and employers on schedules, budgets, and deliverables during project planning. Then we strive to fulfill our obligations in a timely, responsible manner.

Fairness

We respect cultural variety and other aspects of diversity in our clients, employers, development teams, and audiences. We serve the business interests of our clients and employers as long as they are consistent with the public good. Whenever possible, we avoid conflicts of interest in fulfilling our professional responsibilities and activities. If we discern a conflict of interest, we disclose it to those concerned and obtain their approval before proceeding.

Professionalism

We evaluate communication products and services constructively and tactfully, and seek definitive assessments of our own professional performance. We advance technical communication through our integrity and excellence in performing each task we undertake. Additionally, we assist other persons in our profession through mentoring, networking, and instruction. We also pursue professional self-improvement, especially through courses and conferences.

FIGURE 5-1 · ETHICAL GUIDELINES FOR THE SOCIETY FOR TECHNICAL COMMUNICATION
Source: http://www.stc-va.org/fguide.htm

IEEE Code of Ethics

We, the members of the IEEE, in recognition of the importance of our technologies in affecting the quality of life throughout the world, and in accepting a personal obligation to our profession, its members and the communities we serve, do hereby commit ourselves to the highest ethical and professional conduct and agree:

1. to accept responsibility in making engineering decisions consistent with the safety, health and welfare of the public, and to disclose promptly factors that might endanger the public or the environment;

2. to avoid real or perceived conflicts of interest whenever possible, and to disclose them to affected parties when they do exist;

3. to be honest and realistic in stating claims or estimates based on available data;

4. to reject bribery in all its forms;

5. to improve the understanding of technology, its appropriate application, and potential consequences;

6. to maintain and improve our technical competence and to undertake technological tasks for others only if qualified by training or experience, or after full disclosure of pertinent limitations;

7. to seek, accept, and offer honest criticism of technical work, to acknowledge and correct errors, and to credit properly the contributions of others;

8. to treat fairly all persons regardless of such factors as race, religion, gender, disability, age, or national origin;

9. to avoid injuring others, their property, reputation, or employment by false or malicious action;

10. to assist colleagues and co-workers in their professional development and to support them in following this code of ethics.

FIGURE 5-2 · ETHICAL GUIDELINES FOR THE INSTITUTE OF ELECTRICAL AND ELECTRONICS ENGINEERS
Source: http://www.ieee.org/about/whatis/code.html

The Institute of Electrical and Electronics Engineers, a huge international association serving a variety of technology specialists, sets out its guidelines as a list of ten ethical objectives (see Figure 5-2).

Know the codes of conduct that motivate and regulate ethical communication for your company and your profession: you might have to cite such guidelines to justify decisions you make when ethical dilemmas crop up on the job.

Also establish a professional code of conduct for yourself:

- What are your principles?
- What are you willing and unwilling to do?
- What do you expect of yourself?
- Who in your profession do you admire and try to be like?

Ordinarily, codes of conduct assert guiding principles that you must interpret and apply to specific situations. Practice at such applications through cases and scenarios will assist you in making suitable decisions at the appropriate times.

RECOGNIZING UNETHICAL COMMUNICATION

Also essential to communicating ethically is the recognition of ways in which employees and executives alike may be unethical in their communications. Chief among the possibilities are plagiarizing, deliberately using imprecise or ambiguous language, manipulating statistics, using misleading visuals, and promoting prejudice.

Plagiarism and Theft of Intellectual Property

On the job, you may be responsible for the security of five kinds of intellectual property:

1. Copyrightable material: a composition of original material fixed in a tangible medium, such as books, journals, software applications, computer programs, video or audio recordings, or illustrations. This includes materials available in digital files, e-mail messages, and Web pages.

2. Trademark: a display of words or symbols communicated in text, illustrations, or sounds that identify and distinguish the goods and services of a manufacturer or supplier, such as the name or logo of a company.

3. Trade secret: a design, formula, list, method, pattern, or process that offers a competitive advantage over parties who don't have the same information, such as a special recipe.

4. Invention: a new and unique design, device, method, or process that is subject to patent protection.

5. Tangible research property: tangible items created during research related to copyrightable materials, trademarks, trade secrets, and inventions, such as databases, diagrams, drawings, notes, prototypes, samples, and associated equipment and supplies.

Copyrightable material is unique in that for certain purposes (e.g., criticism, news reporting, research, teaching) you have the right to borrow limited portions for presentation or publication without the explicit permission of the owner. If the borrowing is extensive, however, permission is necessary.

On the job, writers will often recycle the words and images from various company documents without identifying the original source. They will readily lift paragraphs from the corporate Web site, for example, to use in a business letter to a potential client, or they will borrow a table from the annual report for insertion into a proposal to a potential funding agency. Such recycling of material is efficient, and it is entirely legal and ethical as long as the participating writers recognize this sharing of effort and no organizational policy disallows it. This is because the company has ownership of the words and images that are being recycled. (If you have doubts about the propriety of such recycling within your company, ask the writer or writers for permission.)

To use materials not generated by your company, however, you must acknowledge the sources of borrowed words, images, and ideas. In the majority of documents you write (such as letters, e-mail messages, and memos), the acknowledgment may be a brief and simple introduction to the borrowed material: for example, "As Dr. Shirley Olson, of the National Institutes of Health, discovered, it's possible to vaccinate mosquitoes to prevent their developing and passing the disease on to human beings."

In formal reports, however, some official system of documentation would be necessary to identify the source of the information and to give full credit to Dr. Olson. Your organization might develop a special style for such source citations or adopt a standard style guide such as *The Chicago Manual of Style* of the University of Chicago Press, the *APA Publication Manual* of the American Psychological Association, or the *MLA Style Manual* of the Modern Language Association. These style guides are periodically revised, so be sure that you have the latest edition. For example, the fifteenth edition of *The Chicago Manual of Style* was published in 2003.

Plagiarism   To use the words, images, or ideas of others without attribution is plagiarism, or a theft of intellectual property. Plagiarism is highly unethical and potentially illegal. Your intentions are immaterial: plagiarism is plagiarism, whether it occurs deliberately or inadvertently. You must, therefore, be especially cautious to avoid plagiarism: your organization could find itself the defendant in a criminal case or a civil suit, and you could lose your job and your reputation.

Note also that material is automatically copyrighted as soon as it is created: it need not carry a copyright notice, and the copyright need not be registered with the U.S. Copyright Office (www.copyright.gov), though material so protected is often less likely to be plagiarized.

Quotations   If you quote a source (including material from digital files, e-mail messages, or Web pages), put the borrowed material inside quotation marks (or display it in a separate indented paragraph) and specify the source. If you paraphrase or summarize, you don't need quotation marks but you still must specify the source. Make sure that the wording and phrasing of even your acknowledged paraphrases and summaries differ sufficiently from the original passage to invalidate any suggestion of plagiarism.

Here, for example, is an original passage, followed by a complete citation of its source:

> Ethylene oxide has a boiling point of 51°F. It is processed as a liquid through the application of pressure. As the temperature of the ethylene oxide increases, the pressure in the feed line will correspondingly increase. At the time of the explosion, the ambient temperature was around 93°F. The feed line was not insulated or cooled. (From United States Environmental Protection Agency, Office of Solid Waste and Emergency Response. 2000. *EPA chemical accident investigation report, Accra Pac Group, Inc. North Plant, Elkhart, Indiana.* EPA 550-R-00-001. Washington, DC.)

Here is a summary that would be considered plagiarism even if the source were cited. Note how this passage changes words here and there but essentially duplicates the original passage:

> The boiling point of ethylene oxide is 51°F. Pressure is applied to process it as a liquid. As the temperature of the ethylene oxide rises, so does the pressure in the feed line. The temperature in the vicinity was roughly 93°F at the time of the explosion, and the feed line was neither insulated nor cooled.

Here is a summary that would be considered ethically appropriate. Note how the order of the sentences has been substantially changed, as well as the words. Nevertheless, the source of the information must be identified:

> Ethylene oxide was pressurized as a liquid in a feed line that was neither insulated from external temperatures nor subjected to any kind of special cooling. Ethylene oxide boils at 51°, but the ambient temperature rose to approximately 93°F. Increased pressure from the boiling chemical inside the feed line caused the line to rupture, resulting in the explosion.

If you are borrowing a substantial portion of the original source (e.g., several paragraphs or a single image), you will have two ethical and legal duties:

1. Acknowledging the source
2. Requesting permission from the owner of the intellectual property

The extensive borrowing of copyrighted material will ordinarily be permitted, but often with restrictions and for a cost. You may contact the copyright owner directly or make your request through a service such as the Copyright Clearance Center (www.copyright.com). If the copyright owner agrees to let you use the material, you must add a note indicating the source and stating that the copy is used with permission. For example:

> From *Ethics in Technical Communication,* by Paul Dombrowski. Copyright 2000 by Allyn & Bacon. Reprinted by permission.

If you change the original material (e.g., summarizing passages or revising illustrations), you would specify the adaptation of the borrowing:

> From *A Legal Primer for the Digital Age,* by TyAnna K. Herrington. Copyright 2004 by Pearson Education. Adapted by permission.

Permission is not needed for material in the public domain (i.e., intellectual property for which copyright protection has expired or material created by agencies of the U.S. government), but such sources must always be acknowledged.

Deliberately Imprecise or Ambiguous Language

Ordinarily, unclear and ambiguous language is a result of the writer's negligence, but it may be a sign of a deliberate effort to mislead or manipulate the reader.

Writers can imply that things are better or worse than they really are through the choice of words. For example, a writer answering an inquiry about her company's voltage generator could reply, "Our voltage generator is designed to operate from the heat of Saudi Arabian deserts to the frozen tundra of Greenland." It may be true that the generator was *designed* that way, but if it *operates* effectively only between Atlanta and Cleveland, the writer has made a false implication without telling a straight lie.

Negative assertions and absolutes are often deceptive. For example, a claim such as "No graphics software does more or costs less than PaintPower" seems to declare that PaintPower does more and costs less. For the claim to be true, however, PaintPower doesn't have to be better or cheaper: it could have the same functions and the same price as other graphics software. Thus PaintPower may be just as good and just as cheap as competing software, but the claim that it's a better and cheaper product is not justified.

Manipulation of Numerical Information

The manipulation of statistics is a leading way to deceive your readers. For example, imagine the writer of a feasibility report who wishes to convey the impression that a certain change in company policy is desirable. She surveys all the workers in the company and finds that 51 percent of the 20 percent who returned the survey favor the change. In the report she writes "A majority of those who completed the survey favored the change." By using *majority*, she makes a stronger case for change than if she reported the exact figure of 51 percent. In addition, by not revealing that this "majority" represents roughly 10 percent of the company's workers, she fortifies the fairly thin support for the policy change. She has not exactly lied, but certainly she has deceived the readers of the report.

Or imagine a mutual fund that leads its market in returns for ten years. In the eleventh year, the original fund manager retires and a new manager is hired. In that year and the next, the fund drops to the bottom tenth of its market in returns. The writer of an advertising brochure for the fund writes the following: "Our fund has been the market leader for ten of the last twelve years." Again, the writer avoids a lie, but the implication that the fund may *now* be the market leader is clearly deceptive.

Use of Misleading Illustrations

Like words, illustrations have the capacity to misrepresent facts and mislead people. For example, only two of a company's hundred employees are African American. The recruiting materials the firm carries to college campuses show a dozen workers doing different jobs, including both minority employees. This communication is unethical because it implies that African Americans constitute almost 17% of the employees—a gross distortion of the real situation. Prospective job candidates would be substantially deceived about the diversity of colleagues and the working environment this company offers.

Or consider a line graph such as Figure 5-3, which depicts seven years of volatile change followed by a period of relative stability.

If the earlier years are trimmed from the line graph, however, the company could be made to look like it only experiences mild ups and downs, as shown in Figure 5-4.

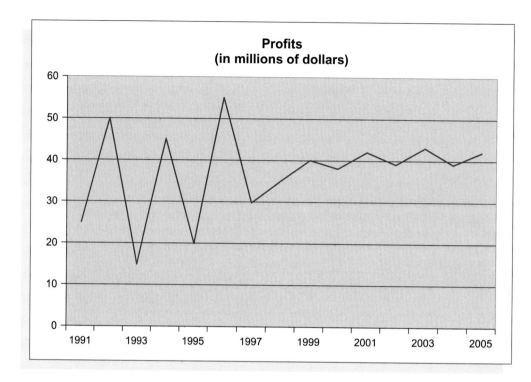

FIGURE 5-3 · A LINE GRAPH WITH THE COMPLETE PICTURE

Promotion of Prejudice

Writers also communicate unethically by voicing prejudice through their choice of words and illustrations. For example, consider the following passage from a company memo:

> One of the constraints that the company operates under is the need to recruit women candidates for managerial positions. Harriet Smith was clearly ineffective as operations manager.

The unmistakable implication here is that women candidates are likely to be underqualified (i.e., having to hire women as managers is "a constraint"). It is also implied that Harriet Smith is representative of all women. It is unlikely that a man who proved to be ineffective as a manager would be considered representative of all men.

If men in the company are always called "Mr." or "Dr." and comparable titles are not used for women (e.g., Mr. William Jones, advertising manager, and Harriet Smith, operations manager), these women seem less credible and authoritative than their male counterparts. If photographs in the company's annual report always show women sitting at desks staring at computer screens and men sitting at big conference tables, the clear message is that men are important executives while women are suited only for clerical positions.

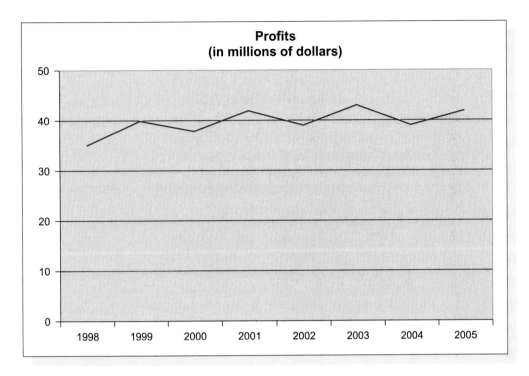

FIGURE 5-4 · THE SAME GRAPH AS FIGURE 5-3 WITH DECEPTIVE OMISSIONS

Your job as a communicator is to make sure that you don't reinforce or inspire prejudice and bigotry. Your ethical obligation is to offer only valid and reliable findings, fair and unbiased analyses, and logically justified conclusions.

ANTICIPATING CONSEQUENCES

In addition to knowing your ethical obligations as a communicator, it is important to consider the consequences of your decisions and actions. That is, even though you know your obligations, by failing to recognize the full implications or impact of what you have written, you may nevertheless have been unethical. A way to make yourself aware of the consequences is to design a fault tree diagram. In your planning or writing, you will perceive various options available to you. As you construct your fault tree, you would sketch each option as a branch, and list the immediate and lasting consequences for each branch. If any of the listed consequences leads to further options, draw another branch showing those options and list their consequences, and so on, until you have exhausted all likely options. Let us illustrate.

Imagine yourself to be a newly graduated civil engineer. You are hired by a land developer to draw up plans for streets and sewage disposal for a large parcel of land on which your client plans to build 45 houses. In walking the parcel, you discover that about half of it is a waste dump filled with trees and other vegetation covered over with several feet of soil. When you draw this to the developer's attention, he tells you that he has dumped debris from other projects on the parcel now slated for development. Upon further questioning, he reveals that he has never sought a county permit for these activities, which means that the dump is an unauthorized land use. You realize that a dump filled with vegetation is a potential source of substantial amounts of highly explosive methane gas. The options you can recommend to the developer boil down to three main categories:

1. Proceed with the development as planned
2. Delay building until the site has been cleared of the dangerous debris
3. Cancel the development plans

To help yourself sort out the consequences of the actions, you diagram the situation, as in Figure 5-5. This fault tree makes it clear that you cannot ethically recommend option 1 because of the dangerous condition that sooner or later could lead to a calamity. Option 2 is ethically acceptable, despite its immediate negative consequence (i.e., the cost of ridding the site of the debris). Option 3 is ethical but has the immediate negative impact of eliminating jobs. If the developer chooses either option 2 or 3, you have likely fulfilled your moral obligation. If the developer decides to proceed with the project, you have a new ethical

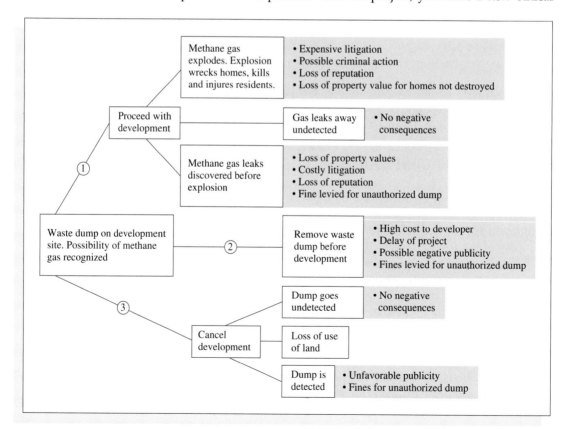

FIGURE 5-5 · FAULT TREE

decision to make. Do you stay quiet, but keep a copy of your report to protect yourself? Or do you "blow the whistle" on the developer to local officials? What are your obligations? What might be the consequences?

APPLYING PRINCIPLES

Cases and scenarios allow you the opportunity to anticipate the kinds of ethical dilemma you might experience on the job and to decide how you might avoid or manage such situations. Cases typically give a detailed narrative, whereas scenarios offer a brief summary. Both are designed to stimulate your thinking and encourage you to practice making ethical judgments. On the job you won't always have time to think through a decision: a boss or client or colleague will ask you to do something that doesn't seem quite right and you'll have to decide on the spot how to address the request. Cases and scenarios heighten your sensitivity to ethics and offer you experience at weighing obligations and consequences.

You might be asked on the job, for example, to

- Shrink the size of safety warnings on the instructions for a new product
- Make recommendations that help your company instead of your client
- Suppress evidence in a feasibility study to make a project look cost-effective
- Describe possible benefits of a policy change as definite benefits
- Write a letter of support for a supplier who has bought you lunch

At times the ethical line is clear, and you'll find it easy to judge the right thing to do. Often, however, you'll find genuine differences of opinion regarding the morality of a specific situation. Was that lunch invitation a kind of bribe? Or a cordial way for the supplier to make a pitch? Is the suppressed evidence truly pertinent, or could it honestly be described as extraneous?

Taking time on the job—over lunch or in meetings—to discuss ethics with your colleagues is a good way to encourage awareness of potential dilemmas and to develop a consensus regarding your moral obligations. Such conversations will also help you to identify possible mentors (i.e., individuals whose opinions you trust and whose behavior you admire) from whom to seek guidance if necessary.

HANDLING UNETHICAL SITUATIONS

If you find yourself asked by a boss, colleague, or client to do something that you don't consider right, don't hesitate to ask questions. Be polite, however. Don't accuse anyone or immediately declare a request unethical. Explain your dilemma and ask for clarification. The other person might be unaware of the difficult moral position his or her request puts you in and might immediately modify the request.

If explanations and negotiations don't satisfy you and if time allows, visit with a supervisor or mentor for guidance. Review the code of conduct of your profession or company for passages that might support or challenge your position.

If investigation and deliberation fail to quiet your moral doubts, explain (in writing, if possible) that you don't feel comfortable doing X, but you could do Y. That is, identify both the thing you can't do (the apparently unethical action requested), but also a thing that you could do (an option you believe to be ethical).

If asked to justify your decision, cite the appropriate passages of your profession's or company's code of conduct. That is, make it clear that the behavior requested would seem to be proscribed by your profession or your company. That is, your position should not be interpreted simply as a refusal by you to carry out a directive.

If you see something occurring on the job that you don't think is right, again, discuss the situation with your supervisor or mentor (in writing, if possible). If this step fails to satisfy you that the activity is justified, consider your ethical obligations, including to your profession and to the public. Always make the decision that you could live with if your choice were made public.

A final piece of advice. Keep in mind that you bring two important credentials to a job: a knowledge of your field and a reputation for integrity. If your reputation for integrity is lost, your knowledge of your field won't matter because you can't be trusted.

■ ■ ■ ■ ■ ▨ ▨ Exercises ▨ ▨ ▨ ■ ■ ■ ■ ■

Further exercises related to the content of this chapter and all documents employed within the following exercises can be found on the book's companion Web site.

1. For the cover of this year's annual report to shareholders, your organization would like to use a photograph of a farm in North Dakota that was published in the book *Taking Measures across the American Landscape* (Yale University Press, 2000). According to the Copyright Clearance Center (www.copyright.com), what would be the cost of your using this photograph? Write a memo to your boss explaining why your organization must request permission to publish this photograph.

2. What ethical dilemmas have you encountered at a job? How did you manage such dilemmas? Did a supervisor or corporate code of conduct offer you sufficient guidance? In a memo to a company's chief ethics officer (or similar position), describe one such experience and make recommendations for changes in the firm's policies and practices.

3. Your immediate supervisor would like you to insert on your company's Web site a 15-second video clip from the 2003 film *The Corporation*. He has a copy of the film on a DVD that he bought himself. Visit the U.S. Copyright Office (www.copyright.gov) to determine the pertinent guidelines regarding such borrowing. Write a memo to your supervisor explaining the copyright guidelines and the importance of compliance.

4. You are the manager of a growing division at your company. Last week you hired five new associates who will start at the beginning of the month. This morning, however, you received a disturbing telephone call from your district supervisor:

Supervisor: We're instituting a change of policy immediately regarding insurance benefits for new employees.

You: What's the change? You know I have five new people starting on the first.

Supervisor: The coverage start date will be delayed for sixty days. The savings for us are huge.

You: You're saying they won't have insurance for the first two months they're on the job?

Supervisor: Yes, that's right.

You: Whose idea was this? It doesn't seem right at all. And it's certainly no way to inspire people's loyalty or confidence in their new company.

Supervisor: The insurance coordinator advised it because of the lower costs. The premiums were killing us and it was either this or raise the deductibles and copayments for all employees.

You: My new hires are exempt, aren't they? They've already accepted their job offers.

Supervisor: No, this is effective immediately. Their offer letters didn't discuss the details of insurance, only that it's available. And it will be, but not for sixty days.

You: Are you serious? They're going to think we—or I—tricked them. I'm going to have to notify them of this new policy immediately: Several did have competing offers and might like to change their minds.

Supervisor: No, don't. They're all young: they're not worried about insurance. This would never be a deciding factor for them in choosing a job. If you bring it up and make a big fuss about it, you'll just make them irritated. They'll find out about the insurance soon enough and by that time they'll be in their new offices and occupied with their jobs and meeting all their new colleagues and . . .

You: But . . .

Supervisor: Look, you could lose the positions. And you'll be desperate for all five of them next month if you're going to meet the deadline on the library project.

Do you write a letter to your five new hires notifying them of the delay in their insurance coverage?

If yes, write the letter. Also write a brief e-mail message to your supervisor explaining why you thought it was necessary to advise the new hires of the policy change.

If no, how would you answer the following e-mail message:

I discovered this morning that my insurance coverage doesn't start for sixty days. Why wasn't this explained earlier? Did you know about this?

5. Radon is an odorless, radioactive gas produced by the breakdown of uranium in the soil. Exposure to radon at sufficient levels can cause lung cancer. The U.S. Surgeon General considers radon to be second only to smoking as a cause of lung cancer in the United States. Concerned residents frequently hire radon removal contractors to test for radon levels in their houses and, if necessary, to install systems to remove the pollutant.

You obtain summer employment with a radon-removal company that tests for radon levels in houses and, if necessary, installs radon-removal systems. The company's name is Darling Radon Removal and the owners are Benjamin and Gloria Darling. Typically, the contractor tests the house for radon and then presents a proposal to the homeowner detailing any work determined to be necessary and naming a price. To obtain more information about radon and its reduction, you read a government booklet titled *Consumer's Guide to Radon Reduction.* From this reliable source, you learn that the most expensive radon-reduction systems are needed for houses that are built either on concrete slabs or with basements. Systems for such houses can run as high as $2,500. Houses built over crawl spaces can almost always obtain adequate reduction by increasing the ventilation of the crawl space, a measure that seldom costs more than $500. You realize that you have been helping Mr. Darling install expensive systems suitable for basement and slab-constructed houses in houses having crawl space only.

You look at a proposal being presented to a homeowner of a house with crawl space construction. In the proposal, you find that Mr. Darling has recommended suction depressurization, a system normally used under basements or slabs. It requires an expensive installation of pipes and fans in the soil under the house to trap and suck away radon. Your employer offers no alternatives to this system. In the proposal, Mr. Darling justifies the suction depressurization system with this statement: "Suction depressurization is the most common and usually the most reliable radon-reduction method." From your research on radon, you know this is a true statement.

What should you do? Write a memorandum to your instructor describing your conclusions and any actions you plan to take.

6. In collaborative groups of four or five people, discuss the following problem, using the bulleted questions as an aid.

Thelma Thomas has been working on her new job as systems analyst in Oglethorpe Consulting for three months. It is her first job out of college, and she is enjoying it. She has made friends with Jim Brown, whose workstation is next to hers. At lunch time, Jim frequently eats at his workstation while playing electronic games at his computer. One day Thelma mentions to Jim that she wonders how he can afford such a variety of expensive games. Jim smiles, and says, "No problem."

Later that day, Thelma gets an e-mail message from Jim directing her to a bulletin board (BB) that he maintains. Logging on to the BB, she finds that it is a repository of copyrighted business and game software, as well as free shareware, that any user of the BB can download without cost. There are warnings on the BB that any users should maintain complete confidentiality about the board's existence.

Upon asking her coworker where all the software comes from, Thelma learns that Jim set the BB up a year ago and has gradually established a network of people who share any software they obtain. She asks Jim if he thinks the distribution of copyrighted software is unethical.

Jim smiles and says, "No problem." He adds, "Some of the people right here at Oglethorpe Consulting download business stuff from the BB that they need on the job. It saves the time it takes to fill out the requisitions to get the firm to buy it. Hey, the firm monitors all our e-mail; they must know what's going on."

When Thelma frowns, Jim says, "Hey, don't use it if you don't want to, but keep quiet about it, OK?"

- Is Jim being unethical? Why? Why not?
- Is anyone being harmed? Who?
- Are the people who download software from the BB being unethical? Why? Why not?
- What is the company's stake in what is going on?
- What should Thelma do?

Compose a memo to the instructor that summarizes your group's discussion and gives its conclusions.

7. In collaborative groups of four or five students, develop a code of conduct for your class. Identify the ethical obligations of both the instructor and the stud`ents. What are the principles that will guide your behavior in the class? What are the ethical issues that are likely to occur in the class?

Write a separate memo describing your group's efforts to compose the code of conduct. Which portions were easiest to write, and which proved a challenge? Which issues were readily decided, and which caused friction? How did you choose which issues to cover and which to ignore?

Write a separate memo explaining why you believe your group's code of conduct would be appropriate and effective for your class. How would you achieve compliance with its ethical guidelines? Would you inspire obedience or punish disobedience or both? Why would your class be willing to adopt your group's code of conduct?

Scenario

Paul Miller looks at his watch. He needs to leave for the airport to catch a plane. Paul has been in Honduras for three days, trying to close a deal to purchase a small local textile plant. He made the trip after initial fax contact and some third-party mediation to establish contact with Miguel Mujarez, who wants to sell the plant. Paul's friend Raul Peña told him that business dealings require a solid relationship with the Mujarez family because so many employees in the company have worked for the plant for decades. The plant, which has been in the Mujarez family since the early 1950s, is very much a family operation.

Paul scheduled four days for the trip. He spent half a day with the plant manager, two days and evenings with Miguel Mujarez, one day with a cousin, Juan Mujarez, and one day with the plant employees. Paul, anxious to get a positive response from Miguel, wants to sign the contract and return to Atlanta.

At dinner, on his last night in Honduras, Paul asked Miguel point-blank when they could complete the deal. Miguel acted as if he hadn't heard the question and continued to discuss one of his sons' feats in soccer. Then he asked Paul about universities in Atlanta, since he hoped to interest his oldest son in attending a school there. He spoke at length about the fine workers at the plant and asked numerous questions about Paul's family and other matters that Paul felt were not relevant. The discussion seemed to be going in circles.

On the final day of Paul's visit, Miguel was supposed to meet him at the office at 10:00. It's now after 12:30, and Miguel is not there. Paul has asked the secretary when Miguel will arrive. She said "Soon" several times but finally acknowledged that late afternoon was more like it because Miguel's parents had arrived in town. Paul tries to control his frustration: if the secretary knew in the morning that Miguel would not be coming to the office for many hours, why hadn't she told him? Why wait until only a short time remains before he must leave?

Juan Mujarez arrives at that moment, expresses great happiness at seeing Paul, and says he hopes all the parties can discuss the sale in a few weeks. Paul quickly tells Juan that he came to Honduras to complete the deal. Juan expresses regret and begins discussing an appropriate time for Paul's next visit.

On the way to the airport, Paul recalls the experience of a friend who had not completed a business transaction with a Japanese firm that had just moved to Ohio until the company had been in Ohio for nearly a year. Paul decides that he cannot spend any more time on this matter. Miguel obviously does not want to sell the company soon, and Paul cannot determine a price that will persuade Miguel to sell. He has wasted nearly a week in trying to close the sale. He admits, wearily, that he just doesn't know how to do business in Central America.

CHAPTER SIX

Writing for International Readers

In business when you think about your audience while planning a report, letter, proposal, or e-mail message, you must realize that because technology now links cities and countries around the world, some or all of your readers may not have English as their first language. As the diminution of trade barriers enables U.S. organizations to do business throughout the world, moreover, business and technical organizations have become increasingly international. Communication, however, is culture specific: the culture of those with whom you wish to communicate determines how you will write; and you cannot write to people in other countries the same way you write to people in the United States. In this book, we primarily emphasize strategies for developing business and technical documents for U.S. readers. In this chapter, however, we show a procedure for planning documents that will be effective with readers in other cultures. In today's global market, you must be able to communicate with people everywhere.

A number of books have been written about protocols for conducting business in other countries and business. At the end of this chapter, we list over sixty of the growing number of useful and interesting books that will help you to understand how culture determines the proper ways to meet and do business: greeting individuals according to the traditions of other countries, the appropriateness of shaking hands, the design and presentation of business cards, proper format for business letters and reports, proper deportment during dinner parties, gift giving, and the significance of holidays and colors are just a few of the matters that you need to learn about.

The study of international communication forms a separate area of study, but here we provide guidelines for written business communications that will enable you to make the transition between the U.S. culture and other cultures. Some of the examples we provide here (and in Chapter 12, on correspondence) will help you see how culture influences the preparation of documents for readers of different traditions.

ESTABLISHING A PERSPECTIVE ON INTERNATIONAL COMMUNICATION

Designing effective written business communications for readers in other countries requires that you approach the development of international communication documents from multiple perspectives, which we shall discuss in this section.

The Fatal Communication Error: Assuming that the United States is the greatest country in the world. This assumption often leads people to think they can do and say whatever they want, with business associates in other cultures automatically following their lead because the U.S. way is obviously the best way.

No perspective could be more detrimental to successful communications with people from other cultures. Outside the United States, you must learn to operate by another set of rules.

Cultures vary, and communicating successfully with people from other cultures requires that you play by their "ground rules." When your audience is an individual or an organization in another country, you need to carefully analyze this audience to understand its perspective.

Everyone thinks his or her culture is "the best," but no one culture is inherently superior to any other. You may not like many of the characteristics of the cultures of people with whom you need to communicate, but you must respect the differences between cultures and the perspectives of readers from outside your own tradition. Always bear in mind that the U.S. culture differs dramatically from most cultures in the world.

UNDERSTANDING READERS FROM VARIOUS CULTURES

Anthropologists tell us that cultures differ in a number of specific ways. As a result, readers in different cultures have different expectations. Cultural anthropologists such as Geert Hofstede (1997, 2000), Fons Trompenaars (1993), Edward Hall and Mildred Hall (1990), and Lisa Hoecklin (1995) have isolated a number of cultural characteristics—differences in values—that enable us to understand differences among cultures. We will discuss eight major values that shape these differences, values that are important for our purposes because they can affect written communication. We will show you example documents that illustrate how values affect communications. To be an effective global communicator, you must first know how characteristics of the U.S. culture shape written documents prepared for distribution in the United States. Understanding specific cultural differences that affect communication is a good basis for understanding how to communicate with business people the world over.

Table 6-1 rates a number of major countries on four of the value dimensions measured by Geert Hofstede, who is perhaps the leading expert on intercultural differences. The higher the score, the more a particular value is stressed in the culture.

Individualism versus Collectivism: Valuing Either Individuals or Groups

In the United States, individualism remains a predominant cultural characteristic. As Lisa Hoecklin (1995, 35) states:

> Individualism is a concern for yourself as an individual as opposed to concern for the priorities and rules of the group to which you belong. The majority of the people in the world live in societies where the interests of the group take precedence over the interests of the individual. In these societies, the group to which you belong is the major source of your identity and the unit to which you owe lifelong loyalty. For only a minority of the world's population do individual interests prevail over group interests.

Individualism drives all other U.S. cultural characteristics. As Table 6-1 shows, the United States tends to be the most individualistic country in the

Table 6-1 ■ Cultural Dimension Scores of Representative Countries				
Country	**Individualism**	**Power Distance**	**Uncertainty Avoiding**	**Masculinity**
United States	91	40	46	62
Australia	90	36	51	61
Great Britain	89	35	35	66
Canada	80	39	48	52
Italy	76	50	75	70
Belgium	75	65	94	54
Denmark	74	18	23	16
Sweden	71	31	29	5
France	71	68	86	43
Ireland	70	28	35	68
Norway	69	31	50	8
Switzerland	68	34	58	70
Germany	67	35	65	66
South Africa	65	49	49	63
Finland	63	33	59	26
Austria	55	11	70	79
Israel	54	13	81	47
Spain	51	57	86	42
India	48	77	40	56
Japan	46	54	92	95
Argentina	46	49	86	56
Iran	41	58	59	43
Jamaica	39	45	13	68
Brazil	38	69	76	49
Arab countries	38	80	68	53
Turkey	37	66	85	45
Greece	35	60	112	57
Philippines	32	94	44	64
Mexico	30	81	82	69
East Africa	27	64	52	41
Portugal	27	63	104	31
Malaysia	26	104	36	50
Hong Kong	25	68	29	57
Chile	23	63	86	28
West Africa	20	77	54	46
Singapore	20	74	8	48
Salvador	19	66	94	40
South Korea	18	60	85	39
Taiwan	17	58	69	45
Peru	16	64	87	42
Pakistan	14	55	70	50
Indonesia	14	78	48	46
Venezuela	12	81	76	73
Panama	11	95	86	44
Guatemala	6	95	101	37

Source: Geert Hofstede, *Cultures and Organizations: Software of the Mind* (New York: McGraw-Hill, 1991).

world. Reverence for individualism expresses itself in a number of major ways, which can be generalized as follows:

- In the United States, children learn to think in terms of "I."
- Individual achievements receive more accolades than team achievements, and even in team efforts, specific team members are usually singled out for outstanding contributions.
- Common sayings like "every man for himself," "winner takes all," "be independent," and "look out for number one" illustrate the pronounced individualism in the United States.
- Emphasis is placed on the individual's responsibility for his or her own destiny.

Another common theme you have probably heard is that people fail because they didn't work or try hard enough. Only within the past two decades have American students and white-collar employees learned to work as teams. Americans tend to separate their business lives from their personal lives. In the U.S. business culture, how one feels about an individual should not influence business decisions involving that person. Americans tend to be direct and to the point. They are hard driving, pragmatic, and competitive in work and often in recreation. It is often said that Americans live to work. Because of the Puritan influence that has long dominated U.S. culture, a strong work ethic is highly valued. Thus, work—success through work—often takes precedence over family and friendships. Because of the importance of success, an individual's self-worth is often bound up with his or her career achievements.

Group-valuing (collectivist) cultures consist of tight social networks in which individuals identify closely with their families and business organizations. They are motivated by the group's needs and achievements. The individual's success is valued as it reflects the success of the group. Groups make most decisions. Employees in collectivist societies act according to the interest of the group, which may not always mesh with the individual's desires. Self-effacement, along with deference to the interest of the group, is the standard. Earnings may be shared with relatives. Promotions are based on seniority within the groups. Often relatives of employees are hired, since nurturing relationships among those in the group are seen as more important than benefiting from the talents of outsiders.

In collectivist cultures, people's business and personal lives merge. Colleagues are polite, formal, and indirect, and concerns about the welfare of the group overshadow the success and needs of any one person. Collectivist cultures allow individuals to be expressive within the group, even though formality and deference are valued within the group and to individuals outside it. Many collectivist societies value family welfare over business issues. That western cultures are more individualist than Latin American, Pacific Rim, and Third World countries is illustrated by the patterns discernible in the ratings of Table 6-1.

Implications for Communication   Written communications are valued in the United States because they often document individual actions. Because the U.S.

culture is heterogeneous, written documents are very important tools for ensuring precise understanding and compliance. In contrast, group-valuing organizations tend to prize oral communication over written communication. When you are writing to individuals within a group-valuing culture, you will want to

- Focus on how the issues you are discussing reflect on the organization and the actions of the group;
- De-emphasize yourself; avoiding excessive use of "I" and focusing on establishing rapport with the organization, not specific individuals;
- Emphasize your relationships with the group before launching into a discussion of the business you wish to transact.

Documents prepared in collectivist cultures may not be as explicit or as detailed as those typically produced in the United States. Because collectivist cultures value "group think," individuals in groups share values and ideas: they do not have to illustrate every idea and document every fact by explicit verbal and numerical communication. In collectivist societies, the group decides the action, and the fact-finding process occurs in an environment that is relatively rule free. However, the more heterogeneous the group, the more explicit documents will need to be to insure that readers grasp the message.

Separation of Business and Private Relationships

How a culture treats and values relationships is critical in understanding that culture and in determining how best to communicate with people in that culture. The treatment of relationships is directly connected with a society's emphasis on groups or individuals, as collectivist (group-valuing) cultures are generally more relationship oriented than individualist cultures. For example, in the United States, personal relationships are usually kept separate from business relationships. Business decisions are based on business information only. People are expected to keep their private life and their business life separate, and attempts to establish or preserve this boundary are respected. An opaque, objective deportment is considered to be appropriate for doing business.

In cultures that value individualism, people have more open public space, but they closely guard their private space. The U.S. approach to business is direct, open, rapid, and extroverted. Many people outside the country perceive the U.S. tendency always to come quickly to the point as harsh and abrasive. Americans, who separate work and private lives, tend to view business relationships differently from personal relationships. In business dealings, business objectives receive priority. Personal relationships are viewed as a necessary, brief prelude to initiating and completing a business transaction.

In collectivist cultures (e.g., in Mexico, in the Pacific Rim nations), people do not separate public and private lives. One unabashedly influences the other. To representatives of cultures that separate business from personal affairs, however, collectivist-oriented business persons appear indirect, noncommittal, and evasive. In a collectivist culture, promotions are often made on the basis of family or friendship rather than competence. Nepotism is rampant in many organizations, and business is conducted in accordance with family needs. A culture that does not distinguish business from private relationships bases its notion of efficiency not on time to completion but on how well one understands others. In such a culture, you can expect to spend extensive time building relationships with individuals in the company with which you want to do business. Family needs often determine business decisions.

Implications for Communication Unlike U.S. communications, which focus on business, communications in relationship-oriented cultures de-emphasize business in favor of the relationships among the individuals transacting the business. Cultures that do not separate business and private relationships expect communications that are formal, reserved but positive, and indirect in discussing business issues. Emphasizing the relationship between you and your business organization and your reader will be paramount. An extremely efficient business presentation may be perceived as inappropriately direct, and your ideas may be rejected accordingly.

Power Distance between Social Ranks

Inequalities exist in any society, and some people have more power, respect, status, and wealth than others. Hofstede defines "power distance" as the degree of closeness, or interdependence, that exists among members of organizational hierarchies:

- Do superiors consult subordinates about decisions?
- Do employees feel comfortable in disagreeing or questioning superiors' decisions?
- Is interdependence of authority evident in supervisor–subordinate relationships, and if so, to what extent?

In short, power distance is measured in terms of the prevalence of ranks or levels of authority. Column 3 of Table 6-1 shows how Hofstede rates various cultures on power distance.

In high-power-distance cultures, employees manage their work according to their superior's specifications, and authoritarian attitudes prevail. Inequalities among people are both expected and desired. Hierarchies in organizations are pronounced; the powerful have privileges that the less powerful do not have, and subordinates expect to take orders. In high-power-distance cultures, employees expect their supervisors to make unilateral decisions. Employees do as they are told without asking questions. Superiors are authoritarian figures. Disagreeing with "the boss" is unacceptable. High-power-distance cultures are characterized by steep organizational pyramids and close supervision of employees. In high-power-distance cultures, age is a positive factor and a major qualification for leadership roles, and formality and politeness in communications are extremely important. One is never openly aggressive.

In low-power-distance cultures, the individual may freely follow his or her own preferences and criticize management. Inequalities among people are minimized, subordinates are consulted, and decentralization of responsibility is popular. Employees have upward mobility, and teamwork is valued because the less and the more powerful depend on each other.

You were told in Chapter 3 to determine the relationships between you and your reader, to choose your content and tone in terms of that relationship. This is because power, in U.S. businesses, is allocated differently depending on the size and type of organization. The power distances within an organization tell you how open, direct, or formal you can be in stating your ideas. The United States is generally a midlevel-power-distance country. Because of the emphasis on the responsibility of the individual and the individual's responsibility for his or her destiny, Americans like to be involved in decision making. Central decision making is accepted, but employees who are not leaders expect to be heard. Great differences in rank are expected, but those who have rank ideally have earned it through extreme individual effort, success in achieving business goals, and hard work. While hierarchies in organizations are dominant, U.S. organizations are moving toward "flatter" organizations and participatory management.

In many of today's U.S. organizations, strong leaders must also be "team players." Youth is often more revered than age. Low-power-distance business cultures and business organizations have flatter organizations, more team decision making, fewer autonomous "bosses," and more decisions coming from group (committee) recommendations. The variations in power distance in the U.S. culture explains why the country has a score of 40 in Table 6-1. Understanding the extent of teamwork in a particular U.S. company, as well as an international company, can be critical to your success in dealing and working with that company.

Implications for Communication   In high-power-distance cultures, using correct forms of address can be extremely important: you must know who should receive a report or letter, the title or rank of that person, and all individuals whose names should appear on the distribution list. Establishing the correct tone in addressing the intended readers is thus important in establishing the correct "distance" between writer and reader. Therefore, tone in documents prepared for readers in high-power-distance cultures who hold relatively superior positions should be quite formal. In contrast, in preparing documents for readers in low-power-distance cultures, strict recognition of business hierarchies and the use of formal address gain less favor. The style of the message can be more casual.

Universal or Relative View of Truth

Cultures also differ in how they perceive truth. A universal view of truth means that what is true can be discovered, defined, and applied in all situations. In other cultures, truth is relative. It changes depending on the needs of the situation or the group affected by a given decision. In many collectivist cultures, relationships and the needs of people in the organization (many who may be relatives or family members) are more important than the objective truth of a situation. The United

States exemplifies what Trompenaars calls a universalist culture in which "truth" exists; clear differences exist in "right" and "wrong," and universal rules of behavior exist and are considered to apply to everyone. "Rules" should be laid down in strictly worded agreements and contracts. Once defined, the "rules" govern business and behavior. This concept of truth evolves from the Puritan roots of the United States: truth exists; it should transcend and guide the actions of individuals, and at times it may take precedence over the immediate needs of people.

"Particularist" cultures, in contrast to "universalist" cultures, believe that truth is relative. What is "true" and what should be done depend on a particular situation. Human relationships are more important than rules, and provisions of written contracts may be suspended or ignored if situations arise that make these provisions undesirable. A written contract is less important than human issues that may arise to affect the contract. In particularist cultures, that is, people are more important than contracts.

Implications for Communication In a universalist culture, writers are advised to be as specific and concrete as possible. Clarity and precision in format, language, and meaning are valuable. In a culture in which truth is relative, comments may be less direct and more dependent on the situation. The message may appear vague. Oral communications may be more significant than written communications. Rules that apply in one business situation may cease to apply when circumstances change. Documents to be distributed in collectivist cultures should always take into account the impact of the situation on the group.

Whether the Entire Message Is Contained in the Text

In the United States, documents are expected to "contain" the complete meaning. The "truth" of the situation must be contained in the text because texts document facts and human actions. These conventions are the hallmark of a text-oriented culture. Written agreements and statements are very important. What you say you will do means little if you do not put your promises in writing. Detailed documents are expected. Conditions not included in a written document are not recognized as applicable, and parties to a contract are not legally bound by obligations not spelled out in the text.

In other cultures, the "meaning" of a business situation or a document comes from the people and the human issues involved in a given decision. The meaning of a document—such as a contract—depends on the situation. The document itself may be ignored, even if it is a contract. In such cultures, schedules are flexible; being late to meetings or social engagements is expected and acceptable. Establishing relationships is seen as more important than doing business. Thus, business days are structured, and documents are less direct than U.S. documents.

Implications for Communication The text of U.S. documents is expected to contain all facts necessary to allow readers to arrive at solid business decisions. Contracts are considered binding. In nontextual cultures, the language is suggestive, oblique, and theoretical; documents themselves are often wordy, tending to focus on organizational situations rather than pristine factual details. What a document ultimately "means" may be a function of the circumstances under which it was prepared. Business obligations may not be clearly or completely stated in the text.

Whether Uncertainty Is to Be Avoided or Accepted

Another important difference focuses on how a culture tolerates uncertainty. (How cultures vary on this value is presented in column 4 of Table 6-1.) Members of cultures that avoid uncertainty appear to be anxiety-prone people who perceive the uncertainties inherent in life as threats that must be fought. Employees formed by these cultures fear failure, take fewer risks, resist change, and place a premium on job security, career patterning, and company benefits. Managers are expected to issue clear instructions, and subordinates' initiatives are tightly controlled. Employees in cultures that dislike uncertainty accept formal procedures, wide power distances within hierarchies, and highly structured organizations. In societies that dislike uncertainty, people tend to exhibit high stress and anxiety levels, to need structured environments, to believe that time is money, and to believe in the value of hard work.

In contrast, people reared in cultures that accept uncertainty are more likely to take each day as it comes. Conflict and competitiveness can be used constructively, and dissent is tolerated. Needs for written rules and regulations are relatively few, and rules that turn out to be unrealistic or unenforceable can be easily changed. Time is seen as a framework for orientation; precision and punctuality are not paramount. Emotions are not shown, and rules that are highly restrictive are avoided. The United States scores about midway on the continuum of avoiding and accepting uncertainty.

Implications for Communication In *cultures* that accept uncertainty, written documents may be less problematic than they are in cultures that seek to avoid uncertainty, where written texts are valued for documentation and governance purposes. In *companies* that dislike uncertainty, precisely written documents, forms, tables, graphs, procedures, policies, and style sheets are valued because they promote uniformity and clarity. While tolerance for uncertainty varies in businesses in the United States, fear of litigation is driving more companies toward insistence on precision in documents.

The Power and Value of Time

Another cultural value that affects communication is the value a culture places on time. The United States is one of the most time-conscious cultures in the world. People in this country value not only productivity, but also efficiency in process and product. Effective use of time—doing more and more work in less and less time—is a cherished U.S. ideal. Whereas many collectivist cultures value relationships before work, here we tend to value work before relationships. Many cultures consider relationships with friends and family, the need to enjoy each day, and the time spent in building and maintaining relationships to be more important than efficient execution of work. Cultures that value time usually value productivity. Cultures that value relationships place people before business. In these cultures, efficient use of time is less valuable than the slower paced focus on relationship building and weaving business with relationships.

Implications for Communication   When you address members of cultures that value relationships over productivity, emphasize relationships with the persons with whom you are doing business. Make business secondary to the relationship. In contrast, in preparing documents for U.S. readers, emphasize the business goal: be precise, direct, and complete. Make goals, expectations, and commitments known.

Masculine versus Feminine

In many cultures, men, rather than women, serve in positions of authority. Occupations tend to be segregated by gender, and inequality of the sexes is often accepted. Only recently have women as well as men been expected to pursue and succeed in careers. Cultures designated by Hofstede as masculine value, achievement, aggressiveness, competitiveness, and financial success; high job stress is a given. Hofstede's feminine cultures, in contrast, feature less occupational segregation by gender. Women occupy well-paid jobs, and the work environment shows less stress, more awareness of individuals' personal needs, and more concern for family and social issues. Feminine cultures value nurturing relationships, consensus, compromise, and negotiation. Column 5 of Table 6-1 shows how cultures rate on this dimension. Clearly, the United States tends to be a masculine culture.

Implications for Communication   American women are well advised to avoid assuming a domineering stance when working with men reared in cultures outside in which males assume positions of authority. When American women write to men in overtly masculine cultures, the tone should be formal and polite. Avoiding immediate, direct discussion of business issues is imperative. Establishing rapport with the individual within the organization is critical. In general, communications for feminine cultures should focus on relationships, while communications for non-U.S. masculine cultures should be assertive and decisive, with an emphasis on the business transaction.

CONSIDERING CULTURE IN THE PLANNING PROCESS

Since the most important influences on how you plan and then draft your document are your audience and your purpose, it is essential to understand to whom you are writing and why. As you consider these factors in planning communications for international readers, you will want to answer the following questions. Because many of these questions pertain to U.S. documents, you can see how effective use of international communications requires that you broaden your perspective as you plan a document.

- To whom is this message directed?
- What do you know about the reader? Age? Interests? Education? Job responsibilities? Title?
- Have you met the reader in person?
- What is the reader's attitude toward you, and how does he or she perceive the topic?
- What are the reader's particular characteristics, as gleaned from written messages or personal encounters?
- If the document is being directed to a reader in a non-U.S. culture, what are its characteristics?
  - Is it oriented toward the individual or toward the group?
  - Do people separate their business and private relationships?
  - Does this culture value success of individuals or success of groups?
  - Do people see truth as universal or as relative to particular circumstances?
  - Is the entire message–and what it means—usually contained in the text?
  - Do the people value time and efficiency?
- How well are the members of your target audience able to read English?
- What is the situation that has led to your need to write this document?
- What purpose do you hope to achieve? What do you want to happen as a result of this document?
- Based on the broad value characteristics of the culture, what choices do you need to make about the following?
  - Structure (deductive or inductive)
  - Organization of ideas
  - Degree of specificity about business purpose
  - Type of information disclosed about you and your organization
  - Quantity of detail presented about you and your organization
  - Style
    - Sentence length
    - Word choice
    - Address protocols
    - Tone
    - Formatting techniques
    - Graphics

In short, considering culture is simply another dimension of considering your audience and the context in which your message will be read. As the guidelines in Table 6-2, suggest, your answer to the foregoing questions will affect how you write to any audience.

Table 6-2 ■ Guide for Designing International Written Business Communications

Non-Western Business Characteristics	Western Business Characteristics
Prefers face-to-face communication *Use a style in written documents that is conversational reflecting oral, narrative discourse; nonlinear discourse*	Insists on written communication as documentation *Structure documents by standard business protocol; use paragraphs; linear discourse*
Values intuitive, aesthetic written communication *De-emphasize structured response and use of headings/ page design*	Values analytical, logical, denotative written communication *Emphasize structured, logical messages; use headings/ page design*
Values indirect communication *De-emphasize "main point" or place it at the end of the message* *State message obliquely*	Values direct communication *Highlight main point; begin with most important ideas first* *State point clearly and objectively*
Values slower pace of communication *Conciseness not valued* *Include comments about nonbusiness issues* *Use opulent style*	Values rapid, efficient pace of communication *Conciseness valued* *Minimize nonbusiness remarks* *Use clear, brief, simple style*
Values formal communications *Emphasize use of formal titles, forum, and audiences for messages*	Values informal communications *De-emphasize titles, consistent decorum, and occasion for communication*
Focuses more on the situation *Focus on the issues as these are bound up with the welfare of the organization*	Focuses more on the ideas involved in the situation *Focus more on issues as business decisions only*
People work near one another or share space and responsibilities *Understand that messages may be shared with many in the group*	Individuals are assigned separate spaces/specific duties *Direct messages toward specific individuals having specific tasks*
Content reflects interdependencies crucial to intra-/inter organizational communication Decision making via consensus *Focus message on all groups that are affected by the message*	Interdependence, which is mistrusted in communications, is de-emphasized Decision making via designated leaders *Focus on hierarchies in the organization and those receiving copies; focus on those responsible for acting on the message*
Content and presentation are conciliatory Harmony should be maintained and confrontation avoided *Control tone to allow everyone to save face in unpleasant situations; avoid blame on individuals or groups*	Content and presentation are argumentative or at least very direct Honesty in communication is desirable *State the issues squarely but tactfully* *Use design to highlight facts and desired outcomes*
Organizational relationships are highest priority Relationships more important than rules *Emphasize company background and rationale for business actions*	Organizational tasks/goals are highest priority; rules and goals before relationships. *Emphasize recommendations, procedures, policies in written form*
Long-term organizational agendas *Discuss business in terms of larger human and company concerns*	Short-term organizational agendas *Emphasize action required and time tables for achieving goals*
Clear distinctions between surface communication and business objective *Messages should be oblique; building relationships carefully and futuristically with business in the background* *Elegance more important than brevity*	Less distinction between business objective and surface communication *Messages should aim to capture and summarize the desired goal of the communique; ambiguity in goals not desirable* *Conciseness more important than elegance*

(continued)

Table 6-2 ■ **Guide for Designing International Written Business Communications** *(Continued)*

Non-Western Business Characteristics	Western Business Characteristics
Agreements easily modified and ignored	Agreements legally binding
Perceived truth may be modified by changing realities; it is not confined to written discourse Documentation in relationships	Perceived truth is defined by written agreements; facts have one interpretation Documentation in written documents
Truth is a product of several perspectives: written, spoken, felt; messages often evade issues, are closed and introverted *Messages may be formal, but they are revocable*	Truth is one textualized statement having one meaning contained in the text *Messages are to the point; often abrasive* *Messages are binding; they contain truth, which can be defined and agreed upon by all parties*
Approach is theoretical and indirect; specific implementation is omitted *Goal required may be hard to explain; linear prose, more undifferentiated text*	Approach is practical; action-oriented, problem-solving approach is used *Goal should be visually clear; bulleted lists, numbered steps*
Time and deadlines are relative and flexible *Do not emphasize time in the message*	Deadlines and commitments are firm *Always include specific time requirements*
High-context communication; work and life intertwined; emphasis on the group responsibility shared by groups *Use message to link aspects of the culture; personal/ public/corporate lives shared* *Tone of document: subjective and oblique*	Low-context communications; work and life separate, emphasis on individuals, responsibility directed toward specific individual(s) *Keep message strictly business; personal life is separate from business life* *Tone of document: objective, direct, and precise*

Note: Written communication strategies are set in italics

EXAMPLE INTERNATIONAL DOCUMENTS

To see how cultural characteristics can affect document design, consider the following example documents, which were written by natives of the United States, Mexico, and Pakistan.

Figure 6-1 exemplifies a typical U.S. business letter: direct, concise, and focused. The main point of the letter appears in the opening paragraph. The closing paragraph indicates the action required and the time constraints. The middle paragraphs convey essential information, and the content develops in a logical sequence with no digressions. The style is informal but efficient: that is, the "medium is the message." Because of a cultural preoccupation with efficient use of time, commitment to achievement, and aggressive business practices, most U.S. business letters are direct and concise. A business letter does not bring up personal issues, even if the writer knows the reader in a nonbusiness context.

U.S. business letters usually follow the development scheme illustrated in Figure 6-2.

Water Authority
District 4

CITY OF GLENDALE
4000 First Street, Room 202
Glendale, TX 77863

March 30, 2001

Mr. Harold Jonas
9515 Richmond Way
Houston, TX 77381

Dear Mr. Jonas:

The City of Glendale Water Authority regrets the continuing problems with insufficient water supply to River Forest Business Park. Irregular pumping has occurred because of faulty release mechanisms, which have not been functioning when water storage sinks below a specified point. We have ordered a new release regulation system, which suppliers have assured us will arrive by April 20. As soon as the systems arrives, installation will begin. The system should be operating within five to seven working days.

Water regulation is a complex process because of the need to conserve water while guaranteeing that residents of every area served by every pump station are served without interruption. For the past three months, to avoid overfill of storage tanks, we have assigned technicians the responsibility of visually monitoring the refill boundary indicators. Once they see that storage capacity is approaching the refill zone, they manually activate the pumping stations. This method has worked well during the recent rainy months, when water requirements were low. Manual operation has allowed us to examine every aspect of the pump system to determine exactly where the problem occurs when the equipment fails to activate in time to avoid water shortages to your neighborhood. During the recent drought, however, manual release has not been efficient, as technicians are often not able to react quickly enough.

Although our analysis of the water system has been lengthy, we have been able to locate the problem and order new equipment. By locating and replacing only the faulty devices, we will save Glendale over $1 million dollars. Unfortunately, the testing process has been lengthy and at times inconvenient for residents of subdivisions dependent on Station 3. However, the problem should be eliminated no later than May 1. We will appreciate your patience as we attempt to implement a long-term solution while conserving both tax money and water.

Sincerely,

M H Marks

M. H. Marks
Director

FIGURE 6-1 · U.S. BUSINESS LETTER

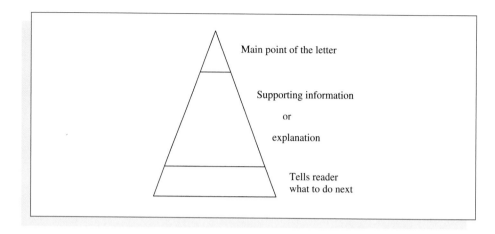

FIGURE 6-2 · SCHEME OF U.S. BUSINESS LETTER

In Figure 6-3, a Pakistani business letter, note the formal, courteous style and politely stated point. As Table 6-1 indicates, Pakistan is a collectivist country with a high score in uncertainty avoidance. Thus, a letter that seeks to collect a debt uses an extremely tactful approach. The main news occurs in the second paragraph. The writer attempts to build a solid case—asking the reader to examine the invoice amounts—before venturing to request payment. Note the courteous closing. Compare the differences in letter format between this letter and Figure 6-1, which uses U.S. block style. The Pakistani business letter follows the development scheme of Figure 6-4.

In Figure 6-5, a business consultant in Mexico City introduces his services to a U.S. company that has recently opened an office in Mexico City. The consultant, a Mexican national, has been working through a friend at the U.S. consulate. In Mexico, as in many Latin American countries, you must establish relationships and credibility with friends of those with whom you wish to do business before directly contacting the individuals in the target company. Thus, the reader will know about the writer, and the writer quickly alludes to the mutual friend.

Because Mexico is a collectivist culture in which relationships are considered more valuable than time, business correspondence uses a more relaxed approach. This characteristic of Mexico's culture expresses itself in the inexpediency of the sentences themselves: the content avoids directness, conciseness, and efficiency. The liquid quality of the language is, in an important sense, preserved by the masculine focus of the culture, which exudes a formal bravado. Note that the letter suggests no time frame and no plan of action. The style is ornate, effusive, aesthetic—certainly not concise or direct—and more complex; the sentences, less efficiently structured. The ideas presented seem more theoretical and less pragmatic. Many non-Western collectivist business cultures will

❖-❖

TELEGRAMS: RUNYON   * * *   PHONE: 8632137
CABLES   : Tomichi   ▼   ▼   Telex: 41212 RUNYON
                          * * *   Fax: 00 82 24 683-9215

Runyon Finnly & Co. Pakistan (PVT.) LTD
Merchants, Engineers & Contractors
IMPORT REGISTRATION NO. K0156
EXPORT REGISTRATION NO. W003971

OUR REF h/1299/TR-54   G.P.O. BOX NO451
YOUR REF_____   KARACHI 74200 (PAKISTAN)
                                  Date Jan. 23, 2001

The Chief Executive   for the kind attention of
Contractors, Ltd.
Jamson Square Building,   _____
Karachi

Dear Sir,

RE: PENDING BILLS

We draw your kind attention to the long pending Bills which we know you will understand our need to discuss with you. Our business relationship has occurred for many years and for that we are grateful:

1) Bill No. DRM/65/2158 dt. 8. 8. 94	Rs..50,000.00
2) " " JM/JC/443 " 30. 11.95	Rs., 45,000.00
3) " " GM/PE/215/7 " 12.10.95	Rs. 2,950.00
	97,950.00

During our visit last month, your Mr. Vibras very kindly promised to settle the above Bills by the first week of Jan. The Bills are pending for more than a year and in addition we are in urgent need of funds. We, therefore, request you to please send us your Cheque in settlement of our above Bills soon after receipt of this letter.

Your co-operation is solicited,

faithfully yours,
for RUNYON FINNLY & CO, PAKISTAN (PVT) LTD.

Asad Paravichi
Asad Paravichi
DIRECTOR

FIGURE 6-3 · PAKISTANI BUSINESS LETTER

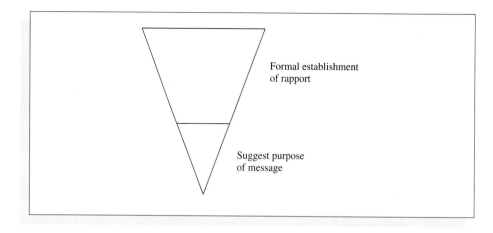

Formal establishment
of rapport

Suggest purpose
of message

FIGURE 6-4 · SCHEME OF PAKISTANI BUSINESS LETTER

allude to the purpose about midway through the letter. However, the business issue is obliquely stated. Figure 6-6 shows the development scheme of this letter, which also differs from the U.S. letter (Figure 6-1) in format.

WRITING BUSINESS COMMUNICATIONS TO READERS IN OTHER CULTURES

If you find that you will be communicating with people from other cultures, particularly in a business context, be sure that you research the business etiquette for each country. These procedures include

- understanding the use of business cards,
- dressing appropriately,
- making proper introductions and greetings,
- knowing how business decisions are made,
- knowing what topics to discuss and to avoid in all conversations, business or social.

An increasing number of books and videos, as well as information on the Web, can help you understand the perspectives of people in other cultures.

Initially, however, remember that what is acceptable in written communications in the United States will likely be unacceptable in many other countries. Knowing this fact is an important first step in learning how to communicate about technical and business issues with people in other countries.

Even though you may be able to use English to address readers in another culture—English is becoming the international technical language—you should attempt to assess aspects of that culture and its associated values that

13 Agosta de 2001

Lic. Felix Ortiz Fraga
General Director
Pintural Chihuahua, S.K.A. de C.V.
44 Salgado, Monterrey, Nuevo Leon

Estimado Lic. Ortiz:

Because of the relationship that exists between us, our mutural desire to enable the effective pursuit of business in Mexico City, it is my pleasure to greet you and present for your consideration the special services in Consulting and Executive Training that is offered by Grupo Empresarial SIA, which has as its object to support your organization in the achievement of the objectives of competitiveness and leadership that are demanded by the economic and social environment in which we actually live. I am greatly indebted to Lic. Julio Montevezos, whose family has long been associated with my family here in Mexico City. It is indeed an honor to greet you upon your arrival in Mexico City.

Among those services offered by Empresarial SIA, you will encounter the following:

☐ Consulting for diagnosis and analysis of processes subsequent to the implantation of operative models oriented to produce knowledge of Mexican business practices integral to your organization.

☐ Training at an executive level in the development of general abilities, such as updating courses for administrative and technical personnel.

As a complement to what was previously mentioned, we offer specialized support for the analysis, definition and implementation of computer systems, along with the infrastructure necessary for your organization.

We would feel very honored to have the opportunity to discuss personally with you the solutions we would be able to propose to you, and by which we would be in contact with you or the person that you indicate to us, in order to agree upon an interview in this respect.

Atentamente

Ing. Raul Orosco Jeminez

FIGURE 6-5 · MEXICAN BUSINESS LETTER

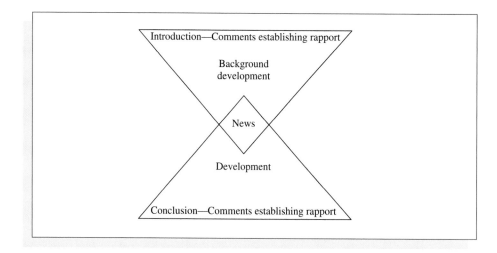

FIGURE 6-6 · SCHEME OF MEXICAN BUSINESS LETTER

may affect communication. For example, for collectivist countries, try to emphasize the group rather than individuals. Attempt to establish rapport with your readers rather than emphasizing the business objective. For high-power-distance countries—those that exemplify and expect differences in ranks—be aware of the possible need for titles, respect for those in powerful positions within the organization, and the status of women in the country.

You can see how these broad cultural characteristics become embedded in written communications by examining a letter written by a female real estate agent in the United States to a Japanese reader who has just moved to Texas.

Situation 1 Katherine Ashcroft, a commercial leasing agent, contacts Keisuke Ashizawa, a Japanese engineer whose team is currently working on a research project in Texas. Mr. Ashizawa, who has discovered that he will need to remain in the United States longer than he had expected, has expressed to his American colleague Kevin Graham, an interest in leasing an apartment. He has also discussed with Kevin a desire to find a local bank. With these requests in mind, Kevin has offered to introduce Mr. Ashizawa to Ms. Ashcroft. The businesswoman would like to work with Mr. Ashizawa (and hopefully other members of his team's research group). After receiving a call from Kevin, she writes the letter shown in Figure 6-7.

Figure 6-7 is concise and polite rather than aggressive. The main issue appears in the final paragraph, and although the writer wishes to become Mr. Ashizawa's real estate agent, the letter only indirectly deals with the business goal. The style is formal, yet the writer attempts to establish rapport with the reader before even mentioning the request. Because the reader is a man, Ms. Ashcroft, who knows that the Japanese culture scores high on the masculine/feminine scale, is careful

March 27, 2001

Mr. Keisuke Ashizawa
1402 Louisiana
Suite 651
Houston, TX 77002

Dear Mr. Ashizawa:

Mr. Kevin Graham, a close friend of our family, has informed me of your arrival in Houston to pursue work for Remco during the next 24 months. We are delighted that you have arrived in Houston during the early spring. Houston is famous for its azalea gardens and flowering fruit trees. Texas is also known for its state flower, the Bluebonnet, which blooms in profusion along the Texas highways. While Houston lacks cherry blossoms, we hope that you will find our azalea gardens an acceptable substitute.

To assist in making your stay in Houston pleasant as well as productive, arrangements have been made for representatives from several Houston banks to contact you. They will be sending you information about their services prior to arranging an appointment. You can then contact those banks that interest you. Or, if these are not acceptable, we would be happy to locate others. My husband, who is senior vice-president of Texas Commerce Bank, is eager to be of service in helping you select whatever financial services you require. We are honored to be able to help you.

Mr. Graham has invited me to meet you during the Offshore Technology Conference, which my husband will also attend. As Kevin has told you, as a residential leasing agent, I am interested in assuring that any visitors to our city receive appropriate accommodations in accordance with their needs. I will look forward to meeting you and hearing more about your work and answering any questions you may have about our culture. We will be pleased to work with Kevin to enable you to see other parts of Texas during the coming months, if you wish.

Sincerely,

Katherine Ashcroft

Katherine Ashcroft

FIGURE 6-7 · REQUEST TO A JAPANESE ENGINEER

to sound appropriately deferential. No timetable is mentioned on any of the issues. Japan is a collectivist culture in which relationships must be established before business can be conducted. Japan also scores high on avoiding uncertainty. Building trust and de-emphasizing the business goal are critical.

CULTURE AND GRAPHICS

Decisions about graphics as well as text should reflect awareness of the culture of your intended readers. For example, do they read left to right or from right to left? This information will help you know how to arrange graphics so that they will be viewed in the proper sequence.

1. If you are unsure about reading habits, consider using arrows with graphics to show the direction that your graphics flow (see Figure 6-8). Or, use numbers on graphics to show the order in which the units should be viewed.

2. Concepts of good and bad can also differ among cultures. In U.S. tables, what is considered acceptable and preferable is usually placed on the right, while in Asian cultures the left-hand side is the more prestigious.

3. Avoid acronymns, abbreviations, jargon, slang, and colloquialisms. Use plain typefaces. Do not attempt to use humor, as ideas of humor vary among cultures.

4. Be sure to use graphics that are internationally recognized (see Figure 6-9).

5. Avoid incorporating into graphics elements that may be interpreted differently in other cultures—animals, religious symbols, national emblems, hand gestures, and colors. For examples, dogs are considered pets in much of the West. To many Asians, they are food. Cattle have an honored position among Hindus in India, but in the United States their meat is eaten.

6. Remember that colors can be particularly problematic, so black and white graphics are usually the safest choice (see Table 6-3).

7. Because of differences in gender roles among cultures and differences in how body language is interpreted, make graphics of people as gender-neutral

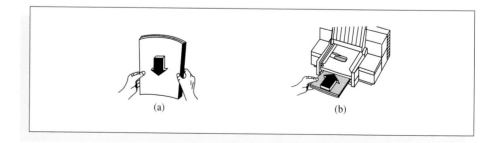

(a)   (b)

FIGURE 6-8 · EXAMPLES OF GRAPHICS USING ARROWS

as possible. Also avoid depicting hand gestures in graphics, as positions of fingers, hands, and arms have different meanings in different cultures.

8. If possible, try to determine how well your readers will know English. For example, if your readers lack fluency, use simple, concrete words and short ("subject, verb, object") sentences. Avoid abbreviations.

9. Watch how you use document design. Many cultures do not use bold-face headings or pay much attention to document design principles. Many Mexican documents, for example, use few headings, and these are underlined rather than set in a second font. Many cultures do not use extensive tables and graphs. Format strategies for correspondence—salutations, dates, subject lines, closings, and titles—also differ among cultures. American word processing programs have helped unify document design through use of common templates among international corporations, but differences still exist and should be respected.

FIGURE 6-9 · EXAMPLES OF INTERNATIONALLY RECOGNIZED GRAPHICS: (A) AIRPORT, (B) FRAGILE, (C) KEEP DRY, (D) BAR, AND (E) HOTEL

Table 6-3 ▪ **Color Associations in Different Cultures**

Culture	Red	Yellow	Green	Blue	Purple
Bolivia		Yellow flowers—contempt			Funerals
Europe and North America	Danger	Caution Cowardice	Safe Sour	Masculine Truth, authority	
Japan	Anger Funerals	Grace Nobility Childish gaiety	Future Youth Energy	Villainy	
China	Joy Festivity	Honor Royalty		Funerals	
Arab countries		Happiness Prosperity	Fertility Strength	Virtue Faith Truth	

▨ FORMAT STRATEGIES IN OTHER CULTURES

Format conventions also differ among cultures. In the following example, you can see how Arabic format conventions surface in a document written in English.

Situation 2 As a follow-up on a contact between the Qatar Department of Agriculture and Water Resources (DAWR) and the marketing manager from the Qatar Foundation Science and Technology Park, the Provost of Texas A&M–Qatar met with the DAWR research group to discuss potential collaborations between the government agency and Texas A&M. The Qatar executive provided a list of immediate research interests. Note, in the excerpt of the document (Figure 6-10) the right-to-left flow of text, a vestige of Arabic writing form. The document was in English but reflects the flow of Arabic text. Sent as a Word file, the file operated right to left.

Potential Research Collaborations Suggested by the Qatar Department of Agriculture and Water Resources – May 2004

I. REMOVING SODIUM AND CHLORIDE FROM SALINE IRRIGATION GROUNDWATER

Introduction 1.

Statistics reveal that more than 70% of the wells (the main source of irrigation in Qatar) have electrical conductivity (EC_w) greater than 3 dS/m. The chemical analysis of groundwater show that sodium (Na) is the most abundant cation and chloride (Cl) the most abundant anion. The concentration of each exceeds 20 mequiv/L in most of the irrigation groundwater. There are a number of potentially undesirable impacts on soils and crops resulting from using saline water for irrigation. The potential irrigation problems include soil salinity, infiltration rate of water in the soil and specific ion toxicity of sodium and chloride. According FAO guidelines for interpretations of water quality for irrigation if EC is > 3 dS/m and Cl > 10 mequiv/L and SAR > 9, there will be severe restrictions on use for irrigation.

Research Work 2.

To avoid detrimental effects on soils and crops, research is needed to develop an economic method for reduction the concentration of Na and Cl to reach the permissible safe limits.

FIGURE 6-10 · FORMAT EXAMPLE (QATAR)

II. DEVELOPMENT OF COOLING SYSTEM IN GREENHOUSES

Introduction 1.

To extend the growing period of vegetables (i.e., from May to September), cooling inside greenhouses is required. Refrigeration or air conditioning cannot be applied economically for cooling greenhouses under arid conditions. The only available technology seems to be evaporative cooling systems. Pad and fan cooling is the mostly widely spread evaporative cooling systems. However, this system has some disadvantages:

Need desalinated water to avoid clogging of pads and this is not economical; ☐
Very low cooling at high humidity (humidity in Qatar is high all the year round); ☐
High temperature gradient from pad to fan. ☐

Technology to Be Developed 2.

It is required to develop technology for improved and economic cooling systems.
Improve the pad and fan cooling system to allow the use of the available brackish groundwater. ☐
Investigate the possibility of lowering the humidity of incoming air before entering the greenhouses. ☐
Investigate the most appropriate combination of shading with ventilation and/or cooling needs. ☐

III. ASSESSMENT AND OPTIMISATION OF GROUNDWATER RECHARGE THROUGH INFILTRATION IN DEPRESSIONS – PHASE I

In Qatar the protection and improvement of the valuable groundwater resources is of highest importance. In this context, especially the recharge of groundwater by rainwater (surface runoff) is a key factor. In recent years, the realization of this concept has already been undertaken by the construction of some 350 infiltration wells in depressions all over the Qatar Peninsula. However, since recharge wells generally are very vulnerable to the accumulation of sediments as well as clogging of the well, the efficiency of the recharge process has significantly declined, causing the loss of precious water by evaporation. Hence, the assessment and optimisation of groundwater recharge through infiltration wells in depressions is essential.

Phase I includes five steps:

Survey of infiltration wells and surrounding depressions .1
Evaluation of different infiltration systems and maintenance works .2
Pilot implementation of optimised infiltration systems .3
Realization of the developed concepts at the most promising sites .4
Detailed reporting .5

FIGURE 6-10 · *CONTINUED*

A FINAL WORD

A number of reference books are available that provide extensive guidance on where to place dates, inside addresses, and reference lines in international documents. These books can also be helpful in explaining what titles to give individuals. These details are important, but none are as important as how you develop the content of the letter. The word order of English, its emphasis on concreteness and brevity as demonstrated in Chapter 4, can be construed as rude, insulting, and improperly aggressive by readers in other cultures. Particularly in Romance language cultures, sentences are more fluid, more complex, and the meaning more oblique. Understanding how to capture in English the tone expected by your international reader is a challenge. Remember that culture determines rhetoric.

But don't assume that all people from a given culture think alike. Culture will often provide the dominant filter through which your international readers view the world; however, it will rarely be the only filter. Sometimes a person's profession (or religion, or political philosophy, or disability, or the fact of being fat or thin left or right handed) will matter more than how he or she reads or writes a document. That is, culture remains a vital but never the only influence, and people of the same culture are still quite capable of thinking and communicating in different ways. Also keep in mind that every effort you make to adapt to another culture will likely be matched by an effort to adapt to yours. The key is respect for different traditions.

Germans prefer details and background information. The French like a formal and authoritative approach. U.S. readers like everything short and direct. The Japanese like instructions that are accurate but polite. Middle Eastern readers value grandiloquent, florid prose, and impassioned style. In China, theory is stated, and readers are expected to determine details. However, the important point is to remember that just as we are steeped in our culture, people from other countries are steeped in theirs. To communicate effectively with them, we must have some understanding of other cultures and of some of the ways that our traditions diverge from customs prevalent elsewhere in the world.

GUIDES TO DOING BUSINESS IN CULTURES AROUND THE WORLD

An increasing number of books and videotapes are available on the differences between the United States and other cultures around the world. Understanding some of these differences, as they apply to doing business, will help you understand how to communicate, orally and in writing, with individuals in other countries. Videotapes that discuss the meaning of gestures and body language can be particularly helpful in illustrating such differences. The following books provide concise, useful information about different cultures and will help you understand how culture affects communication styles and methods.

GENERAL

Axtell, Roger E. *Do's and Taboos Around the World.* 3rd ed. New York: Wiley, 1993.

———. *Do's and Taboos of Hosting International Visitors.* New York: Wiley, 1990.

———. *Do's and Taboos of Using English around the World.* New York: Wiley, 1995.

———. *Gestures: The Do's and Taboos of Body Language around the World.* New York: Wiley, 1998.

———., Tami Briggs, Margaret Corcoran, and Mary Beth Lamb, eds. *Do's and Taboos around the World for Women in Business.* New York: Wiley, 1997.

Bell, Arthur H. *Intercultural Business.* Hauppauge, NY: Barron's, 1999.

Bosrock, Mary Murray. *Put Your Best Foot Forward, Europe: A Fearless Guide to International Communication and Behavior.* St. Paul, MN: International Education Systems, 1995.

———. *Put Your Best Foot Forward, Mexico/Canada: A Fearless Guide to Communication and Behavior / NAFTA.* St. Paul, MN: International Education Systems, 1995.

———. *Put Your Best Foot Forward, Russia: A Fearless Guide to International Communication and Behavior.* St. Paul, MN: International Education Systems, 1995.

Bremmer, Katharina, et al. *Achieving Understanding: Discourse in Intercultural Encounters.* London: Longman, 1996.

Chaney, Lilliam H. *Intercultural Business Communication.* Upper Saddle River, NJ: Prentice Hall, 2000.

Chan-Herur, K.C. *Communicating with Customers around the World: A Practical Guide to Effective Cross-Cultural Business Communicating.* San Francisco: Au Monde Intl. Publishing Co., 1994.

Cremer, Rolf. *The Tongue of the Tiger: Overcoming Language Barriers in International Trade.* River Edge, NJ: World Scientific, 1998.

Doing Business Internationally: The Resource for Business and Social Etiquette. Princeton, NJ: Princeton Training Press, 1997.

Dunning, Sanjyot P. *Doing Business in Asia: The Complete Guide.* New York: Lexington Books, 1995.

Fitzgerald, Helen. *How Different Are We? Spoken Discourse in Intercultural Communication: The Significance of the Situational Context.* Buffalo, NY: Clevedon: Multilingual Matters, 2003.

Gannon, Martin J. *Understanding Global Cultures: Metaphorical Journeys through 23 Countries.* 2d ed. Thousand Oaks, CA: Sage, 2001.

Hall, Edward T., and Mildred Reed Hall, *Understanding Cultural Differences: Germans, French, and Americans* Yarmouth, ME: Intercultural Press, 1990.

Hodge, Sheia. *Global Smarts: The Art of Communicating and Deal Making Anywhere in the World.* New York: Wiley, 2000.

Hoecklin, Lisa. *Managing Cultural Differences: Strategies for Competitive Advantage.* Wokingham, UK: Addison-Wesley, 1995.

Holden, Nigel. *Cross-cultural Management: A Knowledge Management Perspective.* New York: Financial Times Prentice Hall, 2002.

Irvin, Harry. *Communicating with Asia: Understanding People and Customs.* St. Leonards, NSW, Australia: Allen & Unwin, 1996.

Kohls, L. Robert. *Survival Kit for Overseas Living: For Americans Planning to Live and Work Abroad.* Yarmouth, ME: Intercultural Press, 2001.

Koole, Tom. *The Construction of Intercultural Discourse: Team Dicussions of Educational Advisers.* Amsterdam: Rodopi, 1994.

Morrison, Terri. *Dun & Bradstreet's Guide to Doing Business around the World.* Paramus, NJ: Prentice Hall, 2001.

Morrison, Terri, Wayne A. Conaway, and George A. Borden. *Kiss, Bow, or Shake Hands: How to Do Business in 60 Countries.* Holbrook, MA: B. Adams, 1994.

Nelson, Carl A. *Protocol for Profit: A Manager's Guide to Competing Worldwide.* London: International Thomson Business, 1998.

Novinger, Tracy. *Intercultural Communication: A Practical Guide.* Austin: University of Texas Press, 2001.

O'Hara-Devereaux, Mary, and Robert Johansen. *Globalwork: Bridging Distance, Culture, and Time.* San Francisco: Jossey-Bass, 1994.

Reynolds, Sana. *Guide to Cross-Cultural Communication.* Upper Saddle River, NJ: Pearson Prentice Hall, 2004.

Sabath, Ann Marie. *International Business Etiquette. Europe: What You Need to Know to Conduct Business Abroad with Charm and Savvy.* Franklin Lakes, NJ: Career Press, 1999.

Schrank, Jeffrey. *I'm Normal, You're Weird: Understanding Other Cultures. Stage Fright Productions.* Videorecordings. Barrington, IL: Magna Systems, 1998.

Store, Craig. *Figuring Foreigners Out: A Practical/Guide.* Yarmouth, ME: Intercultural Press, 1999.

Thomas, David C. *Essentials of International Management: A Cross-cultural Perspective.* Thousand Oaks, CA: Sage, 2002.

Ting-Toome, Stella. *Communicating Across Cultures.* New York: Guilford Press, 1999.

Trompenaars, Fons, and Charles Hampden-Turner. *Riding the Waves of Culture: Understanding Diversity in Global Business.* 2d ed. New York: McGraw-Hill, 1998.

Walker, Danielle Medina. *Doing Business Internationally: The Guide to Cross-cultural Success.* New York: McGraw-Hill, 2003.

AFRICA

Hamlet, Janice D., ed. *Afrocentric Visions: Studies in Culture and Communication.* Thousand Oaks, CA: Sage, 1998.

BRAZIL

Novinger, Tracy. *Communicating with Brazilians: When "Yes" Means "No."* Austin: University of Texas Press, 2003.

CHINA

An Introduction to Chinese Cultural Values: The Other Pole of the Human Mind. Videorecording presented by Intercultural Resource Corporation in cooperation with the Intercultural communication Summer Institute. Newtonville, MA: Intercultural Resource Corporation, 1996.

Foley, J.A., et al. *English in New Cultural Contexts: Reflections from Singapore.* Singapore Institute of Management. New York: Oxford University Press, 1998.

Gao, Ge, and Shella Ting-Toomey. *Communicating Effectively with the Chinese.* Thousand Oaks, CA: Sage, 1998.

Li, Jenny. *Passport China: Your Pocket Guide to Chinese Business, Customs & Etiquette.* San Rafael, CA: World Trade Press, 1996.

Schnell, James A. *Perspectives on Communication in the People's Republic of China.* Lanham, MD: Lexington Books, 1999.

Wang, Mary M., et al. *Turning Bricks into Jade* [electronic resource]: *Critical Incidents for Mutual Understanding among Chinese and American.* 2000.

GREECE

Broome, Benjamin J. *Exploring the Greek Mosaic: A Guide to Intercultural Communication in Greece.* Yarmouth, ME: Intercultural Press, 1996.

INDIA

Joshi, Manoj. *Passport India: Your Pocket Guide to Indian Business, Customs & Etiquette.* San Rafael, CA: World Trade Press, 1997.

INDONESIA

Cole, Gregory. *Passport Indonesia: Your Pocket Guide to Indonesian Business, Customs & Etiquette.* San Rafael, CA: World Trade Press, 1997.

JAPAN

Abecasis-Phillips, J.A.S. (John Andrew Stephen). *Doing Business with the Japanese.* Lincolnwood, IL: NTC Business Books, 1994.

Goldman, Alan. *Doing Business with the Japanese: A Guide to Successful Communication, Management, and Diplomacy.* Albany: State University of New York Press, 1994.

Gudykunst, William B. *Bridging Japanese/North American Differences.* Thousand Oaks, CA: Sage, 1994.

Haru, Yamada, *Different Games, Different Rules: Why Americans and Japanese Misunderstand Each Other.* New York : Oxford University Press, 1997.

Hendry, Joy. Understanding Japanese Society. 2d ed. London and New York: Routledge, 1995.

———. Wrapping Culture: Politeness, Presentation, and Power in Japan and Other Societies. Oxford: Clarendon Press, 1993.

Hodgson, James D. *Doing Business with the New Japan.* Lanham, MD: Rowman & Littlefield, 2000.

March, Robert M. *Working for a Japanese Company: Insight into the MulticulturalWorkplace.* Tokyo: Kodansha International, 1992.

Nishiyama, Kazuo. *Doing Business with Japan: Success Strategies for Intercultural Communication.* Electronic resource. Honolulu: University of Hawaii Press, 2000.

———. *Welcoming the Japanese Visitor: Insights, Tips, Tactics.* University of Hawaii Press, 1996.

Pinelli, Thomas E. *A Comparison of the Technical Communication Practices of Japanese and U.S. Aerospace Engineers and Scientists* 1996. NASA Technical memorandum 111924.

———. Emerging Trends in the Globalization of Knowledge; the Role of the Technical Report in Aerospace Research and Development (1997), NASA Tech. memo. 112581.

KOREA

Kim, Un-yong. *A Cross-cultural Reference of Business Practices in a New Korea*. Westport, CT: Quorum Books, 1996.

MEXICO

Doing Business in Mexico. New York: Price Waterhouse World Firm Services, 1995.

Gordon, Gus, and Thurmon Williams. *Doing Business in Mexico: A Practical Guide*. New York: Best Business Books, 2002.

Leppert, Paul A. *Doing Business with Mexico*. Fremont, CA: Jain, 1996.

Malat, Randy. *Passport Mexico: Your Pocket Guide to Mexican Business, Customs & Etiquette*. San Rafael, CA: World Trade Press, 1996.

McKinniss, Candace Bancroft. *Business in Mexico: Managerial Behavior, Protocol, and Etiquette*. New York: Haworth Press, 1994.

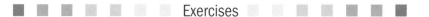 Exercises ■ ■ ■ ■ ■ ■ ■

Further exercises related to the content of this chapter and all documents employed within the following exercises can be found on the book's companion Web site.

1. Price Waterhouse has published a series of books on doing business in countries across the globe. Many of these have been published since 1994: South Africa, Cayman Islands, Czech Republic, Japan, Barbados, Belgium, Botswana, Chile, Canada, Denmark, Egypt, France, Hong Kong, Sweden, People's Republic of China, and Zimbabwe. Write a report, choosing two or three countries from the books in this series available in your library. Explain the differences in marketing strategy among the countries you choose to investigate.

2. Another series published by various branches of the U.S. government, and available from the Government Printing Office in Washington, DC, focuses on important aspects of individual countries. For example:

Heitzman, James, and Robert L. Worden. *India: A Country Study*. 5th ed. Washington, DC: Federal Research Division, 1996.

Hudson, Rex, ed. *Chile: A Country Study*. 5th ed. Washington, DC: PO, 1998.

Metz, Helen Chapin, ed. *Iraq: A Country Study*. 4th ed. Washington, DC: Federal Research Division, Library of Congress, 1990.

Shinn, Rinn S., ed. *Italy, a Country Study*. 2d ed. Foreign Area Studies, the American University. Washington, DC: Headquarters, Department of the Army, 1987.

Watkins, Chandra D. *Marketing in Kenya.* Washington, DC: U.S. Department of Commerce, International Trade, 1992.

Worden, Robert L., Andrea Matles Savada, and Ronald E. Dolan, eds. *China: A Country Study.* Washington, DC: Headquarters, Department of the Army, 1988.

Chose two or three different books in this series. Books are available on over 100 different countries. Write a report comparing and contrasting these countries in major categories of your choice.

Collaborative Projects

3. Prepare a written report on the challenges associated with doing business in a specific country. Focus your report on issues such as the following: management styles, corporate culture, negotiation style, social values, economics, and political systems. Allow each team member to choose and focus on one issue. After each person has completed research, come together as a group. Decide how you will prepare each segment in the written report. Prepare each segment, then make copies of the segment for each team member. Following discussion of each team member's findings, write, as a team, a summary of the findings.

The following works provide useful, more advanced information on culture and its effects on business, economics, and politics than the general guides to doing business in various countries:

Caroll, Raymond. *Cultural Misunderstandings: The French–American Experience.* Chicago: University of Chicago, 1988.

Child, John. *Management in China during the Age of Reform.* Cambridge, UK: Cambridge University Press, 1994.

Clegg, Stewart, and S. Gordon Redding. *Capitalism in Contrasting Cultures.* New York: de Gruyter, 1990.

Curry, Jeffrey. *A Short Course in International Negotiating: Planning and Conducting International Commercial Negotiations.* San Raphael, CA: World Trade Press, 1999.

Durlabhji, Subhash, and Norton E. Marks, eds. *Japanese Business: Cultural Perspectives.* Albany: State University of New York Press, 1992.

Kline, John M. *Foreign Investment Strategies in Restructuring Economies: Learning from Corporate Experience in Chile.* Westport, CT: Quorum Books, 1992.

Jain, Subhash C. *Market Evolution in Developing Countries: The Unfolding of the Indian Market.* New York: International Business Press, 1993.

Kato, Hiroki, and Joan S. Kato. *Understanding and Working with the Japanese Business World.* Englewood Cliffs, NJ: Prentice-Hall, 1992.

Maccoby, Michael, ed. *Sweden at the Edge: Lessons for American and Swedish Managers.* Philadelphia: University of Pennsylvania Press, 1991.

Saik, Yasutaka. *The Eight Core Values of the Japanese Businessman: Toward an Understanding of Japanese Management.* Binghamton, NY: International Business Press, 1999.

Simons, George F., Carmen Vázquez, and Philip R. Harris. *Transcultural Leadership: Empowering the Diverse Workforce.* Houston, TX: Gulf Publishing, 1993.

Soufi, Wahib Abdulfattah, and Richard T. Mayer. *Saudi Arabian Industrial Investment: An Analysis of Government–Business Relationships.* New York: Quorum Books, 1991.

Whitley, Richard. *Business Systems in East Asia: Firms, Markets, and Societies.* London: Newbury Park; Thousand Oaks, CA: Sage, 1992.

——. *Divergent Capitalisms: The Social Structuring and Change of Business Systems.* New York: Oxford University Press, 1999.

Wilson, Peter W., and Douglas F. Graham. *Saudi Arabia: The Coming Storm.* Armonk, NY: M. E. Sharpe, 1994.

REFERENCES

Hall, Edward T., and Mildred Reed Hall. *Understanding Cultural Difference.* Yarmouth, ME: Intercultural Press, 1990.

Hoecklin, Lisa. *Managing Cultural Differences: Strategies for Competitive Advantage.* Reading, MA: Addison-Wesley, 1995.

Hofstede, Geert H. *Cultures and Organizations: Software of the Mind.* London: McGraw-Hill, 1997.

——. *Culture's Consequences.* Thousand Oaks, CA: Sage, 2000.

Horton, William. "The Almost Universal Language: Graphics for International Documents," *Technical Communication*, 4th quarter, 1993, 682–693.

Trompenaars, Fons. *Riding the Waves of Culture.* 2d ed. London: the Economist Books, 1993.

PART TWO

Techniques

Part II moves beyond the basic concepts you will need to develop effective technical writing. All techniques depend on the importance of audience and purpose in planning any document. Part II builds on these fundamental concepts by explaining ways to develop effective reports: how to gather, evaluate, and document information needed to achieve your purpose with your readers; how to format documents that will be easy for readers to access and use; how to create and arrange text that will best present ideas you need to convey; how to select and develop the standard elements of technical reports; and how to create tables and other visuals to enhance readers' access to information. How you apply each technique will depend on your readers, your purpose in writing, and the context in which your writing will be received and used.

■ ■ ■

Scenario

You are the supervisor on a major construction project—a new city jail. This is your first assignment of a building of this size or complexity, and you are eager to make a good impression on your bosses.

You have hired a number of subcontractors for various portions of the project. The subcontractors, in turn, have hired workers (e.g., electricians, plumbers, carpenters, painters) to complete specific jobs. Today you learned, quite by accident, that a number of the workers do not have green cards. If you don't report them, you could be breaking the law. You don't know the precise penalties for employing undocumented workers, but if their immigrant status were discovered, the publicity for city officials and your company could be disastrous. You could easily lose your job. If you cause the workers to lose their jobs, however, you will be depriving their families of crucial income. Moreover, a delay in construction could cause you to miss the promised completion date, obliging your company to pay damages to the city. A delay won't look good on your record.

You would like to speak to your bosses, but you decide that it would be foolish to go in to such a meeting without a better understanding of the issue, the law, and your professional ethical responsibilities. You need information—a lot of it and soon.

CHAPTER SEVEN

Gathering, Evaluating, and Documenting Information

On the job you will often find yourself in situations requiring good research skills. You will need to know what questions to ask, where to look for answers, and how to evaluate the responses you receive from various sources. The ability to gather credible information efficiently will make you a more productive and valuable member of your organization.

Keep in mind that research is a continuous process of asking questions and receiving answers that prompt more questions. Whenever possible, cycle through this process again and again until you are sure that you have all the pertinent and reliable information necessary to make confident decisions or to take appropriate actions.

ASKING PRODUCTIVE QUESTIONS

Efficient research starts with systematic questioning. You have to determine as specifically as possible what you already know and what you don't yet know about your subject. Consider these six questions:

- ⬛ What personal experience do I have regarding this subject?
- ⬛ What have I read about this subject?
- ⬛ What have I heard from friends or colleagues about this subject?
- ⬛ What specific questions would I ask a specialist about this subject?
- ⬛ Why is this subject important to me, my business, or my client?
- ⬛ What keywords would I use to investigate this subject in a library's catalog or on the Web?

These questions will give you a good start on your research by helping you to inventory your existing knowledge of the subject and to identify the gaps in your understanding. As you proceed with your research, you will be able to refine and revise your research questions, and the answers will lead you to new questions and new answers.

For example, as you examine the issue of hiring undocumented workers, several good questions might come to mind:

- ⬛ What are my legal responsibilities as a citizen?
    As the supervisor on this construction project?
    As a representative of my company?
- ⬛ What is my company's responsibility here?
- ⬛ What is its potential liability?
- ⬛ Has a situation like this ever occurred at my company? If so, how was it resolved?
- ⬛ Does my company's code of conduct address this issue? If so, what does it advise?
- ⬛ What are my professional responsibilities as a civil engineer?
- ⬛ Does my professional association's code of conduct address this issue? If so, what does it advise?

LOOKING FOR ANSWERS

Once you know what you're looking for, you must decide where to start looking. Ordinarily, you will start with the most readily available sources and keep investigating until you have answered all your questions, exhausted your sources of information, or run out of time (and often you will run out of time).

A wide variety of information sources are typically available to you.

Interviews

Interviewing subject specialists is a highly efficient way of researching. You ask specific questions and receive answers tailored to the questions. Whether the interview is conducted in person, over the telephone, by letter, or through e-mail, good preparation is essential to getting full and pertinent answers to your questions. Subject specialists might be colleagues and supervisors inside your organization, notable scholars in the field, skilled practitioners from industry, or informed citizens of the local community.

First, do enough background reading on your subject to allow you to ask sophisticated questions. Subject specialists will typically appreciate your preparation. On the other hand, if you could have answered your questions by doing basic research, subject specialists are likely to think you are wasting their time. Consider sending your questions to interviewees ahead of time so that they may prepare thorough answers.

Second, find out as much as possible about your subject specialists:

- What qualifies them to advise you in this area—their education, their job experience, both?
- What have they said or written about this subject?
- What is their particular approach to the subject?
- What is their potential bias?
- What is their reputation within the profession?

Third, compose your list of questions, carefully targeting the areas of expertise of your subject specialists. For example, don't ask a practicing engineer about the origins of engineering ethics; instead, ask the president of the professional association of engineers or a scholar who has specialized in this topic. Ask a practicing engineer about his or her experiences with ethical dilemmas. Don't ask a police officer about immigration regulations; ask the director of the local immigration office or an attorney whose practice is built around immigration law. Ask the police officer about his or her experiences in working with the Immigration and Naturalization Service in matters involving undocumented workers.

Fourth, politely request the interview. Keep in mind that subject specialists won't be entirely forthcoming unless they understand your purpose. Identify the topic and explain how and when you would like to conduct the interview, tell why you chose this individual or group to interview, and let the party or parties know what you will do with the information you receive. Request permission to quote your subject specialists; with a personal or telephone interview, ask permission to record the session to assure the accuracy of your quotations and allow your interviewee to review your draft.

Guidelines for Drafting Questions for Specialists

- Create questions that require explanations or evaluations instead of a simple yes or no answer. Instead of "Does your professional association prohibit the use of undocumented workers?" ask "What does your professional association advise about the hiring of workers who do not have green cards?"
- Ask follow-up questions to encourage elaboration. Solicit details and examples. Follow "What are the hiring risks for local contractors?" with a question such as "How likely is it that a local contractor will be caught?" or "When was the last time a local contractor was caught?"
- Ask follow-up questions for clarification whenever you don't understand an answer. If possible, restate the answer in your own words and ask your subject specialist if your understanding is correct. For example, "So you believe the hiring of undocumented workers by local contractors is a fairly common practice. Is that right?" Here the yes/no question is necessary to prompt your source to confirm or correct your interpretation.
- Early on, include questions that acknowledge your subject specialist's expertise. By demonstrating that you have prepared for the interview by acquiring a basic understanding of the subject, such questions will encourage the specialist to give candid and comprehensive answers. For example, you could say "In your 1999 article, 'Ethics in Civil Engineering,' . . . why did you adopt this position?" Or you might ask "You've been an INS agent for five years. In that time, how has . . .?"
- Keep your interview focused, always steering the comments of the subject specialist to your topic. Intercept the specialist if he or she drifts from the topic. For example, "That's a good observation about the legal requirements. But let me ask you more about professional ethics. How did . . .?"
- Following the interview, write a thank-you letter or e-mail message to each of your subject specialists.

Newsgroups

Newsgroups are electronic communities of people from all over the world who exchange information about a common interest or affiliation by posting questions and answers to online bulletin boards. You will find one or more newsgroups to answer almost every question you might ask. For example, if you have a question regarding the immigration regulations of the United States, you could try the misc.immigration.usa newsgroup or alt.visa.us newsgroup. To find a newsgroup appropriate to your research, visit a directory site on the Web such as www.cyberfiber.com.

Newsgroups are either moderated or unmoderated. In a moderated newsgroup, all messages are reviewed before being posted to the bulletin board. Acting as the group's editor, the newsgroup's moderator will accept timely, relevant messages for publication on the bulletin board, while declining to publish submissions that repeat earlier posts or are irrelevant to the group or otherwise of little merit. In unmoderated newsgroups, all messages are posted directly to the bulletin board without filtering or editing by a moderator.

CHECKLIST

Guidelines for Doing Research via a Newsgroup

- **Upon joining a newsgroup, observe the discussion before you start to ask questions.** Remember that you're joining a conversation that is already in progress. It is polite to listen for a while before speaking. By briefly "lurking" in this way, you will develop a clearer understanding of the purpose of the newsgroup, the nature and style of messages on the bulletin board, and the appropriate way to ask and answer questions.

- **If the newsgroup offers archives of previous messages or FAQs (frequently asked questions), review this material.** This will familiarize you with the major topics of conversation and keep you from raising a subject or asking a question that has already been discussed thoroughly.

- **Compose a clear and specific subject line to your message.** You want to be certain that participants who really know something about your topic will notice your message; you also don't want to waste the time of participants whose expertise lies elsewhere.

- **Keep your message to the point.** Don't ask newsgroup participants to scroll through paragraphs of unnecessary information to locate the questions you are asking or the answers you are offering. If you are asking a question, be as specific as possible. If you are answering a question, copy only the pertinent passages of the original question and try to give a brief but thorough answer.

> - **Don't engage in "flaming."** Deliberately provocative and insulting comments disrupt the collaborative community that the newsgroup is designed to establish. If you consider a message genuinely offensive, comment offlist either to the contributor of the message or to the moderator of the newsgroup.

World Wide Web

On the World Wide Web (or the Web), billions of pages of information await your visit. For example, if you were investigating information on the professional ethics of civil engineers, you could visit the sites of the following professional associations:

American Concrete Institute *www.aci-int.org*

American Institute of Steel Construction *www.aisc.com*

American Society of Civil Engineers *www.asce.org*

American Society for Testing and Materials *www.astm.org*

American Water Works Association *www.awwa.org*

International Code Council *www.iccsafe.org*

Associated General Contractors of America *www.agc.org*

Civil Engineering Research Foundation *www.cerf.org*

Institute of Transportation Engineers *www.ite.org*

International Iron and Steel Institute *www.worldsteel.org*

With millions of sites and billions of pages, the Web is the biggest library of information resources available. The entire government of the United States, for example, is accessible through the Web (see www.firstgov.gov). But how do you know what's out there? And how do you find what you're looking for? To help you navigate the Web, a variety of search engines are available, such as Google, Alta Vista, Excite, InfoSeek, Lycos, and Yahoo. A search engine is a research service that scours the Internet looking for sites with keywords pertinent to your subject and lists the sites it discovers, usually according to relevance to your request as you stated it.

A simple search looks only for the presence of keywords (e.g., *ethics*). A focused search uses plus signs and minus signs to narrow the search. For example, a search for *civil engineering ethics* asks for all files containing at least one of the three words: *civil, engineering,* or *ethics*. A search for + *civil* + *engineering* + *ethics* narrows the focus to the files containing all three words: *civil, engineering,* and *ethics*. A search for − *civil* + *engineering* + *ethics* asks for all files containing both *engineering* and *ethics* but not the word *civil*.

Each service is a little different from the others—more or less comprehensive, more or less specialized. You might wish to try several search engines to see which directs you most quickly to the most appropriate sources for your subject or field. Also keep in mind that search engine firms are often paid to give priority to specific sites in their listings: that is, while one search engine might have been paid to list a site among its first five, a different search engine might list the same site tenth or twentieth in relevance to your subject.

Also available are metasearch engines such as Dogpile (dogpile.com), MetaCrawler (www.metacrawler.com), and Search.com (www.search.com). A metasearch service will submit your keywords to several search engines simultaneously, thus offering a kind of one-stop shopping for your research.

Libraries

If possible, visit a library and discuss your subject with a research librarian. Librarians are familiar with the library's resources—both paper and electronic— and are trained to assist visitors in finding answers to research questions. Librarians won't find the answer for you, but they will help to point you in the right direction.

On the job, however, you typically won't have time to make a physical visit to an outside library. A virtual visit will often have to do. In addition to the physical copies of books, journals, indexes, newspapers, and specialized encyclopedias, more and more libraries are subscribing to electronic sources of information, permitting you to do your library research simply by accessing the library's Web site. By acquiring and organizing the electronic versions of research materials, your local library thus opens the door to a world of information that might otherwise be unavailable to you.

Electronic Library Catalog The electronic library catalog is a listing of all the books, periodicals, and miscellaneous materials such as maps, films, and audio and video recordings that a library houses. Traditionally, the library catalog consisted of rows of chests containing drawers full of index cards listing the institution's entire inventory in alphabetical order by author, by title, and by subject. Today, most libraries have computerized their catalogs, permitting quick electronic searches by author, title, subject, and keyword (see Figure 7-1). For example, you might do a subject search using the words *engineering ethics*. Your search would identify several promising books and list all the information that you would need to locate the books in the library (see Figure 7-2).

Each item in a library has a specific numeric identifier known as a call number. Call numbers operate on one of two systems: the Dewey decimal system, dividing materials into 10 categories (see Figure 7-3), or the Library of Congress system, dividing materials into 21 categories (see Figure 7-4). Maps of your library show where the different categories of books are located.

Enter a search word or phrase:
Select a search option: Keyword   Author   Title   Subject
Limit your search by: Languages: Material types:
Enter a keyword or keyword phrase:
Select a search option: Keyword   Author   Title   Subject
(Selecting the keyword option permits the use of qualifiers.)
Limit your search: Select one or more databases:
Languages: Material types:

FIGURE 7-1 · SEARCHING THE ELECTRONIC LIBRARY CATALOG

---

* Brief Record Hitlist * Full Record Hitlist * Refine Search *
Holdings Display * MARC Display * Download Full Citation for Record *

Previous Record   **Next Record**

Record # 4

Title:	Thinking like an engineer: studies in the ethics of a profession / Michael Davis.
Author:	Davis, Michael, 1943-
Call number:	TA157 .D32 1998
Publisher:	New York: Oxford University Press [1998]
Subject heading(s):	Engineering ethics.
Description:	xii, 240 p.; 25 cm.
Notes:	At head of title: Association for Practice and Professional Ethics. Includes bibliographical references (p. 227–235) and index.
ISBN:	*0195120515 (alk. paper)
DBCN:	ALS-6049
Holdings:	

Holdings

Location	Call number	Material	Status
Main Library–Stacks	TA157 .D32 1998 c.1	Book	Available

FIGURE 7-2 · LISTING FOR A BOOK IN AN ELECTRONIC LIBRARY CATALOG

```
000    General knowledge (encyclopedias and other reference works)
100    Philosophy
200    Religion
300    Social sciences
400    Language (linguistics)
500    Pure sciences (mathematics, chemistry, physics)
600    Applied sciences (engineering)
700    Arts (music, painting, athletics)
800    Literature
900    History and geography
```

FIGURE 7-3 · DEWEY DECIMAL CLASSIFICATION SYSTEM

```
A    General works
B    Philosophy and religion
C    Auxiliary science of history
D    History of Europe, Asia, and Africa
E    History of United States
F    History of United States local, Canada, Latin America, and South
     America
G    Geography and anthropology
H    Social sciences
J    Political science
K    Law
L    Education
M    Music
N    Fine arts
P    Language and literature
Q    Science and mathematics
R    Medicine
S    Agriculture
T    Technology
U    Military science
V    Naval science
Z    Library science
```

FIGURE 7-4 · LIBRARY OF CONGRESS CLASSIFICATION SYSTEM

Knowing how materials are classified in your local library allows you to go to a specific section of the library to browse the shelves. Often through simple browsing you will discover materials that you might never have otherwise located—a useful book, for example, that because of its unusual title failed to show up during your search of the electronic catalog.

Indexes While the electronic library catalog identifies the titles of the magazines and journals in your library, it can't tell you the titles of the specific articles published. To locate that information, you go to electronic indexes such as the Online Computer Library Center's FirstSearch, which includes both WorldCat (a database of millions of materials in thousands of libraries worldwide) and ArticleFirst (a database of over 15,000 journals in a wide variety of fields). FirstSearch also gives you access to specialized databases such as the following:

- **Applied Science and Technology Abstracts:** abstracts of applied science and technology research
- **Business Dateline:** index of articles on regional business activities in over 500 magazines and newspapers
- **GPO Monthly Catalog:** index of all new publications of the Government Printing Office
- **Index to Legal Periodicals & Books:** index of articles in over 600 legal journals
- **Newspaper Abstracts:** abstracts of articles in 30 national and regional newspapers
- **New York Times:** abstracts of all articles in the *New York Times*
- **Periodical Abstracts:** abstracts of articles in over 2,000 journals in a variety of fields
- **Social Science Abstracts:** abstracts of articles in over 400 journals in anthropology economics, geography, political science, law, and sociology
- **Wilson Business Abstracts:** abstracts of articles in over 300 business magazines

For example, your research on illegal immigration would uncover a citation of a promising publication from the U.S. Department of Justice (see Figure 7-5) titled "Administrative Decisions Under Employer Sanctions, Unfair Immigration-Related Employment Practices, and Civil Penalty Document Fraud Laws."

Or by using Web of Knowledge, you could simultaneously search the *Science Citation Index, the Social Science Citation Index,* and the *Arts & Humanities Index* to locate pertinent journal articles and the sources of information cited in such articles. For example, a search of the subject *impact of illegal immigration* would lead you to the article "Does Border Enforcement Protect U.S. Workers from Illegal Immigration?" published in *Review of Economics and Statistics* in 2002. This article includes 42 citations of sources and is itself cited in a 2003

```
GOVDOC NO:    J 1.103:
AUTHOR:       United States. Dept. of Justice.
TITLE:        Administrative decisions under employer sanctions, unfair
              immigration-related employment practices, and civil penalty
              document fraud laws
PLACE:        [Washington, DC]:
PUBLISHER:    The Dept. For sale by the U.S. G.P.O., Supt. of Docs.
YEAR:         1993, 9999
PUB TYPE:     Serial
LANGUAGE:     English
FORMAT:       v.; 24 cm.
FREQUENCY:    Irregular
NUMBERING:    Vol. 3 (Jan. 1992 to Dec. 1993)
NOTES:        Title varies slightly. Decisions and orders of the Chief
              Administrative Hearing Officer, of the U.S. Dept. of Justice
              Executive Office for Immigration Review
SUBJECT:      Alien labor, legal status, laws, etc. United States Cases.
              Emigration and immigration law United States Cases.
              Illegal aliens, employment. United States Cases.
              Alien labor certification. United States Cases.
              Alien labor certification, corrupt practices. United States
              Cases.
ALT TITLE:    OCAHO
              United States. Dept. of Justice. Administrative decisions
              under employer sanctions & unfair immigration-related
              employment practices laws
              United States. Dept. of Justice. 8 USC 1324 ... proceeding
OTHER:        United States. Dept. of Justice. Executive Office for
              Immigration Review. Office of the Chief Administrative Hearing
              Officer.
ITEM NO:      0717-C-20
STOCK NO      U.S. Govt. Print. Off., Supt. of Docs., Mail Stop: SSOP,
              Washington, DC 20402-9328
OCLC NO:      37879193
```

FIGURE 7-5 · CITATION OF A GOVERNMENT PUBLICATION IN FIRSTSEARCH

article (with 24 citations) from the journal *Demography* titled "Do Amnesty Programs Reduce Undocumented Immigration?" Both articles and all their cited articles are potential sources of information.

Or for news and information in the fields of business, law, or medicine, you might do a search through Lexis Nexis. For example, a search of the subject *illegal immigration and construction industry* would uncover a pertinent 2004 U.S. Newswire article titled "Immigration Proposal May Alleviate Construction Workforce Shortage." According to this article, the Associated General

www.oup.com/us/houp

Contractors of America is supporting a proposal to alleviate a shortage of skilled labor in the construction industry by allowing undocumented workers to keep their jobs and apply for legal residence. Finding this source would encourage you logically to visit the website of the AGC (www.agc.org) for more information. In addition, a search of *illegal immigration and employer liability* would uncover a 2004 *National Journal* article titled "The Risks of Guilty Knowledge: Arrests of Illegal Aliens Working for Wal-Mart's Outside Contractors Raise Employer Liability Issues." This article explains immigration law in easy-to-understand language and advises companies using independent contractors to stipulate their compliance in writing.

EVALUATING ANSWERS

With all the information that is available to researchers today, the difficulty lies less in finding answers to your questions as in determining which answers to trust.

With interviews, for example, you have quick access to the newest information on a subject, such as the findings from yesterday's experiment, but you sacrifice the review process. You receive the information directly but without a good gauge on the accuracy of the subject specialist's findings or the potential bias of his or her conclusions and recommendations. You could be receiving erroneous or misleading information, but unless you have a separate method for verifying the findings or judging the reputation of a subject specialist, you won't know.

With newsgroups and the Web, similarly, you have the blessing of the widest possible participation in the creation and distribution of information: everybody's ideas and opinions are readily available for your consideration. The curse of such wide participation, however, is that misinformation is easily distributed by unreliable or unscrupulous sources.

With books and journal articles, on the other hand, the information you receive is often considered credible because it has been through the peer review process. Before publication of the material, scholars in the field, editors, and publishers examine the typescript—usually several times—to identify any incorrect or inconsistent material, which the authors are then asked to eliminate or correct. While peer-reviewed information has high credibility, the process itself may last as long as a couple of years. The information you receive, as a consequence, is never the newest information available.

Ideally, you will have several sources of information that will help you to achieve a credible balance of the most authoritative, most up-to-date information available. If you receive identical information from books and interviews alike, for example, you likely have found good answers to your questions.

In addition, a variety of aids are available to assist you in judging the credibility of your sources. For example, erroneous information distributed in a newsgroup is often challenged by other newsgroup participants. So don't accept the earliest answer to your question in a newsgroup: wait to see if others support or dispute the posting.

In judging a book, check the book reviews written by specialists in the field to determine their opinion of the book's credibility. For articles, check the Letters to the Editor feature in later issues of the magazine or journal to see whether readers have questioned any findings or conclusions. With published research, a good gauge of a source's credibility is the frequency with which it is cited by other scholars in the field.

To assist you in evaluating Web sites, rating services are available. For example, Librarians' Index to the Internet (lii.org) is a directory of carefully reviewed Web sites. Each site listed has been assessed by at least two librarian-editors to ensure that the information is current, correct, written in a coherent style, freely available, compliant with copyright and fair use guidelines, linked to additional credible sources, created by identified and reliable authors, efficiently organized, and easily navigated.

A search of the LII Web site using the keywords *illegal + immigration,* for example, yields the following two sites:

Center for Immigration Studies (CIS)

CIS "is a non-profit, non-partisan research institute which examines and critiques the impact of immigration on the United States." The site publishes news, articles, reports, and publications on immigration issues and lists a variety of immigration topics which reflect concerns of the U.S. government, such as terrorism and national security.

http://www.cis.org/

and

Los Trabajadores/The Workers

A companion site to a PBS program that "brings to life the vivid contradictions that haunt America's dependence on and discrimination against immigrant labor." The site includes an overview of day labor and day laborers, a brief interview with the filmmaker, a day labor quiz, and a nice selection of recommended books, articles, Web sites, and other related resources. Also available in Spanish.

http://www.pbs.org/independentlens/theworkers/

Similarly, a site such as About.com offers guided access to the Web. A trained legion of subject specialists directs you to pertinent sites with credible information and monitors additional resources such as discussion groups and bulletin boards.

Check the date of the site and the author/editor or sponsoring agency.

Also note the domain in the site address:

.aero	aviation community
.biz	businesses
.com	businesses
.coop	cooperatives
.edu	educational institutions
.gov	government agencies
.info	information about individuals, organizations, products, services
.int	international organizations
.mil	military
.museum	museums
.name	individuals
.net	individuals, organizations, and internet service providers
.org	noncommercial organizations
.pro	certified professionals

For example, information on immigration law from a .gov site would usually be considered authoritative, but information from a .com or .org site, which could be biased, incorrect, incomplete, or out of date, would have to be verified.

Ultimately, however, the decision to trust a source is yours. If you are using and distributing information, it is your reputation as a professional—your judgment as a researcher—that is at risk. To assess the credibility of your sources and the reliability of their answers to your questions, consider the following sets of guidelines.

Interviews

- What are the interviewer's credentials? Does he or she have appropriate and pertinent education and job experience?
- What is the individual's reputation in the field? Is he or she considered to be a national, regional, or local authority on the subject?
- Do you know of a trustworthy specialist in the field who has recommended this individual as a source of information?
- Did you notice bias in the individual's comments? Was he or she cautious and careful in offering explanations and opinions or impulsive and imprecise?
- Were the individual's opinions supported with sufficient and plausible evidence?
- Did the expert tell you anything that is contradicted by your other sources?
- Did the expert tell you anything that is verified by your other sources?

Newsgroups

- Who participates in this newsgroup? Is it restricted to specialists in the field or accessible to the public? Do the participants have appropriate and pertinent education and job experience?
- What is the newsgroup's reputation? Is it widely considered a good source of credible information?
- Do you know of a trustworthy specialist in the field who has recommended this newsgroup as a source of information?
- Did you notice bias in the answers to your question? Are the participants cautious and careful in offering explanations and opinions, or impulsive and imprecise?
- Did participants support their opinions with sufficient and plausible evidence?
- Did participants tell you anything that is contradicted by your other sources?
- Did participants tell you anything that is verified by your other sources?

Web Sites

- Who operates the Web site? Does the site display the credentials of the organization or individual so identified? Are those credentials appropriate and pertinent?
- What is the site's reputation? Do you know of a trustworthy specialist in the field who has recommended the site as a source of information?
- Do you notice bias in the information? Is the site designed to advertise? Or does it offer fair and impartial coverage of its topic?
- Does the site offer plausible evidence to support statements? Does it try to educate or entertain?
- How timely is the information? Does the site display the date it was last modified?
- Is any of the information at the site contradicted by your other sources?
- Is any of the information at the site verified by your other sources?

Books and Articles

- Who wrote the book or article? Does the author have appropriate and pertinent education and job experience?
- What is the reputation of the journal or book publisher? Do you know of a trustworthy specialist in the field who has recommended this book or article as a source of information? Is this journal widely considered to be a good source of credible information? Does this publisher ordinarily publish highly regarded books?
- How often is this book or article cited in other publications on this subject?

- Do you notice bias in the book or article? Is the author cautious and careful in offering explanations and opinions, or impulsive and imprecise?
- Does the author present sufficient and plausible evidence to support all statements?
- How timely is the information?
- Is any of the information in this book or article contradicted by your other sources?
- Is any of the information in this book or article verified by your other sources?

CITING SOURCES

If you use information from interviews, newsgroups, Web sites, books, or articles for a document you are writing or a presentation you are delivering, you have a moral and a rhetorical obligation to acknowledge your sources. Your moral obligation is to give appropriate credit to the individuals who deserve it—the people who composed the words, created the illustrations, or developed the ideas you are borrowing. Citing sources also serves the rhetorical function of bolstering the credibility of your investigation because it allows you to attribute your findings to pertinent subject specialists.

The intentional or unintentional borrowing of words, images, or ideas without attribution is plagiarism, a highly unethical and potentially illegal practice. You must conscientiously avoid plagiarism: you could lose your job and your reputation, and you and your organization might have to address criminal charges or a civil suit.

Depending on the rhetorical situation, you may choose a formal or informal method for acknowledging your sources. In a formal system of citation, your references will require specific and consistent formatting, as illustrated in *The Chicago Manual of Style* or *Publication Manual of the American Psychological Association*. A formal system of citation is particularly important if your document or presentation might be used by others to conduct subsequent research. The consistent formatting of citations will assist researchers in locating the sources you used.

If your document or presentation will be used by others chiefly to make decisions or take actions (i.e., by individuals managing information instead of creating information), a formal system of citation may be unnecessary. If informal citation is appropriate, you might acknowledge your sources by simple tagging:

- According to Lew Pauley's *Immigrants and Immigration,* . . .
- In a recent interview, Timothy Cooper of the Immigration and Naturalization Service told me that . . .
- Information from the Department of Labor (www.dol.gov) indicates that . . .

Keep in mind that the citation of sources offers you the opportunity to demonstrate that your research has been fair and thorough. If your list of sources, for example, omits a major book on your topic or the Web site of the organization you are investigating, your readers would have cause to doubt the validity of your findings. Similarly, if all your information comes from a single source or a single kind of source (e.g., all interviews), readers would likely consider your research to be biased or incomplete. Your list of sources is often a good indicator of the quality of your investigation and the merits of your conclusions and recommendations.

■ ■ ■ ■ ■ ■ ■　Exercises　■ ■ ■ ■ ■ ■ ■ ■

Further exercises related to the content of this chapter and all documents employed within the following exercises can be found on the book's companion Web site.

1. Choose a topic of importance to your major or minor. Familiarize yourself with the topic by reading the information available in one of the online encyclopedias such as www.encyclopedia.com. After you have completed this introductory reading, devise a series of questions about the topic that you would like to ask a subject specialist.

2. Interview a professional in your major or minor. List the criteria according to which you chose this individual and assessed the credibility of his or her answers to your questions.

3. Find a newsgroup that might offer answers to questions about your topic. How did you locate this newsgroup? Who are its participants? How active is it? How helpful are the participants? How credible is the information that is posted?

4. To locate information on the Web, enter a series of keywords on four different search engines. Note the similarities and differences in the listings of the four search engines? Which sites appear on all four lists? Which sites appear on only one list? Are the same sites listed higher or lower on different lists?

5. Use the rating service of Britannica.com to submit the same series of keywords regarding your subject. Which Web sites are listed by this service? Which sites identified by the Brittanica editors as credible sources were also listed by the search engines? Do you agree or disagree with Britannica.com's evaluation of the sites?

6. By interviewing professionals in your major or minor or by researching in the library, identify a noted authority on a subject of your choice. What speeches or publications are the basis for this individual's reputation as a highly authoritative scholar? Locate copies of his or her publications. How readily available are such publications? How often is this individual cited by other scholars in the field? Give examples of such citations from books, magazine and journal articles, or Web sites.

Scenario

Juanita Hinojosa in the payroll office must notify the managers of the company's twenty-three departments about a major change in the handling of time sheets and paychecks for hourly employees. She writes and distributes the following memo.

TO: All Departments

FROM: Juanita Hinojosa

SUBJECT: Time Sheets for Hourly Employees

On Friday, June 1, the payroll office will start using a new system for the submission of time sheets for hourly employees. Because of staff and budget reductions, the payroll office will no longer pick up time sheets from each department. Time sheets must be delivered to the payroll office on Friday of each week. Please make sure that all your time sheets are delivered to the payroll office before 3:00 P.M. every Friday. Departments with over 100 employees must submit their time sheets by 1:00 P.M. The payroll office will close at exactly 3:00 P.M. on Friday to allow time to validate and sort the time sheets for delivery to data processing by 5:00 P.M. Data processing staff will enter data from the time sheets, calculate the appropriate amount of pay, print the checks, and deliver the paychecks to the payroll office on Monday morning at 9 A.M. Each department may pick up payroll checks for its hourly employees on the following Monday at 10 A.M. Please note that payroll checks will no longer be delivered to individual departments.

On Monday, June 4, Juanita receives angry telephone calls from six department managers, complaining that the payroll office failed to pick up the time sheets from their department. None of the callers remembers having received Juanita's memo. Juanita is mystified and irritated because she made the extra effort to see that every manager received a paper copy of the memo as well as a high-priority e-mail message.

How can document design strategies be used to improve this memo? Assuming that everyone received this memo, why didn't people remember reading it?

CHAPTER EIGHT

Designing and Formatting Documents

As you have seen, effective writing requires a number of composing strategies that help you to meet the needs of your audience. But effective writing is more than just putting words on the page, using correct sentences organized in logical paragraphs. To be effective, your document must also be visually intelligible.

With the ever increasing capabilities of software to change the appearance of text, to incorporate illustrations, even to include animation and sound, you have many choices in how your document will look on paper or online. This chapter will help you design both paper and online documents.

UNDERSTANDING THE BASICS OF DOCUMENT DESIGN

A reader's earliest impression comes from the appearance of the document, not its content. A dense page of long paragraphs will often discourage or annoy a reader. A page designed to direct readers to important information, however, may add to the persuasiveness of your position or help your readers to find what they need and understand what they find.

Figure 8-1 is a vacation policy for an organization. Figure 8-2 is a revision of that policy that incorporates principles of document design discussed in this chapter. Which would you choose to read? Notice how the headings, list, and table all help the reader to see immediately the organization of the writer's points and to grasp the important information quickly.

These five principles will help you plan your document's visual design:

- ▪ Know what decisions are yours to make.
- ▪ Choose a design that fits your situation.
- ▪ Plan your design from the beginning.
- ▪ Reveal the design to your readers.
- ▪ Keep the design consistent.

Know What Decisions Are Yours to Make

Many companies have a standard format or template for reports, letters, proposals, e-mail messages, and Web sites. Before you develop your document, determine your design requirements. Don't use a format just because it's convenient (see Figure 8-3). If you think the template you are supposed to use isn't appropriate for your audience or message, find out who makes decisions on such issues and make a case for your desired changes.

Keep in mind also that you don't have to use exotic or sophisticated software applications to apply basic document design. A typical word processing program has all the functions you will need to create a visually effective document, like Figure 8-4.

Joint Practice 27: Vacation Days for Management

General

The purpose of this Joint Practice is to outline the vacation treatment applicable to management employees.

Eligibility

Vacations with pay shall be granted during the calendar year to each management employee who shall have completed six months' employment since the date employment began. Vacation pay will not be granted if the employee has been dismissed for misconduct. Vacation allowed will be determined according to the following criteria: (a) One week's vacation to any such management employee who has completed twelve months of service but who could not complete seven years of service within the vacation year. (b) Two weeks' vacation to any such management employee who has completed twelve months of service but who could not complete seven years of service within the vacation year. Two weeks will be allowed if the employee initially completes six months' service and twelve months' service within the same vacation year. (c) Three weeks' vacation to any management employee who could complete seven or more but less than fifteen years' service within the vacation year and to any District level employee who shall have completed six months' employment within the vacation year. (d) Four weeks' vacation to any management employee who completes fifteen or more but less than twenty-five years' service within the vacation year and to Division level employees who shall have completed six months' employment within the vacation year. (e) Five weeks' vacation to any management employee who completes twenty-five or more years of service within the vacation year and to Department head level and higher management who shall have completed a period of six months' employment within the vacation year.

The foregoing criterion is Net Credited Service as determined by the Employees' Benefit Committee. Where eligibility for a vacation week under (a) or (b) above first occurs on or after December 1 of a vacation year, the vacation week may be granted in the next following vacation year if it is completed before April 1 and before the beginning of vacation for the following year. When an authorized holiday falls in a week during which a management employee is absent on vacation, an additional day off (or equivalent time off with pay) may be taken in either the same calendar year or prior to April 1 of the following calendar year. When the additional day of vacation is Christmas Day, it may be granted immediately preceding the vacation or prior to April 1 of the following calendar year.

FIGURE 8-1 · ORIGINAL POLICY

Choose a Design That Fits Your Situation

Don't make your document any more complex than required. You don't need a table of contents or a glossary for reports that are under five pages. Add appendix material only if necessary and useful to your readers.

You'll impress readers most by providing just the information they need in a way that makes it easy for them to find and understand it. Most people skim technical and business documents selectively, looking for sections that are relevant to their needs. They try to grasp the main points quickly because they are

Joint Practice 27: Vacation Time Allowed Management Employees

The following schedule describes the new vacation schedule approved by the company. This schedule is effective immediately and will remain in effect until a further update is issued.

Vacation Eligibility

1. Vacation with pay shall be granted during the calendar year to each management employee who has completed 6 months' service since the date of employment. Employees who have been dismissed for misconduct will not receive vacation with pay.

Net Credited Service	Eligible Weeks
6 months–12 months	1
12 months–7 years	2
7 years–15 years and to District level with 6 months service	3
15 years–25 years and to Division level with 6 months service	4
25 years or more and to Department head or higher management with 6 months service	5

Net Credited Service is determined by the Employee Benefits Committee.

2. If eligibility occurs on or after December 1 of a vacation year,

   • vacation may be granted in the next following year if it is taken before April 1.

3. If an authorized holiday falls in a vacation week,

   • an additional day may be taken in either calendar year or before April 1 of the following year.

4. If the additional day of vacation is Christmas Day,

   • it may be taken immediately preceding the vacation or before April 1 of the following year.

FIGURE 8-2 · REVISED POLICY

busy. Remember that your readers get paid to make decisions and take actions, not to read documents. The more time they must use to read your document, the less productive they (and their company) are and the less cost-effective you (and your document) are.

Similarly, users working with a computer program are not likely to read the entire user's manual. They go to the manual when they have a specific problem

With the substantial growth in computing in the College of Engineering during the past decade, the issue of linking the departments through a computer network has become critical. The network must satisfy a number of criteria to meet the needs of all of the engineering departments. We first state these criteria and then discuss them individually in detail.

To adequately serve both faculty and student needs in the present environment, the network must be able to handle the number of computers currently in use. In addition, the system must be able to expand and link in additional computers as the number of computers increases over the next few years. The different types of computers that the departments presently possess must all be linkable to the network, and the types of computers that are scheduled for purchase must also be able to be connected to the network. The network should permit the transfer of files in both text and binary form in order to facilitate student access to files and collaborative exchange among faculty and research associates. The network must also have adequate bandwidth in order to handle the expected traffic. Finally, the network must permit both students and faculty to link to the existing national networks.

Each department currently has both computer laboratories for students and computers that are associated with faculty research projects. The various departments possess different numbers of computers. The Aeronautical Engineering Department at present has 27 computers, while Civil Engineering has 12. The Electrical Engineering Department has the most in the College with 46. Mechanical has 22, and Nuclear Engineering, the smallest department in the College, presently has 7. This means that the entire College presently has 114 computers which will need to be networked.

In order to meet their different needs, each department has focused on the purchasing of computers with differing strengths. The computers provide for faculty and advanced students to program in a variety of languages including Pascal, C, and Fortran.

The page with just text looks dense and uninviting.

Readers can't tell at a glance what the text is about.

FIGURE 8-3 · AN EXAMPLE OF POOR FORMATTING

or need instructions for a specific task. They want the instructions to stand out on the page or screen. Look at Figures 8-5 and 8-6. The numbered steps in Figure 8-6 make for quick reading and easy understanding relative to the long and confusing paragraph in Figure 8-5.

Plan the Design from the Beginning

Before you start writing, carefully consider how you will organize and display your information. Ask questions like the following:

- How will people use the document? Will most people read it from beginning to end? Will they want to skim it and grab the main points without reading more? Will they want to jump to a specific topic? Even if they read

Large headings make the topics and structure obvious.

A bulleted list makes the points more memorable.

Each item in the list becomes the heading for a subsection. The subsection headings are also bold but smaller than the main section heading.

The shorter line length makes the text easier to read and makes the headings stand out.

The numbers are much clearer in a table.

The footer on every page reminds readers of the overall topic.

With the substantial growth in computing in the College of Engineering during the past decade, the issue of linking the departments through a computer network has become critical. The network must satisfy a number of criteria to meet the needs of all the engineering departments. We first list these criteria and then discuss them individually in detail.

What must the network do?

To serve both faculty and students, the network must be able to

- handle the number of computers currently in use
- link different types of computers
- expand as the number of computers increases
- link to the national networks
- transfer and store both text and binary files

The network must also have adequate bandwidth to handle the expected traffic.

Handling the number of computers currently in use

Each department has both computer laboratories for students and computers that are associated with faculty research projects. The following table shows the number of computers in each department at the end of the last fiscal year.

Aeronautical Engineering	27
Civil Engineering	12
Electrical Engineering	46
Mechanical Engineering	22
Nuclear Engineering	7
Total	114

Linking different types of computers

In order to meet its different needs, each department has focused purchasing on machines with different strengths.

A Proposal to Install a Computer Network   **page 3**
for the College of Engineering

FIGURE 8-4 · THE PAGE FROM FIGURE 8-3, REFORMATTED

the document through once, will they want to come back later and find a specific point quickly?

- How familiar are your readers likely to be with the subject of the document? How much support may they need in understanding and navigating the information?

Drawing Product Help

To draw a box:

Decide where to put one corner of the box and move the mouse so that the cursor is in that position on the screen. Press and hold the left mouse button, sliding the mouse along the diagonal of the box which will appear on the screen as you move the mouse. When the box is the desired size, release the mouse button.

FIGURE 8-5 · INSTRUCTIONS IN PARAGRAPH STYLE
This format is difficult to follow, both on screen and on paper.

Drawing Product Help

To draw a box:

1. Decide where to put one corner of the box.

2. Move the mouse so that the cursor is in that position on the screen.

3. Press and hold the left mouse button, sliding the mouse along the diagonal of the box.

   The box appears as you move the mouse.

4. When the box is the size you want, release the mouse button.

FIGURE 8-6 · INSTRUCTIONS IN LIST FORM
Instructions formatted like this are easier to follow.

- Do your readers come to this kind of document with certain expectations about how the information will be organized and exhibited?
- Will most people see this document on paper or on a computer screen?

If people will skim a document, for example, a table of contents and headings on every page will help them find information quickly. (The rest of this chapter includes techniques for developing effective designs that help people find what they need.)

If readers don't know much about the subject of the document, a glossary of keywords and abbreviations may be advisable. Illustrations might also be useful.

If readers are unfamiliar with documents of the type you are creating, they might benefit from a simple and explicit design that avoids potentially confusing variations. Experienced readers, however, may have rigid design expectations.

Web site users, for example, ordinarily assume that underlined or colored words and phrases are active links and will be annoyed if they click on type so styled and are not transferred to a matching link.

If the document is going to be read on a computer screen, you may have both more constraints and more choices than you would for a document to be printed on paper. We read more slowly on screen than from paper, so limiting the amount of information and leaving space between paragraphs or list items is crucial in an online document. Illustrations and color may be easier and less expensive to include in an online document than in a paper document, however.

Reveal the Design to the Readers

Research on how people process information shows that readers cannot make sense of information unless they can see how it is organized. That is, as we read, we try to make sense of the passage we're reading, to see how the passage fits with previous passages, and to determine what it contributes to the entire document. The more difficult these tasks, the less readable the document.

Tables of contents and headings reveal the organization, scope, and direction of your document and give readers a clear overview. Using headings (at least one on every page) in a memo will show the structure and logic of the discussion and help readers recognize, remember, and retrieve your major points. Longer reports must have headings and a table of contents. In online documents, a contents listing of page titles is like the headings in paper documents.

Keep the Design Consistent

Consistency in design is essential to easy reading. When you have considered your audiences, the content, and the ways that people will read and use your document, you can develop a page design that will work well for your situation. Once you have decided on the appropriate page design, don't change it arbitrarily. You want your readers to know immediately when they are beginning a new section, and when they are in another part of the same section because they recognize the differences in design of each level of heading. Look again at Figure 8-4. Is it obvious where the major section starts? Is it obvious that the two subsections are parallel?

A good way to achieve this consistency is by identifying the different types of information in your document and using the Styles function of your word processing program to duplicate the design.

1. Think about all the types of information you will need to display (e.g., paragraphs, quotations, lists, examples, equations, formulas and various levels of headings).
2. Plan a design that always shows the same type of information in the same way throughout your document. The design could include:
   the type size;
   the typeface, or font;

placement of an element on the page;
whether the text has a border (also called a line or rule) over or under it;
whether the headings are bold or italic;
the amount of space that comes before and after a heading;
the style of the text following each kind of heading, and so forth.

3. Use the Style function of your word processing program to label and fix the design of each type of information.

Figure 8-7 shows how document design can reveal the content and the relationships among the sections of a letter. Any document, including routine letters and memos, will benefit from the use of a consistent design.

CHECKLIST

Designing E-mail Messages: On the job, you will probably write and receive more e-mail messages than any other kind of document. If you design your messages for quick and easy reading, your recipients will be able to respond readily and appropriately to your instructions, questions, and requests. Here are five simple guidelines to follow:

1. **Keep your messages brief.** E-mail is especially effective for brief messages that the recipient will read and reply to quickly. Long, scrolling messages with extensive detail are often better relegated to attachments.

2. **Use short paragraphs.** Short paragraphs separated by white space encourage quick reading and make it easy to perceive and retrieve the chief points of your message.

3. **Use the subject line to effectively specify your message.** A clear and specific subject line will preview your message, making reading easier and aiding understanding. You will also be helping recipients to sort and find your messages later, especially if you write a separate message for each topic.

4. **Use headings to identify the sections of your message.** Headings make it easier to skim your message for its chief points and assist in later retrieval of specific information.

5. **Avoid decorative colors and backgrounds.** While mountains and beaches are lovely to view and vivid colors are eye-catching, such decorations are unlikely to clarify or reinforce your message and often make reading more difficult. Text is easiest to read if it is in a dark color (preferably black) and displayed against a light and solid background (preferably white). Keep in mind that the objective of your e-mail message is to focus your reader's attention on pertinent words and illustrations.

Wilson, Wilson, and Fitch
2202 Winding Parkway, Suite 400
Glendale, Arizona 85320

September 14, 2005

Mr. Nick Marshall
Vice President
Multi-Tech Company
34454 Meadows Avenue
Glendale, AZ 85320

SUBJECT: Tax Treatment of Moving Expenses for Employees

Dear Mr. Marshall:

The moving expenses of your employees are regarded as itemized deductions subject to several requirements and limitations. Your employees should have no problem meeting the requirements for deductibility, but they should be informed of the limitations that apply to these expenses.

Conditions for Eligibility

1. **Distance Test.** The distance between the old residence and the new residence and the new place of employment must be at least 35 miles farther than the distance between the old residence and the new place of employment. Because your employees will be moving across the country, they will meet this test.

2. **Minimum Period of Employment after the Move.** There is a 39-week minimum period of employment following the move. This minimum should have no effect on your employees as long as they continue to work for your company.

Limitations of Expenses

Once the foregoing conditions have been met, the moving expenses qualify as itemized deductions. However, some of these expenses are limited by specific dollar amounts, depending on whether the expense is direct or indirect.

Direct Expenses

Expenses directly associated with moving to the new residence are not limited, except to say that they must be reasonable. Direct expenses include

- Traveling from the old residence to the new residence
- Moving all household goods to the new location.

FIGURE 8-7 · LETTER FORMAT THAT APPLIES DOCUMENT DESIGN PRINCIPLES

Mr. Nick Marshall
September 14, 2005
Page 2

Indirect Expenses

House hunting and temporary living expenses are limited to a total of $1,500 as a deduction.

1. **House hunting expenses** are all expenses incurred while you are actually looking for a house or dwelling. You must be working for your new company and looking for a house to use these expenses as indirect expenses.

2. **Temporary living expenses** include food and lodging expenses you incur after you move to the area of the new place of employment but before you move into a permanent residence.

   Note: Temporary living expenses will be allowed during a maximum 30-day period only.

3. Residence expenses are costs you incur in selling the old residence and/or costs incurred in locating a new one. Examples of these expenses include

   • Closing costs
   • Real estate commissions
   • Expenses necessary in acquiring or settling a lease.

Total deductible expenses in the indirect category, including house hunting costs, temporary living expenses, and residence expenses are limited to $3,000. Any amount incurred in excess of this amount cannot be taken as an itemized deduction.

If you need any further clarification, please call me at 303 444-5609.

Sincerely,

Kelly Jones

Kelly Jones, C.P.A.
Wilson, Wilson, and Fitch

FIGURE 8-7 · *CONTINUED*

DESIGNING EFFECTIVE PAGES AND SCREENS

Visually effective pages and computer screens are designed on a grid, to let readers know where to look for information. Because they have space inside the text, around the graphics, and at the margins, they look uncluttered, and information is easy to locate. Line length and margin widths are chosen to help people read easily. The following suggestions will help you develop visually effective pages and screens:

- Use space to frame and group information.
- Set the spacing for easy reading.
- Use a medium line length.
- Use a ragged right margin.

Use Blank Space to Frame and Group Information

Don't think of blank space as wasted space. It is a critical element in design for both paper and screens because it makes information easier to find and read. Look at Figures 8-8 and 8-9. Which do you think is easier to read?

You can incorporate space into documents in several ways, but one critical location is at the margins. Clear and generous margins enclose and contain the material presented and keep the page or screen from looking crowded and chaotic, making your information look organized and coherent.

If your document will be read on paper, also think about how it will be bound. If you are putting your work in a binder, be sure to leave room for the binding so that holes won't be punched through the text. Similarly, think about whether a reader will want to punch holes in a copy later or put the work in a binder. Figure 8-10 shows how these guidelines appear proportionately for margins on a standard 8½-by-11-inch page:

top margin	1 inch
bottom margin	1 inch
left margin	1 inch, if material is not being bound; 1½ inches, if material is being bound
right margin	1 inch

If you are going to photocopy on both the front and back of the page, leave space for the binding in the left margin of odd-numbered pages and in the right margin of even-numbered pages. Word processing programs allow you to choose mirror margins that alternate for right-hand (odd-numbered) and left-hand (even-numbered) pages. If you cannot set alternating margins, set both the right and the left margin at about 1½ inches to allow for binding two-sided copies.

Graphic designers call margins *passive space* because margins define only the block of the page or screen that readers should look at. Graphic designers know that *active space*—the space inside the text—makes the real difference in designing effective pages or screens.

TO: All Department Heads

SUBJECT: New Copy Procedures

A recent study of our copy center request procedures indicates that we are not ful-filling copy requests as efficiently as possible. A number of problems surfaced in the sur-vey. First, many requests, particularly large orders, are submitted before the copy center opens. Others are submitted after the copy center closes. As a result, the copy center has an enormous backlog of copy orders to fill before it can begin copy orders submitted after 8:30 A.M., when the center officially opens. This backlog may throw the center two or three hours behind schedule. All copy requests throughout the day then require over two hours to complete. By 2:00 P.M., any copy requests submitted may not be filled that day. If large orders arrive unexpectedly even a routine copy request may take two days to complete.

To remedy the situation, we will change to the following copy request procedure beginning Monday, February 7. The copy center will close at 3:00 every afternoon. Two work-study employees will work at the center from 3:00 until 5:00 to complete all orders by 4:00. If you submit copy requests by 3:00, the center will have them ready by 4:00. In short, all requests will be filled the day they are submitted. However, do not leave copy requests after 3:00, as these will not be processed until the following day. However, we guarantee that if you leave your request for copies with us between 8:30 and 3:00, you will have them that day.

Requests for copies of large orders—over 100 copies of one item, single/multiple copies of any document over 25 pages, of front/back photocopying of one item up to 50 copies—will require that a notice be given the copy center one day in advance. That way, the center can prepare for your copy request and be sure to have it ready for you. Copies of the request form are attached. Please complete one of these and send it to Lynda Haynes at the copy center so that she can schedule all big jobs. If you submit a big copy request without having completed the form, your request will be completed after other requests are complete.

Allow plenty of time for routine jobs—at least two hours, and three if possible. Begin-ning February 7, give all copy requests to the receptionist at your office number. Be sure you attach complete instructions. Give your name, your phone number, and your office number. State the number of copies required and any special instructions. Specify staples or clips, color paper, and collation on multipage copies.

Pick-up procedures also change February 7. All copy jobs, after they are complete, will be placed in each department's mail box. No copies will be left outside the copy cen-ter after closing time. No copies will be left with the receptionist. Large orders that will not fit mail boxes will be delivered to your office.

If you have questions about this new procedure, please contact Lynda Haynes at 2257.

Long lines and the uninterrupted flow of text obscure the new procedures.

FIGURE 8-8 · MEMO THAT IGNORES FORMAT GUIDELINES

TO: All Department Heads   DATE: January 27, 2005

FROM: Lynda Haynes

SUBJECT: **New Procedures for Ordering Copies from the Copy Center**

EFFECTIVE DATE: MONDAY, FEBRUARY 7, 2005

To handle orders more quickly and efficiently, the Copy Center is changing its procedures. Please inform everyone in your department and ask them to follow these new procedures.

Large Orders and Routine Requests

First, you must decide if you have a large order or a routine request. A large order is

- more than 100 copies of any item
- more than 50 copies of any item to be copied two-sided (front/back)
- single or multiple copies of any document over 25 pages.

Procedures for a Large Order

1. Fill out one of the attached Requests for Copying a Large Order forms.
2. Send the completed form to Lynda Haynes at the Copy Center at least one day in advance of the day you need the copying done.

   That way, Lynda can schedule big jobs, and you will avoid delays in getting your copying completed.

Procedure for a Routine Request

1. Attach complete instructions to your request. Include
   - your name, phone number, and office number
   - the number of copies you need
   - for multiple-page copies: instructions on collating and staples or clips
   - any special instructions, such as paper color

2. Give all copy requests to the Copy Center receptionists.

3. Allow 2 to 3 hours for your order to be filled.

NOTE:
Routine requests left between 8:30 A.M. and 3:00 P.M. will be processed by 4:00 P.M. on the same day.

The Copy Center will close at 3:00 P.M. Orders left after that time will be processed the next day.

Copy Pick Up Procedures

Copies will be delivered to your department's mailbox. If the order is too large for your mailbox, it will be delivered to your office.

If You Have Questions . . .   Contact Lynda Haynes at ext. 2257.

The information is in the order in which users need it.

The page with lots of white space is easier to read.

The headings in bold break the text into meaningful sections.

FIGURE 8-9 · A REVISION OF FIGURE 8-8

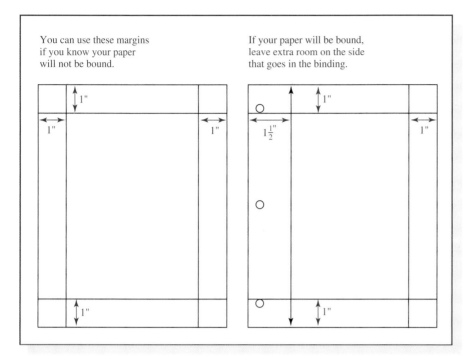

FIGURE 8-10 · PAGE LAYOUTS SHOWING MARGINS

These margins are for 8½-by-11-inch paper. (Some people prefer a larger, 1½-inch margin at the bottom of the page.)

Here are three techniques to bring active space to your pages and screens:

- Use headings frequently (and at least once per page or screen). Put them above the text or to the left of the text and put space before them.
- Use bulleted lists to emphasize three or more parallel points. Use numbered lists for steps in instructions. Lists are often indented inside the text, and items may be separated from each other by a blank space.
- Separate paragraphs with an extra blank line, or indent the first line of each paragraph. In online documents, make your paragraphs even shorter than you would in paper documents so that there will be space even in a small window. Online, one instruction or one short sentence may make an appropriate paragraph. Look again at Figures 8-5 and 8-6. The space is an active design element that makes the instructions in Figure 8-6 much easier to follow than the instructions in Figure 8-5.

Space the Lines of Text for Easy Reading

It's customary in paper documents to use single spacing. (For a brief letter or a one-paragraph memo, double spacing is appropriate.) When you use single spacing, insert an extra line between paragraphs or text units as in Figures 8-2 and 8-8.

Drafts of documents submitted for review and editing are often double-spaced to give writers and editors more room in which to write corrections and notes. When you use double spacing in drafts, you need to be able to show where new paragraphs begin. Either indent the first line of each paragraph or add an extra line between paragraphs.

For documents that will be read on the computer screen, use single spacing, with an extra line inserted between paragraphs. Double spacing is rarely used for continuous text on screen because a typical screen holds only about one-third of what a paper page holds.

Set the Line Length for Easy Reading

Long lines of text (over 15 words) make readers lose their place in moving from the end of one line (right margin) back to the beginning of the next line (left margin). Short lines (under 5 words) are also difficult to read because readers are almost continuously shifting their eyes from the right margin to the left margin of the next line with little time for moving across each line. Figure 8-11 illustrates why both long and short lines of text are difficult to read.

The number of words that fit on a line depends on the size and style of type that you are using. If you have one column of text on a page or screen, try to keep the lines of text to about ten to twelve words. In a format with two equal columns, keep each column to about five to seven words.

Use a Ragged Right Margin

The first line of a paragraph of text either starts at the left margin (*block style*) or is indented two or five spaces (*indented style*). This book uses indented style for paragraphs other than those following a displayed heading. Letters and memos often use block style. Figure 8-1 is an example of the indented style for paragraphs. Figure 8-7 is in block style.

> Long lines of type are difficult for many people to read. Readers may find it difficult to get back to the correct place at the left margin. The smaller the type, the harder it is for most people to read long lines of type.
>
> Very short lines
> look choppy
> on a page
> and make
> comprehension
> difficult.

FIGURE 8-11 · LINE LENGTH
Very long lines and very short lines are hard to read.

This text is justified only on the left. It has a ragged right margin. This is the preferred style for most online documents and documents that are printed on desktop printers.

This text is justified on both the right and the left. This is typical of books and other materials printed on high-resolution printers. If the computer cannot put the space in evenly across the rows, the text will include "rivers" of white that make reading difficult.

This text is justified on the right, but not on the left. This is highly unusual, except sometimes in headings.

This text has each line centered. Centering is sometimes used for headings and title pages.

FIGURE 8-12 · THE FOUR WAYS TO LINE UP (JUSTIFY) TEXT

Although text is almost always lined up on the left margin, it is sometimes also aligned on the right margin, creating a tidy rectangle of text. The text of this book, for example, is aligned on both the left and the right margins. Most of the examples in the figures in this chapter, however, are aligned on the left but not on the right. The technique of making all the text align exactly on both the left and the right margin is called *justifying the text.* If the text is aligned on the left but not on the right, it has a *ragged right margin.* Figure 8-12 shows the four ways of aligning text.

Be careful if you decide to justify type. Think about the purpose and audience for your work. Justified type gives a document a formal tone. Unjustified (ragged right) type gives a document a more friendly, informal feeling. Justified text is often more difficult to read: by making every line the same length, you eliminate a visual signal that helps readers both to keep their place and to find the next line of text. Reading onscreen is more difficult than reading from paper, and writers usually want to make their online documents look friendly and inviting. Therefore most online documents have ragged right margins.

CHOOSING READABLE TYPE

Today's technology gives you a wide array of design choices for the letters, numbers, punctuation, and other characters in your text. The various designs are usually referred to as *type, typefaces,* or *fonts.* Different companies often have style guidelines that dictate which fonts to use for official reports, letters, manuals, and online documents. If you are writing a document that is not covered by the style guidelines, you are likely to have to make decisions about what to use.

These six suggestions will help you choose type that is easy to read:

- Choose a legible type size.
- Choose a typeface (font) that is appropriate for the situation.
- Use special typefaces sparingly.
- Use highlighting effectively.
- Use a mixture of cases, not all capitals.
- Use color carefully.

Choose a Legible Type Size

Type is measured in points. A point is 1/72 of an inch, so 36-point type is about one-half inch high. Figure 8-13 shows type in seven different sizes.

Ordinarily, 10-, 11-, and 12-point type works well for regular text. You may want to use 9- or 10-point type for headers, footers, and footnotes. You can then use larger sizes for titles and headings.

In writing other online documents, make sure the type is large enough to be viewed by readers sitting in front of their monitors. Experiment yourself with different fonts in different sizes to see which is most legible.

The design and spacing of the letters also affect how much room the font takes up on a page or screen and how readable it is. Type in one font may look much tighter and smaller than type in the same point size in another font because of the way the letters are designed and how they are spaced when typed next to each other. Figure 8-14 shows four fonts, all in 13-point type.

This is 8-point type.

This is 10-point type.

This is 12-point type.

This is 14-point type.

This is 18-point type.

This is 24-point type.

This is 36-point type.

FIGURE 8-13 · TYPE COMES IN DIFFERENT SIZES

The quick brown fox jumped over the lazy dog.	Times New Roman
The quick brown fox jumped over the lazy dog.	Century Schoolbook
The quick brown fox jumped over the lazy dog.	Arial
The quick brown fox jumped over the lazy dog.	Courier New

FIGURE 8-14 · DIFFERENT FONTS
The same text in the same point size but in different fonts takes up different amounts of space on the page.

FIGURE 8-15 · STYLES OF TYPE
Type comes in two major styles: serif and sans serif (without serifs).

Serif Typefaces	Sans Serif Typefaces
Times	Helvetica
Palatino	Futura Light
Garamond	Franklin Gothic
Schoolbook	Helvetica Narrow
Courier	O C R A

FIGURE 8-16 · EXAMPLES OF TYPEFACES
Some traditional and new typefaces (fonts) available on a standard computer.

Choose a Font That Suits Your Document

In choosing a font for your document, you will need to consider several issues. Chief among your considerations are the two major categories of type design: *serif type* and *sans serif type.* Figure 8-15 shows the difference between them. Figure 8-16 illustrates fonts in both type families.

Historically, serifs were thought to be better for continuous reading because they seemed to aid the reader's eye in moving horizontally across the lines of text. Thus, books and long paper documents were and are still usually printed with serif type. Sans serif type, however, is often as easy to read, especially by those who are used to it. Sans serif type is also appropriate for visuals with oral presentations because viewers are at a distance from the words and are reading single lines at a time.

Experts are still discussing whether serif or sans serif type is better for online documents. Many writers prefer sans serif typefaces for their online documents because, with the low resolution of computer screens, serifs can make individual letters look fuzzy. Also, sans serif type, like ragged right margins, gives a document a clean and contemporary look.

Once you decide on a font for your document, use the same font throughout the document so that your pages look consistent. If you are using a serif font for the continuous text, however, it is permissible to use a sans serif font for headings, tables, figure captions, graph legends, and headers or footers, etc.

Keep in mind that any arbitrary changes in font will be distracting and confusing to readers, who will likely interpret the difference in font as indicating a qualitative difference in the information displayed. If there is no such difference, don't suggest there is by changing the font.

Use Special Typefaces Sparingly

Highly decorative and unusual typefaces (see Figure 8-17) are effective for getting a reader's attention but not for continuous reading. Such typefaces draw the reader's eye to themselves and are thus less effective in communicating content bearing substantial amounts of information. They are avoided for technical reports and online documentation, but might be used judiciously for flyers, brochures, circulars, and invitations, which require less reading.

THIS FONT LOOKS LIKE OLD-FASHIONED COMPUTER WRITING.

This might be appropriate for invitations.

Use outline fonts sparingly.

This looks like Ye Olde Font.

FIGURE 8-17 · SPECIAL TYPEFACES
Unusual typefaces should be used with caution and only in appropriate situations.
Source: Fonts from Arts & Letters. Computer Support Corporation, Dallas, TX 75244.

Use Highlighting Effectively

There are several ways to bring your reader's attention to specific words or passages that you want to emphasize:

using boldface
using italics
using <u>underlining</u>
changing the color of the type or the color of the background
putting the text in a box

using extra spacing
changing the type size

As you design your documents and decide on highlighting techniques, keep these three points in mind:

- Don't use too many different techniques.
- Don't use any technique too often.
- Use each technique consistently.

Don't Use Too Many Different Techniques   If you use too many different kinds of highlighting, your readers will be distracted by the excess visual information. For example, instead of using bold, underlining, italics, and color for a word you would like to emphasize, maybe just bold or just bold and italics will do the job. In design, less is often more.

Don't Use Any One Technique Too Often   Whole paragraphs underlined, set in boldface or italics, or printed a different color discourage reading. If every word is emphasized, it is impossible for readers to distinguish the more important from the less important. Use highlighting techniques sparingly for genuine impact.

Figure 8-18, a memo that uses excessive underlining in an attempt to emphasize the major points, does not invite reading. In contrast, Figure 8-19 shows how white space, bold headings, and a judicious use of bold in the text make the memo more inviting and easier to read and, especially, make the major points more obvious.

Use Each Technique Consistently   Highlighting helps readers quickly grasp the distinctions between different elements in the text. Once you have chosen a technique for highlighting a particular kind of information, use that technique consistently for that kind of information so that readers won't be confused. If you use italics to identify words listed in a glossary, for example, do so throughout and don't use italics for any other purpose. If you decide to set off cautions and warnings with rules above and below the text, use that technique for all cautions and warnings and not for anything else.

TO: David Stewart   DATE: February 18, 2005

FROM: Kathy Hillman

SUBJECT: Short Course Request from Ocean Drilling

Because I will be away on a three-week teaching assignment, I would appreciate your handling the following request, which came in just as I was preparing to leave today.

Randy Allen, director of the offshore drilling research team, would like a short-course in writing offshore safety inspection reports. He would like the short-course taught from 2–4 P.M. <u>Monday–Friday afternoons, beginning week after next.</u> The class must be scheduled then, as the team leaves the following week for their next research cruise.

The drilling research team spends <u>two weeks each month on cruise.</u> After they return, they have one week to complete their reports before briefing begins for the next research expedition. Because of their rigid schedule, <u>they cannot attend our regularly scheduled writing classes.</u>

Allen says that the cultural and educational backgrounds of the team are varied. Five of the ten regular researchers are native Europeans who attended only European universities. Of the remaining five, two have degrees from U.S. institutions, and three attended Canadian universities. <u>As a result of their varied educational backgrounds, their reports lack uniform handling of English and organization.</u> All the researchers have expressed interest in having a short review of standard English usage so that their reports to management will be more uniform.

<u>Sarah Kelley</u> says she can develop a class for the drilling team. We have materials on reports, style, and standard usage in the files. She can work with Ocean Drilling to determine the best report structure and develop a plan. These <u>items can be easily collected and placed in binders.</u> We also have <u>summary sheets</u> on each topic that will be good reference aids when the researchers write their reports following their cruise.

<u>Sarah will contact you Monday morning. If her teaching the class meets with your approval, please give Randy Allen a call, at extension 721, before noon. He has a staff meeting scheduled at 1:30 and would like to announce the short course then. In fact, if the course cannot be scheduled this month, it cannot be taught for seven months because of off-season cruise schedules. Randy wants this course before the team begins a series of four reports during the off-season.</u>

Please arrange a time for Sarah to meet with Randy so they can go over several previous reports. Sarah wants to be sure that what she covers in the course is what they need.

If you need to talk to me about this request, I will be staying at the Hyatt in New Orleans.

FIGURE 8-18 · MEMORANDUM THAT OVERUSES UNDERLINING AND LACKS HEADINGS

TO:	David Stewart	DATE:	February 18, 2005

FROM: Kathy Hillman

SUBJECT: Request from Offshore Drilling Team for a Special Short Course

ACTION
REQUIRED: **Decision from you by Monday, February 21 at noon**

Because I will be away on a three-week teaching assignment, I would appreciate your handling the following request, which came in just as I was preparing to leave today.

Offshore Drilling Wants a Special Short Course

Randy Allen, director of the offshore drilling research team, would like a short course in how to write offshore safety inspection reports. He would like the short course taught from 2 to 4 P.M. Monday to Friday afternoons, beginning the week after next. Randy wants this course before they begin a series of four reports during the off-season.

They Cannot Attend Our Regular Writing Classes

The class must be scheduled at the time Randy has requested because the team leaves the following week for their next research cruise. The team spends two weeks each month on cruise. After they return, they have one week to complete their reports before briefing begins for the next research expedition. Because of this rigid schedule, they cannot attend our regularly scheduled writing classes. In fact, **if the course cannot be scheduled this month, it cannot be taught for seven months** because of the off-season cruise schedules.

The Offshore Drilling Team Needs Help with Their Writing

Randy says that the cultural and educational backgrounds of the team are varied. Five of the ten regular researchers are native Europeans who attended only European universities. Of the remaining five, two have degrees from U.S. institutions, and three attended Canadian universities. As a result of their varied educational backgrounds, their reports lack uniform handling of English and report organization. All the researchers have expressed interest in having a short review of standard English usage so that their reports to management will be more uniform.

Sarah Kelley Can Develop the Class

Sarah Kelley says she can develop a class for the offshore drilling team. We have materials on reports, style, and standard usage in the files. She can work with the ocean drilling group to determine the best report structure and develop a plan. These items can be easily collected and put in binders. We also have summary sheets on each topic that will be good reference aids when the researchers write their reports following their cruise.

Please Decide by Monday Noon and Call Randy Allen—Extension 721

Sarah will contact you Monday morning. If you approve of her teaching the class, please call Randy Allen at extension 721, before noon. **He has a staff meeting scheduled for 1:30 and would like to announce the short course then.**

Please arrange a time for Sarah Kelley to meet with Randy Allen so they can go over several previous reports. Sarah wants to be sure that what she covers in the course is what they need.

FIGURE 8-19 · REVISION OF FIGURE 8-18

CAPITAL LETTERS GIVE US NO CLUES TO DISTINGUISH ONE LETTER FROM ANOTHER. THEREFORE, THE LETTERS BLUR INTO EACH OTHER VERY QUICKLY, AND WE WANT TO STOP READING. LOWERCASE LETTERS GIVE US CLUES TO THE SHAPES OF THE WORDS, AND WE USE THOSE SHAPES AS WE READ.

Capital letters give us no clues to distinguish one letter from another. Therefore, the letters blur into each other very quickly, and we want to stop reading. Lowercase letters give us clues to the shapes of the words, and we use those shapes as we read.

FIGURE 8-20 · A COMPARISON OF TEXT IN ALL CAPITALS AND TEXT IN MIXED CASE

Use a Mixture of Cases, Not All Capitals

Don't use all capitals for text. As Figure 8-20 shows, the use of all capitals slows reading because we use the different sizes and designs of letters to help us recognize words. For example, the lowercase letters include twelve with conspicuous ascending and descending strokes that give words a distinctive appearance, whereas uppercase letters are all the same height and leave words looking relatively similar.

A sentence in all capitals also takes up about 30 percent more space than the same sentence in lowercase letters. Mixed case is especially important for online documents because space is always at a premium on screen and text is already more difficult to read because of low resolutions. In addition, the use of all capitals in e-mail, offends many readers, who interpret it as intent to shout.

Use Color Cautiously and Consistently

On paper, avoid printing the text in color. Black ink on white paper provides the high contrast between paper and ink that is necessary for easy reading.

Colored paper is a good choice if it genuinely adds to the effectiveness of your document. For example, companies might use light gray or light blue paper for major reports and proposals but white paper for routine letters and memos, or paper of one color for the main section of the report and another color for the appendices, or colored sheets as dividers between sections of reports. If you use colored paper, choose a light color to keep the contrast between ink and paper as high as possible.

You may be able to print a document in more than one color. If so, use color judiciously. Color works well in headings and in illustrations such as statistical graphs and photographs, but it doesn't work well for continous text on paper. This book, for example, uses blue for headings and other elements on the page, but the regular text is black. Note that printing paper documents in color incurs additional cost.

When planning to use color in a printed report or other document, think about what may happen to the document in the future. If it is likely to be photocopied, make sure that color is not the only indicator of a particular feature, because the color will be lost in photocopying or will be reproduced poorly. Keep in mind that readers who happen to be color-blind cannot distinguish certain colors. If your document will circulate in a number of different countries, you will also need to determine the significance of your color choices to your international audiences (refer again to Chapter 6, Table 6-3).

In online documents, color is free, but this does not mean you should use it excessively or arbitrarily. You want readers to pay attention to your information instead of all the pretty colors on the screen, which in any event might be lost or modified when the document is printed out. Keeping a high contrast between the text and the background is also important on screen. Avoid using color for long passages of text, and don't put text on highly textured or illustrated backgrounds that make it difficult to read.

The key to all effective document design is a page or screen that is uncluttered and styled consistently. Striving for simplicity goes a long way to achieving that objective.

HELPING READERS LOCATE INFORMATION

To help your readers find what they need quickly and make sense of the material easily, you have to plan a useful structure, and show that structure to your readers. In the preceding sections of this chapter, we showed you how to use page layout and fonts to make your document clear and easy to use. In this section, we present three ways to help readers find information easily by providing clues to the document's overall structure:

- Write descriptive, informative, and persuasive headings.
- Design distinctive headings.
- Use page numbers and headers or footers.

Write Descriptive Headings

Headings are the short titles that label each section and subsection of your document. Even brief documents, such as memos and letters, can benefit from headings. Compare Figures 8-19 and 8-18 to see how useful headings can be, even in a brief memo.

Headings are the map to your document, identifying the key topics and revealing the direction of thought. These five suggestions will help you write useful headings:

- Use concrete language.
- Use questions, verb phrases, and sentences instead of nouns alone.

- Use standard headings if readers expect them.
- Make sure that the headings at each level are parallel in structure (e.g., all beginning with a verb).
- Make sure the headings match the table of contents.

Use Concrete Language Generic headings such as *Introduction* or *Conclusion* give no indication of the topic you are discussing. Headings should be specific to your document, and should reveal both the subject and your claims. Readers should be able to get a clear sense of your overall message by reading only your headings.

Use Questions, Verb Phrases, and Sentences instead of Nouns Alone The best way to write headings is to put yourself in your readers' place. Will readers come to your document with questions? Then questions will make good headings. Will they come wanting instructions for doing tasks? Then verb phrases that match the actions you advise will make good headings. Will they come seeking knowledge about a situation? Then statements of fact about that situation will make good headings.

In addition, avoid headings that are individual nouns or strings of nouns: such headings are often perceived as ambiguous. For example, a heading such as "Evaluation Questionnaire Completion" makes it impossible to predict the kind of information that this section will offer. Much clearer would be a heading such as "How Do I Complete the Evaluation Questionnaire?" or "What Is the Deadline for Completing the Evaluation Questionnaire?" or "Who Must Complete the Evaluation Questionnaire?"

Figure 8-21 shows how effective questions, verb phrases, and statements can be as headings.

Headings like these in actual documents were shown earlier in this chapter: Figure 8-9 (two levels of headings on the same page) and Figure 8-19 (all headings are sentences, so that a busy reader can skim through them and grasp the important messages in the memo).

Use Standard Headings If Readers Expect Them You may be working on a document for which readers expect to see a certain set of headings in a certain order, as in a standard proposal format. In that case, organize your material in the order your readers expect and use the customary headings. Figure 8-22 shows the headings used in a typical proposal format.

Make Sure the Headings at Each Level Are Parallel Like list items, headings in a document should be parallel. Parallelism is a very powerful tool in writing. See for yourself the difference parallelism makes by comparing the two sets of headings in Figure 8-23.

Make Sure the Headings Match the Table of Contents To check how well your headings tell your story and to check how well you've maintained parallel structure in headings, use your word processing program to create an outline

Questions are useful as headings in a brochure.
>   What does the gypsy moth look like?
>   How can we protect trees from gypsy moths?
>   How often should we spray?

Verb phrases are useful in instruction manuals.
Verb phrases can be gerunds, like these:
>   Adding a graphic
>   Selecting the data
>   Choosing type of graph to use
>   Adding a title

Verb phrases can be imperatives, like these:
>   Make your attendance policy clear.
>   Explain your grading policy.
>   Announce your office hours.
>   Supply names of texts to be purchased.
>   Go over assignments and their due dates.

Short sentences are useful in memos and reports.
>   Our workload has doubled in the past year.
>   We are also being asked to do new tasks.
>   We have logged 560 hours of overtime this year.
>   We need three more staff positions.

FIGURE 8-21 · DIFFERENT STRUCTURES YOU CAN USE FOR EFFECTIVE HEADINGS

Project Summary	Facilities and Equipment
Project Description	Personnel
Rationale and Significance	Budget
Plan of Work	

FIGURE 8-22 · STANDARD HEADINGS FOR A PROPOSAL FORMAT

view or a table of contents for your draft document. Both in print and online, the table of contents will consist of these headings. In a print document, readers can use the table of contents to locate a particular section. They know they're in the right place if the heading for that section matches the wording in the table of contents. The same is important online, since readers mostly navigate documents by jumping directly from a heading in the contents to a screen of information. Users will be confused and annoyed if the heading on the screen doesn't match the heading they clicked on in the table of contents.

Nonparallel Headings	Parallel Headings
Graph Modifications	Modifying a graph
Data selection updating	Changing the data
To add or delete columns	Adding or deleting columns
How to change color or patterns	Changing the color or patterns
Titles and legends can be included	Adding titles and legends

FIGURE 8-23 · NONPARALLEL AND PARALLEL HEADINGS
Headings that use the same sentence structure—parallel headings—are easier for users to follow.

Design Distinctive Headings

Headings do more than outline your document. They also help readers find specific parts quickly, and they show the relationship among the parts. It is helpful to readers to design headings that can be easily distinguished from the text; similarly, each level of heading must be easy to distinguish from all the other levels. The print document with four levels of headings shown in Figure 8-24 illustrates the use of boldface to distinguish all headings from the text, and the use of type size, capitalization, and position on the page to distinguish the heading levels.

These six suggestions will help you design distinctive headings:

- Limit the number of heading levels.
- Create a pattern for the headings and stick to it.
- Match size to importance.
- Put more space before a heading than after it.
- Use headings frequently.
- Consider using numbers with your headings.

Limit the Number of Heading Levels Don't make the hierarchy of levels more complicated than it needs to be. Headings are supposed to be an aid to reading, not an obstacle. Having too many levels of headings may create confusion. Paper documents don't need more than four levels of headings. If your copy seems to require five or more levels, consider dividing the material into two chapters. Two levels of headings will suffice for online documents: readers see much less at one time online than in a paper document.

Create a Pattern for the Headings and Stick to It Although your choices depend in part on the technology you are using, you almost certainly have several options for showing levels of headings. Figure 8-24 demonstrated a variety of ways to show different levels of headings. You can combine these to create the pattern for your headings. For example, you can change size, position, *and* capitalization to show the different levels of headings. Look again at Figure 8-24 to see how the writer has combined size and position to create a

---

Controlling Soil-Borne Pathogens
in Tree Nurseries

Types of Soil-Borne Pathogen and Their Effects on Trees

Simply stated, the effects of soil-borne pathogens...

....

The soil-borne fungi

At one time, it was thought that the soil-borne fungi......................................

Basiodiomycetes. The Basiodiomycetes are a class of fungi whose species

....

Phycomycetes. The class Phycomycetes is a very diversified type of fungus.
It is the ....

The plant parasitic nematodes

Nematodes are small, unsegmented...

....

Treatments and Controls for Soil-Borne Pathogens

....
....
....

FIGURE 8-24 · FOUR LEVELS OF HEADINGS IN A REPORT

pattern in which the level of each heading is obvious. To see a consistent pattern of headings in a complete report, look ahead to Figure 16-3.

Match Size to Importance Changing the type size is one way to indicate levels of headings. If you use different type sizes, make sure that you match the size to the level of importance. If the headings are different sizes, readers expect first-level headings to be slightly larger than second-level headings, second-level headings to be slightly larger than third-level headings, and so on, as in Figure 8-24. The lower-level headings can be the same size as the text, but no heading should be smaller than the text. That would violate readers' expectations.

Put More Space before a Heading Than after It Headings announce the topic that is coming next in your document. Therefore, you want the heading to lead the reader's eye down the page or screen into the text that follows. One way to do that is to have more space, on the page or screen, before the heading than after it, as in Figures 8-4 and 8-9. In this way, the heading and its accompanying text constitute a visible chunk of information.

If you are going to use a rule with the heading, consider putting it *above* the heading rather than below it. A rule above the heading creates a "chunk" that includes both the heading and the text that it covers. A rule above the heading also draws the reader's eye down into the text that follows instead of up and away from that text.

Use Headings Frequently Frequent headings help readers know where they are in a document at all times. In a report, you probably want a heading for every subsection, which might cover two or three paragraphs. In general, in print, you want to have clues to the text's arrangement on every page; online, you should have a heading on every screen or window. On a Web site page, you want to keep each topic short and give each topic a heading.

Consider Using Numbers with Your Headings In many companies and agencies, the standard for organizing reports and manuals is to use a numbering system to differentiate levels of headings. Figure 8-25 shows the three most commonly used systems:

■ The traditional outline system
■ The century-decade-unit system (often called the Navy system)
■ The multiple-decimal system

These systems allow you to designate a section elsewhere in the report by citing the number of its heading. Numbering systems, however, have several disadvantages. In all these systems, if you want to add or remove a section, you have to renumber at least part of the report. Unless your software does this automatically, renumbering is tedious and highly susceptible to error. In addition, readers often have trouble following these numbering systems, especially if there are more than three levels. The multiple-decimal system is particularly difficult for most people to use. For example, by the time you get to Section 1.1.1.1 (and some government reports go to a fifth level, 1.1.1.1.1), you may have lost track of what the first-level heading (Section 1) actually was.

If you are not required to use a numbering system, it is better not to institute one. Instead you can show the hierarchy of heading levels distinctly with changes in type size and position, as demonstrated in examples in this and other chapters.

However, many government agencies and many companies use one of these numbering systems, so it pays to be familiar with them. If you number your headings, of course, you must also number the corresponding entries in the table of contents.

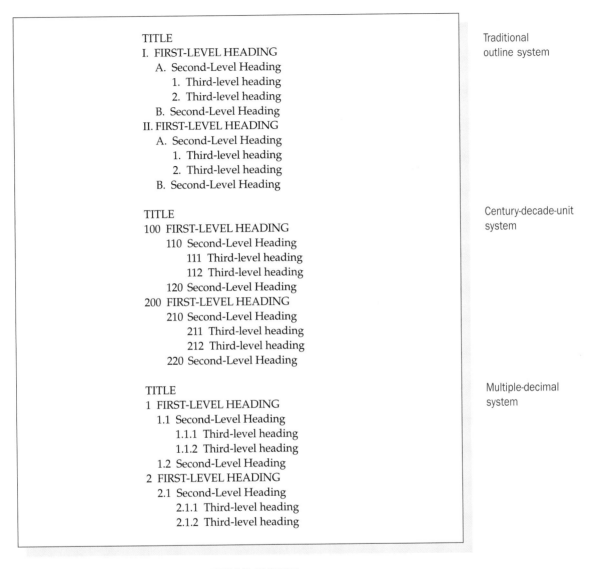

FIGURE 8-25 · THREE TYPES OF NUMBERING SYSTEM

Use Page Numbers and Headers or Footers

In addition to clearly worded and visually accessible headings, page numbers and running headers and footers are important aids to efficient reading.

Number the Pages Page numbers help readers keep track of where they are and provide easy reference points for finding sections or passages in a document. Always number the pages of paper drafts and final documents.

Inserting page numbers may be unnecessary for documents that will be used online, since most word processing programs keep track automatically of the number of pages and typically display this information in the bottom margin of the document window. If readers are likely to print your online document, however, they will certainly appreciate your inclusion of page numbers. In slide shows identifying each slide with a numbering notation like this

Slide 1 of 10

will help the audience to track your oral presentation.

Short manuscripts and reports that have little prefatory material mostly use arabic numerals (1, 2, 3). The convention is to center the page number at the foot of the page or to put it in the upper outside corner (upper left corner for left-hand pages, upper right corner for right-hand pages). Always leave at least one double space between the text and the page number. Page numbers at the bottom of the page often have a hyphen on each side, like this:

- 17 -

As reports grow longer and more complicated, the page-numbering system also may need to be more complex. If you have material that comes before the main part of the report, it is customary to use small roman numerals (i, ii, iii) for that material, reserving arabic numbers for the body of the report.

In a report, the introduction may be part of the prefatory material or the main body. The title page isn't numbered but is counted as the first page. The page number following the title page is 2 or ii.

You also have to know whether the document will be printed on one side of the paper or both. If both sides will carry printing, you may have to number several blank pages. New chapters or major sections usually start on a right-hand page. The right-hand page always has an odd number. If the last page of your first chapter is page 9, for example, and your document will be printed double-sided, you have to include one blank page 10 so that when the document is printed, copied, and bound, your second chapter will start on a right-hand page, 11.

The body of a report is usually paginated continuously, from page 1 to the last page. For the appendixes, you may continue this, or you may change to a letter-plus-number system. In that system, the pages in Appendix A are numbered A-1, A-2, and so on, Appendix B is numbered B-1, B-2, and so forth. If your report is part of a series, or if your company has a standard report format, your page numbering should match that of the series or standard format.

Numbering appendixes with the letter-plus-number system has several advantages:

▪ It separates the appendixes from the body. Readers can tell how long the body of the report is and how long each appendix is.
▪ It clearly shows that a page is part of an appendix and which appendix it belongs to. It makes pages in the appendixes easier to locate.

- It allows the appendixes to be printed separately from the body of the report. Sometimes the appendixes are ready before the body of the report
  has been completed, and being able to print them first may save time and
  help you meet a deadline.
- It allows the changes in pagination of either an appendix or the body
  without requiring changes in the other parts.

Include Headers or Footers   In long documents, give identifying information
at the top or bottom of each page: information at the top is a **header;** information at the bottom is a **footer.** Organizations often have a standard format for
headers and footers. A header formatted to show author's name, title of report,
and date might look like this:

Jane Fernstein   Feasibility Study   June 2005

With a header like this, the page numbers would likely appear in the footer.

A typical header for a letter might show the name of the person receiving
the letter, the page number, and the date. It might look like this:

Dr. Jieru Chen   -2-   June 16, 2005

or

Dr. Jieru Chen
Page 2
June 16, 2005

Figure 8-7 used this type of header.

Headers and footers rarely appear on the first page of a document because
first pages already carry identifying information like the title, author, recipient,
and date. Word processing programs allow you to start headers and footers on
the second page.

DESIGNING WEB SITES

Web pages offer a separate series of design challenges. Visitors to a site are likely
to skim pages quickly, looking for links to pertinent information. Their expectations include extensive uses of color, illustrations, video clips, interactivity, and
dynamic individualizing of the information displayed.

If you design your pages to accommodate the special reading habits of site
visitors, the Web could be a highly efficient way for your organization to distribute information about its goods and services. For example, you might
develop a site that functions like a billboard, promoting your organization's
image. Or you could create a site like a catalog, with pages that identify your
company's products and permit customers to make purchases by credit card.
You could also create a site like a technical manual, offering instructions on

how to assemble, maintain, or service your company's products. To such sites, you might add "Contact us" links that allow users to e-mail their questions or complaints to a customer service department.

And the multimedia capability of the Web gives you the power to design dynamic billboards, catalogs, and manuals. For example, if your site uses sound and animation, your customers could see and hear your products in operation. You might speak directly to your customers through a brief audio-video clip, directing their attention to your newest product, highlighting special items, or personally guaranteeing their satisfaction. Animated illustrations make it possible to demonstrate difficult or dangerous procedures, minimizing the potential ambiguity of words and static diagrams. The possibilities are almost unlimited.

Creating the Site

To create a Web site, you will need access to a computer that has a direct Internet connection, the server software necessary to establish a Web service, and the space available for storage of Web pages. A variety of Internet service providers offer access to such Web servers.

You will also need authoring software. Web pages often are written using HyperText Markup Language (HTML), coding that identifies the textual elements of each page, such as titles, headings, lists, and block quotations. You could do the HTML coding yourself or save time with authoring software such as Dreamweaver or Microsoft's Front Page. Such programs apply HTML to your word-processed files by translating your designated formatting features to HTML tags. Even if you have authoring software, knowing the basics of HTML will help you to interpret the coding of pages you would like to imitate and permit you to do a little quick editing of your pages whenever necessary. The World Wide Web Consortium (W3C) offers a free guide to basic and advanced HTML coding (http://www.w3.org/MarkUp).

In creating your site, consider the following guidelines.

1. **Explore and experiment.** Navigate the Web for a while, locating pages you would like to imitate or designs you would like to adapt for your site. The almost unlimited variety of sites offers a good stimulus for your creativity.

2. **Strive for a consistent design for all the pages at your site.** This will make your site more memorable and will help visitors to recognize when they have entered and when they have exited your site.

3. **Make your site inviting but simple.** Don't litter your pages with meaningless colors, icons, drawings, and photographs: focus people's attention on information, not decorations.

4. **Focus your efforts on the home page.** This is where visitors will usually enter your site. Make this page especially inviting and easy to navigate. If your home page is unattractive or confusing, visitors will exit your site without investigating further.

5. **Give visitors opportunities to interact with your pages.** Remember that a Web site is a dynamic medium, intrinsically different from the static paper

page. Visitors don't simply want to read or view your Web pages: they want to interact with your pages—to click on links and to access audio or video clips. They want to act over the Web, ordering merchandise, downloading software, making reservations, completing applications, and so on.

6. Be sensitive to the cultural differences of your international audience. Keep in mind that the Web is used by people all over the world. Avoid idioms, slang, and biased language. If appropriate to your audience, create multiple versions of your site in different languages.

7. Make your site accessible to visitors with disabilities. W3C has adopted guidelines on accessibility (http://www.w3.org/WAI/) that encourage barrier-free design and equivalent presentation (e.g., offering a textual description of pictorial or auditory materials).

Designing the Pages of the Site

In designing your pages, keep in mind the following common practices.

1. Include a complete menu of links at the top and bottom of each page. Make it as easy as possible for visitors to navigate your pages with a minimum of scrolling.

2. Avoid excessive links in the running text. Each link in the running text is a distraction from your message, requiring readers to make a choice—exit the page or continue reading. Give the complete menu of links at the top and bottom of the page and only crucial links in the running text.

3. Include identifying information on each page. At the bottom of every page, place a copyright notice (e.g., © 2005 ABC Corporation), the date of the latest revision (e.g., *last revised November 1, 2005*), and your e-mail address. Visitors to your site will want to know how to credit information they find there, how up to date the information is, and how questions or comments can be registered.

4. Choose a light, solid color for the background of your pages. Text displayed on dark or patterned backgrounds can be attention getting, but it is also more difficult to read. Remember that the background is just that: background. If the background is a distraction, it is ineffective and inefficient. Try to focus your reader's attention instead on the foreground—the genuine information on your pages.

5. Adjust the length of your pages to your information. Ordinarily, pages on the Web are relatively short: approximately one to three screens of information. If the material on a page is a long list, however, readers might prefer to scroll through the multiple screens of information and avoid the wait that accompanies a change of pages.

6. Keep illustrations small. Large illustrations are impressive but often take a long time to load. For indifferent visitors and people with slower browsers or connections, that long load time is annoying. Instead, include only a thumbnail-size illustration that interested visitors could click on in order to load the larger version.

7. Restrict animation to video or audio-video clips that the visitor specifically clicks on to view. Continuous motion (e.g., a blinking word) is highly distracting, reducing the reader's ability to pay attention to the other information on the page. Just because you can animate a page doesn't always make it a good idea.

8. Use only two levels of heading. On a scrolling page, readers often have difficulty keeping track of the organization and hierarchy of information.

9. Minimize the use of italics. Italics are difficult to read on the screen.

10. Use bold type selectively and consistently. Bold type is attention getting and identifies information of special importance, but it loses its power of emphasis if used excessively or arbitrarily.

11. Edit and proofread carefully. Errors in grammar and punctuation are distracting and diminish the credibility of your site.

12. Check the design of your site from a variety of computers and monitors. A page that looks good or loads quickly on your computer might load slowly on a different computer or appear distorted on another monitor. In designing your pages, therefore, consider the likely visitors to your site. What kind of hardware and Internet connections will they have? Should you design your pages for viewing on high-speed computers and 21-inch monitors or slower machines and smaller screens?

Maintaining the Site

Once you have designed your site, you must maintain it. Remember the following guidelines.

1. Register your site with a variety of search engines. This will maximize the number of people who can locate your site and see the information and services you have to offer.

2. Keep updating your pages. Give visitors to the site a reason to return regularly. Offer new information and new opportunities to interact with your pages.

3. Periodically check all your internal and external links to see that each is still active. (Internal links connect to other pages of your site; external links connect to pages at other sites.) Dead links annoy your visitors and jeopardize the efficiency and credibility of your site. Ordinarily, internal links break accidentally because of mistakes in the HTML coding while someone is updating or editing the pages. External links break if a Web site closes or changes its address (e.g., because it switches to a different Internet access provider). A variety of programs are available to check links.

4. If possible, avoid changing the address of your site. Changing your address will cause dead links at all the other sites with links to your pages. In addition, earlier visitors to your site who noted your location will not be able to return. If you must change your address, as soon as possible notify visitors to the existing site of the coming change. Include a link to your new site, and keep the existing site open for at least one month while you direct visitors to your new address. E-mail this address to appropriate discussion lists, and register it with appropriate search engines.

TESTING YOUR DESIGN

Both in print and online, the way your information looks is critical to its readability and usability. The choices you make about blank space, typography, and the wording and position of headings will help or hinder readers as they try to find and understand the information they need.

A particularly notorious and important example of ineffective information design is the Palm Beach County, Florida, "butterfly" ballot of the 2000 elections (see Figure 8-26). To vote for the first presidential candidate listed (George W. Bush), people were supposed to punch the first hole; to vote for the second candidate (Al Gore), they had to punch the third hole. Many voters who thought they were completing their ballots correctly realized later that their ballot did not go to the candidate of their choice. Why not? Consider the environment in which the ballots were used. After waiting for a long time to vote, aware that other impatient people were lined up behind you, would you be attentive to the intricacies of ballot design or would you try to hurry?

FIGURE 8-26 · "BUTTERFLY" BALLOT
Source: © Greg Lovett/The Palm Beach Post

In addition, if you think you know how to vote, do you expect to make mistakes?

Clearly, people who intended to vote for the Republican ticket on the ballot (or the Natural Law ticket) were more likely to vote correctly with or without paying full attention to the details of the graphics of the ballot. To vote correctly for any other candidates on this ballot required a voter to exercise a higher degree of mindful attention. Some voters whose ballots did not go to the candidate of their choice because they had failed to exercise that higher level of attention blamed themselves.

For the most part, however, people blamed the designer of the ballot, who had not done a proper usability test to determine whether all voters would have equal likelihood of completing the ballot correctly. Looking at the ballot and deciding it "looks" clear or asking colleagues (or the public) to review the design isn't enough. Testing the ballot with real voters under conditions that simulate the voting process is the right and ethical way to do it. Certainly it would have been less expensive, relative to all the subsequent legal challenges, the on and off counting of ballots, and the truly disastrous publicity that resulted.

You don't have to test the design of every document you write on real users: you won't have the time or the money to do so. But usability testing is especially important if your document:

■ Is of a kind you've never written before
■ Is on a topic of great importance or complexity
■ Will be used by a large number of readers
■ Might be used in ways that could differ widely or lead to serious mistakes

Planning the Usability Test

A valid test of usability includes good planning of four factors: (1) time, (2) money, (3) location, and (4) participants.

In the time available, you will need to determine how many users you can test, how many tests can be run at the same time, and how long each test can be. Keep in mind that not all users will tolerate long testing periods. Three to five users will ordinarily be enough to give you meaningful test results.

You will need money for both testing materials (such as copies of the document), but also for possible rewards for the users to thank them for participating in the test. The rewards don't have to be expensive or elaborate; they should, however, demonstrate your appreciation and assure subsequent cooperation (if necessary, e.g., for a second round of testing, a follow-up interview, or a later survey).

You also must choose a location for the testing. This might be an artificial setting that is convenient for you or a field setting that duplicates the real environment in which your document will be used.

You must also determine the testing methods, including your objectives. What do you want to know about the document from your user?

- Is it easy for the reader to find information?
- Is it easy for the reader to learn from the information provided?
- Is it easy for the reader to recover from a mistake or misunderstanding?
- Does the document invite the reader's attention?
- Does the document keep the reader's attention?
- Does the reader think the document is user friendly?
- Does the reader think the document is credible?
- Does the reader think the document is helpful?
- Which sections of the document are the reader's favorite and least favorite?

The measurements you will use are also a consideration in your planning. How will you know that the document is or isn't usable? Will you consider the time needed to read each section, time needed to recover from a mistake, time occupied in searching for information, number of tasks completed (or not completed) in a given time, number of mistakes made in answering a questionnaire, number of assists required to find information or answer questions? Your testing methods must also include decisions on the portions of the document you will test and on the script you will follow while conducting the test.

Finally, you will need to identify ideal subjects for the usability test and invite their participation. You will want individuals as similar as possible to the likely users of your document.

Conducting the Test

To obtain valid and reliable results, the test must be conducted consistently for each user. Ordinarily, document testing includes the following steps.

1. Greet each participant.
2. Explain the testing process to him or her.
3. Remind the participant that the object of study is the document, not the participant, and that there are no right or wrong answers.
4. Remind the participant that he or she may stop at any time.
5. Verify that the participant is comfortable before starting.
6. Introduce the test.
7. Observe the participant carefully but unobtrusively during the test.
8. Assist the participant only if necessary.
9. Interview the participant after the test.
10. Thank the participant for his or her cooperation.

Interpreting and Revising

Usability testing, like other forms of beta testing, will typically offer unexpected insights on the design of your product, in this case, a document. Readers will notice things and experience difficulties that you would never have imagined. The results of your testing will allow you to make important adjustments to the typography, blank space, headings, and headers and footers. Revisions derived from usability testing will almost always make your document more effective for more readers.

☑ Planning and Revision Checklist

You will find the Planning and Revision Checklist that follows Chapter 2, Composing, and the section entitled Planning the Content in Chapter 3, Writing for Your Readers, valuable in planning and revising any presentation of technical information. The following questions, which summarize the key points in this chapter, specifically apply to document design.

General Questions

Planning

- Have you considered how people will use your document?
- Have you checked on the software and hardware that you will use to prepare both drafts and final copy? (Do you know what options are available to you?)
- Have you found out whether you are expected to follow a standard format?
- Have you thought about how you will make the organization obvious to your readers? (What will you do to make it easy for people to read selectively in your document?)

Revision

- Is your document clean, organized, and attractive?
- Is your text easy to read?
- Will your readers be able to find a particular section easily?
- If your document is supposed to conform to a standard, does it?
- Have you tested your document with representative readers to determine its usability?

Questions about Using Space Effectively

Planning

- Have you set the margins to give enough white space around the page, including space for binding, if necessary?
- Have you planned which features to surround with extra white space, such as lists, tables, graphics, and examples?
- Have you set the line length and line spacing for easy reading?
- Have you decided how you are going to show where a new paragraph begins?
- Have you decided whether to use a justified or ragged right margin?

Revision

- Have you left adequate margins? Have you left room for binding?
- Is the spacing between the lines and paragraphs consistent and appropriate?
- Have you used white space to help the reader find information?
- Have you put white space around examples, warnings, pictures, and other displayed elements?
- Can the reader tell easily where sections and paragraphs begin?
- Is the margin ragged right? If so, is this choice effective? Would the document benefit from greater formality? If the margin is justified, is this choice effective? Does the document need to look less formidable? Is the text still easy to read?

- Have you used lists for steps in procedures, options, and conditions?
- Did you test representative readers to ascertain their perceptions of the spacing of material on the page or screen?

Questions about Making the Text Readable

Planning

- Have you selected a type size and typeface that will make the document easy to read?
- Have you planned for highlighting? Have you decided which elements need to be highlighted and what type of highlighting to use for each?
- Do you know whether you can use color? If this is allowed, have you planned which color or colors to use and where you will use color in the document?

Revision

- Is the text type large enough to be read easily?
- Have you been consistent in using one typeface?
- Have you used uppercase and lowercase letters for the text and for most levels of headings?
- Have you used highlighting functionally? Is the highlighting consistent? Does the highlighting make important elements stand out?
- Did you ask representative readers to test the text for readability?

Questions about Making Information Easy to Locate

Planning

- Have you planned your headings? Have you decided how many levels of headings you will need? Have you decided on the format for each level of heading?
- Will the format make it easy for readers to tell the difference between headings and text? Will the format make it easy for readers to tell one level of heading from another?
- Have you decided where to put the page numbers and what format to use?
- Have you decided on headers and footers (information at the top or bottom of each page)?
- Have you found out whether you are expected to use a numbering system? If numbering is required, have you found out what system to use and what parts of the document to include?

Revision

- Have you reread the headings critically? Are they informative? Unambiguous? Consistent? Parallel?
- Can readers get an overall picture of the document by reading the headings?
- Is the hierarchy of the headings obvious?
- Is it clear at a glance what is heading and what is text?
- If readers want to find a particular section quickly, will the size and placement of its heading help them?
- Are the pages of a paper document numbered?
- Are there appropriate headers or footers?
- If you are using a numbering system, is it consistent and correct?
- Did you test the ability of representative readers to locate information easily?

▪ ▪ ▪ ▪ ▪ ▪ ▪ ▪   Exercises   ▪ ▪ ▪ ▪ ▪ ▪ ▪ ▪

Further exercises related to the content of this chapter and all documents employed within the following exercises can be found on the book's companion Web site.

1. Figure 8-27 is a first draft of the instructions for the GrillWizard.

You are a friend and neighbor of the owner and founder of Fierce Products, a new company in Lubbock that manufactures the GrillWizard. Fierce Products is a family-owned company with forty-five employees. The owner (and spouse of the author of the instructions) has hired you to revise the design of the document to assure safe and efficient operation of the product as well as to convey a positive impression of the quality of Fierce Products.

After you design this document, you will return it to the company owner for final approval. Fierce Products is ready to release the Grill-Wizard to market. As soon as the instructions are ready, the product will be boxed and shipped. The sooner you submit your design, the sooner Fierce Products can start making money.

2. Figure 8-28 is a public notice about severe acute respiratory syndrome (SARS). This document was designed by the Centers for Disease Control and Prevention as a pdf file that could be downloaded from the CDC Web site (www.cdc.gov) for subsequent printing and distribution. Evaluate the design of this document. How is it effective and ineffective? Consider especially its use of color, typography, and spacing.

3. Conduct a usability test on the CDC's public notice on SARS (Figure 8-28), using at least three participants. Design the test so that you might determine the following:

· Readers' ability to find information
· Readers' ability to learn information
· Document's ability to get readers' attention
· Document's ability to keep readers' attention
· Readers' perception of the document's credibility

Report the results of your testing in a memo to the director of the CDC with your suggestions for revision of this and similar public notices on SARS.

4. Assume that the memo in Figure 8-29 was sent to instructors at your school to advise them that they have been selected to administer some teaching evaluation forms that the university wants to test. The main part of the memo, paragraph 2, provides instructions on how to administer the evaluations. The recipients of this memo are not expecting it and are only slightly familiar with the new teaching evaluation, a copy of which is attached to the memo.

GrillWizard

Fast and efficient frying and cooking with propane gas. Light the burner and instantly you have a hot 100,000 BTU continuous flame.

The **GrillWizard** will help you eliminate fish and other lingering cooking odors from your home.

The **GrillWizard** is used for fast frying of fish, potatoes, onion rings, chicken, vegetables, donuts. Substitute water for oil and it's also great for shrimp, crab, and lobster boils as well as steaming clams.

This cooker is completely portable with all parts easily assembled and disassembled for compact transportation and storage, yet it weighs only 40 lb.

When the control valve is open, a full 100,000 BTU of powerful heat prepares cooking oil in 3 minutes for frying. Adjust the heat with a touch of the control valve.

Operating Instructions

Place grill on level ground. Insert tapered end of tubing into the hole in the base of the grill. (CAUTION: Make sure the cooker is level and the burner is facing up.)

Attach grill connector to propane cylinder.

Completely open propane valve.

Slightly open control valve at grill connector and light cooker at top of tube immediately (CAUTION: **DO NOT stand directly over cooker when lighting burner.)**

Adjust control valve for desired flame height.

When finished cooking, always close both grill connector valve and propane cylinder valve completely.

The **GrillWizard** works with any size of propane tank cylinder and all will give off the same amount of heat. A 20 lb cylinder will provide approximately 6 hours of cooking time if valves are completely opened.

The intense heat produced by the **GrillWizard** allows you to fast-fry all foods. The cooking oils of conventional fryers drop in temperature as food is added, but the **GrillWizard** maintains its temperature with just a quick touch of the control valve.

The **GrillWizard** has been designed for easy care. However, keep all dirt and foreign objects out of connectors, hose, valves, and openings. Failure to do so could cause obstruction of gas and greatly diminish the effectiveness of the **GrillWizard.**

CAUTION: If you suspect leaks, **DO NOT** light unit before checking.

For outdoor use only.

After washing pan with soap and water, dry thoroughly and coat the entire pan with cooking oil on paper towel to prevent rusting.

FIGURE 8-27 · INSTRUCTIONS FOR THE GRILLWIZARD

GUIDELINES AND RECOMMENDATIONS

Interim Guidelines about Severe Acute Respiratory Syndrome (SARS) for Persons in the General Workplace Environment

The Centers for Disease Control and Prevention (CDC) is investigating the spread of a respiratory illness called severe acute respiratory syndrome (SARS). CDC has issued two types of notices to travelers: advisories and alerts. A ***travel advisory*** recommends that nonessential travel be deferred; a ***travel alert*** does not advise against travel, but informs travelers of a health concern and provides advice about specific precautions. CDC updates information on its website on the travel status of areas with SARS (www.cdc.gov/ncidod/sars/travel.htm) as the situation evolves.

SARS is an infectious illness that appears to spread primarily by close person-to-person contact, such as in situations in which persons have cared for, lived with, or had direct contact with respiratory secretions and/or body fluids of a person known to be a suspect SARS case. Potential ways in which infections can be transmitted by close contact include touching the skin of other persons or objects that become contaminated with infectious droplets and then touching your eyes, nose or mouth.

Workers, who in the last 10 days have traveled to a known SARS area, or have had close contact with a co-worker or family member with suspected or probable SARS could be at increased risk of developing SARS and should be vigilant for the development of fever (greater than 100.4° F) or respiratory symptoms (e.g., cough or difficulty breathing). If these symptoms develop you should not go to work, school, or other public areas but should seek evaluation by a health-care provider and practice infection control precautions recommended for the home or residential setting; **be sure to contact your health-care provider beforehand to let them know you may have been exposed to SARS so arrangements can be made, as necessary, to prevent transmission to others in the healthcare setting**. For more information about the signs and symptoms of SARS, please visit CDC's website (www.cdc.gov/ncidod/sars/). More detailed guidance on management of symptomatic persons who may have been exposed to SARS, such as how long you should avoid public areas is available at the exposure management page.

As with other infectious illnesses, one of the most important and appropriate preventive practices is careful and frequent hand hygiene. Cleaning your hands often using either soap and water or waterless alcohol-based hand sanitizers removes potentially infectious materials from your skin and helps prevent disease transmission.

The routine use of personal protective equipment (PPE) such as respirators, gloves, or, using surgical masks for protection against SARS exposure is currently not recommended in the general workplace (outside the health-care setting).

For more information, visit www.cdc.gov/ncidod/sars or call the CDC public response hotline at (888) 246-2675 (English), (888) 246-2857 (Español), or (866) 874-2646 (TTY)

May 8, 2003 | Page 1 of 1

DEPARTMENT OF HEALTH AND HUMAN SERVICES
CENTERS FOR DISEASE CONTROL AND PREVENTION
SAFER · HEALTHIER · PEOPLE™

FIGURE 8-28 · CDC SARS GUIDELINES

DATE: November 11, 2005

FROM: Karen Jones
                 Associate Director of Testing

TO: Faculty

As you know, the university has made every effort to see that teaching evaluations, which are given once a year, are as accurate a reflection as possible on the effectiveness of your teaching. We know that this goal is your desire. To help us better achieve this goal, we are launching a pilot program to test a new kind of evaluation. You were one of 50 faculty who agreed to test the new evaluation system. Because you are getting this memorandum, you are one of the faculty chosen for the trial evaluations. After you receive your scores, we will send you a response form to allow you to express your views on the evaluation. We will then set up an interview with you so that we can more fully discuss your views of the accuracy of the results and changes you think should be made.

When you receive the questionnaire, a copy of which is attached, we want you to do a number of things. Please announce that the questionnaire will be given and urge students to attend class that day. If some students are absent the day you give the questionnaire, give those students a questionnaire the next class period. For the trial questionnaire, it is imperative that every student in the trial sections complete a questionnaire. Have someone else administer the questionnaire—either a colleague or your department secretary. Be sure you are not present while the students are completing the questionnaire. Have the person who is monitoring the questionnaire collect all of them and place them back in the envelope. These should be sealed in front of the students. The person monitoring the questionnaire should return these to the testing office (104 Haggarty) immediately after the test. The tests should be left at the test desk, which is the first desk on the right after you enter the office. Give the test to the clerk in charge of the trial test evaluation. Her name is Micki Nance. She will be there from 8–11 and 1–4 every class day during the test week, which will be the first week in December (December 5–9). Sign the sheet to indicate that you have returned your trial test. You will receive your printout by the first week in February. When you receive your printout, it will include a date that tells you when we will want talk to you further. The response card, indicating your feelings about the accuracy of the trial evaluation, should be completed and returned immediately.

If you have any questions, please call Sammy Carson at ext. 9912.

FIGURE 8-29 · MEMO FOR EXERCISE 4

TO: Plant Engineering Managers   DATE: September 14, 2005

FROM: John Bridgers
           District Superintendent

SUBJECT: Company Education Policy

Here is the company's policy on education, which many of you have asked about. Please keep it for your files for reference.

Policy 44.7. Advanced Education and Training. This policy applies to all employees except technicians and maintenance personnel. In order to encourage management personnel to achieve increasing professional competence in their disciplines and to enhance advancement potential, personnel who register for credit at the undergraduate or graduate level in accredited institutions will be reimbursed for tuition costs, registration fees, required textbooks, lab equipment, and other required materials upon completion of these courses. Certification that the specific course(s) will enhance the employee's professional growth must be provided by the employee's direct supervisor and countersigned by the supervisor's superior, unless the employee's supervisor holds the rank of vice president. Successful completion, defined as a grade of B or higher, must be attained in any course before the employee can apply for compensation. Costs of travel to the institution and costs of nonrequired materials such as paper, photocopies, and clerical help will not be reimbursed nor submitted to the company clerical workers. Submission to the Training Division of all receipts for all expenses, approval of the direct supervisor that the course fills the requirements of this policy, and documentation of successful completion are required before reimbursement will be permitted by the Training Division. Supervisors may allow release time for their employees to enroll in credit courses when work schedules permit. Release time is encouraged only when scheduled meetings of extremely important courses occur during regular working hours. If possible and necessary, personnel may be required to make up working time outside normal working hours. If the credit college course can be taken outside the individual's normal working hours, no release time will normally be allowed. To receive reimbursement, personnel should submit Training Division Form 6161 to the Training Division in accordance with the instructions on that form.

FIGURE 8-30 · MEMO FOR EXERCISE 5

Revise the design of this memo to help teachers administer the evaluation correctly and to let them know what they should do afterward.

5. Revise the memorandum in Figure 8-30, which is to be sent to plant engineering managers, so that it clearly explains the company's policy for paying for continuing education courses.

6. The instructions memorandum shown in Figure 8-31 was intended to explain how to delete security codes. Use principles of document design to make the memo easier to read and understand. This

TO: SWMTR II Departmental Coordinators and Time Reporters

FROM: SWMTR System Administrator

SUBJECT: Deletion of Security Codes

This memorandum reminds time reporters to delete their security codes when they are no longer reporting payroll time data, informs new time reporters to establish unique security codes, and emphasizes the importance of using innovative passwords.

Security codes should be deleted when a time reporter leaves a group or is no longer responsible for inputting payroll. When new time reporters are assigned, they must obtain new security codes. The security codes of a prior time reporter should not be used by the new time reporter. These security code procedures are documented in the Time Reporter Training Manual, Section IV.

Security codes may be added or deleted by sending a completed SW4570 form (Security Code Request) to SWMTR Administrator at One Palm Center, 17-A-2, St. Louis, Missouri 63101. Payroll numbers that are to be added or deleted to a security code must also be included on the SW4570. Also submit an SW4570 if you are not receiving your verification reports. Include your current security code, check the "Add Payroll Number" Box, and list all payroll numbers for which you are responsible under "New Payroll Numbers."

The last five positions of the security code are referred to as the "password." In SWMTR Memorandum 860721, dated March 4, 2005, it was stated that passwords should be made up of completely random numbers. For security purposes, you should not use a sequence of consecutive digits or the same digit repeated five times. The inputters database shows that approximately 22% of the time reporters are using this type of password. Passwords should be made more discrete so that the number schemes are not easily broken.

Additional information or questions should be referred to your local SWMTR II departments coordinator of your local payroll office.

FIGURE 8-31 · MEMO FOR EXERCISE 6

memo will be used as a reminder—to help employees remember how to handle deletion of security codes. It will be read carefully by those who do not know how to delete security codes.

7. Visit the Web site of the leading professional association in your field. Look specifically for information on student membership. How long does it take you to locate this information? Is the site easy to navigate? Is the design of the site engaging? What is your impression of the association from its Web site? Does it seem like a professional association you would want to join? Why or why not?

Scenario

Hector Varner, a new entry-level engineer at KMJ Contractors, finds that he has been asked to write a report summarizing what a project group has done on analyzing a new road repair sealant to determine cost and effectiveness. The senior project manager, Koren Melano, asks Hector: "Have you ever written a report?"

"Just academic reports," Hector responds. "I know that all reports have specific parts, and I think I know how to do those. The problem is, I have a hard time getting started."

"I suggest," responds Koren, "that you make a list of what the group discussed. Summarize what was said about each topic. Check the minutes of the meetings. Those should help you. Whatever you decide to do, write a report that people will read. Don't just throw it together and expect everyone to read it. This report should help us know where we are after four weeks of discussion. We will need this report to help us decide what to do next in considering new sealants. The group needs to have the report several days before our next meeting."

This chapter will help you consider ways to create and manage text, a skill you will need when you begin planning and writing reports.

CHAPTER NINE

Creating and Managing Text

In developing a report, first know the topic. You then collect information from research and interviews, and assimilate what you already know about your topic. Next, think about how you will organize the information you are collecting:

- ▨ How you will arrange the ideas
- ▨ Where you will place material within specific sections
- ▨ How you will decide "what goes where"

Then, you will need to develop the content within each section. Considering arrangement of content as you are planning it helps you think in an organized way about generating your content as you collect information. Organization is the heart and soul of effective writing.

COLLECTING AND GROUPING INFORMATION

As you gather and research information, try grouping your material and notes into specific categories. Label the categories, and then begin your report. You may want to develop your report around main sections—introduction, followed by information categories ordered in terms of report purpose and reader needs. Open and name the file and begin. For example,

Introduction: State the purpose of your report—what you expect to accomplish, what you want your readers to know. You can add more information later. Stating your purpose at the beginning of the draft helps you stay focused.

Category/topic 1: phrase describing the issue you want to present

Category/topic 2: another descriptive phrase

Category/topic 3: another descriptive phrase

Once you have determined your report information categories, begin inserting information under each topic or subject category. You can combine, revise, and delete ideas later. Focus initially on grouping pieces of information under the appropriate topic headings.

This method will help you organize your information, and you can watch your ideas grow to meet the report purpose. As you insert information, you can see how a topic or category develops; rearrangement of topics, paragraphs, and sentences is easy. Composing electronically allows you to write notes to yourself and to use different colors and fonts for text you may want to move, delete, or revise. This method helps anyone who may be trying to work on a report, answer phone calls, manage e-mail messages, and deal with other routine business events throughout the work day. In short, you can arrange, insert material, save it, and add other material as time permits.

Situation 1 Allen Harper was asked to visit a research facility that is likely to be for sale soon. Dick Crandall, the VP for operations, wants a brief overview on the usability of the building as a corporate training facility. Harper visits the building, takes notes on major usability features that he knows will interest Crandall, and writes a brief report for the operations manager, who is known for wanting only important facts and recommendations. He begins with a simple report purpose statement, followed by a bulleted list of the facility's features likely to be of interest to Crandall. Harper uses notes to himself (in color) as he drafts. Harper concludes with a brief recommendation of what he believes should be done, if Crandall thinks the building should be purchased. The report (see Figure 9-1) responds to the information needs and reading (skimming) style of a specific reader.

PLANNING CONTENT DEVELOPMENT

The design of your reports will depend not only on the purpose of the document but on the kind of report required and the information needs of your readers.

Reports with Standard Arrangement Patterns

Some kinds of reports have fairly standard arrangement patterns, such as the empirical research report, described in detail in Chapter 15. Empirical research reports usually begin with an introduction, then move to a review of existing research, statement of methods used in the current experiment, exact procedure used, results of the research, discussion of results, and conclusion with appropriate recommendations for future research.

Progress reports, discussed in Chapter 16, usually have basic sections that present work completed and work remaining, problems encountered (with suggested solutions), and perhaps funds expended for the period covered by the report.

Often, the plan of the document you will write is predetermined. Many government funding agencies require that all reports follow a specific plan, and content may be inserted into a template and submitted online. Many organizations also have a standard plan for their policies—sections to be included and the order in which they are to appear. Policy and procedure manuals usually follow the same organization scheme.

Often, however, as in Situation 1, you must decide how to set up a report to present your research or the information you need to convey. Generally, you have two basic choices: topical arrangement and chronological arrangement.

Topical Arrangement   In topical arrangement, you present your ideas in logical order and inclusively: the report should supply the information the reader needs. For example, in the following example report on disease management of citrus fruit, the writer groups information about specific citrus diseases and uses a parallel arrangement, to describe the main diseases.

Introduction—Description of Treatments for Citrus Diseases

Disease #1—Melanose

· Description
· Factors to be considered before the application of fungicides for melanose control

    Table I. Chemical Controls for Melanose

TO: Dick Crandall   DATE: October 9, 2004

FROM: Allen Harper

SUBJECT: Space Assessment of RAMP, 6004 Highway 7

As you requested, I toured the RAMP research facility during my trip to Atlanta October 4–8, 2004. Based on your questions about using the facility for corporate training, I have concluded the following.

Conclusions

- RAMP offers us the space we need and appears to be in excellent condition. Modifications will be necessary, but those required can be done in stages. We can focus on Building 1, then renovate Building 2 and 3 as we want to. **The RAMP is composed of two main buildings and one small building (Buildings 1, 2, and 2A on the attached site map).**

- **Each small office has two Internet connections. The main lecture hall has one Internet connection and no projection equipment.** [Check on price of wireless for the entire building. DC will probably want to know ASAP.]

- **The current RAMP buildings are divided into offices for start-up projects. Some of the RAMP income is derived from the space leased by small start-up companies. Cost: $22/sq foot.** [Is this price low/high?]. [Call Dave Redding about laboratory contracts. Cheryl Kempe has his phone number.]

- **The debt on RAMP is slightly over $12M with approximately $5M dues in April of 2007. (See p. 4 of the Notes to Consolidated Financial Statements.)** [Get current balance owed from Fiscal. Perhaps delete for now? Call Debbie in Fiscal.]

- For us to use RAMP for training, instead of research, the entire first floor of Building 1 would need to be reconfigured. We could relocate some of the start-up companies to Buildings 2 and 2A, which would preserve the income.

- An attached sketch shows how renovation of the first floor of Building 1 might look. Note:

  Eight office sites for trainers

  Two computer labs

FIGURE 9-1 · RESPONSE TO SITUATION 1

Space Assessment of RAMP—2

Five training rooms—two with breakout rooms off the main area.

One tiered lecture room

Conversion could be made gradually, depending on how many training rooms are needed ASAP. The attached sketch, included along with the original plan, shows one redesign plan.

Assessment of the RAMP Building

- The facility seems to be in good repair. Maintenance seems to be satisfactory. I found no indication of leaks.

- HVAC system needs to be inspected.

- Wireless capability could be easily installed in the main lecture room.

- New projection equipment needs to be purchased. [Cost?]

- Main lecture hall could be converted into a tiered lecture room. The two rooms on either side of the lecture hall could be eliminated to further expand the main lecture hall. The kitchen would have to be moved or eliminated—perhaps moved to Building 2.

- Renovation costs will be high, close to $1M (according to rough estimates by Gavin Newberry—see attached e-mail answer to my query) even if only the first floor of Building 1 is refurbished. Current office furniture is usable, but new furniture, IAV equipment, and computers would need to be purchased for training and meeting rooms. Closets/storage would have to be added to all training rooms. Floor coverings would need to be replaced.

- Location—RAMP is easily accessible from three main arteries.

- Estimated remodeling time: 4 months.

Attachments: 3

FIGURE 9-1 · *CONTINUED*

Disease #2—Greasy Spot

· Description
· Factors to be considered in the management of greasy spot

Table II: Chemical Controls for Greasy Spot

Disease #3—Foot Rot

· Description
· Factors to consider in managing the disease

Table III. Chemical Controls for Foot Rot

Disease #4—Citrus Nematode

· Description
· Sampling instructions to determine presence of citrus nematode

Table IV: Citrus Nematode Counts Considered Low, Medium, or High at Specific Times during the Growing Season

Conclusion: Factors to consider before applying nematicides

Table V: Chemical Controls for Nematodes (Summary)

Having decided on this arrangement, the citrus researcher can begin inserting information beneath each topic heading.

The parallel arrangement and the use of tables allows readers to move quickly through the report and enables readers to compare the various citrus diseases. The book's companion Web site has the full report, which is designed to be easily read by growers who need to be able to diagnose these typical citrus diseases.

Chronological Arrangement Some topics can be presented by time. You explain or present information sequentially, in the order in which it occurred. The following outline of a literature review of a topic identified as *Cultural Control of the Boll Weevil—A Four-Season Approach—Texas Rolling Plains* illustrates chronological arrangement. This technical report surveys and reviews existing research on the control of boll weevils throughout the agricultural year. Note, too, that the segments use parallel development, and each segment ends with a summary. This approach allows the reader to choose where to begin and how much to read. For example, the reader may wish to read only the Summary and Introduction and the factual summary for each season.

Table of Contents

· Summary
· Introduction
· Spring Cultural Control
    ○ Prepare the land for planting
    ○ Utilize delayed planting

- ○ Use uniform planting
- ○ Summary
- · Summer Cultural Control
  - ○ Shorten the growing season
  - ○ Change the microclimate
    Row direction
    Bed shape
    Row spacing
  - ○ Summary
- · Fall Cultural Control
  - ○ Use harvest-aid chemicals
  - ○ Consider role of planting date
  - ○ Terminate irrigations in August
  - ○ Summary
- · Winter Cultural Control
  - ○ Eliminate the overwintering habitat
  - ○ Modify the overwintering habitat
  - ○ Avoid the overwintering habitat
  - ○ Summary
- · Conclusions
- · Acknowledgments
- · Supporting Research Studies

You can read the full report at our companion Web site.

Reports Designed for Specific Reader Needs

Given your readers' perspective on the topic, what information needs to be placed first? Refer again to Situation 1, which is designed to give that reader what he wants immediately and efficiently. If a reader is likely to resist what you say, you may want to "buffer" critical points with background information that prepares the person for the main "news." If the reader wants "the facts," then begin with "the facts."

Topical arrangement is the most common type of arrangement. If some information is more important than others, you place the information in descending order of importance, usually, to ensure that the most important appears first. For example, in the memo illustrating Situation 1, Allen Harper begins with the item—adequate space—that he believes will be of most interest to his reader, who cares about space and cost. Situation 1 does not lend itself well to a chronological style of reporting.

Persuasive Arrangement and Development

Many times reports advance a point or position through persuasion. Thus, understanding what objections you will need to overcome will be critical to your planning your report and the presentation of your positions. Writers may be required to develop reports for readers who will not welcome the conclusions and recommendations that must be presented. Or, these reports may be prepared for readers who have no preconceived ideas. Each type of report can be designed to anticipate the perspective views of readers. Let's assume, for example, in Situation 1, that after his on-site visit, Harper came to believe that the RAMP building should be seriously considered. He knows, however, that Crandall has all but decided to purchase another site. Crandall told Harper to visit RAMP, to cover all bases, expecting to hear that his decision to disregard RAMP would be questioned. To prepare his boss for the surprise, Harper might begin his report as follows:

As you requested, I toured the RAMP research facility during my trip to Atlanta October 4–8, 2004. Based on my findings, I believe that you should seriously consider the RAMP facility as the site for our CE operation. RAMP offers us what we need in terms of space and location. Renovation costs can be done in stages to reduce cost, and the facility offers us the space and arrangement we need.

In short, awareness of the reader's perspective often determines how you may best ensure that your ideas will be considered in the most favorable light possible.

Figure 9-2, argues that Florida's Indoor Air Quality Programs provide adequate, cost-effective control for indoor pollution. Note that each bulleted item in the abstract is discussed (pp. 2–4); each bulleted point is introduced under Findings. Note the use of document design and highlighting to identify where each of the three findings begins. Major topics occur as topic sentences in the supporting paragraphs. People can read this report selectively.

Document design makes the report inviting. Each paragraph begins with a topic sentence. Note how an explanatory statement introduces each table. After you have studied Chapters 10 and 11, reexamine this report as an example of effective report design and effective integration of graphics to enhance readability.

STRATEGIES FOR DEVELOPING CONTENT

In its most basic form, technical writing explains, describes, or defines concepts. Writers create explanation by several methods: first, they divide the concept to be presented or explained into workable units (such as the reports on citrus disease and the four-season approach to boll weevil control). Technical writing must be carefully and logically partitioned so that readers can follow the content as they move through the text. In addition to partition, you will use other content development methods to explain:

- Definition of terms
- Description
- Background (history) of the concept
- Visual and verbal illustrations

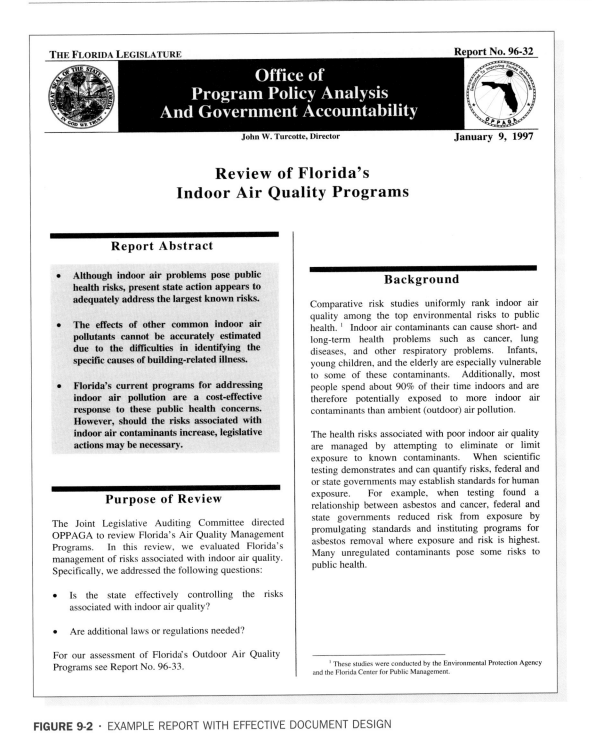

THE FLORIDA LEGISLATURE

Report No. 96-32

Office of
Program Policy Analysis
And Government Accountability

John W. Turcotte, Director

January 9, 1997

Review of Florida's
Indoor Air Quality Programs

Report Abstract

- Although indoor air problems pose public health risks, present state action appears to adequately address the largest known risks.

- The effects of other common indoor air pollutants cannot be accurately estimated due to the difficulties in identifying the specific causes of building-related illness.

- Florida's current programs for addressing indoor air pollution are a cost-effective response to these public health concerns. However, should the risks associated with indoor air contaminants increase, legislative actions may be necessary.

Purpose of Review

The Joint Legislative Auditing Committee directed OPPAGA to review Florida's Air Quality Management Programs. In this review, we evaluated Florida's management of risks associated with indoor air quality. Specifically, we addressed the following questions:

- Is the state effectively controlling the risks associated with indoor air quality?

- Are additional laws or regulations needed?

For our assessment of Florida's Outdoor Air Quality Programs see Report No. 96-33.

Background

Comparative risk studies uniformly rank indoor air quality among the top environmental risks to public health.[1] Indoor air contaminants can cause short- and long-term health problems such as cancer, lung diseases, and other respiratory problems. Infants, young children, and the elderly are especially vulnerable to some of these contaminants. Additionally, most people spend about 90% of their time indoors and are therefore potentially exposed to more indoor air contaminants than ambient (outdoor) air pollution.

The health risks associated with poor indoor air quality are managed by attempting to eliminate or limit exposure to known contaminants. When scientific testing demonstrates and can quantify risks, federal and or state governments may establish standards for human exposure. For example, when testing found a relationship between asbestos and cancer, federal and state governments reduced risk from exposure by promulgating standards and instituting programs for asbestos removal where exposure and risk is highest. Many unregulated contaminants pose some risks to public health.

---

[1] These studies were conducted by the Environmental Protection Agency and the Florida Center for Public Management.

FIGURE 9-2 · EXAMPLE REPORT WITH EFFECTIVE DOCUMENT DESIGN

Findings

While the state has established regulations to manage the most serious indoor air risks, other unregulated pollutants may also pose health risks.

Current regulation reduces risks associated with indoor air contaminants. The scientific community generally recognizes three of the biggest risks to public health resulting from poor indoor air are from tobacco smoke, radon, and asbestos. The U.S. Environmental Protection Agency (EPA) estimates 3,000 annual lung cancer deaths are caused by exposure to second-hand tobacco smoke. In addition, the EPA estimates between 7,000 and 30,000 annual lung cancer deaths from radon exposure. An estimated 1.3 million workers in construction and general industry are exposed to asbestos. In 1994, there were over 17,000 lung cancer deaths attributed to inhalation of carcinogens, such as asbestos, in the workplace.

Federal regulations exist to reduce the risks associated with second-hand tobacco smoke, radon, and asbestos, three high-risk pollutants. Title 41 of the Code of Federal Regulations regulates smoking in public spaces. The federal government also regulates most aspects of asbestos production and use. For example, the Occupational Safety and Health Administration has established standards for worker safety, and the EPA has established procedures to control asbestos emissions. Additionally, the Indoor Radon Abatement Act regulates exposure to radon by establishing a mechanism for testing homes, schools, and federal buildings for radon.

Florida has also developed regulations to reduce health risks caused by exposure to second-hand tobacco smoke, radon, and asbestos. Chapter 386, Part II, F.S., the Florida Clean Indoor Air Act, bans smoking in public buildings, which creates public places free of tobacco smoke. Additional legislation, s. 404.056, F.S., protects the public health by reducing human exposure to radon, a harmful environmental radiation. This legislation establishes a coordinating council on radon protection, initiates a public information program, institutes mandatory testing in schools and day care facilities, and requires notification on real estate documents. Third, s. 255.552, F.S., establishes the Asbestos Management Program to safely manage all asbestos-containing material in state owned buildings.[2] **While other unregulated contaminants pose some risks, those risks cannot be accurately estimated.**

---

[2] See Office of Program Policy Analysis and Government Accountability Report No. 94-49, <u>Review of the Asbestos Management Program Administered by the Department of Labor and Employment Security</u> issued May 24, 1995.

Other common indoor air contaminants such as biological agents and volatile organic compounds also create public health risks. Biological agents are bacteria, fungus, mold, and mildew that usually result from improperly maintained air conditioners and air ducts. Volatile organic compounds (VOCs) are compounds containing carbon and hydrogen resulting from various cleaning solvents, paints, pesticides, and aerosol sprays. Long-term or repeat exposure to these pollutants in high concentrations can lead to severe respiratory and other health problems.

Accurately estimating risks from these lesser known contaminants is complicated by the interaction of multiple elements. Indoor air pollution produced by contaminants such as biological organisms, building materials, cleaning agents, and pesticides are less understood than currently regulated contaminants. These elements may act in combination with other factors such as inadequate temperature or humidity control, to worsen air quality problems. Thus, identifying the cause of indoor air contaminants associated with poor indoor air is very difficult. Consequently, while these contaminants may pose some risks, the state cannot accurately estimate those public health risks.

Given the inability to accurately estimate health risks, state actions have focused on providing technical assistance and guidance rather than instituting new regulation.

The state addresses the lesser-known indoor air contaminants and the risks they pose through several programs. The state's primary vehicle for addressing indoor air problems is the Indoor Air Toxics Program of the Department of Health (formerly the Department of Health and Rehabilitative Services). This program responds to public indoor air complaints, provides employers with guidance for correcting indoor air problems, and provides information and education on indoor air issues. The program received a general revenue appropriation of $360,000 in fiscal year 1994-95 and $336,000 in fiscal year 1995-96. These funds are distributed to county health departments in the form of matching grants. Program officials believe they are generally able to respond to calls for assistance on indoor air problems.

In addition, six other state agencies implement indoor air-related activities. (See Exhibit 1.) These activities range from establishing specifications for building designs intended to reduce the possibility of indoor air contaminants causing health problems, to inspecting and testing sites with suspected air quality contamination. In addition these agencies provide

FIGURE 9-2 · *CONTINUED*

public outreach activities such as producing guidance documents, and providing education and technical assistance. Several agencies offer indoor air consulting and advisement to the general public, employers and building owners and operators. Additionally, agencies have recently created and updated procedures and policies to more adequately address indoor air issues. These agencies have little or no duplication of service or program jurisdiction and collaborate their efforts.

Exhibit 1
Several State Agencies Implement
Indoor Air Related Activities

Agency (Department of)	Agency Role Related to Indoor Air Quality
Health	Administer the Florida Clean Indoor Act
Management Services	Resolve indoor air quality problems in state buildings
Community Affairs	Develop state building codes and standards
Labor and Employment Security	Investigate workplace-related indoor air quality problems
Education	Improve indoor air quality of schools
Corrections	Identify and investigate indoor air problems in correctional institutions
Business and Professional Regulation	Identify and investigate indoor air problems in state-licensed facilities

Source: Developed by the Florida Legislature, Office of Program Policy Analysis and Government Accountability.

Current state measures are consistent with Environmental Protection Agency policy. Presently, the EPA is committed to fully explore the potential for voluntary actions and preventive approaches to control indoor air quality before using additional mandatory programs. EPA policy indicates research and nonregulatory approaches can offer effective strategies for reducing exposure to indoor air pollutants and that mandatory guidelines are premature until the successes of voluntary programs have been evaluated. Current state measures are not relying on additional regulation but attempt to prevent indoor air quality problems through education, voluntary actions, and technical assistance.

Although current state efforts are managing risks, state liability is increasing and should be monitored to insure total workers' compensation claims remain relatively low.

While state actions appear appropriate indoor air quality concerns remain and should be monitored. In 1994, the Legislature required the Department of Management Services, in conjunction with other state agencies, to study indoor air quality and report to them in January 1995. In its report, the Department identified an increasing number of lawsuits, injuries, and financial settlements resulting from poor indoor air as a cause for concern. The Florida Department of Labor and Employment Security is responsible for overseeing workers' compensation claims. Its best estimate is that between 1990 and 1994 the annual number of settled claims related to indoor air pollution increased from 2 to 65.[3] (See Exhibit 2.)

Exhibit 2
The Estimated Number of Workers' Compensation
Claims Related to Indoor Air Quality
Increased Annually Between 1990 and 1994[1]

Year	Cost of Claims (Per Year)	Number of Claims (Per Year)	Average Cost Per Claim
1990	$236,800	2	$118,400
1991	187,800	4	46,950
1992	391,300	25	15,652
1993	243,000	41	5,927
1994	404,490	65	6,223
Average	$292,678	27	$ 10,682

[1] 1994 is the most recent year for which data are available.

Source: Developed by the Florida Legislature, Office of Program Policy Analysis and Government Accountability from Department of Labor and Employment Security data.

---

[3] The Department of Labor and Employment Security's Division of Workers' Compensation does not have a specific category for illness or injury related to indoor air. The information presented in the table is based on the Division's review of indoor air quality claims completed for three nature-of-injury codes; respiratory disorders, all other occupational diseases, and all other cumulative injuries.

FIGURE 9-2 · *CONTINUED*

While indoor air problems have increased workers' compensation claims, total financial payments related to these claims are still low. Payments for claims related to indoor air between 1990 and 1994 are approximately $1.5 million (an average of approximately $300,000 per year) representing approximately 0.03% of the total compensation payments. (See Exhibit 2.) Between 1990 and 1994, the state spent approximately $4.5 billion in total workers' compensation payments or benefits.

Given the relatively low risk of indoor air quality problems in state-owned buildings, extensive testing and modifications would not be cost-effective. It is extremely costly to identify and correct for the risks associated with indoor air contamination. Estimates by the Indoor Air Quality Committee for an initial one-time testing of indoor air quality in all state buildings would exceed $50 million. Furthermore, the estimated cost for annual testing, after the initial testing, would exceed $3.2 million and estimated annual remediation costs would exceed $12.1 million. The Indoor Air Quality Committee, formed at the request of the Legislature, assists the Department of Management Services in determining the extent of indoor air quality problems in state-owned buildings. The Committee is comprised of environmental specialists, architects, engineers, and others with indoor air expertise from state agencies and other entities.

Conclusions and Recommendations

State actions have reduced the risks associated with indoor air contaminants. Identifying and correcting indoor air quality problems for currently unregulated contaminants is both difficult and extremely expensive. Current conditions related to indoor air problems do not appear to warrant additional regulation or costly testing and remediation of state buildings.

We recommend continued use of current indoor air prevention and problem-solving measures being implemented by state agencies. State agencies should continue to monitor indoor air quality problems to ensure that state efforts continue to effectively protect public health. In addition, the Department of Labor and Employment Security should closely monitor worker's compensation claims related to poor indoor air quality.

Agency Response

The Florida Department of Health expressed concerns that it may be premature to conclude that present actions adequately address the largest known health risks from indoor air problems. The Department does agree that continued use of existing prevention and problem solving measures is effective, but indicated that additional financial support is needed to fund additional local county health department responses, increase training, and create a statewide indoor air quality service delivery mechanism.

The Florida Department of Labor and Employment Security, in response to this report, agrees with the findings and recommendations. The Department, however, cautions that the current claims data is only a "best guess" for claims related to indoor air quality and should not be used to define total claims or costs.

This project was conducted in accordance with applicable evaluation standards. Copies of this report may be obtained by telephone (904/488-1023 or 800/531-2477), by FAX (904/487-3804), in person (Claude Pepper Building, Room 312, 111 W. Madison St.), or by mail (OPPAGA Report Production, P.O. Box 1735, Tallahassee, FL 32302).

Web site: http://www.state.fl.us/oppaga/

Project Supervised by: Julie Ferris (487-4256)

Project Conducted by: Bob Dahlstrom (487-9271)
Lyndon Rodgers (487-3805)

-4-

FIGURE 9-2 · *CONTINUED*

- Analogies
- Examples
- Cause/effect analysis
- Comparison
- Contrast

You can deconstruct good technical writing to see how these strategies were used in building explanation or exposition.

For example, consider this portion of a report written for high school physical science students:

Electrical Principles, Terminology, and Safety

Electricity can be defined in several ways. The layman defines electricity as a form of energy that can be converted to light, heat, sound, and motion. Electrical engineers define electricity as the flow of electrons from one atom to another. This flow of electrons is controlled in an electric circuit. The amount of energy produced depends on the number of electrons in motion.

The writer begins with a definition of electricity, moving from basic to advanced.

Understanding how electricity works requires a basic understanding of an atom. The center of an atom, or nucleus, contains positively charged protons. In an electrically neutral of "balanced" state, each atom has an equal number of electrons and protons. The electrons are arranged in orbital layers around the nucleus with a negative charge. A hydrogen atom, for example, contains one proton and one electron. Electrons can be forced to leave their outer orbital layer and attach to the outer ring of an adjacent atom. When extra electrons are attached to an atom, they become negatively charged, while the atom that gives up the electron has become positively charged.

The writer then *describes* the creation of electricity. He *compares* the behavior of electricity with the behavior of the atom. He also uses an *example*.

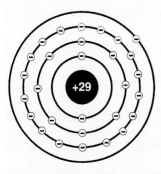

Copper Atom

Another example is the copper atom, which has 29 protons in the nucleus and 29 electrons in the orbiting layers around the nucleus. However, the electrons in outer orbit in a copper atom are loosely held, facilitating the exchange of electrons between copper atoms. Even though electrons have been exchanged, the atoms are still balanced. Each atom of copper has the same number of protons and electrons. Because electrons are loosely held in a copper atom allowing for the exchange of electrons, copper is a good conductor of electricity, while hydrogen is a good insulator.

He then uses another, more complex example, with a drawing to help the reader visualize the example.

The path through which an electrical current flows is called a circuit. A current is the movement of electrons through an electrical conductor. Generators produce and release current the moment it is needed. The complete electrical pathway from the power source through the load and back to the source is known as a closed circuit. As long as electrical equipment is turned on and the circuit is completed or closed, there is a continuous flow of current. If the electrical pathway is

The writer then contin-
ues to explain how
electrical energy
moves through a
circuit. He adds *defini-
tions* as he proceeds,
and uses *cause/effect
analysis.* Note that he
is building his expla-
nation and giving the
reader knowledge of
the terminology need-
ed for understanding
the basics of electri-
city. He uses visuals
to help readers SEE
his description.

not complete, the circuit is said to be an open circuit. For example, when electrical equipment is turned off, the circuit is opened and electron flow stops.

There are two basic types of circuits: series and parallel. A third type is series-parallel, which is a network of circuits that incorporates both series and parallel circuits.

In a series circuit, all the current must flow through each device in the circuit. The current in a series circuit has only one path to flow through. Removing or opening any one device in the series circuit will stop the flow of current for the whole circuit. The same current or amperage flows through each load in a series circuit. Switches, fuses, and circuit breakers are always connected in series.

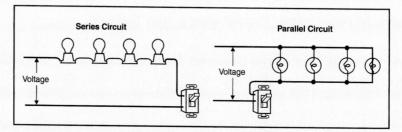

Series Circuit vs Parallel Circuit

A parallel circuit has separate paths for the current to flow through. Parallel circuits provide the same voltage across each load (lights or appliances). Removing one device in the parallel circuit has no effect on the other devices in the circuit. The current flow is divided through each load on the circuit and is equal to the total current from the source. In most cases, except for some Christmas tree lights, appliances are lights are connected in parallel.

Basic Electrical Wiring Skills, 8602-A, Instructional Materials Service, Texas A&M University, 2003.

Note that the writer has built his report as follows:

- Basic definition + examples
- Definition of circuits: open and closed
- Types of circuits: series defined along with examples; parallel defined along with examples; visual aid to depict operational differences between series and parallel circuits

Development by definition and description can be found in other areas, such as the following excerpt from a U.S. Department of Energy Office of Environmental Management that explains various approaches to contain ground contaminants. This technical description, which can be seen in full on our companion Web site, develops each description about the same main topics or categories. Parallel development allows readers to easily compare and contrast the various approaches. The plan used throughout the document is illustrated in the following excerpt.

3.1.5 Synthetic Membranes

Uses/Target Contaminants

Synthetic membranes contain the groundwater, treating no particular target group of contaminants (Figure 3-5). Synthetic membranes are used to contain groundwater, divert groundwater, and/or provide a barrier for the groundwater treatment system.

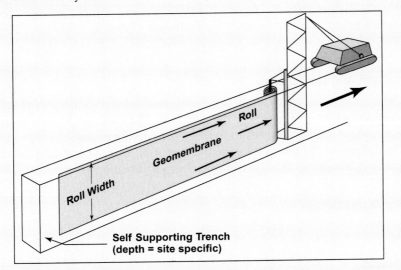

Figure 3-5. Typical use of synthetic membrane as a cutoff wall.

Description

Synthetic membranes used for vertical cutoff walls are generally made from high-density polyethylene; however, other polymers have been used. Membrane sheets can be continuous, but usually finite length panels that interlock are preferred. The final depth of installation is a function of the ability of the trenching technique.

Synthetic membranes can be used either alone or in conjunction with other materials, such as bentonite, grout, cement, or sheet piling. The design may incorporate a double membrane system with a leak detection between the layers.

Five methods have been developed for installation of synthetic membranes. (1) Trenching machine installation involves the excavation of an unsupported trench with the membrane lowered vertically in the trench and progressively unrolled. (2) The vibrated insertion plate method uses panels attached to a steel truss insertion plate. A vibratory pile hammer forces the assembly (insertion plate plus membrane panel) to the desired depth. The insertion plate is withdrawn, leaving the panel behind. Interlock connections are required to join panels together. (3) The slurry supported method involves using a conventionally excavated and slurry trench. The membrane panels are inserted into the slurry to the desired depth using a steel frame. (4) The segmented trench box method uses a modified steel box to support the sides of the trench as the panel is inserted. (5) The vibratory beam method involves a combination of the slurry wall and vibrated insertion plate methods. Generally in this use, the membrane is seen as secondary to the slurry wall.

Availability and Cost

This technology is readily available. The cost of installation varies with site conditions. The greater the depth of installation and the presence of cobbles and boulders will increase the cost of installation. Costs can range from $20/m^2$ to $250/m^2$.

Advantages and Limitations

Synthetic membranes provide a continuity and relative impermeability that conventional grout and cement walls may lack. The various methods of installation provide flexibility in design to meet site-specific needs.

Limitations of the technology include potential leaks at seams, depth limitations, and increased cost in cobble–boulder environments.

Similar Technologies

Bentonite slurry walls, cement-based grout walls, and sheet piling.

References

Environmental Protection Agency, 1991. *Stabilization Technologies for RCRA Corrective Actions Handbook*, EPA/625/4-91/029.

Rumer, Ralph, and James K. Mitchell (eds.), 1995. Assessment of Barrier Containment Technologies, a Comprehensive Treatment for Environmental Remediation Applications, International Containment Workshop, Baltimore.

Source: http://www.em.doe.gov/define/techs/remdes2.html

ORGANIZATION AND CONTENT DEVELOPMENT

Clear organization characterizes good technical writing. Artistic technical writing uses organization to reveal content. As the preceding example and the report examples on the book's companion Web site illustrate, good technical writing unfolds into logical, carefully organized, visually accessible chunks of information. Unlike novels and essays, technical writing should allow readers to enter/exit the document easily, depending on their information needs. Technical documents may be hundreds of pages long, or one page. The approach to developing content is the same.

Our final example of content development uses definition, description, examples, visual illustrations, analysis, comparison/contrast, and analogy—the entire range of development—to introduce the Human Genome Project. The project is divided into specific topics at the beginning. Readers can then click and select any topic. The entire project, complete with updates made after publication of this book, can be found at http://www.ornl.gov/sci/techresources/Human_Genome/publicat/primer/toc.html. For the version current as of, see the book's companion Web site. For online reports particularly, organization becomes critical. Readers scrolling through virtual pages must be able to see, from the beginning, the logical organization of the material they will read.

Primer on Molecular Genetics

More information about genetics and the science behind the Human Genome Project

- Human Genome Project Science——an overview
- Frequently Asked Questions about HGP Science
- Students Page
- *To Know Ourselves*
- *Human Genome News*
- Glossary of Genetic Terms
- Acronyms
- Research in Progress (More technical information

Contents

- **Introduction**
  - DNA
  - Genes
  - Chromosomes
- **Glossary**
- **Mapping and Sequencing the Human Genome**
  - Mapping Strategies
  - Genetic Linkage Maps
  - **Physical Maps**
    - Low-Resolution Physical Mapping
      - Chromosomal map
      - cDNA map
    - High-Resolution Physical Mapping
      - Macrorestriction maps: Top-down mapping
      - Contig maps: Bottom-up mapping
  - **Sequencing Technologies**
    - Current Sequencing Technologies

- about the HGP)
- **Presentation Resources**
  - ○ Genome Videos
  - ○ Genome Audio Files
  - ○ Genome Images
- Home

- ■ Sequencing Technologies Under Development
- ■ Partial Sequencing to Facilitate Mapping, Gene Identification
  - ○ **End Games: Completing Maps and Sequences; Finding Specific Genes**
- **Model Organism Research**
- **Informatics: Data Collection and Interpretation**
  - ○ Collecting and Storing Data
  - ○ Interpreting Data
  - ○ Mapping Databases
  - ○ Sequence Databases
    - ■ Nucleic Acids (DNA and RNA)
    - ■ Proteins
- **Impact of the Human Genome Project**

← Return to Human Genome Project Information

For online reports, like this one, note that the figures may be accessed as links, in contrast to the figure in the DoE's description of groundwater contaminants given earlier. Whether you insert visuals directly into the text or make them available as links depends on the relations of these aids to the text. Online documents usually provide links to graphics from the text. However, if the reader needs to move back and forth between the visual and the text, it makes more sense to place the visual IN the text. If the visuals may change, as they do in the following excerpt from the Human Genome Project Web site, including visuals as links allows those who maintain the report to update the visuals without changing the text. Also, in the report from the Human Genome Project, the figures are not critical to the reader's understanding the information. In this document, note the development of ideas by means of definition, description, visual/verbal illustration, and cause–effect analysis.

Human Genome Project Information

Introduction

The complete set of instructions for making an organism is called its genome. It contains the master blueprint for all cellular structures and activities for the lifetime of the cell or organism. Found in every nucleus of a person's many trillions of cells, the human genome consists of tightly coiled threads of deoxyribonucleic acid (DNA) and associated protein molecules, organized into structures called chromosomes (*Fig. 1: the Human Genome at Four Levels of Detail*).

If unwound and tied together, the strands of DNA would stretch more than 5 feet but would be only 50 trillionths of an inch wide. For each organism, the components of these slender threads encode all the information necessary for building and maintaining life, from simple bacteria to remarkably complex human beings. Understanding how DNA performs this function requires some knowledge of its structure and organization.

DNA

In humans, as in other higher organisms, a DNA molecule consists of two strands that wrap around each other to resemble a twisted ladder whose sides, made of sugar and phosphate molecules, are connected by rungs of nitrogen-containing chemicals called bases. Each strand is a linear arrangement of repeating similar units called nucleotides, which are each composed of one sugar, one phosphate, and a nitrogenous base (*Fig. 2: DNA Structure*). Four different bases are present in DNA: adenine (A), thymine (T), cytosine (C), and guanine (G). The particular order of the bases arranged along the sugar–phosphate backbone is called the DNA sequence; the sequence specifies the exact genetic instructions required to create a particular organism with its own unique traits.

The two DNA strands are held together by weak bonds between the bases on each strand, forming base pairs (bp). Genome size is usually stated as the total number of base pairs; the human genome contains roughly 3 billion bp (*Fig. 3: Comparison of Largest Known DNA Sequence with Approximate Chromosome and Genome Sizes of Model Organisms and Humans*).

Each time a cell divides into two daughter cells, its full genome is duplicated; for humans and other complex organisms, this duplication occurs in the nucleus. During cell division the DNA molecule unwinds and the weak bonds between the base pairs break, allowing the strands to separate. Each strand directs the synthesis of a complementary new strand, with free nucleotides matching up with their complementary bases on each of the separated strands. Strict base-pairing rules are adhered to adenine will pair only with thymine (an A–T pair) and cytosine with guanine (a C–G pair). Each daughter cell receives one old and one new DNA strand (*Figs. 1* and *4: DNA Replication*). The cells' adherence to these base-pairing rules ensures that the new strand is an exact copy of the old one. This minimizes the incidence of errors (mutations) that may greatly affect the resulting organism or its offspring.

Genes

Each DNA molecule contains many genes—the basic physical and functional units of heredity. A gene is a specific sequence of nucleotide bases, whose sequences carry the information required for constructing proteins, which provide the structural components of cells and tissues as well as enzymes for essential biochemical reactions. The human genome is estimated to comprise more than 30,000 genes.

Human genes vary widely in length, often extending over thousands of bases, but only about 10% of the genome is known to include the protein-coding sequences (exons) of genes. Interspersed within many genes are intron sequences, which have no coding function. The balance of the genome is thought to consist of other noncoding regions (such as control sequences and intergenic regions), whose functions are obscure. All living organisms are composed largely of proteins; humans can synthesize at least 100,000 different kinds. Proteins are large, complex molecules made up of long chains of subunits called amino acids. Twenty different kinds of amino acids are usually found in proteins. Within the gene, each specific sequence of three DNA bases (codons) directs the cells' protein-synthesizing machinery to add specific amino acids. For example, the base sequence ATG codes for the amino acid methionine. Since 3 bases code for 1 amino acid, the protein coded by an average-sized gene (3000 bp) will contain 1000 amino acids. The genetic code is thus a series of codons that specify which amino acids are required to make up specific proteins.

The protein-coding instructions from the genes are transmitted indirectly through messenger ribonucleic acid (mRNA), a transient intermediary molecule similar to a single strand of DNA. For the information within a gene to be expressed, a complementary RNA strand is produced (a process called transcription) from the DNA template in the nucleus. This mRNA is moved from the nucleus to the cellular cytoplasm, where it serves as the template for protein synthesis. The cells' protein-synthesizing machinery then translates the codons into a string of amino acids that will constitute the protein molecule for which it codes (*Fig. 5: Gene Expression*). In the laboratory, the mRNA molecule can be isolated and used as a template to synthesize a complementary DNA (cDNA) strand, which can then be used to locate the corresponding genes on a chromosome map. The utility of this strategy is described in the section on physical mapping.

Chromosomes

The 3 billion bp in the human genome are organized into 24 distinct, physically separate microscopic units called chromosomes. All genes are arranged linearly along the chromosomes. The nucleus of most human cells contains 2 sets of chromosomes, 1 set given by each parent. Each set has 23 single chromosomes, 22 autosomes, and an X or Y sex chromosome. (A normal female will have a pair of X chromosomes; a male will have an X and Y pair.) Chromosomes contain roughly equal parts of protein and DNA; chromosomal DNA contains an average of 150 million bases. DNA molecules are among the largest molecules now known.

Chromosomes can be seen under a light microscope and, when stained with certain dyes, reveal a pattern of light and dark bands reflecting regional variations in the amounts of A and T vs G and C. Differences in size and banding pattern allow the 24 chromosomes to be distinguished from each other, an analysis called a karyotype. A few types of major chromosomal abnormalities, including missing or extra copies of a chromosome or gross breaks and rejoinings (translocations), can be detected by microscopic examination; Down syndrome, in which an individual's cells contain a third copy of chromosome 21, is diagnosed by karyotype analysis (*Fig. 6: Karyotype*). Most changes in DNA, however, are too subtle to be detected by this technique and require molecular analysis. These subtle DNA abnormalities (mutations) are responsible for many inherited diseases such as cystic fibrosis and sickle cell anemia or may predispose an individual to cancer, major psychiatric illnesses, and other complex diseases.

OTHER TYPES OF DEVELOPMENT

Writers can use many other approaches to develop reports. For example, you might determine your purpose, think of questions your readers will want answered, and plan your report about the questions. Begin with an introduction; then develop the body of the report to answer the questions. Figure 9-3 shows how this technique was used to develop an information report on pesticide storage. The authors open the report with a boxed list of questions, then move on to introductory material presented as an overview. The body of the report develops around headings that answer the questions. We show the opening three pages of the report to allow you to see what this development method looks like. The full report can be found at the book's companion Web site.

B-6025

TEX★A★Syst

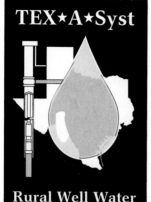

Rural Well Water Assessment

Texas Agricultural Extension Service
The Texas A&M University System

Reducing the Risk of Ground Water Contamination by Improving Pesticide Storage and Handling

B.L. Harris, D.W. Hoffman and F.J. Mazac, Jr.*

1. Do you store pesticides on your land?

2. Do you use or store any agricultural chemicals near a water well?

3. Are chemicals stored on a permeable surface such as wood, gravel or soil, or are chemicals stored on an impermeable surface with no curb?

4. Do you have any chemical containers that are rusted, damaged or leaking?

5. Are chemicals stored in an area where containers could become damaged or where a chemical spill could occur?

6. Are chemicals stored in a location that is unlocked and open to vandalism or to children?

7. Do you fill the sprayer tank directly from a water well?

8. Do you fill a sprayer tank with a hose that does not have a check valve, or put the hose in the tank so that it is below the liquid line during filling?

9. Do you leave the sprayer tank unattended when filling?

10. Do you mix or load chemicals upslope or within 150 feet of a water well?

11. Do you have a concrete pad with a curb to contain spills during the mixing or loading of chemicals?

12. Do you wash the sprayer tank out and dump the rinsate on your land less than 150 feet from a water well?

13. Do you apply pesticides without reading the label first?

14. Has it been longer than 5 years since you attended a pesticide applicator training?

If these questions create doubt about the safety of your management practices, this publication will provide helpful information.

*Professor and Extension Soils Specialist; Research Scientist, Texas Agricultural Experiment Station; and Extension Associate-Water Quality, The Texas A&M University System.

Texas Agricultural Extension Service • Zerle L. Carpenter, Director • The Texas A&M University System • College Station, Texas

FIGURE 9-3 · EXAMPLE REPORT DEVELOPED FROM READERS' QUESTIONS

Pesticide Handling Overview

Pesticides play an important role in agriculture. They have increased farm production and enabled farmers to manage more acres with less labor. Taking voluntary action to prevent pesticide contamination of ground water will help ensure that pesticides remain available for responsible use.

Pesticides work by interfering with the life processes of plants and insects. Some pesticides are also toxic to humans. If a pesticide enters a water supply in large quantities, which could happen with spills or back-siphonage accidents, acute health effects (toxic effects apparent after only a short period of exposure) could occur, depending on the toxicity of the pesticide. Contaminated ground water used for drinking water supplies may cause chronic exposure (prolonged or repeated exposure to low doses of toxic substances). Chronic exposure may be hazardous to humans and livestock.

Normally pesticides are not found in water supplies in high enough concentrations to cause acute health effects, which can include chemical burns, nausea and convulsions. Instead, pesticides usually occur in trace amounts, and the concern is for the chronic health problems that may result from prolonged exposure.

Proper pesticide management on your property is an important step toward preventing ground water contamination. This guide will provide information about the following areas:

1. Pesticide storage
2. Mixing and loading practices
3. Spill clean up
4. Container disposal
5. Other management practices
6. Evaluation table
7. Pesticide Leachability Chart

Pesticide Storage

If stored in a secure, properly constructed location, pesticides pose little danger to ground water. Common sense suggests keeping pesticides out of the way of activities that might knock over a jug or rip open a bag. Short-term storage (during a season) poses a lower risk than year-round storage, but storage for any length of time can be a risk to ground water.

Secondary containment includes an impermeable (waterproof) floor and walls around the storage area. This will minimize the amount of pesticide seeping into the ground if a bulk liquid pesticide storage tank leaks.

If a spill does occur, an impermeable concrete floor should prevent chemical seepage into the ground. Putting a curb around a concrete floor also will prevent chemicals from spreading to other areas.

A mixing or loading pad provides secondary containment during the transfer of pesticides to spraying equipment or nurse tanks.

Building a New Storage Facility

Building a new facility just for pesticide storage may be expensive, but it is usually easier than trying to modify areas meant for other purposes. When building a new facility, keep in mind a few principles of safe pesticide storage.

★ Locate the building downslope and at least 100 feet away from your water well. The distance from the well should be greater if the site has sandy soils or fractured bedrock near the soil surface. The risk of pesticides contaminating ground water is influenced by properties of both the pesticide and the soil on which it is spilled or applied.

★ Drain surface water to a confined area because, in the event of a fire, contaminated surface water can be collected more safely.

★ Locate the mixing and loading area close to your storage facility in order to minimize the distance that chemicals are carried.

★ Provide a well-drained building foundation or secondary containment floor that is high above the water table. The finished soil grade should be 3 inches below the floor and sloped to provide surface drainage away from the building. The subsoil should have a low permeability.

★ Keep large drums or bags on pallets and off the floor. Shelves for smaller containers should have lips to keep the containers from sliding off. Steel shelves are easier to clean than wood if a spill occurs. Store dry products above liquids to prevent wetting from spills.

FIGURE 9-3 · *CONTINUED*

★ Provide a containment area large enough to confine 125 percent of the contents of the largest bulk container, plus the displaced volume of any other storage tanks in the area.

★ Keep the storage area or building locked for security. Preventing unauthorized use of pesticides reduces the chance of accidental spills or theft. Post signs or labels to identify the area as a pesticide storage area. Labels on the outside of the building will give firefighters information about pesticides if they must respond to a fire or a spill. Also, it is a good idea to maintain, in a separate location, a list of the chemicals and amounts stored.

★ Provide adequate road access for deliveries and emergency equipment.

★ Keep pesticides separate to prevent cross-contamination. Herbicides, insecticides and fungicides should be kept on separate shelves or in separate areas.

For information on other factors to consider in the design of a storage facility, such as ventilation, water access, temperature control and worker safety, contact your county Extension office or the Texas A&M University Department of Agricultural Engineering.

Modifying an Existing Storage Facility

Remodeling an existing facility to serve as pesticide storage may be less expensive than building a new facility, but remodeling can be complicated. When existing buildings must accommodate other activities, also using them to store pesticides could compromise the safety of people and the environment. Storing chemicals in a separate facility reduces the risk associated with fire or accidental spills. Never store pesticides inside a wellhouse or in a facility containing an abandoned well.

Even if you decide to improve your current storage building, applying the above principles can be expensive. Compared to the cost of a major accident or a lawsuit, however, storage improvements are a bargain. Also, note that the last five items listed in the section above are important points to remember for existing storage facilities.

The least expensive alternative you may have is to reduce the amounts and types of pesticides stored. If that is not practical, consider

how stored pesticides can be protected. Sound containers are the first defense against a spill or leak.

When modifying a structure, it is important to note that the building should have a solid floor. If liquid pesticides are stored the building also should have a curb. The modified structure should be large enough to hold 125 percent of the contents of the largest full container, plus the displaced volume of any other storage tanks in the area.

When modifying an existing structure, label windows and doors to alert firefighters to the presence of pesticides and other products stored in the structure. It is always a good idea to keep a list of the stored chemicals and amounts in a separate location.

If a fire should occur, consider where the surface runoff water will go and where it might collect. For example, adding a curb around a floor can help confine contaminated water. In making the storage area secure, also make it accessible in order to help get chemicals out in a hurry.

Mixing and Loading Practices

Ground water contamination can result even from small spills in the mixing and loading area. Small quantities spilled regularly in the same place can go unnoticed, but the chemicals can build up in the soil and eventually reach ground water. By mixing and loading on an impermeable concrete surface most spilled pesticides can be recovered and reused.

A Mixing and Loading Pad

Containing pesticide spills and leaks requires an impermeable or waterproof surface for mixing and loading. The surface, or pad, should be large enough to contain leaks from bulk tanks, to hold wash water from cleaning equipment, and to keep spills from transferring chemicals to the sprayer or spreader. (See Figure 1.)

The size of the pad depends on the equipment used. The pad should provide space around the parked equipment for washing and rinsing. Having several rinsate (rinse water) storage tanks allows the user to keep rinsate separate from other chemicals. That way, the rinsate can be used as mixing water on subsequent loads.

FIGURE 9-3 · *CONTINUED*

■ ■ ■ ■ ■ ■ ■ ■   Exercises   ■ ■ ■ ■ ■ ■ ■

Further exercises related to the content of this chapter and all documents employed within the following exercises can be found on the book's companion Web site.

1. Examine the following excerpt from a discussion of U.S. government regulations. Identify the methods the author has used to explain or illustrate regulations.

In terms of federal taxation, regulations are pronouncements issued by the U.S. Treasury Department under authority granted by Congress. Regulations provide guidance to taxpayers by interpreting and explaining the Internal Revenue Code. Regulations can be used by government employees and the general public to ascertain the Treasury's view on tax matters. Regulations fall into two categories: administrative regulations and legislative regulations.

Administrative regulations are issued to interpret the language of the Code and to prescribe all needed rules and regulations for the enforcement of the Code. For example, Administration Regulations 1.616-1 prescribe requirements that must be met to file for an extension to pay tax.

Legislative regulations are used when Congress directs the Internal Revenue Service to fulfill effectively a law-making function. Such authority is granted because in certain technical areas of the tax law, Congress cannot or does not care to address the detailed or complex issues associated with a tax issue. Accordingly, Congress directs the IRS to issue regulations on the matter. An example of this action can be seen in Section 385 of the Code. The original language of Section 385 directed the IRS to generate regulations to distinguish debt from equity in "thinly capitalized" corporations.

When issued, regulations are proposed, final, or temporary regulations.

Regulations are first issued in "proposed" form to allow interested parties to comment in public hearings. After the hearings are complete, usually within thirty days, proposed regulations are withdrawn, modified, or finalized. Even though proposed regulations do not have the force of law, the tax researcher should be aware of these regulations because they are often finalized without changes.

When proposed regulations are finalized, they become final regulations. These regulations, although not issued by Congress, have nearly the force of statutory law. The purpose of final regulations is to explain and illustrate the law as enacted by Congress. When found to be inconsistent with the intent of Congress, Final regulations can and have been held invalid by the courts.

Due to major changes in the Internal Revenue Code, the Treasury Department often issues temporary regulations without holding public hearings. Temporary regulations are issued to provide the taxpayer with immediate guidelines concerning a new provision of the law, such as filing requirements or definitional classification. Temporary regulations are legally binding and should not be confused with proposed regulations. Temporary regulations are fully in effect and must be followed until they are superseded, where proposed regulations are not in effect and are issued for comments and informational purposes only.

2. The Cast Metal Foundation of the United Kingdom maintains a Web site that explains various casting processes. Here is one of those processes: How does the author develop the content?

Resin Shell Moulding

The shell moulding process is a precision sand casting process capable of producing castings with a superior surface finish and better dimensional accuracy than conventional sand castings. These qualities of precision can be obtained in a wider range of alloys and with greater flexibility in design than die-casting and at a lower cost than investment casting. The process was developed and patented by Croning in Germany during World War II and is sometimes referred to as the Croning shell process.

The fundamental feature of the process is the use of fine-grained, high purity sand that contributes the attributes of a smooth surface and dimensional accuracy to mould cores and castings alike. In conventional sand moulding the use of such fine sand is precluded because it would dramatically reduce mould permeability. This has the effect of retarding the escape of air and mould gases, causing short-run castings or castings containing gas defects. However, the distinguishing feature of the shell moulding process is that the mould is literally a shell, being in the region of only 10mm (0.4in) thick. It was the ability to produce such a thin shell mould which made the process a revolutionary development in metal founding. The coincident development of plastics, like Bakelite, which were based on thermosetting resins such as phenol formaldehyde, provided the basis for shell moulding. In shell moulding the fine sand is coated with a thermosetting resin which provides the relatively high strength required, enabling a thin section, or shell, mould to be produced.

The requirement that the mould should accurately replicate the pattern detail and dimensions if a precision casting is to be produced is also met by the shell moulding process. This is achieved because the resin bond is developed whilst the mould is in contact with a heated pattern plate. Furthermore, the mould is separated from the pattern without the need to enlarge the cavity, as is the case in green sand moulding. These features apply equally to the production of cores by the process. A further improvement in casting accuracy can be obtained if zircon sand is used instead of silica sand. That arises because the expansion of zircon sand, caused by the heat of the cast metal, is both lower and more predictable than that of silica sand. Foundry production of castings by the process is comparatively straightforward and the process lends itself readily to close control, with the advantage of consistency in the castings produced.

The advantages and disadvantages of the process are summarised below (18, 19):

Advantages

- lower capital plant costs, when compared with mechanised green sand moulding
- capital outlay on sand preparation plant is not essential
- good utilisation of space
- low sand-to-metal ratio
- mould coatings are unnecessary
- lightweight moulds are produced which are readily handled and have good storage characteristics
- skilled labour is not required
- shells have excellent breakdown at the knockout stage
- lower cleaning an fettling costs
- castings have a superior surface finish and dimensional accuracy, when compared with green sand moulded castings

Disadvantages

- the raw materials are relatively expensive
- the size and weight range of castings is limited
- the process generates noxious fumes, which must be effectively extracted

Source: http://www.castmetalsfederation.com/process.asp?procid=9&name=Resin%20Shell%20Moulding

---

3. Plan a one-page report to describe a major academic program at your college or university. The target audience will be new freshmen who may want to consider this program as a major area of study. You should prepare the report in a way that suits it for placement on a university Web site that describes academic programs. What information will you include? How will you arrange the material? What will your general plan look like?

4. Your professional association has invited its student members to make suggestions for benefits and service that student members would like to see the association offer. Plan a written response to this invitation. What will you suggest? How will you arrange your response? How will you develop the content of your response? Put your plan in writing and then draft a response based on your plan.

Scenario

Your office has been asked to write a summary report of the company's research the past year. You decide to have a team of three people to write the report. You ask each division for a summary of its work and then begin to combine the information.

"What else do we need in this report," one report team member asks? "Obviously we need a cover sheet. What about a table of contents? Do we need a summary? What about a glossary? Not everyone in the company understands everyone else's work. Also, who else will receive a copy? Perhaps we need a summary for distribution to people who may not be interested in the full report."

"We need to plan the report," the second member adds. "We have choices to make in addition to summarizing these area reports."

This chapter will familiarize you with the elements of technical reports and provide guidelines for their use.

CHAPTER TEN

Developing the Main Elements of Reports

Business and technical reports contain a number of standard elements designed to help readers access the material the documents contain. Each element has a different purpose and often targets the needs of different readers. Remember that reports should be developed with a variety of readers in mind. Effective report design reflects (1) the reality of more than one reader and (2) the need to accommodate busy readers who must be able to access information quickly.

Reports are not novels: few are read from cover to cover. Any report, but particularly a long report, should be designed so that readers can easily find what they want to read. The elements of a well-designed report will help readers find their way through the material.

We provide guidelines for developing such major components of reports as prefatory elements, main elements, and appendixes, with examples of each. Not all reports use all elements: which ones you use and how you use them will depend on various factors:

- The style sheet of the business or organization for which you work
- Your topic
- Your intended readers
- Length of the report
- Format—paper or online—in which the report will appear

Once familiar with the development methods of each report element, however, you can observe a range of technical reports and understand the rationale and logic of their form. The book's companion Web site provides several reports for you to examine with these questions in mind: How do the report elements help you access the report? How do the elements differ among these example reports? How does the design of the report elements affect your interest in reading the report?

PREFATORY ELEMENTS

Readers initially access a report from the prefatory elements, which show what the document discusses and tells how it approaches the topic. Long, complex reports that will likely be distributed to many readers require prefatory elements. The letter of transmittal, title page, table of contents, abstract, and/or summary are often your reader's first—and only—experience with the report.

Each of these elements should be written to help readers grasp and accept the rhetorical purpose of the report. In no case should prefatory elements be considered routine report "paperwork" needing only perfunctory writing. Instead, each prefatory element should enforce the rhetorical purpose of the report. No matter how well-written and researched the report content, the effectiveness of the report begins with the effectiveness of its content elements.

Many readers will decide to read the body of the report if they find the prefatory elements compelling.

Letter of Transmittal

The letter of transmittal is addressed to the individual who will initially receive the report. This person is likely to be the individual responsible for routing the report to appropriate readers, who will digest and use the content. Consulting reports usually are accompanied by a letter of transmittal addressed to the CEO of the organization that solicited the consulting work. Or, the transmittal letter is addressed to the person who authorized and requested the information, analysis, work, or recommendations covered in the report.

The letter of transmittal should at least include the following information:

- Statement of transmittal
- Reason for the report
- Statement of subject and purpose of the report

In addition, transmittal letters (or memoranda, for internal reports) may include the following items:

- Background material identifying the larger issue or problem addressed by the report and/or giving a brief history of the issue
- Mention of earlier reports that may be needed
- Information that may be of special interest or significance to the readers
- Specific conclusions or recommendations that might be of special interest to the person to whom the report is addressed
- Financial implications
- Acknowledgments of those who provided help in the project

Some transmittal letters, such as Figure 10-1, serve as the opening segment of the report. The letter of transmittal contains enough information to become the introduction to the report. You may view this entire report at our companion Web site.

Title Page

Title pages perform several functions. They provide critical identifying matter and may contain a number of tags that help readers distinguish the report from others on a similar subject or about specific projects. Many organizations have a standard format for title pages to ensure that the following information appears:

- Name of the company or individual(s) preparing the report
- Name of the company or course for which the report was prepared
- Specific title and subtitle of the report
- Type of report (assessment, proposal, etc.)

United States General Accounting Office
Washington, D.C. 20548

December 6, 2000

The Honorable Rodney E. Slater
The Secretary of Transportation

Dear Mr. Secretary:

In September, we testified before the Committee on Science, House of
Representatives, on the Federal Aviation Administration's (FAA) computer
security program.[1] In brief, we reported that FAA's agencywide computer
security program has serious, pervasive problems in the following key
areas:

- personnel security,
- facility physical security,
- operational systems security,
- information systems security management,
- service continuity, and
- intrusion detection.

We also noted that until FAA addresses the pervasive weaknesses in its
computer security program, its critical information systems will remain at
increased risk of intrusion and attack, and its aviation operations will
remain at risk. These critical weaknesses need to be addressed. To assist
you in bringing this about, we are making recommendations to you based
on the suggestions we made in our September 2000 testimony, which is
reprinted in appendix I.

We performed our work from March through September 2000, in
accordance with generally accepted government auditing standards.
Department of Transportation (DOT) and FAA officials provided us with
comments on a draft of this report; they are discussed in the "Agency
Comments" section.

Recommendations

Given the importance of a comprehensive and effective computer security
program, we recommend that the Secretary of Transportation direct the
FAA Administrator to complete the following actions.

---

[1]*FAA Computer Security: Actions Needed to Address Critical Weaknesses That Jeopardize
Aviation Operations* (GAO/T-AIMD-00-330, September 27, 2000).

Page 1   GAO-01-171 FAA Computer Security Weaknesses

FIGURE 10-1 · LETTER TO A CABINET OFFICER SUMMARIZING A REPORT TO CONGRESS

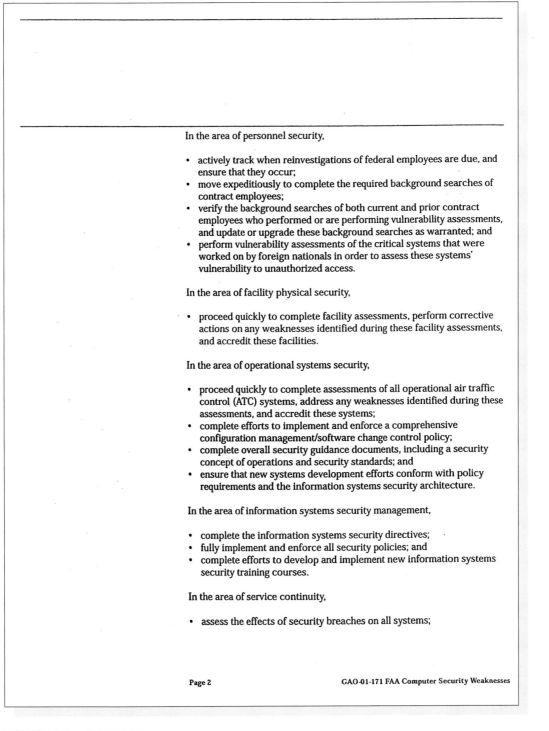

In the area of personnel security,

- actively track when reinvestigations of federal employees are due, and ensure that they occur;
- move expeditiously to complete the required background searches of contract employees;
- verify the background searches of both current and prior contract employees who performed or are performing vulnerability assessments, and update or upgrade these background searches as warranted; and
- perform vulnerability assessments of the critical systems that were worked on by foreign nationals in order to assess these systems' vulnerability to unauthorized access.

In the area of facility physical security,

- proceed quickly to complete facility assessments, perform corrective actions on any weaknesses identified during these facility assessments, and accredit these facilities.

In the area of operational systems security,

- proceed quickly to complete assessments of all operational air traffic control (ATC) systems, address any weaknesses identified during these assessments, and accredit these systems;
- complete efforts to implement and enforce a comprehensive configuration management/software change control policy;
- complete overall security guidance documents, including a security concept of operations and security standards; and
- ensure that new systems development efforts conform with policy requirements and the information systems security architecture.

In the area of information systems security management,

- complete the information systems security directives;
- fully implement and enforce all security policies; and
- complete efforts to develop and implement new information systems security training courses.

In the area of service continuity,

- assess the effects of security breaches on all systems;

Page 2   GAO-01-171 FAA Computer Security Weaknesses

FIGURE 10-1 · *CONTINUED*

- enhance existing contingency plans to address potential systems security breaches; and
- correct inadequacies in facility contingency plans.

In the area of intrusion detection,

- increase efforts to establish a fully operational Computer Security and Intrusion Response Capability that allows for the detection, analysis, and reporting of all computer systems security incidents promptly and
- ensure that all physical security incidents are reported to security personnel.

As you know, the head of a federal agency is required by 31 U.S.C. 720 to submit a written statement on the actions taken on our recommendations to the Senate Committee on Governmental Affairs and the House Committee on Government Reform not later than 60 days after the date of this report. A written statement must also be sent to the House and Senate Committees on Appropriations with the agency's first request for appropriations made more than 60 days after the date of this report.

Agency Comments

We obtained oral comments on a draft of this report from DOT and FAA officials, including representatives of the Office of the Secretary of Transportation, FAA's Chief Information Officer, FAA's Director of Information Systems Security, and FAA's Deputy Associate Administrator for Civil Aviation Security. These officials generally agreed with our recommendations and stated that they are working to implement them. In addition, these officials offered detailed comments, which we have incorporated as appropriate.

We are sending copies of this report to Senator Slade Gorton, Senator Frank R. Lautenberg, Senator Joseph I. Lieberman, Senator John D. Rockefeller IV, Senator Richard C. Shelby, Senator Fred Thompson, Representative James A. Barcia, Representative John J. Duncan, Representative Ralph M. Hall, Representative Steven Horn, Representative William O. Lipinski, Representative Constance A. Morella, Representative Martin O. Sabo, Representative F. James Sensenbrenner, Jr., Representative Jim Turner, and Representative Frank R. Wolf in their capacities as Chairmen or Ranking Minority Members of Senate and House Committees and Subcommittees.

FIGURE 10-1 · *CONTINUED*

We are also sending copies of this report to the Honorable Jane F. Garvey, Administrator of the Federal Aviation Administration, and to the Honorable Jacob J. Lew, Director of the Office of Management and Budget. Copies will also be made available to others upon request.

Should you or your staff have any questions concerning this report, please contact me at (202) 512-6408 or Linda Koontz, Director, Information Management Issues, at (202) 512-6240. We can also be reached by e-mail at *willemssenj@gao.gov* and *koontzl@gao.gov*, respectively. Major contributors to this report are identified in appendix II.

Sincerely yours,

Joel C. Willemssen
Managing Director, Information Technology Issues

Page 4 GAO-01-171 FAA Computer Security Weaknesses

FIGURE 10-1 · *CONTINUED*

- Date of submission
- Code number of the report
- Contract numbers under which the work was performed
- Company or agency logo
- Proprietary and security notices
- Names of contact/responsible individuals
- Descriptive abstract

Figure 10-2 contains many of these items in addition to the signatures of the individuals responsible for the report content. Examine several of the title pages on the book's companion Web site. You will see a wide variety of designs. Regardless of the items included and the format selected, you should do your best to design an attractive title page.

The report title, which must indicate clearly the content of the report, should contain keywords that reflect the crucial content areas as well as the purpose and perhaps the scope and/or type of the report.

Submission Page

Reports may use a submission page, which includes the list of contributors to the report and/or the names and signatures of the authorizing officer or project leader. Submission pages emphasize a point we make throughout this book: reports require accountability. Signatures on a submission page indicate that the authors stand behind the content. The submission page usually either precedes or follows the title page. Or, as exemplified in Figure 10-2, the title and submission pages may be combined.

Table of Contents

The table of contents (TOC) performs at least three major functions. First and most obviously, it indicates the page on which each major topic begins. Second, the TOC forecasts the extent and nature of the topical coverage and suggests the logic of the arrangement and the relationship among the report parts. The TOC contains all major headings used in the report, so these displayed items, like the title of the report, should tell something about the content of the material that follows. Skimming a table of contents should give readers a clear idea of the topics covered, the content presented under each heading, the amount of coverage devoted to that topic, the development of the report, and the progression of information. While you may wish to use as headings "Introduction," "Conclusion," and "Recommendations," avoid "Discussion": it tells your reader nothing about the section's content.

Finally, the TOC should reflect the rhetorical purpose of your report—what you want your readers to know and why the report is important to them. Tables of contents vary, but Figure 10-3 shows how a detailed table of contents can summarize the report. Figure 10-4 shows a shorter table of contents, but the rhetorical purpose of the report clearly emerges.

FHWA-VA-EIS-99-01-F
State Project: 0064-114-F12, PE-102

HAMPTON ROADS CROSSING STUDY
FINAL ENVIRONMENTAL IMPACT STATEMENT &
SECTION 4(F) EVALUATION

U.S. Department of Transportation – Federal Highway Administration
and
Virginia Department of Transportation

Submitted Pursuant to:
42 U.S.C. 4332(2)(c); 23 U.S.C. 128(a);
49 U.S.C. 303(c); 16 U.S.C. 470(f); 23 CFR 450.318

<u>Cooperating Agencies</u>
US Army Corps of Engineers – Norfolk District, US Environmental Protection Agency – Region III,
US Fish and Wildlife Service, National Marine Fisheries Service

_2/28/01_
Date of Approval

Chief Engineer
Virginia Department of Transportation

_3/1/01_
Date of Approval

Division Administrator
Federal Highway Administration

The following persons may be contacted for additional information concerning this document.

Mr. Earl T. Robb
Environmental Administrator
Virginia Department of Transportation
1201 East Broad Street
Richmond, VA 23219
(804) 786-4559

Mr. Ed Sundra
FHWA
400 North Eighth Street
Room 750
Richmond, VA 23240-0249
(804) 775-3337

This Hampton Roads Crossing Study final environmental impact statement has been prepared to determine the impact of a proposed new crossing of Hampton Roads in southeastern Virginia. The primary project purpose is to develop intermodal alternatives that can work together to improve accessibility, mobility, and goods movement in the Hampton Roads metropolitan area to help relieve the congestion that occurs at the I-64 Hampton Roads Bridge Tunnel. Project termini include the I-64 and I-664 interchange in Hampton; the I-64 and I-564 interchange in Norfolk; VA 164 near Coast Guard Boulevard in Portsmouth; and the I-64, I-264, and I-664 interchange in Chesapeake.

FIGURE 10-2 · TITLE PAGE FROM A STUDY PREPARED BY GOVERNMENT AGENCIES (HAMPTON ROADS CROSSING STUDY)

FY 2002 Performance Evaluation Report
of Battelle Memorial Institute

Table of Contents

I. OVERALL SUMMARY/RATING **1**

II. CRITICAL OUTCOMES **3**

1.0 SCIENTIFIC AND TECHNOLOGICAL EXCELLENCE (60%) **3**
1.1 THROUGH 1.4 DOE-HQ PROGRAM OFFICE EVALUATIONS **3**
1.5 CREATE LEADING-EDGE SCIENTIFIC CAPABILITIES TO SUPPORT EVOLVING DOE MISSION NEEDS **5**
1.5.1 Progress against Biomolecular Systems Initiative *6*
1.5.2 Progress against Computational Sciences and Engineering Initiative *6*
1.5.3 Progress against the Nanoscience and Technology *7*
1.6 CREATE AND MAINTAIN STRATEGIC ACADEMIC PARTNERSHIPS THAT STRENGTHEN SCIENTIFIC CAPABILITIES **7**
1.6.1 Impact of the Joint Global Change Research Institute with University of Maryland *7*
1.6.2 Impact of the Joint Research Institute with University of Washington *8*
2.0 MANAGEMENT AND OPERATIONS EXCELLENCE CRITICAL OUTCOME (25%) **10**
2.1 PROVIDE MANAGEMENT AND OPERATIONAL EXCELLENCE IN ACHIEVING KEY CONTRACT PERFORMANCE REQUIREMENTS **10**
2.1.1 Provide ES&H management systems that sustain and enhance Laboratory operations *10*
2.1.2 Performance against Business Management sub-indicators *11*
2.1.3 Sustain and enhance the effectiveness of Integrated Safeguards and Security *11*
2.1.4 Provide management and operational excellence in achieving adequate investment in maintenance and energy conservation efforts *11*
2.2 MAINTAIN AND ENHANCE LABORATORY CAPABILITIES TO MEET CURRENT AND FUTURE MISSION NEEDS **12**
2.2.1 Effective execution of the Information Technology (IT) Infrastructure Strategic Plan to provide the IT infrastructure needed to achieve the 2010 Vision of the Laboratory *12*
2.2.2 Effective execution of the Facilities Strategic Plan to provide the facility space and infrastructure revitalization needed to achieve the 2010 Vision of the Laboratory *12*
2.2.3 Effective execution of the Facilities Strategic Plan to provide the facility space and infrastructure modernization needed to achieve the 2010 Vision of the Laboratory *12*
2.3 PROVIDE INTEGRATED MANAGEMENT SYSTEMS THAT ENABLE EFFECTIVE AND EFFICIENT BUSINESS PERFORMANCE **13**
3.0 LEADERSHIP EXCELLENCE (15%) **17**
3.1 ATTRACT, DEVELOP AND RETAIN THE CRITICAL STAFF NECESSARY TO ACHIEVE SIMULTANEOUS EXCELLENCE IN S&T, OPERATIONS, AND COMMUNITY TRUST **17**
3.1.1 Implement a Laboratory level assessment that establishes a baseline for measuring staff engagement *17*
3.1.2 Achieve and maintain competitive base pay levels for all job classifications *17*
3.1.3 Achieve and maintain a voluntary separation rate (VSR) at or below industry average *18*
3.2 HELP DEFINE AND SHAPE THE FUTURE OF THE REGION BY HELPING ESTABLISH A ROBUST, SUSTAINABLE, REGIONAL ECONOMY **18**
3.2.1 The number of new businesses started or expanded in the local area where Battelle had a material role in their establishment *18*
3.2.2 Effectiveness in providing technical assistance to regional firms *18*
3.2.3 Develop and champion at least one new economic development initiative *19*
3.3 DEMONSTRATE THE RELEVANCE OF THE NATIONAL LABORATORY AND OTHER NORTHWEST RESEARCH INSTITUTIONS TO THE ECONOMIC, SCIENCE AND TECHNOLOGY NEEDS OF THE REGION **20**
3.3.1 Establish meaningful partnerships to link regional needs to science and technology strategies *20*
3.3.2 Focus efforts of the regions' research institutions and business communities and units of government on implementing new S&T solutions to a critical few consensus needs of the Northwest *20*
3.3.3 Initiate efforts to analyze and leverage the Intellectual property (IP) of the major regional research and development institutions *21*

i

FIGURE 10-3 · DETAILED TABLE OF CONTENTS FROM A REPORT OF THE BATTELLE MEMORIAL INSTITUTE

FY 2002 Performance Evaluation Report
of Battelle Memorial Institute

ii

FIGURE 10-3 · *CONTINUED*

I-95/I-395 HOV RESTRICTION STUDY

Table of Contents

FIGURE 10-4 · BASIC TABLE OF CONTENTS FOR A TRAFFIC ENGINEERING STUDY (HOV RESTRICTION)

I-95/I-395 HOV RESTRICTION STUDY

FIGURE 10-4 · *CONTINUED*

List of Illustrations

If a report contains tables, graphics, drawings, photos—any type of visual aid, it's customary to list them in a section entitled "Illustrations" or "Exhibits." The displayed items are usually divided into "tables" and "figures." Specific types may also be listed, however, if you have an array of illustrations: for example, maps, financial statements, photographs, charts, and computer programs. Figure 10-5 shows an example list of illustrations; this one accompanied the high-occupancy-vehicle restriction study whose TOC appeared in Figure 10-4.

Glossary and List of Symbols

Reports dealing with specialized subject matter often include abbreviations, acronymns, symbols, and terms not known to readers outside a particular discipline or subspecialty. Knowledge of your readership should determine whether a glossary or list of symbols is needed. Many reports prepared by government agencies for the general public provide glossaries for the convenience of nonspecialists.

Glossaries are often prefatory elements, but they may also appear at the end of the report. When you first use a symbol or a term included in the glossary, tell your reader where to find the glossary (e.g., "in Appendix Q"). You may wish to place an asterisk (*) by a word or symbol that will be covered in the glossary. Figure 10-6, from the Hampton Roads Crossing Study whose title page was shown in Figure 10-2, is a list of acronyms.

ABSTRACTS AND SUMMARIES

Abstracts and summaries are the most important prefatory items in a report. As noted earlier, the title page, table of contents, abstract, and summary may be the only parts of the report many individuals read. Thus, the prefatory elements should be carefully planned, as each provides a slightly different perspective on the report. While the title page and table of contents outline the content and direction of the report, the abstracts and summaries each provide the essence:

- Topic
- Purpose
- Result(s)
- Conclusion(s)
- Recommendation(s)

Each item can stand alone or be read in conjunction with other prefatory elements. At any point, the reader should be able to turn to the table of contents and

LIST OF TABLES

LIST OF FIGURES

FIGURE 10-5 · LIST OF TABLES AND FIGURES FOR THE HOV STUDY (SEE FIGURE 10-4)

GLOSSARY OF COMMONLY USED ACRONYMS

AASHTO	American Association of State Highway Transportation Officials
ACHP	Advisory Council on Historic Preservation
ADT	Average Daily Traffic
AWOIS	Automated Wreck and Obstruction Information System
BTUs	British thermal units
CAA	Clean Air Act
CAAA	Clean Air Act Amendments of 1990
CALTRANS	California Department of Transportation
CEQ	President's Council on Environmental Quality
CFR	Code of Federal Regulations
CMS	Congestion Management System
CO	Carbon monoxide
CTB	Virginia Commonwealth Transportation Board
CZMA	Coastal Zone Management Act
DEIS	Draft Environmental Impact Statement
DEQ	Virginia Department of Environmental Quality
DOD	U.S. Department of Defense
DOI	U.S. Department of the Interior
EO	Executive Order
ER-M	Effect Range—Median
ESA	Endangered Species Act
FEIS	Final Environmental Impact Statement
FEMA	Federal Emergency Management Administration
FHWA	Federal Highway Administration
FONSI	Finding of No Significant Impact
FTA	Federal Transit Administration
GIS	Geographic Information System
GTIS	Graphic Traffic Information System
HC	Hydrocarbons
HOV	High Occupancy Vehicle
HRBT	I-64 Hampton Roads Bridge Tunnel
HRPDC	Hampton Roads Planning District Commission
HUD	U.S. Department of Housing and Urban Development
ISTEA	Intermodel Surface Transportation Efficiency Act
JRIA	James River Institute of Archeology
LOS	Level of Service
LRT	Light Rail Transit
LWCFA	Land and Water Conservation Fund Act
MIS	Major Investment Study
MMMBT	I-664 Monitor Merrimac Memorial Bridge Tunnel
MMPA	Marine Mammal Protection Act
MPO	Metropolitan Planning Organization
MSA	Metropolitan Statistical Area
NAAQS	National Ambient Air Quality Standards
NEPA	National Environmental Policy Act
NHL	National Historic Landmark
NHS	National Highway System
NMFS	National Marine Fisheries Service

FIGURE 10-6 · EXAMPLE ITEMS FROM GLOSSARY OF ACRONYMS FOR THE HAMPTON ROADS CROSSING STUDY

move swiftly and precisely to a specific segment in the body of the report. The wording of the table of contents should echo throughout an abstract or summary.

Often, the abstract follows or appears on the title page (Figure 10-2). Summaries may also follow the title page. While both contain similar information, the summary usually provides more extensive information than the abstract and may be written for a decision maker whose needs differ from those of readers who want only the essence of a report.

Abstracts are either informative or descriptive. Abstracts, often accompanied by keywords, are prepared for use in online indexes and databases. Keywords can be used to retrieve a report stored in a database. Thus, you should think carefully about the keywords that characterize the content of your report. What words will your readers most likely use in a database search? Abstracts with keywords may be separated from the report but linked to the full report. Based on the information in the abstract, a reader can decide whether to open the full report.

Because of the information overload alluded to repeatedly, well-written abstracts have become critical to report access. The abstract should explain the purpose of the report, the findings, conclusions, and recommendations—anything of significance. Because of the proliferation of abstract services, many readers will see only the abstract of your work. If the abstract clearly shows the relevance of the content and key points, however, some may decide to retrieve or order your complete report.

www.oup.com/us/houp

Informative Abstract

The informative abstract includes the following:

- The research objectives
- The research methods used
- The findings, including principal results
- The conclusions

Recommendations may also be included. Informative abstracts usually range from fifty to five hundred words, depending on the length of the report, and the requirements of the organization disseminating the report and the abstracting service. The informative abstract helps readers decide whether to access the entire report for more thorough examination. Informative abstracts begin with a statement of the report's purpose, as shown in the following abstract of a report in the NASA collection. The remaining sentences give major highlights and conclusions.

1. **Abstract:** This demonstration project assessed the crew members' compliance to a portion of the exercise countermeasures planned for use on board the International Space Station (ISS) and the outcomes of their performing these countermeasures. Although these countermeasures have been used separately in other projects and investigations, this was the first time they'd been

used together for an extended period (60 days) in an investigation of this nature. Crew members exercised every day for six days, alternating every other day between aerobic and resistive exercise, and rested on the seventh day. On the aerobic exercise days, subjects exercised on an electronically braked cycle ergometer using a protocol that has been previously shown to maintain aerobic capacity in subjects exposed to a space flight analogue. On the resistive exercise days, crew members performed five major multijoint resistive exercises in a concentric mode, targeting those muscle groups and bones we believe are most severely affected by space flight. The subjects favorably tolerated both exercise protocols, with a 98% compliance to aerobic exercise prescription and a 91% adherence to the resistive exercise protocol. After 60 days, the crew members improved their peak aerobic capacity by an average 7%, and strength gains were noted in all subjects. These results suggest that these exercise protocols can be performed during ISS, lunar, and Mars missions, although we anticipate more frequent bouts with both protocols for long-duration spaceflight. Future projects should investigate the impact of increased exercise duration and frequency on subject compliance, and the efficacy of such exercise prescriptions.

Keywords exercise physiology, physical exercise, weightlessness, muscular strength, bones

Source S. M. C. Lee,* M. E. Guilliams,* A. D. Moore, Jr.,* W. J. Williams,* M. C. Greenisen, S. M. Fortney, **Exercise Countermeasures Demonstration Project During the Lunar-Mars Life Support Test Project Phase IIA** *TM-1998-206537*, 1/1/1998, pp. 71.

*Krug Life Sciences, Inc.

Source: http://ston.jsc.nasa.gov/collections/TRS/_1998-abs.html

Descriptive Abstract

The descriptive abstract states what topics the full report contains. Unlike the informative abstract, it cannot serve as a substitute for the report itself. The following descriptive abstract, from another document abstracted in the NASA collection, begins with the report purpose and then explains content areas or topics covered in the report.

1. **Abstract:** The purpose of this report is to identify the essential characteristics of goal-directed whole-body motion. The report is organized into three major sections. Section 2 reviews general themes from ecological psychology and control-systems engineering that are relevant to the perception and control of whole-body motion. These themes provide an organizational framework for analyzing the complex and interrelated phenomena that are the defining characteristics of whole-body motion. Section 3 applies the organizational framework from the first section to the problem of perception and control of aircraft

motion. This is a familiar problem in control-systems engineering and ecological psychology. Section 4 examines an essential but generally neglected aspect of vehicular control: coordination of postural control and vehicular control. To facilitate presentation of this new idea, postural control and its coordination with vehicular control are analyzed in terms of conceptual categories that are familiar in the analysis of vehicular control.

Keywords motion, motion perception, perception, control, adaptive control

Source: Gary E. Riccio* and P. Vernon McDonald,** **Multimodal Perception and Multicriterion Control of Nested Systems: I. Coordination of Postural Control and Vehicular Control,** *TP-1998-3703,* 1/1/1998, pp. 76.

*Nascent Technologies, Ltd.
**National Space Biomedical Research Institute.

Source: http://ston.jsc.nasa.gov/collections/TRS/_1998-abs.html

Differences in abstracts are tending to disappear, as some abstracts display both informative and descriptive characteristics.

SUMMARY

The summary, often called "executive summary," targets decision makers or readers who do not have time to read the full report but do need to know its essential issues, conclusions, recommendations, and financial implications. While abstracts concentrate on main information of the report, summaries are broader and may be several pages long rather than one or two paragraphs. Length is determined by the needs of the reader or readers. Figure 10-7 lists possible questions decision makers might ask about a report topic. To plan a summary, consider including the following items, but consider the information you think your readers will need to know:

- Subject and purpose of the project
- Research approach used
- Topics covered
- Essential background—why the report/research was needed
- Results
- Conclusions
- Recommendations
- Cost and/or potential return on investment
- Anticipated implementation problems and suggested solutions

Figure 10-8 presents the executive summary excerpted from the high-occupancy vehicle restriction study whose table of contents appeared in

Problems
What is it?
Why is solution undertaken?
Magnitude and importance?
What is being done? By whom?
Approaches used?
Are they thorough and complete?
Suggested solution? Best? Consider others?
What now?
Who does it?
Time factors?

New Projects and Products
Potential?
Risks?
Scope of application?
Commercial implications?
Competition?
Importance to company?
More work to be done? Any problems?
Required personnel, facilities, and
    equipment?
Relative importance to other projects or
    products?
Life of project or product line?
Effect on company's technical position?
Priorities required?
Proposed schedule?
Target date?

Tests and Experiments
What tested or investigated?
Why? How?
What did they show?
Better ways?
Conclusions? Recommendations?
Implications to company?

Materials and Processes
Properties characteristics, capabilities?
Limitations?
Use requirements and environment?
Areas and scope of application?
Cost factors?
Availability and sources?
What else can be used?
Problems in using?
Significance of application to company?

Field Troubles and Special Design Problems
Specific equipment involved?
What trouble developed? Any trouble history?
How much involved?
Responsibility? Others? Company?
What is needed?
Special requirements and environment?
Who does it? Time factors?
Most practical solution? Recommended action?
Suggested product design changes?

FIGURE 10-7 · WHAT MANAGERS WANT TO KNOW
Source: James W. Souther, "What to Report," *IEEE Transactions on Professional Communication* PC-28 (1985): 6.

Figure 10-4. Note the effective use of format: headings, listing, and concise statements all contribute to the readability of the document. The table of contents and the summary complement each other and provide a structural tool to aid readers in grasping the report purpose and results. In a sense, a summary is an extended abstract: it covers many of the same elements but gives more detail.

Summaries have become increasingly popular. Reports may include a well-developed summary before the actual discussion. The Congressional Research Report on pages 254–55 begins with a summary, which incorporates the introduction, as well.

I-95/I-395 HOV Restriction Study

Executive Summary

The I-95/I-395 **H**igh **O**ccupancy **V**ehicle (HOV) facility is a reversible two-lane freeway, about 27 miles long, between the southern terminus at Dumfries near Route 234 and the northern terminus between Route 27 and Eads Street in Arlington. Beyond this northern terminus, there are separate lanes for northbound and southbound traffic that extend across the Potomac River on the Rocheambeau Bridge. During the HOV-restricted periods, the HOV facility carries more people at higher speeds than the parallel general purpose lanes. The advantages of reliable higher speeds and lower travel times make the HOV lanes attractive to a large number of commuters in carpools, vanpools and buses.

The Virginia Department of Transportation (VDOT) frequently receives comments from citizens, specifically those using the general purpose lanes, that the HOV facility is underutilized. The HOV facility is viewed by some to be inefficient and propose that its restrictions on usage be modified. These concerns led VDOT to have this study undertaken. A technical committee was formed, comprised of representatives from numerous agencies, organizations and private transportation providers, to guide and monitor the study. Alternatives were identified for evaluation that included:

1. Changing the HOV lane occupancy requirements from HOV 3+ to HOV 2+ for either the entire corridor or for a portion of the corridor (e.g., HOV 2+ outside the Capital Beltway and HOV 3+ inside the Beltway),

2. Changing the HOV-restricted times during the morning (AM) and/or afternoon/evening (PM) periods,

3. Providing additional access ramps to/from the HOV facility at appropriate locations, and

4. Providing three (3) HOV lanes inside the Beltway.

In addition, the study team was directed to investigate the potential impacts on HOV lane demand that could result from upcoming construction activities associated with the Springfield I-95/I-395/I-495 interchange improvement project.

The findings of this study are summarized below in terms of the questions that were posed by VDOT and the Technical Committee.

What are the impacts on HOV and transit usage if the HOV 3+ restriction was changed to HOV 2+ for the entire corridor?

This change is projected to result in significantly higher traffic volumes on the HOV lanes, with AM peak hour volume increases of 30 to 50 percent inside of the Beltway. These higher volumes will cause travel speeds to drop by 50 percent, to an average speed of 32 mph from Arlington south to the Beltway. A corresponding degradation in level of

FIGURE 10-8 · EXECUTIVE SUMMARY FOR THE HOV STUDY
Source: http://www.virginiadot.org/projects/studynova-hov395exsumm.asp

service to LOS E or F for this section is also projected. Provision of a third HOV lane inside the Beltway may relieve this congestion; however, the estimated cost to provide a third lane inside the Beltway is approximately $22 million and, in the end, will result in a facility with substandard shoulder and lane widths. By 2010, traffic volumes outside the Beltway will necessitate addition of a third HOV lane at least down to the Fairfax County Parkway. Although traffic volumes are projected to increase dramatically, person movement on the HOV lanes will remain relatively constant due to the fact that the increased traffic volumes on the HOV lanes will come primarily from a breakup of 3 or more person carpools already in the HOV lanes into 2-person carpools. To the extent that persons now traveling in the general purpose lanes divert to the HOV lanes, those trips are likely to be replaced on the general purpose lanes by trips diverting from other roadways.

Increased congestion, decreased speeds, and the need to only have 2 persons in a car on the HOV lanes will also have negative effects on bus and vanpool operators. Private and public bus ridership could decrease by as much as 50 to 80 percent and vanpool riders could be reduced by 60 percent. This will result in lower fare revenues and increased operating costs. A change to HOV 2+ will also significantly diminish the effectiveness of current informal carpool matching, or slugging, activities. This change is projected to result in only modest regional vehicle emission increases, estimated by MWCOG to be on the order of 0.01 to 0.06 percent.

What are the impacts on HOV and transit usage if the HOV 3+ restriction was changed to HOV 2+ outside of the Beltway only?

This change is projected to result in a 30 to 60 percent increase in traffic outside of the Beltway on the HOV lanes, although travel speeds will remain high enough to provide an incentive for HOV use. Traffic volumes inside of the Beltway will decrease by approximately 15 percent as persons currently in 3 or more person carpools split into 2-person carpools to take advantage of the travel time savings outside of the Beltway while not having to continue forming 3 or more person carpools. In the short term, projected increases in traffic volumes will not reduce existing speeds or result in LOS E or F conditions; however, there will be a critical issue related to the transition of 2-person vehicles from the HOV lanes to the general purpose lanes that will need to occur south of the Beltway, at the Newington flyover ramp, which will result in increased congestion at the merge onto the general purpose lanes. In 2010, this problem will be alleviated by the new ramps planned from the HOV lanes to the Beltway, which is now scheduled as the final phase of the improvement program. By 2010, increased vehicle volumes outside of the Beltway will result in LOS E or F conditions on projected between Edsall Road and the Horner Park & Ride Lot exit ramp. Although speeds and travel time are not expected to deteriorate significantly, traffic volumes on this section will be approaching capacity with an increased likelihood of slowdowns and diminished reliability.

Effects of this change on transit and vanpool usage is much less significant than it is with a change to HOV 2+ for the entire corridor, with ridership decreases projected in the

FIGURE 10-8 · CONTINUED

10 to 20 percent range. This change will also be much less damaging to slugging activities, since most slug trips are made to points in the northern portion of the corridor, which would remain at HOV 3+. The estimated effects on air quality are slight, with only a nominal increase in nitrogen oxides projected.

What are the impacts of changing the hours of HOV restrictions during the AM and/or PM periods?

Extending the restricted periods of HOV operations by a half hour, either by starting earlier in the morning at 5:30 AM instead of 6:00 AM, or extending operations later in the evening from 6:00 PM to 6:30 PM, could in the long term serve to increase person movement on the HOV lanes during these half hour periods as demand and congestion increase on the general purpose lanes. However, in the short term, total person trips on the HOV and general purpose lanes combined could decrease during the extended half hour periods by approximately 10 percent. Increases in transit and vanpool ridership are expected on the order of 1 to 3 percent for the peak period. Shortening the restricted period in the morning by ending HOV restrictions at 8:30 AM would probably not have negative impacts in the southern portion of the corridor, but may lead to congested conditions in the HOV lanes in the northern portion of the corridor during this 8:30 to 9:00 half hour period. The upcoming Springfield Interchange construction project, which is scheduled to continue for up to eight years, could accelerate the rate of projected demand growth for the HOV lanes, which would argue for extending the HOV-restricted periods rather than shortening them. Were it not for the interchange construction project, ending the restricted period earlier in the AM could have been warranted, at least in the southern portion of the corridor.

Are new ramps to/from the HOV lanes justified and feasible?

More access points to and from the HOV lanes should enhance utilization of the HOV facility. New access ramps for morning northbound/evening southbound HOV traffic at Seminary Road, Route 123 and the Fairfax County Parkway are all feasible from an engineering perspective but will be costly, $5.2 million, $26.9 million, and $12.2 million, respectively. In lieu of adding a ramp at Seminary Road, one possibility would be to construct a new slip ramp from the northbound HOV lanes to the general purpose lanes at a point between Edsall Road and Route 236 (Duke Street). This new ramp would serve to enhance HOV and bus access to Alexandria and Arlington. In addition, if it could be constructed quickly in conjunction with implementation of the HOV 2+ outside Beltway/HOV 3+ inside Beltway alternative, this ramp would divert traffic from the Newington flyover ramp and remove traffic from the Beltway interchange construction area. Projected volumes for a new Rt. 123 ramp are relatively low, but a new ramp at the Fairfax County Parkway could attract as many as 500 HOV vehicles per hour. Demand estimates for a new ramp at Seminary Road are approximately 200 to 300 HOV vehicles per hour.

FIGURE 10-8 · *CONTINUED*

How will construction of the new Springfield Interchange affect HOV operations?

Considerable delays to vehicles on the general purpose lanes through the Springfield area are expected once the interchange construction project begins. These delays could range from 20 to 60 minutes. An analysis was performed to estimate how many new carpools might be formed given these high levels of delay on the general purpose lanes. It was estimated that traffic volumes on the HOV lanes during the AM and PM peak hours may increase by 50 to 75 percent in Prince William County and by 20 to 30 percent in Fairfax County. These increases represent 150 to 225 vehicles, and 400 to 600 vehicles, respectively. Volume increases inside of the Beltway would be expected to be much lower since over 80 percent of the person trips traveling from outside the Beltway to the Pentagon and downtown areas are already using HOV or transit modes. Vanpool ridership would not be expected to experience similar increases because most vanpool destinations are to the Pentagon and downtown areas. Only a nominal switch to transit would be expected.

I-95/I-395 HOV Restriction Study, Virginia Department of Transportation

FIGURE 10-8 · *CONTINUED*

Arsenic in Drinking Water: Recent Regulatory Developments and Issues

Mary Tiemann
Specialist in Environmental Policy
Resources, Science, and Industry Division

Summary

In 1996, Congress directed the Environmental Protection Agency (EPA) to propose a new standard for arsenic in drinking water by January 1, 2000, and to issue a final standard by January 1, 2001. Congress also directed EPA, with the National Academy of Sciences (NAS), to study arsenic's health effects to reduce the uncertainty in assessing health risks associated with exposure to low levels of arsenic. EPA issued the current standard of 50 parts per billion (ppb) in 1975. In 1999, the NAS concluded that the standard did not achieve EPA's goals for public health protection and recommended that it be tightened as soon as possible. On June 22, 2000, EPA proposed a revised standard of 5 ppb and projected that compliance could be costly for small communities. A question of ongoing scientific debate concerned whether significant adverse health effects occur from ingesting arsenic at very low levels. Because EPA proposed the rule nearly 6 months late, some stakeholders expressed concern that the Agency would not have time to evaluate public comments and complete analyses before issuing a final rule.

On January 22, the final rule, which set the standard at 10 ppb, was published in the *Federal Register* with an effective date of March 23, 2001; public water systems were given until 2006 to meet the new standard. On May 22, EPA extended a previous 60-day delay of the rule's effective date to February 22, 2002 in order to review risk, benefit, and cost issues associated with the rule. The 2006 compliance date for water systems remained unchanged. On October 31, EPA announced that the standard would be 10 ppb. On November 8, Congress approved the conference report to EPA's FY2002 appropriations bill (H.R. 2620), which includes language prohibiting EPA from using the funds to delay the January rule. This report reviews EPA efforts to develop a new arsenic rule and summarizes key provisions and subsequent events.

Source: http://www.cnie.org/nle/crsreports/water/h2o-40.pdf

DISCUSSION OR BODY OF THE REPORT

The main part of the report—the discussion—takes most of the writer's development time. The discussion will explain, in detail appropriate to the context, the readers, and the purpose of the report, the following:

- Why the report was done
- Its objectives
- Its methods
- Its findings
- Its results
- Analysis of results
- Conclusions emerging from results
- Recommendations with respect to the results

The discussion is the guts of the report: without it, effective summaries and abstracts could not be written. The presentation of information in the discussion allows the report writer to draw conclusions and perhaps recommendations. In short, the discussion must support all conclusions and recommendations.

The discussion can also stand alone; it is complete in itself because it begins with an introduction and explains the approach used in developing the report. Many discussions end with a factual summary, which is a concise narrative of the report's findings.

Ironically, however, the main discussion is *the report segment read least*. While most readers will look at the summary, abstracts, and table of contents, few will actually delve into the discussion. Nevertheless, the discussion is the source, the foundation, and repository of documentation for every statement in the abstract and summary. Your conclusions need to evolve from the discussion. Recommendations need to evolve from the discussion and the conclusions.

The readers who need to pay close attention to certain parts will read the introduction, check the table of contents, skip to relevant sections, and then move to the conclusion or factual summary, which pulls together the main results or ideas.

Parts of the Discussion

The main body of the report generally begins with an introduction (or introduction and summary), which forecasts what is to follow in the report. It directs the reader to the subject, purpose, scope of the report, the plan of development, and any additional information the reader may need. Examine the following introduction from a NASA report. How does the introduction prepare you for the report content?

Telecommunications for Mars Rovers and Robotic Missions

William D. Home, Stanford Telecommunications, Inc.
Rolf Hastrup and Robert Cesarone, Jet Propulsion Laboratory, California Institute of Technology

Telecommunications plays a key role in all rover and other robotic missions to Mars both as a conduit for command information to the spacecraft and for the return of scientific data from the instruments and engineering data from the spacecraft. Telecommunications to the Earth may be accomplished using direct-to-Earth links via the Deep Space Network (DSN) or by means of relay links provided by orbital missions at Mars. A number of factors make direct-to-Earth telecommunications for robotic or rover missions at Mars very challenging, especially for small systems [1, 2]. These include the distance between Mars and Earth, the inability of Earth-based systems to regularly communicate to all portions of Mars, the power and mass constraints on systems at Mars along with the substantial power and/or antenna requirements. By decreasing the communications range and providing coverage to virtually all portions of the Martian surface, orbiting missions with telecommunications relay systems can send and receive large amounts of data to and from rover missions and relay these data to and from Earth. These orbiting relay systems are viewed as enabling elements as they provide telecommunications support for multiple missions including extremely constrained missions, such as microrovers or balloons.

With the recent launch of Mars Global Surveyor (MGS) and Mars Pathfinder, the next generation of Mars exploration has begun. Over the next decade, the Mars Exploration Program [3], led by the National Aeronautics and Space Administration (NASA) and the Jet Propulsion Laboratory (JPL), will launch one or more missions to Mars during each of the launch opportunities in 1998, 2001, 2003, and 2005 [4, 5]. Of these, the orbiting missions, starting with MGS, will each carry a UHF communications package that will provide a relay link to and from appropriately equipped landed elements. The relay capability will vary with each mission but will enable and enhance small, low-power telecommunications systems for robotic assets on or near the surface, such as rovers and balloons.

This paper considers the relay capabilities of MGS, the Mars Surveyor Program (MSP) '98 Orbiter, and future orbiters. The discussion will include both link design considerations and multimission support operations necessary for rover and robotic designs based on results from a JPL-led study of relay communications at Mars. In addition to the telecommunications support, the relay systems of the orbiters can also provide navigation and location services to robotic missions.

Source: Prepared for Space Technology, NASA, 1998.

Without a proper introduction, readers will have difficulty following the main discussion. Thus the introduction should always state the report subject, purpose, and plan of development. Some reports place the background and the scope in separate sections that follow the introduction; this practice is feasible if the two items are extensive. What you include in the introduction depends on your readers. If they are awaiting the report, your introduction can be short. However, if the document will be archived and read later by people who know little about its context, you will need to provide a longer, more informative introduction. Avoid long introductions, however: focus on stating the subject and purpose in terms that explain the relevance of the report to readers. But as you plan the introduction, do anticipate your readers—think about who will read this report and when.

An introduction on the long side will benefit from the use of headings to separate its components.

- **Subject.** The report subject should be stated directly: This paper will discuss. . . .
- **Purpose.** The statement of purpose tells the reader why you are writing about the subject: the significance of the topic or its relationship to existing organizational issues.
- **Scope.** The extent of the coverage of the subject should be made clear.
- **Background or history of the topic.** This section tells why the report was written and what is important about the subject matter. If this is one of several reports, how does this report relate to the others?
- **Plan of development.** The topics covered in the report and the order in which they are presented will reprise in paragraph form the listing in the table of contents.

Strategy for Presenting the Discussion

The discussion should be planned around each topic mentioned in the plan of development. This plan should be evident in the Table of Contents. Examine Figure 10-4 and other reports on the book's companion Web site.

Note that the logic of the discussion should be evident from the headings, as these are repeated in the table of contents.

Each paragraph in the discussion should begin with a topic sentence (see Chapter 4), followed by supporting sentences, data, and visuals, arranged according to the conventions outlined in Chapter 9.

Conclusion

Reports end with a statement of the primary issues covered in the discussion. The factual summary gives the essential facts presented without interpretation. The place for interpretation of the factual summary is the executive summary. The discussion begins with an introduction and ends with a factual summary or conclusions and then recommendations, enabling the discussion to stand alone for the convenience of readers who are not immediately concerned with interpreting the material.

Recommendations

Recommendations, if required or needed, emerge from conclusions. However, many reports end with a conclusion. The type of ending depends on the type of report. The longer the discussion, the more useful a factual summary will be to your reader. Many long reports place the factual summary before the conclusions and recommendations.

Appendixes

Appendixes contain documents that support information you include in the discussion. While this material may not be read carefully, it serves an important role. An appendix may contain tables of data, statistical studies, spreadsheets, or any other form of support for the points or arguments you make in your discussion. If you are recommending an action, the inclusion in the appendix of copies of letters or e-mails from others requesting that action provides convincing support for your recommendation. Be sure to refer to such appendix documents in the text—they should not be a surprise at the end.

ONLINE REPORTS

Online reports are becoming increasingly popular as the Web and intranets are used archivally. When a report is destined for online reading and storage, its preparation will differ from that of the standard paper report.

Online reports usually begin with the report title followed by the abstract. The body of the report may follow, or readers may click on a link that takes them to the remainder of the report. In either case, each segment of the report is linked to the table of contents. Readers may scroll down the table of contents and click on report segments that interest them. Figure 10-9 exemplifies a report prepared for online viewing. Note that the main items in the index are linked to the report sections.

Figure 10-10 shows an online report introductory page. Readers may move to each segment of this report from the Environmental Protection Agency, beginning with the prefatory elements in a separate pdf file. A brief descriptive abstract that includes the size of the pdf file introduces each report section.

Recommendation: Page 1 of 4

SB 358 Section 22
Committee Report ICF/MR Oversight
September 28, 2000

Index

Executive Summary
Introduction and Background
Process of Developing Recommendations
Committee Recommendations
Attachments:
    Section 22 of SB 358 (76th Regular Session)
    Committee Member Listing
    Listing of various Federal and State laws and rules impacting ICF/MR regulation
    OAG Opinion JM-14
    Differences between public and private ICF/MR providers

Executive Summary

The Health and Human Services Commission (HHSC) appointed a nine-member committee pursuant to Section 22 of SB 358, 76th Regular Session (Attachment 1) in November 1999. As directed by that law, the committee is comprised of four members representing residents, families and advocates; three members representing providers; and two ex officio members representing state agencies (Attachment 2). The committee's charge is to evaluate the delivery and regulations of services provided to residents in those Texas Intermediate Care Facilities for the Mentally Retarded (ICF/MR) that must comply with the provisions of Chapters 222 and 252, Health and Safety Code. The committee's charge and this report cover only ICF/MR facilities and do not include any facilities operated under waiver program, such as the HCS program.

The Committee recommends that the current ICF/MR licensing, surveying and certification functions currently at the Department of Human Services (DHS) remain there and not be transferred to the Department of Mental Health and Mental Retardation (MHMR). This recommendation is subject to the following three items.

1. DHS will retain specialized staff to audit/review those ICF/MR facilities under its purview and will refrain from using nursing home audit/review staff to assist with ICF/MR audits/reviews.
2. DHS will work with MHMR in developing applicable policies/rules and will submit proposed policies/rules to MHMR for impact statement prior to adoption by DHS.
3. DHS & MHMR will work closely on all facility closures.

These recommendations are discussed more thoroughly in the Committee Recommendations section.

FIGURE 10-9 · EXAMPLE SUMMARY AND INTRODUCTION FOR A REPORT TO A STATE AGENCY
Source: http://www.hhsc.state.tx.us/medicaid/reports/sb358/SB358.htm

Recommendation:

Introduction and Background

The ICF/MR program is the largest institutional program targeted for Texans having profound to mild retardation and developmental disabilities. ICF/MR facilities are operated by both private and public entities and generally have a capacity of six or more beds. Numerous businesses operate the private ICF/MR facilities, while either the State (MHMR) or local governmental entities (i.e., community MHMR centers) operate the public ICF/MR facilities. As of July 2000, there were 719 privately operated ICF/MR facilities having a capacity of 6,879 and a census of 6,631. In addition, as of July 2000 there were 183 public-operated facilities (including state schools, state centers, state operated community services and community MHMR centers) having a capacity of 7,342 and a census of 6,536.

All of the publicly and privately operated ICF/MR facilities participate in the Title XIX (Medicaid) program. Participation in the Title XIX program includes a mandate that the facility complies with both federal and state laws and regulations. A general listing of those various federal and state laws and regulations identified as impacting the ICF/MR facilities is included in this report (Attachment 3).

Both publicly and privately operated facilities must meet state and federal certification and licensure standards. However, publicly operated facilities meeting licensure standards are based on the state's standards for participation and they are exempt from the state licensure requirement based on the Texas Attorney General's Opinion JM-12, dated March 14, 1983 (Attachment 4). As MHMR is not licensed, it is not subject to any licensing penalties. Included in this report (Attachment 5) is a chart DHS prepared for its HB 1396 report that shows the major differences between public and private ICF/MR providers.

DHS has oversight responsibility for all ICF/MR facilities, including both those operated in the public and the private sector. DHS performs annual inspections (called surveys) of all certified and licensed facilities. For the privately operated facilities, this DHS oversight involves a review of the facility against both the operations and licensing standards. For the public facilities, the DHS oversight involves a review against only the standards. In addition, DHS performs ad hoc inspections for other items such as change of ownership or complaints of abuse.

MHMR is charged with ICF/MR program administrative and operational authority (except survey, certification, and licensure) and is designated by HHSC as the Medicaid "State Operating Agency" for that program. In that capacity and as the state MR authority, MHMR sets mental retardation services principles and values and establishes ICF/MR policy.

Process of Developing Recommendations

Section 22 of SB 358 (76th Regular Session) directed the committee to evaluate the impact and feasibility of consolidating at MHMR the following ICF/MR responsibilities:

- Service delivery,
- Licensing,

FIGURE 10-9 · CONTINUED

Recommendation: Page 3 of 4

- Surveying, and
- Regulation

In evaluating that impact and feasibility, that legislation directed the committee to:

- Review the current service delivery, licensing, surveying, and regulation.
- Analyze current consumer and provider satisfaction with service delivery, licensing, surveying, and regulation.
- Determine the impact on residents and providers of consolidating service delivery, licensing, surveying, and regulation at MHMR.
- Determine if the transfer of service delivery, licensing, surveying, and regulation to MHMR from DHS, or a successor agency, to achieve consolidation is in the best interests of the residents.
- Adopt a plan to address the consolidation if the committee determines that the transfer is in the best interests of the residents.

As directed by the previously noted legislation, the committee evaluated the delivery and regulation of the services to residents of ICF/MR facilities regulated under Chapters 222 and 252 of the Health and Safety Code. The committee held four public meetings in Austin between December 1999 and June 2000. During those public meetings, the committee reviewed the current service delivery, licensing, surveying and regulation of ICF/MR facilities.

The committee solicited input from various organizations representing ICF/MR providers, residents, families and advocates. The committee sought input on consumer and provider satisfaction concerning ICF/MR service delivery, licensing, surveying and regulation. Although the lack of funding prevented the committee from undertaking any independent satisfaction studies, the committee did hear comments on those points during the public hearings. In addition, the committee searched for studies that focused on consumer and provider satisfaction. The only recent relevant satisfaction information the committee located was the January 14, 2000 report entitled "The Customer Feedback Questionnaire: Description and Analysis" authored by Philip Salem, Professor of Speech Communications at Southwest Texas State University. This report focused on provider satisfaction and was based on questionnaires DHS, Long Term Care Regulatory (LTC-R) provided to and collected from those entities it regulates. Those entities include both for profit and not for profit nursing homes and ICF/MR facilities. The survey found no significant differences in reported satisfaction among the different reporting entities. That survey indicated that as a whole, the surveyors do a superior job.

Staff from both DHS and MHMR briefed the committee on operations of those agencies related to ICF/MR regulation and oversight. During public meetings, individual committee members briefed the committee on various topics.

Committee Recommendations

The Committee recommends that the current ICF/MR licensing, surveying and certification functions should remain with DHS and not be transferred to MHMR.

FIGURE 10-9 · *CONTINUED*

Recommendation:

The recommendation is subject to the following three items:

- DHS will retain specialized staff to audit/review those ICF/MR facilities under its purview and will refrain from using nursing home audit/review staff to assist with ICF/MR audits/reviews.

There are significant differences between nursing homes and ICF/MR facilities. The two types of facilities serve different populations and are subject to different regulations. The committee believes that it is a disservice to both DHS staff and entities under review, to expect a nursing home surveyor will also be sufficiently versed with ICF/MR regulation to satisfactorily perform reviews of ICF/MR facilities. The committee recommends that DHS retain adequate mental retardation experienced staff at both the central office and regional offices.

The committee also commends DHS for its efforts in standardizing surveys between the various regions and suggests it continue with these efforts. The committee believes that minimal differences should exist in the methods employed by various survey staff and between the different regions performing those surveys. The committee recommends that DHS continue survey staff in-service training to ensure consistency.

- DHS will submit proposed policies/rules to MHMR for impact statement prior to adoption by DHS. DHS will work with MHMR in developing applicable rules/policies.

Since both DHS and MHMR promulgate regulations impacting ICF/MR facilities, it is imperative that the two agencies work in concert to ensure consistent oversight. Toward that end, the committee recommends that prior to adopting any ICF/MR regulations, DHS submit them to MHMR for review and comment. Asking DHS to work with MHMR prior to adopting any ICF/MR regulations helps ensure that the integrity of MHMR's fiscal management remains intact. Equally important, a collaborative relationship between the two agencies helps ensure that DHS regulations reflect the mental retardation services principles and values established by the MHMR Board. In addition, the committee recommends that both DHS and MHMR work on improving communication between those agencies and the public and private sectors.

- DHS & MHMR will work closely on all licensed facility closures.

For all ICF/MR facility closures, both DHS and MHMR should work in concert to ensure the clients' well being is a foremost concern for both short-term and long-term resolutions. Interagency communications should be of the utmost concern in these situations.

Finally, the committee also discussed the possibly of segregating the current single trust fund into several individual trust funds. Currently Nursing Homes, ICF/MR facilities and Assisted Living Facilities pay into the single trust fund. That trust fund is entitled "Nursing and Convalescent Home Trust Fund and Emergency Assistance Funds" and was established pursuant to Section 242.096 of the Health and Safety Code. Some speakers expressed concern about the possibility of cross-subsidization among the current entities paying into the trust fund. The committee did not research either the costs that might be incurred or the benefits that might accrue from implementing this proposal. The committee reached no consensus on this other item other than recommending that the costs and benefits of a potential segregation be further researched.

FIGURE 10-9 · *CONTINUED*

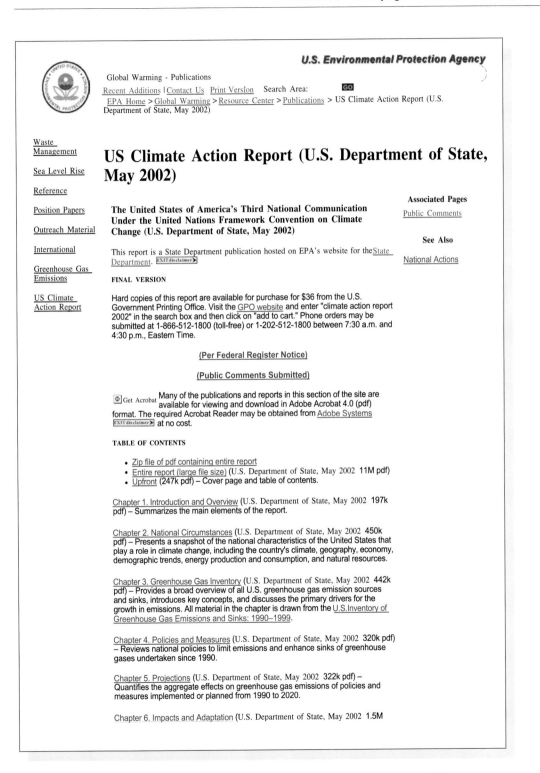

FIGURE 10-10 · ONLINE TABLE OF CONTENTS FOR A U.S. CLIMATE ACTION REPORT

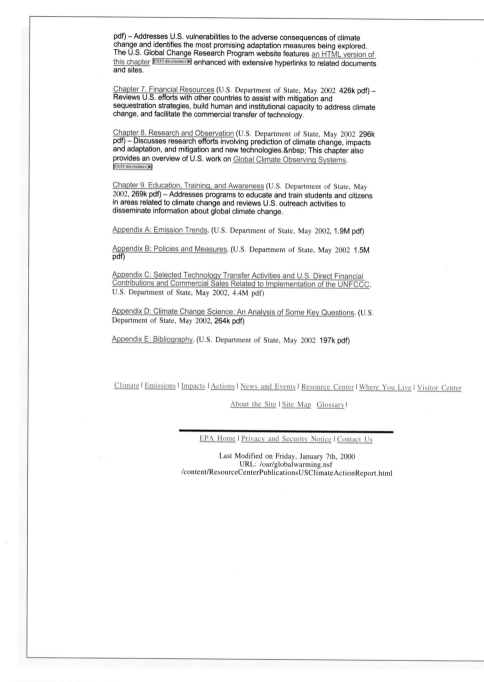

pdf) – Addresses U.S. vulnerabilities to the adverse consequences of climate change and identifies the most promising adaptation measures being explored. The U.S. Global Change Research Program website features an HTML version of this chapter ▣EXIT disclaimer▶ enhanced with extensive hyperlinks to related documents and sites.

Chapter 7. Financial Resources (U.S. Department of State, May 2002 **426k pdf**) – Reviews U.S. efforts with other countries to assist with mitigation and sequestration strategies, build human and institutional capacity to address climate change, and facilitate the commercial transfer of technology.

Chapter 8. Research and Observation (U.S. Department of State, May 2002 **296k pdf**) – Discusses research efforts involving prediction of climate change, impacts and adaptation, and mitigation and new technologies. This chapter also provides an overview of U.S. work on Global Climate Observing Systems. ▣EXIT disclaimer▶

Chapter 9. Education, Training, and Awareness (U.S. Department of State, May 2002, **269k pdf**) – Addresses programs to educate and train students and citizens in areas related to climate change and reviews U.S. outreach activities to disseminate information about global climate change.

Appendix A: Emission Trends. (U.S. Department of State, May 2002, **1.9M pdf**)

Appendix B: Policies and Measures. (U.S. Department of State, May 2002 **1.5M pdf**)

Appendix C: Selected Technology Transfer Activities and U.S. Direct Financial Contributions and Commercial Sales Related to Implementation of the UNFCCC. U.S. Department of State, May 2002, **4.4M pdf**)

Appendix D: Climate Change Science: An Analysis of Some Key Questions. (U.S. Department of State, May 2002, **264k pdf**)

Appendix E: Bibliography. (U.S. Department of State, May 2002 **197k pdf**)

Climate I Emissions I Impacts I Actions I News and Events I Resource Center I Where You Live I Visitor Center

About the Site I Site Map Glossary I

EPA Home I Privacy and Security Notice I Contact Us

Last Modified on Friday, January 7th, 2000
URL: /oar/globalwarming.nsf
/content/ResourceCenterPublicationsUSClimateActionReport.html

FIGURE 10-10 · *CONTINUED*

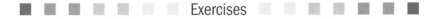

■ ■ ■ ■ ■ ■ ■   Exercises   ■ ■ ■ ■ ■ ■ ■

Further exercises related to the content of this chapter and all documents employed within the following exercises can be found on the book's companion Web site.

1. Examine several of the reports on the book's companion Web site (select reports from Chapters 13 to 17: Sample Documents).
   a. What differences do you see in how reports are developed?
   b. Can you discern the writers' intents in designing the elements of each report?
   c. How do audience and report length affect the design and the plan for each element?

2. Consider the table of contents in Figure 10-4. How would you adjust this TOC for a nonspecialist audience of concerned citizens?

3. Consider the executive summary in Figure 10-8. Use the strategies in Chapter 4 (Achieving a Readable Style) to cut the number of words in this summary by 10 percent without losing critical information.

4. Consider a major report that you are preparing for this or another course. What terms will you need to present in a glossary? Prepare a glossary of those terms and justify each item. Exchange your glossary with a student from a different major. Did the nonspecialist reader understand your definitions?

5. Study the student report "Stem Cell Research: The Debate" located on the book's companion Web site (Appendix B). How well does this report adhere to our development guidelines on the following issues?

   · Writing for your readers
   · Achieving a readable style
   · Gathering, evaluating, and documenting information
   · Designing and managing content
   · Developing the main elements of reports
   · Creating visual aids

Scenario

It's only your third week on the job, but you realize that to do it right, you're going to need a better digital camera. You must write a memo to your boss to justify the request. You don't have much time to write this memo, and you know your boss won't have much time to read it. You could write several paragraphs comparing and contrasting the 1 megapixel camera you inherited with the office with the 6 megapixel camera (with 4× optical zoom lens) that you'd like to have, but you decide displaying the specifications in a simple table would be a lot easier on your boss. And to emphasize the substantial difference in image quality, you choose two pictorial displays that dramatically illustrate the image quality possible on your existing equipment versus the crisp images that the new digital camera would give you. Side by side on the page, the two pictures build a persuasive case: you could never have said it better.

CHAPTER ELEVEN

Creating Tables and Figures

In communicating technical information, you will often need to use illustrations either in addition to words or instead of words to convey your message. Learning how to create effective tables, graphs, diagrams, and drawings will give you additional tools for helping readers to understand your subject.

As in all forms of technical communication, the use of illustrations has ethical implications. In every diagram and drawing, every table and graph, you have a moral responsibility to offer a clear and correct impression of your subject.

CHOOSING ILLUSTRATIONS

How do you determine whether illustrations are really desirable? The answer: consider your rhetorical situation, including both purpose and audience.

Consider Your Purpose

- *What do you want to have happen as a result of the document?*
- *What do you want the reader to do or think after reading the document?*
- *How will illustrations help you to achieve your objective?*

For example, Victor Crowell, the owner of Crowell Heating & Cooling in Rising Star, New Mexico, is writing a recommendation report regarding the installation of new air-conditioning equipment for a local business. He wants the potential client to approve this recommendation quickly, to clinch the sale. Victor decides that illustrations will dramatize the benefits of the new equipment. To help the prospective client to envision the projected installation and to reinforce his own track record with such installations, Crowell uses both a photograph and a bar graph showing the high cooling costs with the current system versus the lower costs projected for the new equipment. Victor believes the two illustrations will focus attention on the persuasive evidence.

Erica Vasquez of Rapid Computers has a different challenge. She is writing a set of instructions for customers on how to install a new graphics accelerator in their computers. She wants such installations to proceed without incident, reinforcing the image of Rapid Computers as easy to operate and upgrade. She wants people to read the instructions and feel confident that they can perform the installations; she doesn't want users to get stuck in the middle of the procedure and, in frustration, call Rapid Computers for technical support. Erica believes that a flowchart, outlining the steps in the procedure, will give customers a comforting overview. She intends to include lots of drawings to show users exactly how to open the case, complete the installation, and close the case.

Consider Your Audience

What are your readers likely to expect? If they are accustomed to receiving illustrated reports, you will not want to risk disappointing and distracting them by failing to deliver. If your audience doesn't expect illustrations, you will need to gauge carefully the positive or negative impact of including graphics. For example, readers of financial reports expect to see tables of sales figures, profits and losses, stockholders' equity, and long-term debt. If your financial report doesn't display this information in tables, your readers probably will be confused and irritated. On the other hand, Web users expect to see graphic art (including animated illustrations). A site that is all paragraphs is considered primitive and uninviting. Readers of résumés, however, still expect words on a page and might interpret illustrations as a distraction to mask a weak set of credentials.

In addition to the expectations of your audience, consider their background, attitude, and environment.

- If your readers have little or no knowledge of your subject, illustrations may create a virtual experience by showing what the subject looks like, summarizing its specifications, highlighting its components, revealing its organization, or displaying its operation.
- If your readers are limited in their ability to read English, illustrations may help them circumvent the difficulties of deciphering words on a page.
- If your readers aren't motivated to spend time on your document, illustrations that make it look more inviting and accessible may encourage them to read it.
- If your audience has little time to read, illustrations may assist by capsulizing information.
- For readers in a distracting environment, illustrations that display information in an eye-catching and engaging manner may serve to focus their attention.

For example, Victor Crowell believes the prospective client will be nicely surprised by the two illustrations. He has noticed that in conversations, the client is often impatient with long explanations and likes to go straight to the point. Instead of offering a boring, all-prose recommendation report, therefore, Victor integrates illustrations that focus on key issues and expedite the job of reading.

Similarly, Erica Vasquez knows that most of her readers have never before installed a graphics accelerator. She knows that to motivate users to read the instructions and try this installation, she must make the job look as easy as possible. And instead of using words alone, thus requiring readers to picture objects and actions for themselves, Erica includes a series of drawings that show users what they need to do and what the work should look like at every step along the way. For the minority of customers who are experienced in such installations, Erica believes a flowchart outlining the steps will provide sufficient guidance. And for the inexperienced majority, the flowchart will make the whole procedure look coherent and systematic.

Having decided to use illustrations, how do you decide which illustrations to use? Once more, you will have to consider your rhetorical situation, chiefly your audience and purpose.

Consider Your Audience Again

Is your audience familiar with the kind of illustration contemplated? Can they interpret the illustration? If your audience is unfamiliar with circle graphs, for example, information displayed in this manner would be ineffective without considerable verbal explanation. If many words are necessary to explain how to interpret one picture, however, you are probably using the wrong kind of illustration for that audience. If your audience is international, you must also be sensitive to cultural variations in illustration interpretations. For example, the convention of using bubbles in drawings to depict what people are thinking (see Figure 11-1) will be readily understood in some cultures but can cause

FIGURE 11-1 · DEPICTION OF THINKING
This drawing is from a research report prepared by the U.S. Department of Justice. Aimed at American audiences, the drawing communicates effectively by using a readily understood convention of bubbles to depict that the school administrator is considering different security techniques. For a person unfamiliar with the practice of using bubbles to imply thought, however, the drawing is almost impossible to decipher.
Source: National Institute of Justice. *The Appropriate and Effective Use of Security Technologies in U.S. Schools: A Guide for Schools and Law Enforcement Agencies.* NCJ178265. Washington, DC: 1999, p. 2.

FIGURE 11-2 · INTERNATIONAL VARIATIONS IN AUTOMOBILE LICENSE PLATES
Automobile license plates differ considerably from one nation to the next in size, shape, and color as well as in the information displayed. With no universal standard for the design, you must adapt drawings and diagrams according to the national origins of your audience.
Source: http://www.worldlicenseplates.com/ July 15, 2003.

confusion in others. International variations in the design of objects such as trash containers, mailboxes, or automobile license plates also can lead to misinterpretation of pictures displaying such objects (see Figure 11-2).

Consider Your Purpose Again

- *If the purpose of your document is to summarize information and substantiate a claim, use a table.*
- *If you wish to exhibit or emphasize information, use a figure.*

Tables are rows and columns of numbers and words that offer readers the details of a subject. If you want your readers to remember or retrieve specific pieces of information, tables will serve your purpose.

For example, Tom Skiludi of National Oil & Gas, headquartered in Pittsburgh, is reporting to the northeast regional supervisor on the year's operations. He knows the supervisor will pass the figures on to the company's chief financial officer. Tom chooses a table as the efficient way to display the information (see Table Exhibit 11-1). Figures come in a wide variety and dramatize different kinds of information (see Table 11-1).

To emphasize or reinforce the meaning of data, start with a table and then use a figure to spotlight the implications of the data. For example, in reporting on the year's operations, Tom considered several possible illustrations, including a line

National Oil & Gas, Northeast Region Operation

	2004	2003
Gas throughput (billion cubic feet)		
Distribution	453.5	427.6
Transmission	659.7	640.2
Gas production (billion cubic feet)	224.5	263.8
Oil production (thousand barrels)	3458.8	3144.9

TABLE EXHIBIT 11-1
This is a simple and clean table, offering specific numerical information. Notice that the words are left aligned while the numbers are aligned on the decimal. White space is used (instead of thin rules) to separate the rows and columns of information.

Table 11-1 ▮ The Purposes of Figures

Illustration	Image	Purpose	Example
Line graph		To show the degree and direction of change relative to two variables	Sales figures over a five-year period
Bar graph		To compare and contrast two or more subjects at the same point in time.	The populations of major cities according to the 2000 census
Column graph		To reveal change in a subject at regular intervals of time	The number of registered voters in your city during the last five elections
Circle graph		To display the number and relative size of the divisions of a subject	The distribution of the population by major

(continued)

Table 11-1 ■ The Purposes of Figures (*continued*)

Illustration	Image	Purpose	Example
Flowchart		To show the sequence of steps in a process or procedure	The process for installing a computer software application
Organizational chart		To map the various divisions and levels of responsibility within an organization	The hierarchy of military officers in the United States Air Force
Diagram		To identify the parts of a subject and their spatial relationship	The rooms of a building
Drawing	 ON/OFF	To exhibit selected features of an object or process	The on/off button on a machine or the direction to turn lever
Photograph		To show what a subject looks like in realistic detail	The scene of a crime

graph to dramatize the company's rising oil production and a pair of circle graphs to highlight the stability in the relative proportions of gas distribution and transmission. He decided, however, on a bar graph because it would readily focus attention on a potentially ominous piece of information: while gas transmission and distribution both increased in 2004, gas production decreased (see Figure 11-3).

Ordinarily, you will use a table to arrange your data for easier analysis and interpretation. Graphics programs will then allow you to select a dramatic way of displaying the data.

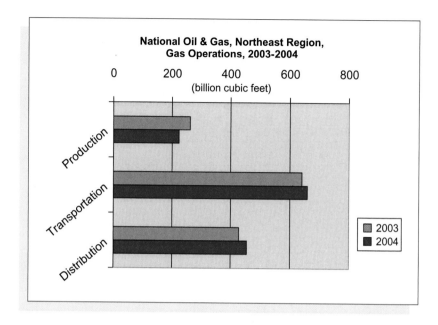

FIGURE 11-3 · BAR GRAPH
Notice that although the 2004 bars exceed the 2003 bars for both Transportation and Distribution, the reverse is the case for Production, which is located prominently at the top of the figure. In addition, the vertical grid lines make it easier to decipher the approximate numerical values depicted by each bar.

Often a variety of illustrations will serve your purpose. Consider, for example, the tables and figures in the U.S. Environmental Protection Agency's investigation report of a chemical accident at a manufacturing facility in Elkhart, Indiana. On page i of this 41-page report, the purpose and audience are identified:

> A key objective of the EPA chemical accident investigation program is to determine and report to the public the facts, conditions, circumstances, and causes or probable causes of chemical accidents that resulted, or could have resulted, in a fatality, serious injury, substantial property damage, or serious off-site impact, including a large scale evacuation of the general public. The ultimate goal of the accident investigation is to determine the root causes in order to reduce the likelihood of recurrence, to minimize the consequences associated with accidental releases, and to make chemical production, processing, handling, and storage safer. This report is a result of an EPA investigation to describe the accident, determine root causes and contributing factors, and identify findings and recommendations.

The report describes the explosion and fire that occurred at the plant on June 24, 1997, while pressurized containers were being filled with ethylene oxide

Physical Properties	Ethylene Oxide
Boiling point	51°F
Liquid density at 68°F	7.25 lb/gal
Flash point (tag open cup)	<0°F
Autoignition temperature	804°F
Lower explosive limit (LEL)	3%
Upper explosive limit (UEL)	100%

TABLE EXHIBIT 11-2
This simple two-column table is easy to read with its bold headings and wide cells. The horizontal lines dividing the rows are unnecessary, and a sans serif type would offer a cleaner display, but readability is satisfactory nevertheless because the text in each cell is quite sparse. The three rows reporting temperatures might be grouped (instead of separated), and the text in the right column might all be right aligned (instead of centered); both changes would promote easy and accurate reading of the three different temperatures and the two different percentages.
Source: U.S. Environmental Protection Agency, Office of Solid Waste and Emergency Response. EPA Chemical Accident Investigation Report, Accra Pac Group, Inc. North Plant, Elkhart, Indiana. EPA 550-R-00-001. Washington, DC: 2000, p. 1.

(EtO, a sterilizer used in hospitals). One employee was killed and 59 employees and firefighters required medical attention.

Among the illustrations in the report are the following:

■ A table identifying the properties of ethylene oxide (see Table Exhibit 11-2)
■ A map of the site of the accident (see Figure 11-4)
■ A series of flowcharts explaining the process of filling pressurized containers (see Figure 11-5)
■ A diagram displaying causes in relation to effects (see Figure 11-6)
■ A series of photographs revealing the damage to the equipment and buildings (see Figure 11-7)

Without the illustrations, the EPA investigators could not have achieved their purpose or communicated effectively with their audience.

Notice, however, the absence of pictures of human beings. There are no drawings of people making mistakes in their administration of the facility or the operation of the equipment and no photographs of the person killed or the many employees injured in the accident. (The fatality is identified on Figure 11-4, the map of the accident site, with a simple X instead of the icon of a human being.) These omissions diminish the importance of the people at this facility, both as contributors to the accident and as its victims. By focusing the viewer's attention on buildings and machines, the illustrations mask the human dimension of the accident.

Accra Pac Accident Site

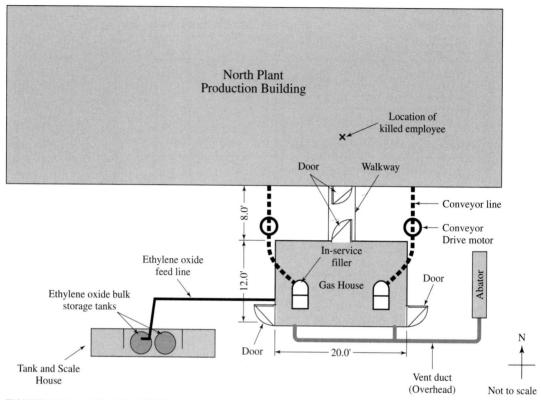

FIGURE 11-4 · MAP OF ACCIDENT SITE

This map of the accident site includes detailed labeling of all pertinent buildings and equipment. The labels are placed inside the buildings whenever possible and are usually horizontal for easier reading. A directional arrow is given to indicate the geographical orientation of the map. A note below the arrow specifies that the map is a pictorial approximation of the relative positions of the buildings instead of a precise scale drawing. Notice the callous labeling of the human fatality, who is denied the dignity of a human icon (much less a portrait photograph) and identified only as "killed employee."

Source: U.S. Environmental Protection Agency, Office of Solid Waste and Emergency Response. *EPA Chemical Accident Investigation Report, Accra Pac Group, Inc. North Plant, Elkhart, Indiana.* EPA 550-00-001. Washington, DC: 2000, p. 4.

As the illustrations in this report demonstrate, you have a variety of ways to help your readers perceive your meaning. In choosing illustrations, always consider your purpose and your audience. Consider illustrations while you are planning your document, but even as you are writing and revising, continue to look for ways to enhance readers' ability to visualize your message.

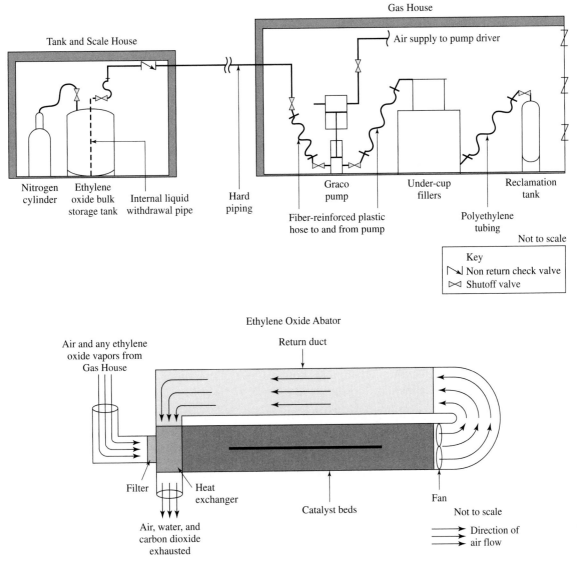

Ethylene Oxide: Simplified Filling Configuration

Gas House

Tank and Scale House

Air supply to pump driver

Key
Non return check valve
Shutoff valve

Nitrogen cylinder | Ethylene oxide bulk storage tank | Internal liquid withdrawal pipe | Hard piping | Graco pump | Under-cup fillers | Reclamation tank

Fiber-reinforced plastic hose to and from pump

Polyethylene tubing

Not to scale

Ethylene Oxide Abator

Return duct

Air and any ethylene oxide vapors from Gas House

Filter | Heat exchanger | Catalyst beds | Fan | Not to scale

Air, water, and carbon dioxide exhausted

Direction of air flow

FIGURE 11-5 · DIAGRAM OF EQUIPMENT OPERATIONS
This series of diagrams offers a simplified depiction of the operation of the equipment that is the subject of the accident investigation. It allows viewers to see the process as it is supposed to occur so that the sources of failure can be readily perceived. Notice that the labels are always horizontal, for ease in reading. The use of identical arrows both to identify components and to indicate direction, however, could cause confusion. The labels of components could be linked to their respective locations with simple lines.
Source: U.S. Environmental Protection Agency, Office of Solid Waste and Emergency Response. *EPA Chemical Accident Investigation Report, Accra Pac Group, Inc. North Plant, Elkhart, Indiana.* EPA 550-R-00-001. Washington, DC: 2000, pp. 6–8.

Gas House Ethylene Oxide Vapor Recovery and Disposal System

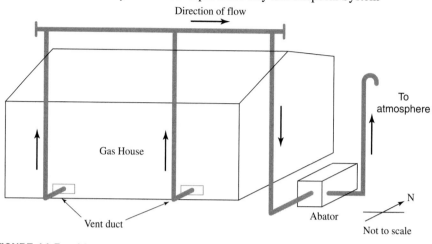

FIGURE 11-5 · *CONTINUED*

CREATING ILLUSTRATIONS

In deciding when and how to use illustrations, remember the following guidelines:

■ **Simplify your illustrations.** Keep your illustrations as simple as possible so that readers have no difficulty understanding your message. Avoid distracting readers with unnecessary details or decorative flourishes.

■ **Use computer applications critically.** Computer graphics software and clip art allow you to include all manner of illustrations. To ensure the effectiveness of your tables and figures, however, you must choose carefully. Graphics software, for example, might create artistic but misleading graphs, and some clip art has a cartoonish style that isn't serious enough or detailed enough to do justice to all subjects. It is your job to choose illustrations that display your information clearly and correctly.

■ **Choose illustrations carefully.** Realize that you usually have a choice of illustrations. If a table or figure does not convey your point quickly and clearly, look for other ways of displaying the information visually.

■ **Title your illustrations.** Give each table and figure a title that clearly indicates the content of the display.

■ **Number your illustrations.** If you use several illustrations in your report, number them. Number the tables and figures separately (e.g., Table 1, Table 2, Figure 1, Figure 2). Place the number and title above a table and below a figure.

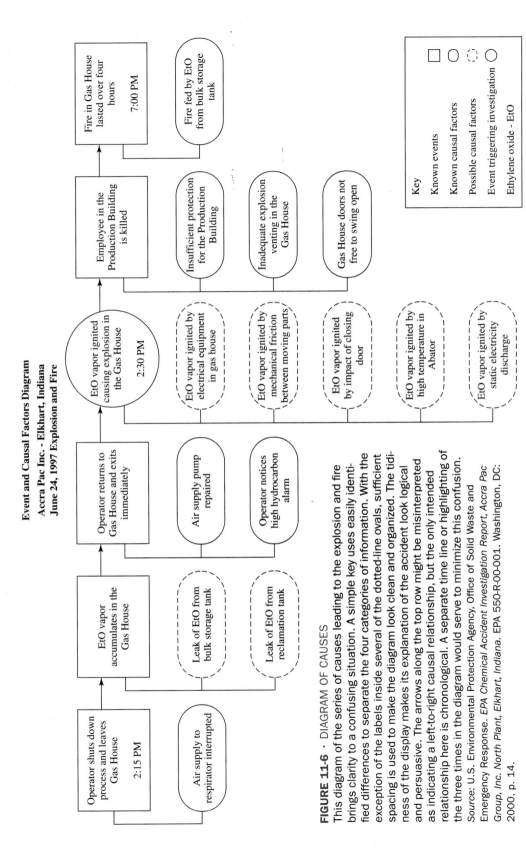

Event and Causal Factors Diagram

**Accra Pac Inc. - Elkhart, Indiana
June 24, 1997 Explosion and Fire**

FIGURE 11-6 · DIAGRAM OF CAUSES

This diagram of the series of causes leading to the explosion and fire brings clarity to a confusing situation. A simple key uses easily identified differences to separate the four categories of information. With the exception of the labels inside several of the dotted-line ovals, sufficient spacing is used to make the diagram look clean and organized. The tidiness of the display makes its explanation of the accident look logical and persuasive. The arrows along the top row might be misinterpreted as indicating a left-to-right causal relationship, but the only intended relationship here is chronological. A separate time line or highlighting of the three times in the diagram would serve to minimize this confusion.

Source: U.S. Environmental Protection Agency, Office of Solid Waste and Emergency Response. *EPA Chemical Accident Investigation Report, Accra Pac Group, Inc. North Plant, Elkhart, Indiana.* EPA 550-R-00-001. Washington, DC: 2000, p. 14.

FIGURE 11-7 · PHOTOGRAPHS OF DAMAGED BUILDING AND EQUIPMENT
This series of photographs reveals the damage to building and equipment in full color, offering viewers a vivid and realistic depiction—a "you are there" experience at the site of the explosion. No drawings would have quite the dramatic impact that the photographs do. Notice the careful labeling of objects in each photograph that serves to direct and focus the viewer's attention.
Source: United States Environmental Protection Agency, Office of Solid Waste and Emergency Response. *EPA Chemical Accident Investigation Report, Accra Pac Group, Inc. North Plant, Elkhart, Indiana.* EPA 550-R-00-001. Washington, DC: 2000, p. 4.

FIGURE 11-7 · *CONTINUED*

FIGURE 11-7 · *CONTINUED*

[*Note:* To avoid the confusion of two numbers for the same displayed item, we have deleted from the table exhibits and example figures the original identifying numbers used in their respective sources.]

■ **Alert your readers.** Always alert your readers to illustrations by referring to them in the text. Every time you refer to the illustration, use the table or figure number (not "the table above" or "the preceding figure").

■ **Position your illustrations strategically.** Each illustration should appear as close to the related passage as possible. Announce the table or figure—what it is or shows—then insert the illustration, and add any explanation your reader will need to fully understand the material. Don't lead readers through a complicated explanation and only afterward mention the illustration. Send them to the illustration immediately; then they can shift back and forth between the explanation and the illustration as necessary.

■ **Consider size and cost.** Calculate the impact of illustrations on the expected length of your document. Illustrations will often increase the size and add to the cost of production and distribution.

■ **Identify your sources.** If you borrow or adapt a table or figure from another source, identify that source below the illustration.

Designing Tables

■ Every column in a table should have a heading that identifies the information it contains. In a table of numbers, include the unit of measurement, such as "miles per hour." For large numbers, add a designation such as "in thousands" or "in millions" to the column heading (and delete the corresponding zeros from the numeric data). Headings should be brief; any additional information can be provided in a footnote to the table. Use superscripts (lowercase letters or numbers), or symbols (e.g., * or †), to indicate footnoted material.

■ If possible, box your table to separate it from surrounding paragraphs.

■ Keep tables as simple as possible. Include only data relevant to your purpose.

■ Consider omitting rules between rows and columns. If white space can be used to separate these units of information, your table will have a less crowded appearance.

Table Exhibits 11-3–11-5 show three ways of presenting tabular data.

Designing Bar and Column Graphs

■ Avoid putting excessive information on a bar or column graph and thereby hindering the reader's ability to decipher it. Consider using a separate graph to communicate each point.

■ Be sure to label the x-axis and the y-axis—state what each measures and the units in which each is calibrated. Readers can't understand your graph if they don't know what it measures or how the results are expressed.

Ranking of State Pedestrian Fatality Rates from All Crashes in 2001				
Rank	State	Pedestrians Killed	Population (Thousands)	Pedestrian Fatality Rate per 100,000 Population
1	New Mexico	72	1,829	3.94
2	Arizona	159	5,307	3.00
3	Florida	489	16,397	2.98
4	South Carolina	108	4,063	2.66
5	Hawaii	30	1,224	2.45
6	Louisiana	98	4,465	2.19
7	Nevada	45	2,106	2.14
8	Delaware	17	796	2.14
9	Texas	449	21,325	2.11
10	Mississippi	59	2,858	2.06
Source: NCSA, NHTSA, Traffic Safety Facts 2001, Table 113				

TABLE EXHIBIT 11-3

This table is effectively designed with clear headings for the entire table as well as each of the rows. Note how bold type and a light background color work together to differentiate the headings in the table from the data. Note also that words are aligned on the left and numbers on the right (or on the decimal point) to allow the easiest possible reading, navigation, and comparison of information. The only exception is the numbers used to designate the ranking, but misreading here is unlikely.

Source: U.S. Department of Transportation, National Highway Traffic Safety Administration. *Pedestrian Roadway Fatalities.* DOT HS 809 456. Washington, DC: 2003, p. 30.

Large Truck Fatalities by Fatality Type and Day vs Night 1996–2000					
Combination Trucks: Day vs Night					
Fatality Type	Day	Night	Total	Day (%)	Night (%)
Occupant					
Single-vehicle	1,000	790	1,790	56%	44%
Multiple-vehicle	11,097	5,969	17,066	65%	35%
Total	12,097	6,759	18,856	64%	36%
Nonoccupant	757	674	1,431	53%	47%
Single-unit Trucks: Day vs Night					
Fatality Type	Day	Night	Total	Day (%)	Night (%)
Occupant					
Single-vehicle	473	169	642	81%	19%
Multiple-vehicle	4,479	917	5,396	83%	17%
Total	4,952	1,086	6,038	82%	18%
Nonoccupant	545	125	670	74%	26%
Source: NHTSA, NCSA, FARS.					

TABLE EXHIBIT 11-4

This table is a mess of crisscrossing lines of various weights, all of which make reading difficult. Bold type is used for first- and third-level headings, but second-level headings and numerical data are displayed in the same style of type. It's also a table with excessive information, reporting both the number and percentage of day and night accidents involving two different kinds of fatality and two kinds of truck. At least two separate tables would make for easier reading: one for the numbers and one for the percentages. As is, the column of totals occurs in the center of the table instead of on the right as we might ordinarily expect it to.

Source: U.S. Department of Transportation, National Highway Traffic Safety Administration. *An Analysis of Fatal Large Truck Crashes.* DOT HS 809 569. Washington, DC: 2003, p. 30.

Recommendations for reporting results of testing for antibody to hepatitis C virus (anti-HCV) by type of reflex supplemental testing performed

Anti-HCV screening test results	Supplemental test results	Interpretation	Comments
Screening-test negative*	Not applicable	Anti-HCV negative	Not infected with HCV, unless recent infection is suspected or other evidence exists to indicate HCV infection
Screening-test positive* with high signal-to-cut off (s/co) ratio	Not done	Anti-HCV positive	Probably indicates past or present HCV infection; supplemental serologic testing not performed. Samples with high s/co ratios usually ($\geq$95%) confirm positive, but <5 of every 100 might represent false-positives; more specific testing can be requested, if indicated
Screening-test positive	Recombinant immunoblot assay (RIBA[fi])-positive	Anti-HCV positive	Indicates past or present HCV infection
Screening-test positive	RIBA-negative	Anti-HCV negative	Not infected with HCV, unless recent infection is suspected or other evidence exists to indicate HCV infection
Screening-test positive	RIBA-indeterminate	Anti-HCV indeterminate	HCV antibody and infection status cannot be determined; another sample should be collected for repeat anti-HCV testing (>1 month) or for HCV RNA testing
Screening-test positive	Nucleic acid test (NAT)-positive	Anti-HCV positive, HCV RNA-positive	Indicates active HCV infection
Screening-test positive	NAT-negative RIBA-positive	Anti-HCV positive, HCV RNA-negative	The presence of anti-HCV indicates past or present HCV infection; a single negative HCV RNA result does not rule out active infection
Screening-test positive	NAT-negative RIBA-negative	Anti-HCV negative, HCV RNA-negative	Not infected with HCV
Screening-test positive	NAT-negative RIBA-indeterminate	Anti-HCV-indeterminate HCV RNA-negative	Screening test anti-HCV result probably a false-positive, which indicates no HCV infection

* Screening immunoassay test results interpreted as negative or positive on the basis of criteria provided by the manufacturer.

TABLE EXHIBIT 11-5

This table is easy to read with effective spacing to separate the cells of information. The sans serif typeface also reinforces the clean and clear display. Nevertheless the table could be more effective. The unnecessary repetition of "Screening-test-positive" in the left column keeps readers from quickly grasping that the table is reporting on only three conditions: positive test, positive test with qualification, and negative test. Thin rules across the table separating the three conditions would also help to make this point immediately clear.

Source: U.S. Centers for Disease Control and Prevention. Guidelines for Laboratory Testing and Result Reporting of Antibody to Hepatitis C Virus. *Morbidity and Mortality Weekly Report* 52 (No. RR-3): 2003, p. 11.

■ For bar graphs, start the *x*-axis at zero and equally space the intervals to avoid distorting the length of the bars. For column graphs, start the *y*-axis at zero and equally space the intervals to avoid distorting the height of the columns.

■ Color can enhance the effect of a graph, but excessive color can reduce comprehension and distort information. Use the same color for all bars or columns that are representing the same items. Avoid using color simply for decorative effect.

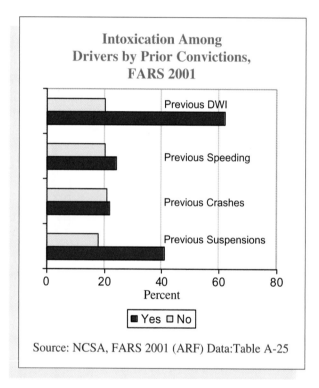

FIGURE 11-8 · BAR GRAPH

This bar graph makes a clear and vivid point: drivers involved in fatal automobile accidents who were intoxicated are more likely to have a history of driving convictions. The bright blue bars used for intoxicated drivers seem to jump off the graph, whereas pale blue bars identify drivers who were not intoxicated. The four categories are also crisply labeled for easy reading and understanding. Nevertheless this bar graph could be better. The title of the figure ought to mention "drivers in fatal automobile accidents" specifically to clarify the scope of the data. Instead of a yes/no legend, the two colors might be identified as intoxicated/not intoxicated. The order of the categories of prior convictions is also arbitrary: a more dramatic, logical, and coherent display would position the bars longest to shortest from top to bottom.

Source: U.S. Department of Transportation, National Highway Traffic Safety Administration. *Alcohol Involvement in Fatal Crashes 2001.* DOT HS 809 579. Washington, DC: 2003, p. 17

■ Make the graph accurate. Computer graphics allow a tremendous range of special effects, but artistic graphs are not always effective or accurate. Three-dimensional bars and columns are often deceptive because readers have difficulty visually comparing the relative lengths of the bars and heights of the columns.

■ Try to write captions or labels on or near the bars. Avoid legends (or keys) that slow reader comprehension. When bars are divided into a great many divisions that cannot be interpreted without consulting a legend, readers are likely to become confused. However, colored bars can be difficult to label: even black text is hard to read on all but the lightest colors. For that reason, many

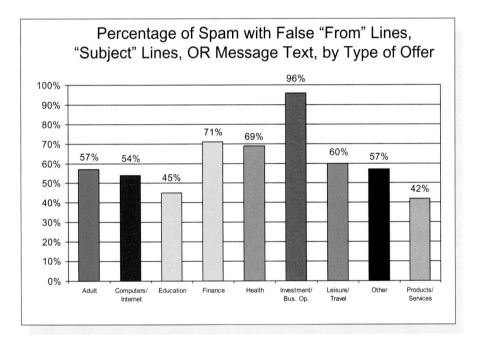

FIGURE 11-9 · COLUMN GRAPH
This column graph would be better designed as a bar graph (i.e., with horizontal bars instead of vertical columns). The series of vertical columns aligned on the x-axis from left to right might be initially perceived as a chronological progression—the same subject displayed at consecutive points in time. In fact, the figure displays different subjects at the same points in time. Displaying the information as a bar graph would avoid any initial misunderstanding and speed the viewer's access to the information. The arbitrary use of color is also distracting, making you pay attention to color instead of quantity. The two-dimensional columns keep the display simple, and the labeling of each column is clear and specific; but the ordering of the columns is alphabetical instead of quantitative (from largest to smallest or vice versa).
Source: U.S. Federal Trade Commission. *False Claims in Spam: A Report by the FTC's Division of Marketing Practices*. Washington, DC: 2003, p. 10.

graphics software programs offer you the use of legends. In short, legends are fine for a maximum of four segments or bars—and if the legend is close to the bars.

▪ For divided bar and column graphs with extensive divisions, distinguish divisions by means of color or shading, not cross-hatching patterns. Cross-hatching often creates distracting optical effects.

▪ Avoid crowding the bars or columns in a graph. Such visual clutter makes a graph difficult to interpret. Using three-dimensional bars or columns will also

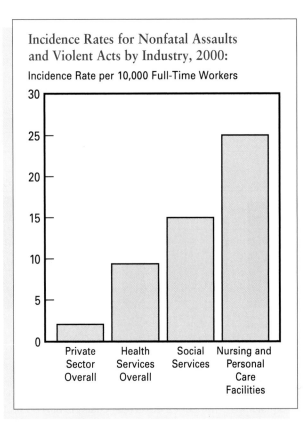

FIGURE 11-10 · COLUMN GRAPH
This column graph is clean and concise. The blue columns are in clear contrast to the white background. The columns also differ enough in size from each other to make the point of the graph easy to decipher (i.e., workers in the health care and social service fields run a substantially higher risk of assault). Numerical labeling of each column and gridlines in the background, as a consequence, are unnecessary.
Source: U.S. Department of Labor, Occupational Safety and Health Administration. *Guidelines for Preventing Workplace Violence for Health-Care and Social-Service Workers.* OSHA 3148. Washington, DC: 2003, p. 1.

reduce the number of categories that can fit in a given space. Effective and inviting graphs leave generous space between the bars or columns.

See Figures 11-8 to 11-11 for examples of column and bar graphs.

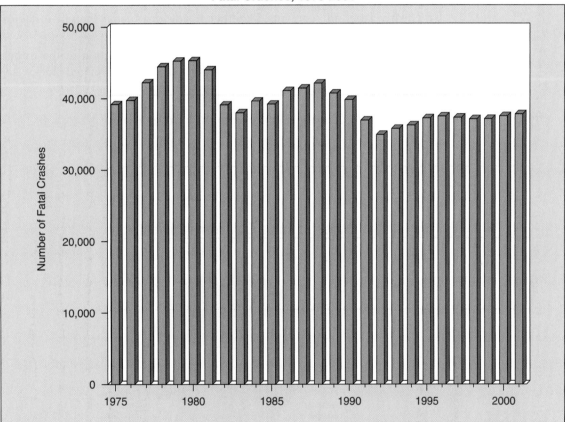

FIGURE 11-11 · COLUMN GRAPH

This graph clearly shows why three-dimensional columns are undesirable. The unnecessary third dimension on each column makes the entire graph look crowded and difficult to decipher. The visual effect is distracting and almost annoying as your eyes continually shift their focus from the light gray and dark gray sides of each column to the narrow and flickering white space between the columns. This effect is exaggerated by the height of the columns. It would be easier for readers to eliminate the redundant information from 0 to 30,000 and break the y-axis (indicating the break with two diagonal slashes through the y-axis just above the zero point). The labeling of the axes is easy to read, but note the inhumanity of the display: only the title alludes to what is being represented—over a million dead human beings.

Source: U.S. Department of Transportation, National Highway Traffic Safety Administration. *Traffic Safety Facts 2001.* DOT HS 809 484. Washington, DC: 2002, p. 14.

Designing Circle Graphs (Pie Charts)

▪ Restrict the number of segments in a circle graph to seven or eight. There is simply a limit to the number of segments into which a circle can be divided before appreciation of the relative sizes is jeopardized. If necessary, create a second circle graph that combines several smaller segments.

▪ Use shading or color to differentiate segments and make them easier to see.

▪ Avoid using three-dimensional circle graphs because they can distort the apparent size of the segments.

▪ Clearly label all segments. Whether inside or outside the circle, labels should run horizontally for easier reading.

▪ As you segment the graph, begin with the largest section in the upper right-hand quadrant. The remaining segments should be arranged clockwise, in descending order.

See Figures 11-12 and 11-13 for examples of circle graphs.

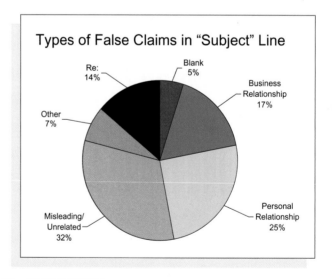

FIGURE 11-12 · CIRCLE GRAPH

This circle graph uses clear labels for all six sections. Note, however, that instead of starting in the upper right quadrant with the biggest piece and progressing clockwise to the smallest piece for easiest reading, the sections are arbitrarily ordered and colored. Information is not displayed in a logical sequence and color isn't used systematically but is merely distracting. Notice, for example, how your eyes go right to the black section of the circle even though that section deserves no special priority on your attention.
Source: U.S. Federal Trade Commission. *False Claims in Spam: A Report by the FTC's Division of Marketing Practices.* Washington, DC: 2003, p. 6.

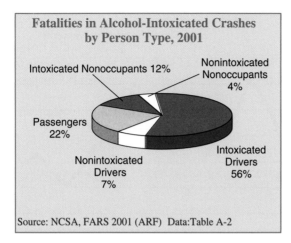

FIGURE 11-13 · CIRCLE GRAPH

While this circle graph is nicely labeled, the sections offer a distracting third dimension. Instead of matching the size of the segment to the lightness or darkness of coloring (i.e., the larger the segment, the lighter or darker the color), colors are assigned arbitrarily (and, inexplicably, blue and white are used twice). The graph starts with the largest segment in the upper right quadrant, but thereafter the rationale of the ordering collapses, leaving it to the viewer to sort the segments by size. Visually and logically, the graph is much more complicated than it needs to be.

Source: U.S. Department of Transportation, National Highway Traffic Safety Administration. *Alcohol Involvement in Fatal Crashes 2001.* DOT HS 809 579. Washington, DC: 2003, p. 11.

Designing Line Graphs

■ Label each axis clearly. Like bar and column graphs, line graphs must have clearly labeled scales to show the variables being measured. Ordinarily, the independent variable is placed on the horizontal *x*-axis, and the dependent variables are placed on the vertical *y*-axis (and, in a three-dimensional graph, on the diagonal *z*-axis).

■ Choose the scale of each axis to show the appropriate steepness of the slope of the line. Typically, the scales start at zero, with the intervals equally spaced on each axis.

■ In designing line graphs, be sure to choose the spacing for each axis so that the steepness (slope) of the line accurately measures the actual trend suggested by the data. Computer graphics will allow you to adjust the intervals on the *x*- and *y*-axes, but your job is to decide whether the slope of the graph accurately depicts your data or gives a distorted impression.

■ Avoid using more than three data lines on one graph unless they are spaced apart and do not overlap. Graphs with several intersecting data lines are usually difficult to interpret.

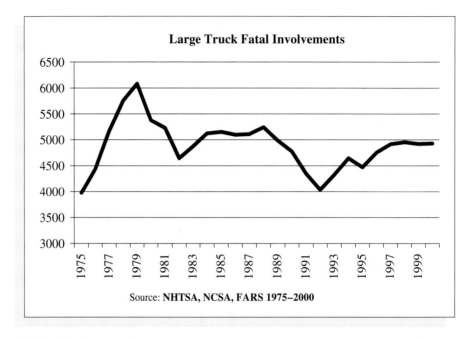

FIGURE 11-14 · LINE GRAPH

This is a simple, clean line graph. Both scales are readily interpreted, and the single data line is plainly visible. Notice, however, that the y-axis starts at 3000 instead of 0. The labeling of the x-axis would be easier to read if it were on a 45- or 60-degree angle instead of a 90-degree angle.

Source: U.S. Department of Transportation, National Highway Traffic Safety Administration. *An Analysis of Fatal Large Truck Crashes.* DOT HS 809 569. Washington, DC: 2003, p. 4.

■ Keep the data lines on your graph distinctive by using different colors or styles for each line.

■ If possible, label each data line. Avoid legends (or keys) that slow reader comprehension.

See Figures 11-14 to 11-16 for examples of line graphs.

Designing Flowcharts

■ Make the flowchart as simple as the process itself. If a process is simple, the flowchart should progress in a single direction, usually top to bottom or left to right. Complicated designs that spiral and zigzag imply a more complicated process.

■ Use the same shape in the same size for all equivalent steps or phases, but different shapes for steps or phases of a different kind (e.g., circles for the stages

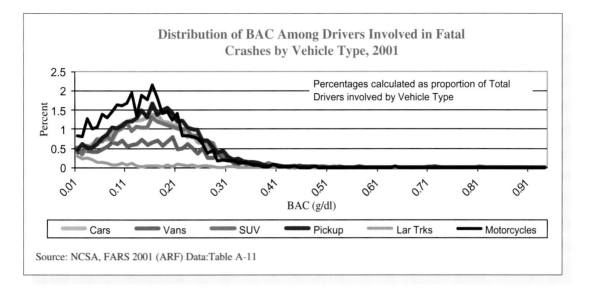

FIGURE 11-15 · LINE GRAPH

This line graph uses color coding to differentiate its six data lines. Given that the data lines intersect at multiple points, the color is essential for easy reading. Notice that black—offering the highest contrast and thus greatest visibility—is used to denote the most significant data line. The primary color of blue is used for the data line that is second in importance, but red is used for fifth instead of third. The abbreviation of BAC in the label for the x-axis is appropriate, but might be spelled out (blood alcohol concentration) in the title of the figure, especially for readers who are skimming the report.

Source: U.S. Department of Transportation, National Highway Traffic Safety Administration. *Alcohol Involvement in Fatal Crashes 2001.* DOT HS 809 579. Washington, DC: 2003, p. 9.

in researching a document, squares for the stages of writing, and diamonds for the production stages of printing and binding).

■ Label each of the steps or phases.

■ If space allows, put the labels directly on the part; if not, attach the label to the subject with thin rules. Avoid using arrows for the purpose of labeling because they imply action.

■ Use a clean sans serif type for the labels to maximize legibility.

■ Position all the labels on the horizontal so that the viewer doesn't have to rotate the page or screen to read the labels.

■ Connect each step or phase in the sequence to the next step or phases with clear directional arrows.

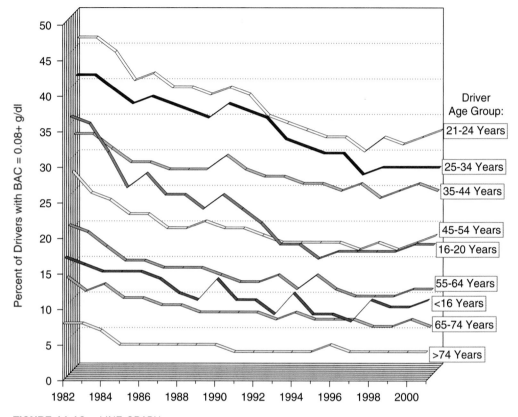

FIGURE 11-16 · LINE GRAPH

This line graph uses a decorative and distracting third dimension that serves only to make the graph difficult to read. Instead of using a complicated legend for its nine different data lines, it labels each line but positions the labels on the right side of the graph. This forces people to read each data line from right to left instead of left to right. The labels might be more useful if made smaller and positioned closer to the left side of the graph.

Source: U.S. Department of Transportation, National Highway Traffic Safety Administration. *Traffic Safety Facts 2001.* DOT HS 809 484. Washington, DC: 2002, p. 37.

Laboratory Algorithm for Antibody to Hepatitis C Virus (anti-HCV) Testing and Result Reporting

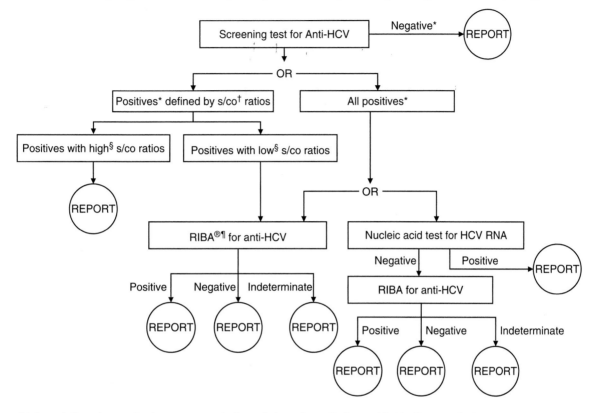

* Interpretation of screening immunoassay test results based on criteria provided by the manufacturer.

† Signal-to-cutoff ratio (s/co).

§ Screening-test–positive results are classified as having high s/co ratios if their ratios are at or above a predetermined value that predicts a supplemental-test–positive result ≥95% of the time among all populations tested; screening-test–positive results are classified as having low s/co ratios if their ratios are below this value.

¶ Recombinant immunoblot assay.

FIGURE 11-17 · FLOWCHART

This flowchart uses rectangles for the testing and circles for the resulting report. Notice also the easy-to-understand top-to-bottom direction of the illustration. Each stage is given a legible label in a sans serif type. Uppercase letters are appropriately used to highlight the two occurrences of "OR" that identify alternative stages, but the uppercase letters are quite unnecessary for labeling the already distinctive circles.

Source: U.S. Centers for Disease Control and Prevention. Guidelines for Laboratory Testing and Result Reporting of Antibody to Hepatitis C Virus. *Morbidity and Mortality Weekly Report* 52 (No. RR-3): 2003, p. 9.

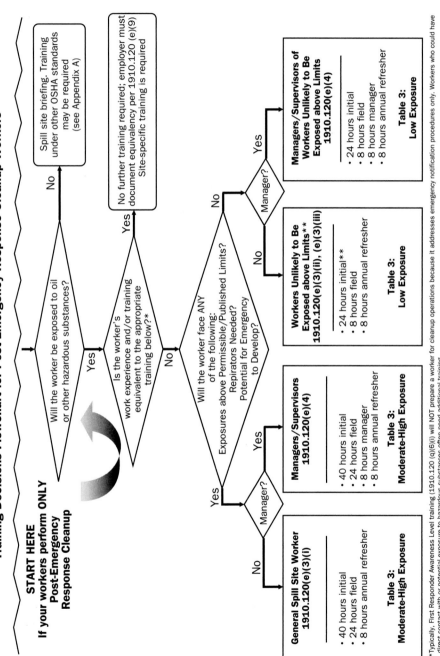

Training Decisions Flowchart for Post-Emergency Response Cleanup Workers

START HERE
If your workers perform ONLY Post-Emergency Response Cleanup

Will the worker be exposed to oil or other hazardous substances?

No → Spill site briefing. Training under other OSHA standards may be required (see Appendix A)

Yes ↓

Is the worker's work experience and/or training equivalent to the appropriate training below?*

Yes → No further training required; employer must document equivalency per 1910.120 (e)(9) Site-specific training is required

No ↓

Will the worker face ANY of the following:
Exposures above Permissible/Published Limits? Respirators Needed? Potential for Emergency to Develop?

Yes ↓   **No ↓**

Manager?

- **No →** General Spill Site Worker 1910.120(e)(3)(i)
  - 40 hours initial
  - 24 hours field
  - 8 hours annual refresher

  Table 3: Moderate-High Exposure

- **Yes →** Managers/Supervisors 1910.120(e)(4)
  - 40 hours initial
  - 24 hours field
  - 8 hours manager
  - 8 hours annual refresher

  Table 3: Moderate-High Exposure

Manager?

- **No →** Workers Unlikely to Be Exposed above Limits** 1910.120(e)(3)(ii), (e)(3)(iii)
  - 24 hours initial**
  - 8 hours field
  - 8 hours annual refresher

  Table 3: Low Exposure

- **Yes →** Managers/Supervisors of Workers Unlikely to Be Exposed above Limits 1910.120(e)(4)
  - 24 hours initial
  - 8 hours field
  - 8 hours manager
  - 8 hours annual refresher

  Table 3: Low Exposure

*Typically, First Responder Awareness Level training (1910.120 (q)(6)(i)) will NOT prepare a worker for cleanup operations because it addresses emergency notification procedures only. Workers who could have direct contact with or potential exposure to hazardous substances often need additional training.

**If you need to train workers for a specific spill and for tasks that involve minimal exposure (e.g., beach cleanup workers) you may be able to use the reduced training provision OSHA describe in its compliance directive. CPL2-2.51. This directive applies in limited circumstances. See the directive and Table 1B.

FIGURE 11-18 · FLOWCHART

This flowchart uses diamonds for the questions but two different shapes for the answers: round-corner rectangles and square-corner rectangles. Notice also that the diamonds differ in size, as do the rectangles, giving viewers extra visual variables to decipher and adding unnecessarily to the complexity of the illustration. Except for the bulleted lists and the footnotes, all the type is centered, adding to the viewer's difficulty. The clear top-to-bottom direction of the illustration, however, makes the decision process easy to follow.

Source: U.S. Department of Labor, Occupational Safety and Health Administration. *Training Marine Oil Spill Response Workers Under OSHA's Hazardous Waste Operations and Emergency Response Standard.* OSHA 3172. Washington, DC: 2001, p. 10.

■ Connect reversible or interactive steps or phases with double-headed arrows.

■ Connect recursive or cyclical steps or phases with circular arrows.

■ Connect optional steps or phases with dotted-line arrows.

See Figures 11-17 and 11-18 for examples of flowcharts.

Designing Diagrams

■ Keep the diagram as simple as possible, avoiding unnecessary details or distracting decorations and focusing the viewer's attention on the key features or parts of the subject.

■ Label each of the pertinent parts of the subject.

■ If space allows, put the labels directly on the part; if not, attach the label to the subject with thin rules (never arrows).

■ Use sans serif type for the labels to maximize legibility.

■ Position all the labels on the horizontal so that the viewer doesn't have to rotate the page or view the screen with tilted head to read the labels.

Editing Photographs

■ Keep the photograph as simple as possible, focusing the viewer's attention on the key features or parts of the subject.

■ Exercise caution in editing the photograph. A photograph in a viewer's mind is a representation of reality, and ethical communicators strive to meet that expectation. Cropping a photograph to close in on a subject and eliminate distractions in the background is standard practice; but avoid adding to, erasing, or changing the size or color of objects in a photograph.

■ If appropriate or necessary for the viewer's understanding, insert labels for each of the pertinent parts of the subject to direct the viewer's attention.

■ If space allows, put the labels directly on the part; if not, attach the label to the subject with thin rules (arrows imply action).

■ Use sans serif type for the labels to maximize legibility.

■ Position all the labels on the horizontal so that the viewer doesn't have to rotate the page or view the screen with tilted head to read the labels.

See Figures 11-19 and 11-20 for examples of effective and ineffective photographs, respectively.

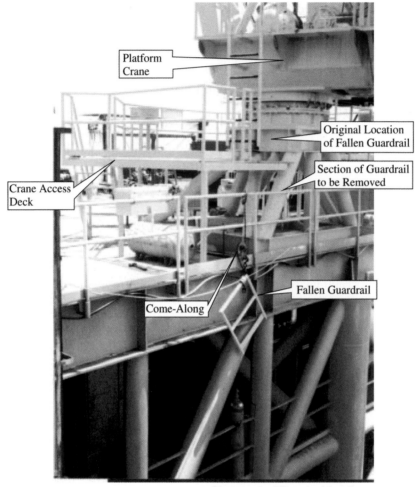

Platform Crane

Original Location of Fallen Guardrail

Section of Guardrail to be Removed

Crane Access Deck

Fallen Guardrail

Come-Along

Photograph of Post-Accident Scene

FIGURE 11-19 · WELL-PRESENTED PHOTOGRAPH
This photograph carries clear labels to identify each of the pertinent pieces of equipment at the accident site.
Source: U.S. Department of the Interior, Minerals Management Service. *Investigation of Fall and Fatality, Main Pass Block 140, May 19, 2000.* MMS 2001-042. Washington, DC: 2001, p. 13.

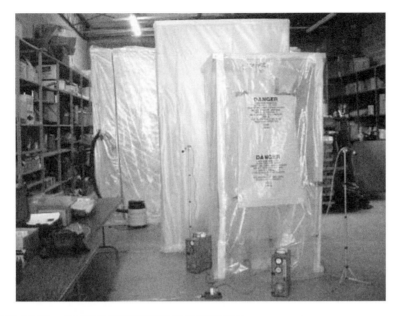

FIGURE 11-20 · POORLY PRESENTED PHOTOGRAPH
This photograph makes viewers look at a lot of distracting detail instead of focusing their attention on essential information. The containment system is nicely centered in the photograph but on the left and right are extraneous shelves and tables loaded with supplies and equipment. In addition, the essential components of the containment system are unlabeled, making it difficult for viewers to notice all the items in the photograph that constitute the containment system. A truly effective photograph would display the containment system in a virtually empty room with either labels or a detailed caption identifying each of the essential components.
Source: U.S. Environmental Protection Agency, Office of Pollution Prevention and Toxics. *Pilot Study to Estimate Asbestos Exposure from Vermiculite Attic Insulation.* Washington, DC: 2003, p. 68.

DESIGNING ILLUSTRATIONS ETHICALLY

Displaying information ethically requires careful choices in the design of illustrations.

For example, the scale of the *x*- and *y*-axes on a line graph has a significant impact on the data display. In designing a graph, you ordinarily start the *x*- and *y*-axes at 0. Exceptions are possible if beginning at some other point (i.e., "suppressing the zero") will not distort information. If several line graphs are to be compared and contrasted, it would be unethical to suppress

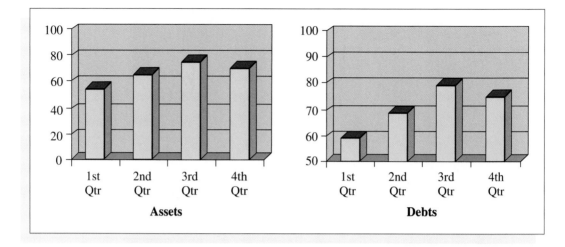

FIGURE 11-21 · DECEPTIVE PAIRED COLUMN GRAPHS
Positioned side by side, the two graphs create a distorted impression. All the columns in the debts graph are smaller than the corresponding columns in the assets graph, giving the incorrect impression that throughout the year, debts were smaller than assets. The assets graph starts the *y*-axis at 0, but the debts graph suppresses the zero, and the percentages on the *y*-axis of the assets graph are given in increments of 20, whereas the debts graph uses increments of 10. Readers will be deceived unless they notice these differences in the two scales. The three-dimensional columns are distracting and hinder a clear comparison of the two graphs.

the zero on some graphs but not on others: readers might focus only on the data lines and overlook the difference in the starting points of the *x*- and *y*-axes (see Figures 11-21 to 11-25).

Pictographs (graphs using pictorial images) can be deceptive if the art is likely to distract the viewer or if the graphic information is subject to be misinterpreted, as in the first pictograph of Figure 11-26.

Using distorted graphs, however, isn't the only error that will result in the creation of unethical illustrations. It is unethical to create a drawing in which a product seems to show features it doesn't really have. It is unethical to design a flowchart that disguises a procedure's complexity by making things look relatively simple. It is unethical to stage or doctor a photograph to create a positive or negative impression of your subject that isn't fully justified.

In addition, if you illustrate situations that entail people, you must strive to be sensitive to their humanity. For example, to use a circle graph to depict the human beings killed in various kinds of automobile accidents (as Figure 11-13 does) or to ignore the human beings killed in a gas explosion (as Figure 11-27 does) genuinely diminishes the dignity of the victims. To illustrate this

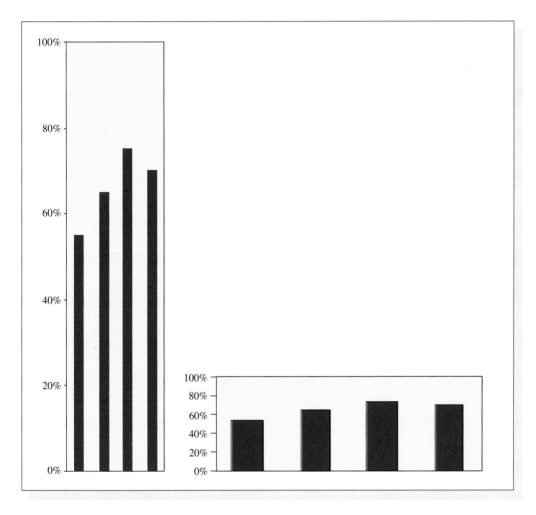

FIGURE 11-22 · DECEPTIVE COLUMN GRAPHS
It is also possible to distort graphic information by stretching one axis and shrinking the other, thereby making the differences among the data points either more or less apparent. Tall and narrow columns exaggerate differences in the data, while short and wide columns disguise such differences.

information ethically, you might insert (with permission of the family) a portrait photograph of a victim, thus offering a vivid reminder of the tragic impact on human beings of accidents, negligence, and other causes of death or injury. In certain cases, however, it is desirable to avoid illustrations altogether, conveying delicate information with words only.

The unethical use of illustrations damages your credibility and hinders your audience's understanding of your subject. (You may want to refer to Chapter 5, Writing Ethically, to help you remain aware of issues that pertain to the ethical presentation of technical information.)

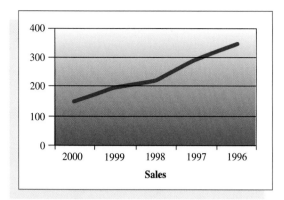

FIGURE 11-23 · DECEPTIVE LINE GRAPH
Readers who look quickly at this graph of sales figures might come away with the impression that sales are rising. They might not notice that the time line on the x-axis is reversed; that is, instead of the earlier date on the left and the later date on the right as one would ordinarily expect, the horizontal scale starts with 2000 and proceeds backward in time to 1996. Sales, in fact, are falling. The thick data line and the small labels on the x-axis contribute to the deception, as does the color: green would suggest money to American audiences. In addition, the graduated shading of the background (darker at the bottom and lighter at the top) lifts the reader's eye and reinforces the misperception of rising sales.

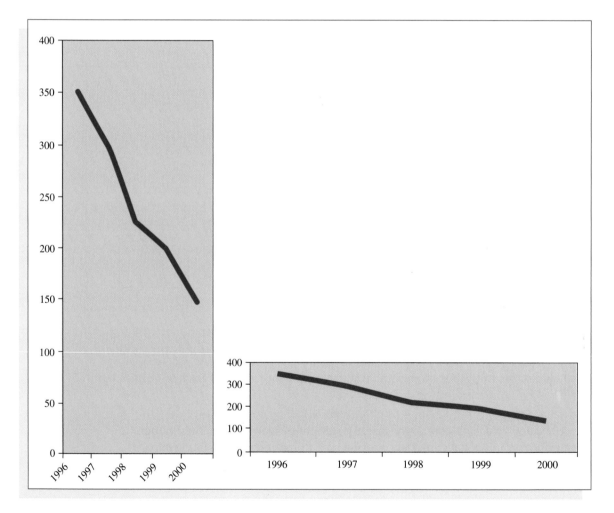

FIGURE 11-24 · MORE DECEPTIVE LINE GRAPHS
It is also possible to distort graphic information by stretching one axis and shrinking the other, thereby making the data line either more vertical (depicting a rapid and dramatic change) or more horizontal (depicting a slow and moderate change).

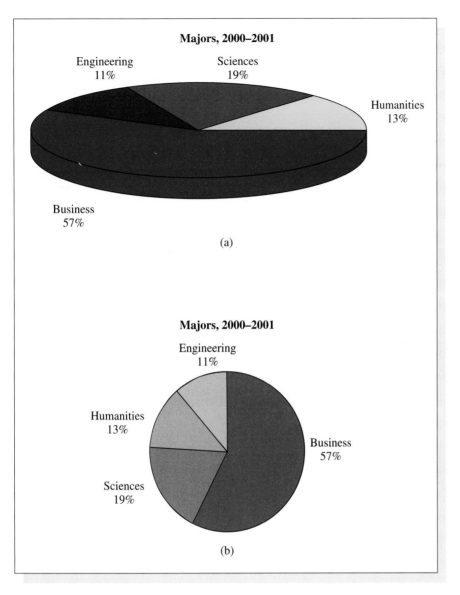

FIGURE 11-25 · DECEPTIVE VS CLEAR CIRCLE GRAPHS

The first circle graph (a) is quite deceptive: even though the labeling of each section is correct, the visual impression is that the Humanities major is the smallest group instead of the second smallest group (i.e., engineering is really the smallest). The light coloring of the humanities major section versus the dark coloring of the other sections distorts the relative size of the light section: the brown, blue, and green sections essentially collapse so that the reader perceives a large dark section versus a small light section. Also disrupting the reader's understanding of the relative size of each section are the slight elevation of the circle graph and the three-dimensional perspective as well as the arbitrary ordering of the sections. The second circle graph (b), on the other hand, offers a clear and unambiguous display of information because of its two-dimensional view with a systematic ordering of the sections (largest to smallest) and a systematic coloring scheme (from darkest to lightest).

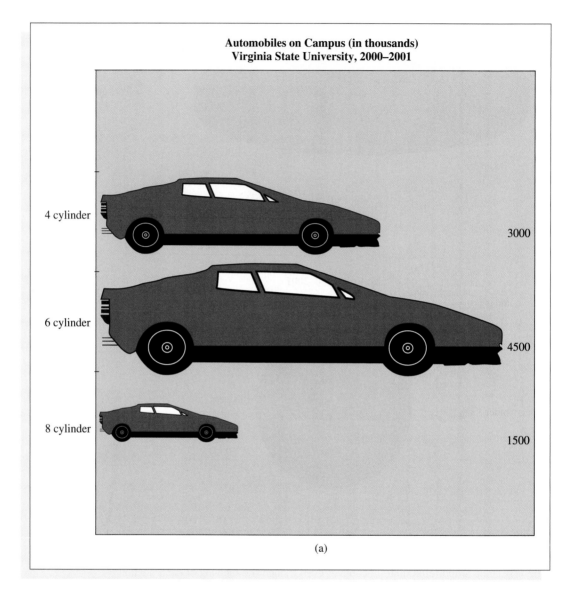

FIGURE 11-26 · DECEPTIVE VS CLEAR PICTOGRAPHS

The first pictograph (a) is potentially deceptive. Even though the numerical information is clearly specified in words, the visual impression is distorted. The picture for the 6-cylinder category is both three
times wider and three times taller than the picture for the 8-cylinder category, making it nine times
larger instead of only three times larger. Similarly, the picture for the 4-cylinder category is twice as wide
and twice as tall as the picture for the 8-cylinder category, making it four times larger. The visual message, as a consequence, contradicts the verbal message. Since the visual message (bright red automo-
biles) is more conspicuous than the verbal message (in small, light type), the reader could easily misunderstand this illustration. To avoid such confusion, design the visual message to reinforce the verbal
message by using one picture for a specific number (e.g., 1000) and duplicating that picture (or portions of it) to depict multiples of that number, as shown in the second example (b).

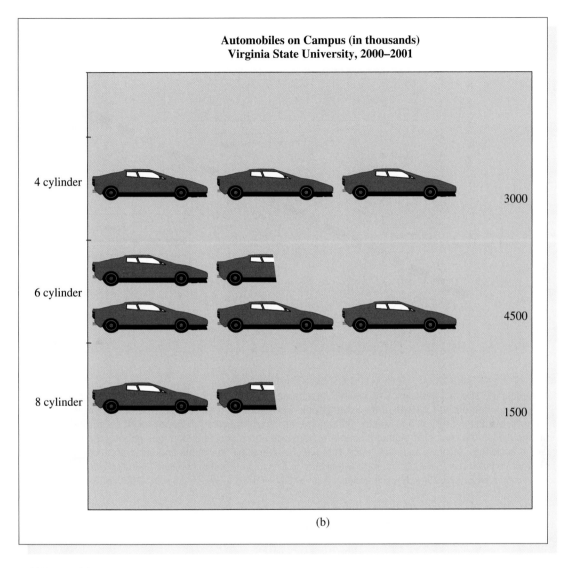

FIGURE 11-26 · *CONTINUED*

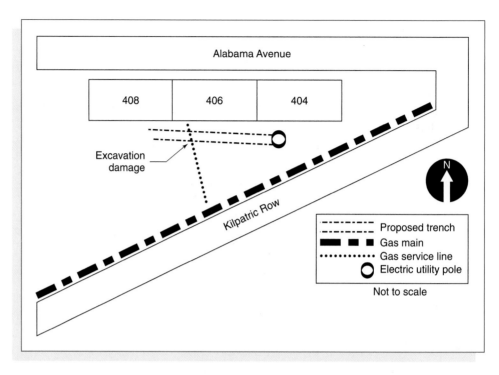

FIGURE 11-27 · UNETHICAL MAP: SCHEMATIC OF AN ACCIDENT AREA

This map depicts the site of a gas explosion caused by a damaged service line. The emphasis here is on the three buildings that were demolished instead of the three people who were killed and the six who were injured. The locations of streets, buildings, and gas and electric utilities are all clearly specified. A viewer might be surprised that human beings died in the explosion because none are displayed at the accident location. Ignoring the human consequences of accidents, failures, and negligence, however, is insensitive to victims and survivors alike.

Source: National Transportation Safety Board. *Natural Gas Service Line Rupture and Subsequent Explosion and Fire, Bridgeport, Alabama, January 22, 1999.* Pipeline Accident Brief NTSB/PAB-00/01. Washington, DC: 2000, p. 3.

☑ Planning and Revision Checklist

Planning Illustrations

- How important are illustrations to your presentation?
- How complex are your illustrations likely to be?
- How expert is your audience at reading illustrations?
- What kinds of illustration is your audience familiar with?
- Are you working with some concepts that can best be presented visually and others that seem to call for a combination of words and graphics, as well as concepts that do not require to be illustrated with visual aids for a particular audience?

- Do you have any definitions that should be presented visually in whole or in part?
- Do you have any processes or algorithms that could be depicted visually in a flowchart?
- Will you be presenting information on trends or relationships? Should some of this information be presented in tables and graphs?
- Do you have masses of statistics that could be summarized in tables?
- Do you need to depict objects? If so, what is it about the objects that must be displayed? Should you try to focus attention on specific aspects of the objects? Do you require the realism of photographs?
- What are the design conventions of your illustrations?
- How much prose explanation of your illustrations are you likely to need?

Revising Illustrations

- Are your illustrations suited to your purpose and audience?
- Do your illustrations communicate information ethically?
- Are your illustrations effectively displayed and easy to find?
- Are your illustrations numbered and labeled?
- Do your verbal and visual elements complement each other?
- Are your illustrations genuinely informative instead of simply decorative?
- When necessary, have you helped your readers to interpret your illustrations with commentary or annotations?
- Will your readers easily understand the processes you have displayed visually?
- Have you included necessary units of measure in your tables and graphs?
- Are your tables simple, clear, and logical? Are the numbers in your tables aligned correctly on the decimal points?
- Do your graphs need a grid for more accurate interpretation? Have you avoided the use of keys or at least kept them simple? Have you plotted your graphs according to the usual conventions: independent variable horizontally, dependent variable vertically? If you have used a suppressed zero, will this aspect of the presentation be obvious to your readers?
- Have you acknowledged the sources for borrowed or adapted tables and figures?

■ ■ ■ ■ ■ ■ ■ Exercises ■ ■ ■ ■ ■ ■ ■

Further exercises related to the content of this chapter and all documents employed within the following exercises can be found on the book's companion Web site.

1. Analyze the effectiveness of the line graph and accompanying table in Figure 11-28. How would you revise the two illustrations to clarify the visual display of information?

2. Critique the ethicality of the design of Figure 11-29. How would you revise this display to minimize the possibility of deception?

3. Examine several technical publications (e.g., brochures, manuals, pamphlets) in your field of study. Choose examples of four effective or ineffective illustrations. Write a report for majors in your field that analyzes the effectiveness or ineffectiveness of each illustration. Develop conclusions and recommendations on the use of illustrations in your field. Be sure to integrate a copy of each illustration with your analysis of it.

4. Examine the various kinds of illustration used in several of the professional journals in your major field of study. Which types of illustration do you ordinarily find? Which types don't you find? Which occur most often? Which occur least often? Reproduce sample tables and figures to demonstrate the conventions for illustrations in your field. Report your findings in an oral presentation to your class.

5. You are the director of marketing for Howell & Field, Inc., which raises turkeys in Texas, Oklahoma, Nebraska, and Kansas. You receive a call from Rita Elizondo, Howell & Field's vice president of operations, who wants you to give an oral sales report to the company's board of directors. Your presentation should take approximately 10 minutes. Prepare several illustrations to explain Howell & Field's sales figures for 2000–2004.

Sales (thousand of pounds)

	2000	2001	2002	2003	2004
Texas	90.4	98.6	99.8	104.6	97.6
Oklahoma	60.3	62.4	60.6	58.4	54.8
Nebraska	58.4	60.6	65.6	67.8	72.4
Kansas	40.8	48.5	52.5	56.3	58.9

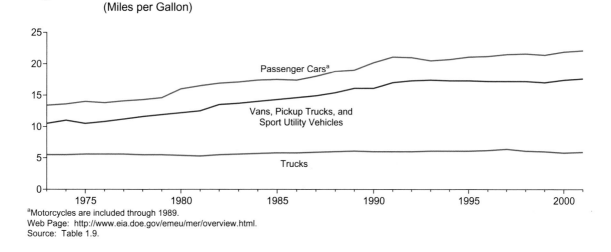

Figure 1.9 Motor Vehicle Fuel Rates
(Miles per Gallon)

[a]Motorcycles are included through 1989.
Web Page: http://www.eia.doe.gov/emeu/mer/overview.html.
Source: Table 1.9.

Table 1.9 Motor Vehicle Mileage, Fuel Consumption, and Fuel Rates

	Passenger Cars[a]			Vans, Pickup Trucks, and Sport Utility Vehicles[b]			Trucks[c]			All Motor Vehicles[d]		
	Mileage (miles per vehicle)	Fuel Consumption (gallons per vehicle)	Fuel Rate (miles per gallon)	Mileage (miles per vehicle)	Fuel Consumption (gallons per vehicle)	Fuel Rate (miles per gallon)	Mileage (miles per vehicle)	Fuel Consumption (gallons per vehicle)	Fuel Rate (miles per gallon)	Mileage (miles per vehicle)	Fuel Consumption (gallons per vehicle)	Fuel Rate (miles per gallon)
1973	9,884	737	13.4	9,779	931	10.5	15,370	2,775	5.5	10,099	850	11.9
1974	9,221	677	13.6	9,452	862	11.0	14,995	2,708	5.5	9,493	788	12.0
1975	9,309	665	14.0	9,829	934	10.5	15,167	2,722	5.6	9,627	790	12.2
1976	9,418	681	13.8	10,127	934	10.8	15,438	2,764	5.6	9,774	806	12.1
1977	9,517	676	14.1	10,607	947	11.2	16,700	3,002	5.6	9,978	814	12.3
1978	9,500	665	14.3	10,968	948	11.6	18,045	3,263	5.5	10,077	816	12.4
1979	9,062	620	14.6	10,802	905	11.9	18,502	3,380	5.5	9,722	776	12.5
1980	8,813	551	16.0	10,437	854	12.2	18,736	3,447	5.4	9,458	712	13.3
1981	8,873	538	16.5	10,244	819	12.5	19,016	3,565	5.3	9,477	697	13.6
1982	9,050	535	16.9	10,276	762	13.5	19,931	3,647	5.5	9,644	686	14.1
1983	9,118	534	17.1	10,497	767	13.7	21,083	3,769	5.6	9,760	686	14.2
1984	9,248	530	17.4	11,151	797	14.0	22,550	3,967	5.7	10,017	691	14.5
1985	9,419	538	17.5	10,506	735	14.3	20,597	3,570	5.8	10,020	685	14.6
1986	9,464	543	17.4	10,764	738	14.6	22,143	3,821	5.8	10,143	692	14.7
1987	9,720	539	18.0	11,114	744	14.9	23,349	3,937	5.9	10,453	694	15.1
1988	9,972	531	18.8	11,465	745	15.4	22,485	3,736	6.0	10,721	688	15.6
1989	10,157	533	19.0	11,676	724	16.1	22,926	3,776	6.1	10,932	688	15.9
1990	[a]10,504	[a]520	[a]20.2	11,902	738	16.1	23,603	3,953	6.0	11,107	677	16.4
1991	10,571	501	21.1	12,245	721	17.0	24,229	4,047	6.0	11,294	669	16.9
1992	10,857	517	21.0	12,381	717	17.3	25,373	4,210	6.0	11,558	683	16.9
1993	10,804	527	20.5	12,430	714	17.4	26,262	4,309	6.1	11,595	693	16.7
1994	10,992	531	20.7	12,156	701	17.3	25,838	4,202	6.1	11,683	698	16.7
1995	11,203	530	21.1	12,018	694	17.3	26,514	4,315	6.1	11,793	700	16.8
1996	11,330	534	21.2	11,811	685	17.2	26,092	4,221	6.2	11,813	700	16.9
1997	11,581	539	21.5	12,115	703	17.2	27,032	4,218	6.4	12,107	711	17.0
1998	11,754	544	21.6	12,173	707	17.2	25,397	4,135	6.1	12,211	721	16.9
1999	11,848	553	21.4	11,957	701	17.0	26,014	4,352	6.0	12,206	732	16.7
2000	11,976	547	21.9	11,672	669	17.4	25,617	4,391	5.8	12,164	720	16.9
2001[P]	11,766	532	22.1	11,140	633	17.6	26,431	4,491	5.9	11,800	692	17.1

[a] Motorcycles are included through 1989.
[b] Includes a small number of trucks with 2 axles and 4 tires, such as step vans.
[c] Single-unit trucks with 2 axles and 6 or more tires, and combination trucks.
[d] Includes buses and motorcycles, which are not shown separately.
P=Preliminary.
Notes: Geographic coverage is the 50 States and the District of Columbia.

Web Page: http://www.eia.doe.gov/emeu/mer/overview.html.
Sources: • **Passenger Cars: 1990-1994:** U.S. Department of Transportation, Bureau of Transportation Statistics, *National Transportation Statistics 1998*, Table 4-13. • **All Other Data:** • **1973-1994:** Federal Highway Administration (FHWA), *Highway Statistics Summary to 1995*, Table VM-201A. • **1995 forward:** FHWA, *Highway Statistics*, annual, Table VM-1.

FIGURE 11-28 · GRAPH AND TABLE FOR EXERCISE 1

General Aviation (noncommercial)
Per 100,000 aircraft-hours flown

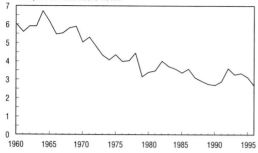

Air Carriers (5-year moving averages)
Log scale

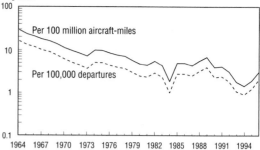

Per 100 million aircraft-miles

Per 100,000 departures

Light Trucks: Occupants
Per 100 million vehicle-miles

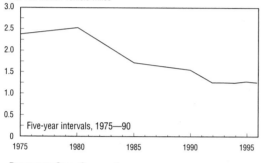

Five-year intervals, 1975—90

Large Trucks: Occupants
Per 100 million vehicle-miles

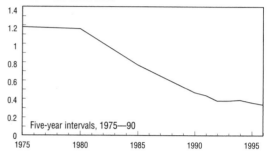

Five-year intervals, 1975—90

Passenger Cars: Occupants
Per 100 million vehicle-miles

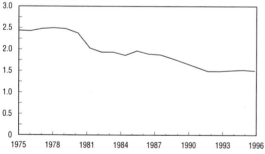

Motorcycles: Riders
Per 100 million vehicle-miles

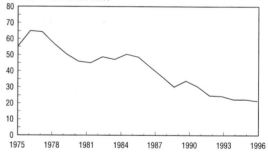

Railroad: Passengers, Employees, Contractors and Other Nontrespassers, and Trespassers (excludes grade-crossing fatalities)

Per 100 million train-miles

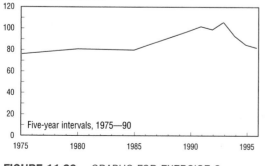

Five-year intervals, 1975—90

SOURCES: General aviation For 1960—74, data include air taxi. Data from U.S. Department of Transportation, Federal Aviation Administration, *FAA Statistical Handbook* (Washington, DC: 1960—74). For 1975—96: National Transportation Safety Board, *Annual Review of Aircraft Accident Data, General Aviation* (Washington, DC: Annual volumes). For all other modes: U.S. Department of Transportation, Bureau of Transportation Statistics, National Transportation Statistics 1998, available at http://www.bts.gov/ntda/nts.

FIGURE 11-29 · GRAPHS FOR EXERCISE 2

6. After your sales presentation, Elizondo asks you to write a report for a group of prospective investors from Australia. She specifically wants you to make projections for 2005–2009 based on the 2000–2004 figures. Design the illustrations that you will include in your written report.

7. Aeschylus Corporation, which designs 3-D computer games, is looking for new employees, especially individuals able to conceive and design state-of-the-art gaming software. It has been a small private corporation for five years and hopes to achieve a major expansion. The president of Aeschylus, Robin Pierce, has asked your organization, Creative Consultants, to develop a brochure for prospective new employees, incorporating appropriate illustrations. You've interviewed Pierce and have gathered quite a bit of information on the company that you believe will be pertinent:

History	Founded in 1996, in Kansas City, Missouri, by Robin Pierce, B.S. in Computer Science, 1992, University of St. Louis; MBA, 1994, Missouri University
Sales	2000: $200,000; 2001: $750,000; 2002: $2 million; 2003: $7 million; 2004: $16 million
Employees	Specialists in design (3), graphics and animation (4), programming (5), video and movie editing (3), production (2), quality assurance (1), documentation (1), sound (1), and music (1)
Products	6 current products, looking to diversify, especially in action–adventure, newest product: Street Maniac (driving game); biggest seller: Trojan War (fantasy role-playing)
Facilities	State-of-the-art computer equipment, serves both PC and Macintosh systems
Location	1224 Howard Avenue, Kansas City: new building, spacious offices
Salary and benefits	Competitive with industry; excellent medical coverage, 4 weeks vacation each year

Design a brochure for Aeschylus Corporation that incorporates as wide a variety of illustrations as possible.

8. In designing your brochure for Aeschylus Corporation, you are surprised by a moral dilemma. Robin Pierce is willing to hire people with disabilities and would like you to include that information in the brochure. At this time, Aeschylus has no permanently disabled

employees. However, programmer Jacqueline Brown, who injured her back in a skiing accident, is temporarily restricted to a wheelchair. In a meeting with your design group, Pierce suggests including a photograph of all the Aeschylus employees (with Brown in the wheelchair) to show prospective employees that people with disabilities are welcome. You mention the idea to your staff photographer, Kishor Mitra, but he objects, claiming such a picture would be deceptive and thus unethical. As project leader, you have to determine which side should prevail and then write one of two memos: either to convince Pierce, your client, that such a photograph would be wrong, or to persuade Mitra, your employee, to set up the photograph.

PART THREE

Applications

Technical writing emerges in a variety of forms—correspondence, internal memoranda, and e-mail; reports, proposals, instructions, procedures, and policies—to list a few of the common types of writing done routinely in business, academic, and research organizations. Each forms the basis for communication within organizations. The effectiveness of these applications requires that you apply basic concepts and techniques studied in Chapters 1 through 11.

Throughout your career, you will likely encounter many additional kinds of technical writing, such as specifications, technical manuals, and financial analyses. We believe that your understanding and practicing the applications of fundamental rhetorical principles and techniques to the development of basic types of writing inherent to organizations will prepare you for any type of writing you may be asked to do. Fundamental principles and techniques are critical to the effectiveness of each of the documents discussed in our final chapters.

■ ■ ■

Scenario

You've just hung up after a telephone conversation with a client who called you to complain about a delayed shipment. You promised to investigate. And she promised to give you 48 hours.

The wise thing to do at this point is to summarize your conversation in writing. You send the client a brief e-mail message that presents your understanding of the conversation and print out a copy for your file. This written record of your promise to the client and your client's promise to you will confirm your expectations and vice versa. If you have misunderstood the telephone conversation, or if she did not say precisely what she meant, your message will immediately bring to light the discrepancy and create the opportunity to eliminate it. For example, she might have said "within 48 hours," clearly expecting your call before that deadline; but if all you heard was "48 hours," you might have thought that you had 48 hours to investigate and would be safe if you called in your report shortly thereafter. While a difference of only minutes might be involved, such a misunderstanding could jeopardize the business relations between your two organizations. A timely exchange of written messages could avoid later confusion and disappointment.

In this situation, you don't have a lot of time, so e-mail is the quick and inexpensive choice to communicate your message, but you could also write a letter and either fax it to your client or use overnight delivery service.

CHAPTER TWELVE

Planning Correspondence and E-Mail

- **Determining Your Purpose**

- **Analyzing the Audience**

- **Composing Letters, Memos, and E-Mail**

- **Finding the Appropriate Style**
  Direct versus Indirect Style
  Conversational Style

- **Special Considerations for E-Mail**

- **Special Considerations for International Correspondence**

- **Keeping Copies of Correspondence**

  Exercises

On the job you will face a wide variety of situations that require you to write e-mail messages, letters, or memos. Oral communication—in person or by telephone—will dominate your day-to-day interactions, but much of your communication will also be written.

You will meet with people to emphasize the importance or urgency of your message and to personalize the communication. A face-to-face conversation offers the greatest opportunity to establish a close human relationship: the parties see each other, offer greetings, and occupy the same space at the same time. A full exchange of verbal and visual information is possible, with ability to ask and answer questions.

You will telephone if a face-to-face conversation is impractical or if, having established a close relationship, you nevertheless wish to emphasize the importance or urgency of a communication. Telephone conversations offer a limited opportunity to personalize the communication because only the parties' voices are available (unless the call allows video as well as audio transmission), so be aware of the tone of voice you use. You do, however, retain the ability to ask and answer questions. There is, of course, no mechanism for generating automatically a written transcript of the conversation.

You will write to create a record of decisions or promises. You will write to offer details that people might misinterpret or misunderstand if received orally. You will write because your audience expects certain information to be delivered in writing. And you will write whenever you would like to receive a written reply.

DETERMINING YOUR PURPOSE

On the job you will compose correspondence for a variety of purposes, such as to report, inquire, or complain. In determining your purpose, ask yourself the following questions:

- What do I want to have happen as a result of my letter, memo, or e-mail?
- What do I want my reader to think or do after reading my message?
- Given the result that I want (and guided by ethical considerations), what must I do in my letter to achieve that result?

Table 12-1 displays several purposes of correspondence, the desired result, and common strategies used for achieving it.

Consider, for example, the following situation. Maria Montez is a second-year engineering student at Florida Technological Institute. She is enrolled this semester in a technical writing course. A major project for the course is a presentation to the class on a critical issue in her major field of study. Maria is aware of the difficulties that engineers encounter because Americans continue to use the traditional "pounds and inches" system of weights and measures instead of adopting the metric system, which is the international standard. She believes the American rejection of the metric system arises from a failure to educate people effectively. Her research leads her to Professor Nicholas Hanson at

Table 12-1 ■ Purpose of Correspondence

Purpose	Desired Result	Common Strategies
Reporting	The reader takes the right action or makes the right decision.	• Use headings to identify the sections of your report. • State the subject and purpose of your message. • If your report or recommendation has been requested, identify the requesting individual and the circumstances leading to the request. • Summarize your report. Emphasize the key points important to the decision-making process as well as conclusions and recommendations. • Offer the details that support your conclusions and recommendations. Integrate illustrations as necessary. • Specify your conclusions and recommendations.
Inquiring (solicited)	You obtain the desired information, and your reader is willing to provide you additional information if you request it.	• Identify the advertisement or offer that solicited your inquiry. • Identify yourself and establish your need for the information. • Request the information. Specify the precise product or process in which you are interested. If appropriate, explain how the company might make a sale to you, thereby prompting a quick and complete reply to your inquiry.
Inquiring (unsolicited)	You obtain the desired information, and your reader is willing to provide you additional information if you request it.	• Identify yourself. • State clearly and specifically the information or materials that you want. • Establish your need for the requested information or materials. • Tell the recipient why you have chosen him or her as a source for this information or material. • Close courteously. Never write "thank you in advance," a phrase that presumes the reader will comply with your request.
Replying to inquiries	The reader receives all the information requested and doesn't have to write again to request more information.	• If possible, answer the questions in the order in which they appear in the inquiry. • Repeat enough of the original questions to remind the reader of the subject and identify which question you are answering. • Try to be as complete as possible, to assist your reader fully and to avoid the need for a second exchange of correspondence on the subject. • If appropriate, refer the reader to additional sources of information on the subject.

(continued)

Table 12-1 ■ **Purpose of Correspondence** *(Continued)*		
Purpose	**Desired Result**	**Common Strategies**
Complaining	The reader makes the adjustment that you desire.	• Be firm but polite. • Assume that the organization will try to correct the situation fairly. • Be specific about the problem and the inconvenience or injury you have suffered as a result. • Provide necessary documentation (e.g., dates, times, places, names, product numbers). • Encourage the organization to make a fair adjustment. • If possible, suggest the adjustment you would like the organization to make.
Offering adjustments	The reader is satisfied with the adjustment you offer and maintains his or her relationship with your organization.	• Be friendly. • Focus on keeping the customer's goodwill. • Express regret about the problem and thank the customer for bringing it to your attention. • Explain the circumstances that caused the problem. • Describe specifically what the adjustment will be. • Resolve any special problems that may have accompanied the complaint. • Close politely.
Refusing adjustments	The reader accepts your refusal and maintains his or her relationship with your organization.	• Be friendly and polite. • Express regret about the problem and thank the customer for writing. • Explain the reason for the refusal in detail. Show the customer that you have given the complaint serious consideration. Prepare the reader for the negative decision. • State the refusal clearly but kindly. If possible, offer a partial or alternative adjustment. • Close the letter in a friendly way, leading the reader's attention away from the negative decision. Focus on the future instead of the past. Never apologize for the negative decision or solicit the reader's understanding of your reasons for reaching it.
Thanking	The reader feels appreciated for his or her efforts and is inclined to assist you again on a subsequent project.	• Be brief. • Be explicit in your expression of gratitude. • Identify specifically the reader's contribution. • Describe the positive impact of the the reader's efforts. • Close politely.

the University of Cleveland, whose name is cited in the professional literature as a leading advocate of the metric system. He has also written a journal article on the economic benefits that would arise from implementation of a single international standard for weights and measures. But none of the articles available really address the question of how to teach Americans the metric system.

Maria decides that she will write a letter to Professor Hanson to solicit his advice. She is a little pressed for time: It's February 1 and the presentation is scheduled for April 3. Professor Hanson's journal article gives his e-mail address at the University of Cleveland. But Maria decides a formal letter would be better, chiefly because she wants the professor to know that she has taken the time to compose her request and hasn't just hurried off a quick e-mail message to him. She's asking him to take the time to answer her questions and believes he will take her more seriously and give her more cooperation if she invests some time of her own in this request.

Since Professor Hanson is a busy man, Maria gets right to the point and immediately states her purpose for writing. She also decides to use short paragraphs to make the letter easy to skim through quickly. She proceeds to explain who she is. To give herself credibility, she specifically mentions that she has been reading journal articles on this subject. And she explains to Professor Hanson why she has chosen to contact him for information, slipping in a nice compliment while doing so. She then concisely asks her questions. Finally, she expresses her appreciation and, by including her e-mail address, tries to make it as easy as possible for Professor Hanson to reply. And Maria knows he is likely to reply a lot sooner by e-mail.

Professor Hanson is impressed with Maria's letter (see Figure 12-1). He is pleased to offer his advice to this serious student, and he appreciates being able to save time by using e-mail for his reply.

Because Maria is a student, Professor Hanson isn't sure how much she already knows about the metric system. He chooses to avoid technical language, but he includes a lot of examples, listing them to make them easier to read. Although he is pleased to answer this one letter, he doesn't want to be bothered by additional questions, especially if answers are readily available from other sources. He has a looming deadline on a research project and just doesn't have time to answer questions from students who aren't enrolled in his courses. He doesn't offer to provide additional information, therefore, referring Maria instead to a pertinent source.

Maria is grateful for the information that Professor Hanson has provided (see Figure 12-2). It gives her oral presentation a good sense of direction: she will compare the two methods of teaching the metric system and demonstrate the advantages of learning to think in metric weights and measures. She is also surprised that her earlier search on the Web using the keyword phrase "metric system" didn't identify the U.S. Metric Association. She will visit its Web site to see if it has teaching materials available.

Maria composes a quick e-mail message to Professor Hanson (see Figure 12-3). She wants him to know she appreciates his assistance so that he'll be willing to help again the next time a student writes to him with a question.

Maria Montez
1225 55th Street
Gatesville, FL 32039
mmontez@fti.edu

Professor Nicholas Hanson
Department of Civil Engineering
University of Cleveland
Cleveland, OH 44125

Dear Professor Hanson:

I am writing to request information about effective ways to teach people the metric system of weights and measures.

I am a second-year student at Florida Technological Institute. For a course in technical writing that I am currently taking, I am giving a presentation on the proper way to educate Americans in the use of the metric system. The presentation is scheduled for April 3.

In the journals in which I have been researching the subject, you are frequently mentioned as a major authority in the field. Would you be kind enough to give me your opinion about how metrics should be taught?

The specific aspect of this question that concerns me is whether metric measurements should be taught in relation to present standard measurements, such as the foot and pound, or whether they should be taught independently of other measurements. I see both methods in use.

Any help you can give me will be greatly appreciated, and I will, of course, cite you in my presentation. For your convenience, I include my e-mail address: mmontez@fti.edu.

Sincerely yours,

Maria Montez

Maria Montez

FIGURE 12-1 · MARIA'S LETTER OF INQUIRY

Dear Ms. Montez:

The question of how to teach people about the metric system concerns a great many educators. I suppose it's inevitable that those familiar with the present system will be tempted to convert metric measurements to ones they already know. For example, people will say, "A kilogram, that's about two pounds."

In my opinion, however, such conversion is not the best way to teach metrics. Rather, people should be taught to think in terms of what the metric measurement really measures, to associate it with familiar things. Here are some examples:

* A paper clip is about a centimeter wide.

* A dollar bill weighs about a gram.

* A comfortable room is 20 degrees C.

* Water freezes at 0 degrees C.

* At a normal walking pace, we can go about 5 kilometers an hour.

These are the kinds of association we have made all our lives with the present system. We need to do the same for metric measurements. As in learning a foreign language, we really learn it only when we stop translating in our heads and begin to think in the new language.

I hope this answers your question. You can get valuable materials about metrics by writing to the U.S. Metric Association, 10245 Andasol Avenue, Northridge, CA 91325 (fax: 818-368-7443; http://lamar.colostate.edu//~hillger).

Sincerely,

Nicholas Hanson

FIGURE 12-2 · PROFESSOR HANSON'S E-MAIL REPLY TO MARIA'S LETTER OF INQUIRY

Dear Professor Hanson:

Thank you for your quick reply to my questions. Your advice has been a great help as I prepare my oral presentation.

I also appreciate the reference to the U.S. Metric Association. I will check its Web site for more information.

Sincerely,

Maria Montez

FIGURE 12-3 · MARIA'S E-MAIL THANK-YOU

ANALYZING THE AUDIENCE

In composing correspondence, consider your readers:

- Who are they?
- What do they know already?
- What is their purpose in reading your letter or memo?
- How receptive are they likely to be to your message?

On the job, you will write to individuals and to groups. And even if you are chiefly addressing a single reader, you may also be directing copies of your message to others. In such cases, you must think about the knowledge and experience of both your primary and secondary readers. If, for example, those receiving copies will be unfamiliar with the topic under discussion, you will have to incorporate background information into the message or add attachments.

In another common situation, you may be explaining a technical problem and its solution to a colleague with technical knowledge equal to yours.

If, however, you are sending a copy of your message to your boss, who lacks that technical expertise, you would be wise to start your discussion with a summary (see Chapter 10). Fill the boss in quickly on the key points and tell him or her the implications of your message.

In addition, consider why people might be reading your message. You have a purpose in writing your letter, memo, or e-mail message, but readers have purposes as well. For example, one person may be reading to evaluate your recommendation and accept or reject it. Your purpose then must be to provide enough information to make that evaluation and decision possible. Or, if your purpose is to explain why you missed a deadline, the reader will be reading to determine whether your reasons are valid and acceptable. You will then need to provide enough information to justify your position.

You also need to determine your audience's attitude toward you, your subject, and your purpose:

- Do you have a good relationship with your audience?
- Do they ordinarily accept or reject your explanations and recommendations?
- Is your audience positively or negatively disposed to your subject and purpose?

If you are registering a complaint, for example, you need to know how such a communication is likely to be received. If the audience is typically unsympathetic or doesn't know or trust you, you will have to provide substantial evidence to build a persuasive case. A sympathetic audience, inclined to trust your judgment, will probably require less evidence.

Consider the following situation. Morad Atif, the manager of building services at Accell, Inc. would like to purchase new reversible power drills for a crew of twenty-five technicians. He has examined the options available and decided on a drill that he believes is suitable, but he needs approval from the company's financial officer, Tamika Williams. He has discussed the subject with Williams, but a written request is still required. Williams likes to have good records.

Morad knows that buying the drills isn't as much the issue as the choice of drills. The previous manager was fired for accepting gifts from equipment suppliers. Morad is new on the job and he believes that suspicion still lingers: he'll have to prove that he is trustworthy and that his recommendations are credible.

Morad adopts a direct style, getting right to the point of the memo in the opening sentence. He knows that he has to come across as straightforward. He mentions the price and briefly states why the new equipment is needed. In the paragraphs that follow, he'll focus on the choice of drills, using a bulleted list to give the evidence visual emphasis. He'll also start the list with a reason that might be especially persuasive to the financial officer: cost. And he'll close the list with a reason that might also be pertinent: prevention of loss of equipment. For the convenience of his reader, he decides to list the price information, since Williams will later need these figures for the company's financial records.

To prompt a quick decision, Morad closes the memo with a question and a tentative deadline (see Figure 12-4). Otherwise, he worries, his request might not be given priority.

COMPOSING LETTERS, MEMOS, AND E-MAIL

How do you decide whether to use a letter, a memo, or e-mail? Consider the following factors.

- *What is the usual practice of your organization?*

In some organizations, for example, day-to-day business updates often are communicated by e-mail, but policy changes are distributed through memos, and letters are used only to convey messages to and from the district supervisor. Ordinarily, you will adopt the usual practices of your organization.

- *What is the relative efficiency or practicality of each communication medium?*

E-mail is a high-speed, low-cost substitute for traditional paper correspondence. Potentially, e-mail has the immediacy of oral conversation while allowing users to keep an electronic record or generate a paper one. Your message goes from your computer to your recipient's computer within seconds and without the delays associated with the printing, copying, addressing, mailing, and sorting of paper memos and letters. And answers to e-mail messages often arrive within minutes instead of days or weeks.

ACCELL. Inc.

313 Slide Avenue, Portland, OR 97194-2113, USA, 1-800-555-6234
www.accell.com

DATE: September 5, 2005

TO: Tamika Williams, Financial Officer

FROM: Morad Atif, Manager of Building Services

SUBJECT: Request for Approval on Purchase of Power Drills

I am writing to request your approval of the purchase of 25 reversible power drills for my technicians. The total purchase price including tax is $810. The new drills will considerably speed the crew's ability to do its job while reducing fatigue and injury.

Company sales representatives recommended several models for the specific requirements of my technicians, and I examined the Towson 7190, Bedford 23088, Decker 404, Joppa 9420, and Stark 560.

I decided on the Towson drill for the following reasons:

- The Towson drill is the least expensive of the five considered.
- The variable speed trigger offers versatility.
- The double reduction gears guarantee additional torque for bigger and tougher jobs.
- The locking mechanism is recessed to avoid accidental lock-on, thus providing extra safety.
- The chuck key (used to change drill bits) comes on a plastic clip attached to the power cord, making it convenient to change drill bits and virtually eliminating the chance that a technician will lose this important item.

Here is the price information:
25 Towson 7190 drills @ $30/drill 5 $750
8% tax 5 $60

total 5 $810

To meet my assigned productivity objectives, I would like to make this purchase immediately. Would it be possible to receive your approval by Friday, September 7?

FIGURE 12-4 · MORAD'S MEMO REQUESTING APPROVAL

E-mail is also a less expensive medium for correspondence. Because e-mail messages are composed, transmitted, received, and stored electronically, the savings for organizations are substantial: for example, the rising costs of paper and postage are minimized; fewer file folders and file cabinets are required for sorting and storing correspondence; and the costs of labor (for distributing and filing messages on paper) are eliminated.

E-mail communication, however, has its risks. In the rapid exchange of ideas that e-mail encourages, you might fail to exercise care in choosing your words and phrases. You might write something that isn't quite what you intended, leading to a message that is incorrect, ambiguous, or impolite. Such a message could damage your credibility and your relationship with your audience. E-mail might be as quick as oral conversation, but readers can't hear your voice or see your face: they have only written words with which to judge the accuracy, civility, and sincerity of your message.

In addition, because e-mail exists only as you type it on the computer screen, you might be tempted to think of it as less important or official. Keep in mind, however, that e-mail is written communication: it has the potential permanence and legal significance of paper correspondence. E-mail requires thoughtful composing, editing, and proofreading.

Finally, though e-mail is *potentially* as quick as oral conversation, it isn't *necessarily* as quick. For example, the message you send today may not elicit a response today. It may be read tomorrow, next week, next month, or never. The absence of a guaranteed timely response to e-mail messages can be frustrating when someone you need to hear from is slow to reply.

Letters and memos address subjects of sufficient importance to justify the extra time, effort, and cost involved in preparing and transmitting a paper version of your message. Once printed, the letter or memo must be delivered, at a minimum, passing through your organization's internal mail delivery system: it is picked up, sorted, and distributed—a labor-intensive process. If the recipient is outside your organization, postage must be added to the cost of your message.

If time is a factor, however, you could fax your letter or memo. This process also allows you to exchange simple drawings and copies of paper documents with handwritten comments or signatures. Also a possibility is electronically scanning such materials and attaching the file to your e-mail message—like sending a cover letter with a report clipped to it.

- *What is your rhetorical situation?*

If your organization has no consistent practices regarding correspondence or if your situation is extraordinary, you must return to a consideration of your audience, your subject, and your purpose in writing:

▪ **What is your purpose?** Are you writing to inquire? To complain? To thank? Is your message informative or persuasive? A letter or memo often seems more careful and deliberate, more official and authoritative, than e-mail, and thus usually more credible and persuasive.

▨ **Who is your audience?** Ordinarily, letters and e-mail go to people outside the company or organization, while memos and e-mail are distributed internally. Within organizations, letters are usually reserved for special circumstances that require formal communication, such as announcing one's resignation. In addition, the more important your audience is or the less familiar you are to your audience, the more likely you are to choose the formality of a letter.

FINDING THE APPROPRIATE STYLE

After analyzing your audience and considering your purpose, you have to decide on a suitable writing style.

Direct versus Indirect Style

In the United States, the preferred style of correspondence is direct communication: news, explanation, and closing. To Americans, this style is candid and efficient: getting to the point immediately with the critical information, offering explanation or clarification, and ending politely. If the audience will be pleased with the news, the direct style is appropriate. Direct communication is also the right choice for situations of urgency, offering the earliest and easiest access to critical information.

The indirect style is usually the better choice, however, in two rhetorical situations: when you anticipate a negative reaction to your news and when you are writing to an international audience. If the news is likely to disappoint or irritate your audience, or if your message is of little or no urgency, the indirect style will cushion the negative impact of the news. And international audiences often prefer the indirect style, considering it more civilized and courteous. In the indirect style, you establish a human relationship with the audience before discussing the news. Ordinarily, you begin graciously, provide whatever background information or reasonable arguments are necessary to help the audience understand and accept the news, report the news itself, and close politely.

Consider the following situation. Nicholas Cooper is the president of Cooper & Cooper, a hog processing corporation. Laura Pauley, the president of Red River Farming, a major hog supplier, has called to request that the contract price on hogs be raised. Because of increased labor costs, rising feed prices, and new environmental and agricultural regulations, Red River Farming is losing money at the existing price.

Cooper considers this request and would like to fax a reply as soon as possible. While he is sympathetic to the plight of Red River Farming, his firm has also experienced a couple of rough years because the public consumption of hog products—bacon, ham, sausage—seems to be dropping. He'd like to be candid with Pauley, but he would also like to avoid antagonizing a major supplier. He drafts a response using the direct style (see Figure 12-5).

Cooper & Cooper

414 Williams Street, Lawrence KS 66045-1200

1-800-555-1919

www.cooper&cooper.com

Laura Pauley, President
Red River Farming
P.O. Box 299
Amarillo, TX 78404-0299

Dear Ms. Pauley:

I have received your letter requesting an immediate increase in the contract price that Cooper & Cooper pays for hogs from Red River Farming.

I regret that we will be unable to raise the contract price. According to our contract, which expires on December 31, 2005, you receive 25 cents per pound. This is the highest amount we pay to any of our other suppliers, and some receive less.

In general, the price of hogs and the demand for pork products has been declining instead of rising. To remain competitive in the market we have to buy our hogs at a cost that reflects market conditions.

We will begin negotiations later this year on a new contract. However, if hog prices remain as is or decrease further, a price increase in the new contract seems unlikely.

Sincerely,

Nicholas Cooper

Nicholas Cooper
President

FIGURE 12-5 · NICHOLAS'S LETTER USING A DIRECT STYLE

Upon reviewing the draft, Cooper decides to try indirection. Next (see Figure 12-6) he starts by thanking Pauley for her letter without mentioning its subject. He emphasizes the good relationship of the two companies and compliments Red River Farming. It isn't until the second paragraph that he addresses Pauley's request, using sympathetic language such as "I understand." His justification for not raising the price is the same as that used in

---

Cooper & Cooper

414 Williams Street, Lawrence KS 66045-1200

1-800-555-1919

www.cooper&cooper.com

Laura Pauley, President

Red River Farming

P.O. Box 299

Amarillo, TX 78404-0299

Dear Ms. Pauley:

Thank you for your letter of August 10, 2005. We have long appreciated our association with your company. Your hog shipments have consistently been of high quality, reflecting the expertise and efficiency of your company's operations.

I understand that the current contract price we pay for your hogs no longer seems satisfactory to you. This price, however, is the same as the price we pay our other suppliers. In fact, in several cases, it is a higher price.

To continue buying hogs from your company, we must remain competitive in a difficult market. As you know, the demand for pork products has been declining instead of rising. I'm sure you understand our dilemma as we understand yours.

Our current contract expires at the end of this year. We will begin negotiations on a new contract in a few months. At that time we will know more accurately than we do now where hog prices are heading.

In the meantime, we hope that our long relationship will continue to the mutual profit of both our companies.

Sincerely,

Nicholas Cooper

Nicholas Cooper

President

FIGURE 12-6 · NICHOLAS'S LETTER USING AN INDIRECT STYLE

the direct version, but here Cooper uses two sentences, thus making his point more emphatic. In the third paragraph, he solicits Pauley's understanding of his position by means of expressions such as "As you know" and "I'm sure you understand." In the fourth paragraph, he raises the subject of contract negotiations, as in the direct letter, but without the pessimistic statement that could leave Pauley dispirited or hostile. Cooper decides to be positive: there's always a chance that things could change. He closes by once again emphasizing the relationship of the two companies. The message of this indirect letter is the same as in the direct version, but the refusal is implied instead of explicitly stated. That is, in the letter shown in Figure 12-6, Cooper hasn't exactly said "no" but he has previewed all the reasons why he may have to.

Keep in mind that "direct" and "indirect" are opposite points on a continuum of correspondence, and the letters, memos, and e-mail messages you write will usually be more or less direct, more or less indirect, according to the rhetorical situation. For example, your audience might be both American and international. Or your information might require both immediate decisions and a series of later actions from your audience. And your message may please some of your readers but disappoint others. Your job will be to analyze carefully your entire rhetorical environment.

Conversational Style

In your correspondence, aim for a simple conversational style.

Everything we said about style in Chapter 4, Achieving a Readable Style, applies especially to letters, memos, and e-mail. Professionals receive so much correspondence every week that they truly can't afford to be paralyzed by messages that are difficult to read or understand. You must create for them messages with short paragraphs, lists, simple sentence structure, and common words. Above all, avoid fancy language and the formality of the passive voice.

And avoid clichés. They'll make your letter seem formulaic, filled with canned expressions instead of specifically suited to your audience, subject, and purpose. And clichés will make you sound like a pompous official instead of a caring and articulate human being.

Instead of:	Write:
We beg to advise you that . . .	I'd like you to know that . . .
We are in receipt of your letter that . . .	I received your letter that . . .
It is requested that you send a copy of the specified document to our office.	Please send me a copy of your latest progress report.

In short, ask yourself whether you would or could say in conversation what you have written. If you know you never would say it, don't write it. Restate it in simpler language for a more readable style.

Finally, always focus on the human being reading your letter, memo, or e-mail. Develop a you-attitude, using the word "you" more often and "I" less often. Try to see things from the reader's point of view. Suppose, for example, you were writing a letter of job application to a prospective employer. You might write:

```
I believe that my employment with XYZ Corporation will
be a great learning experience for me and allow me to
develop my skills as a mechanical engineer.
```

Here you are seeing the subject of employment from your point of view, emphasizing the benefits to you. Your reader, however, is more interested in learning how XYZ Corporation will profit by hiring you, how you will contribute to XYZ's objectives and operations. Adopting the you-attitude, you might write:

```
My training as a mechanical engineer will support your
mission at XYZ Corporation to design environmentally
friendly automotive products. Specifically, my recent
studies in high-temperature superconductivity will help
you to develop state-of the-art motors and generators.
```

Here, you are seeing the subject from XYZ's perspective, offering details that demonstrate your understanding of the organization.

Finally, keep your correspondence concise. Avoid overwhelming or intimidating readers with more information than they want or need. Remember that your readers get paid to take actions and make decisions based on the information you have provided. The longer you keep people reading, the longer they must delay making a decision or taking an action.

While brevity in letters, memos, and e-mail is good, you should avoid seeming brusque or impatient. Occasionally, a longer message gives a better impression, especially if your news is disappointing. In such situations, people appreciate your taking the extra time to explain in detail.

SPECIAL CONSIDERATIONS FOR E-MAIL

Because e-mail has characteristics of both oral and written communication, of both informal and formal communication, writers adopt a variety of styles for their e-mail messages. While e-mail has the structure of a memo (with its designations of FROM, TO, and SUBJECT), it is usually written as a business or personal letter, with a salutation at the beginning of the message and the writer's "signature" at the end.

If your relationship with the reader is strictly professional, you might compose e-mail as though it were a business letter, starting with a greeting such as "Dear Dr. Smith:" or "Dr. Smith:" and closing with a "Sincerely," and your name, title, and organization. If your relationship is both professional and personal (e.g., if you've talked face to face or by telephone), you might adopt a friendlier greeting such as "Dear Bill," or "Bill," omit the "Sincerely," and close with a short e-mail signature. For a message that is strictly business, you might omit the salutation altogether and proceed directly to the news of your message.

Whichever opening or closing you choose, consider also the etiquette of your e-mail message.

■ **Be polite.** Never compose and send e-mail when you're irritated or discouraged; you might put on record something you'll later regret. Consider also your reader's feelings, and be sensitive to them. Certain words may be perceived as insulting or offensive. To minimize misunderstandings, especially avoid satiric, ironic, or sarcastic comments. If you believe your words might be misinterpreted, you might incorporate emoticons to signal your attitude: for example, :-) for a smile, ;-) for a wink, or :-(for a frown. Keep in mind, however, that emoticons are often perceived as frivolous and would diminish the professional quality of an e-mail message intended for a supervisor or a high-level business contact.

■ **Never write a message you wouldn't want others to see.** Any recipient can easily copy or forward your e-mail messages to your boss, your colleagues, or the local newspaper. Also, your e-mail correspondence may be legally monitored by your organization, which may claim your messages as its intellectual property. In addition, like the paper correspondence of your organization, your e-mail messages may be subject to subpoena and legal review.

■ **Respect the privacy of e-mail messages.** Exercise discretion before copying and distributing an e-mail message without the sender's permission. The sender might have intended portions of a message for your eyes only.

■ **Answer your e-mail promptly, especially requests for information.** Your reader may be unable to take action or make a decision before receiving your reply.

■ **Keep your messages brief.** Don't ask your reader to scroll through paragraphs of unnecessary information to locate the news of your message. If you are replying to a previous message, copy only the pertinent passages of the original message.

■ **Send one message per topic.** By limiting each message to a single issue or question (identified clearly in the subject line), you assist the recipient in sorting e-mail messages and retrieving specific messages later by keywords. If your message addresses several topics, you dilute the attention that each will receive and raise the possibility that topics of lesser urgency or importance will go unnoticed.

■ **Keep your paragraphs short.** Short paragraphs organize your message visually and simplify reading.

■ **Edit and proofread carefully.** While typographical errors are characteristic of many e-mail messages, numerous typos may diminish your credibility and distract from your message. Audience analysis is important here: some readers will regard such errors as unimportant, but others won't. If your relationship with your audience is strictly professional, apply strict standards to your grammar, spelling, and punctuation.

■ **Develop a signature file giving full contact information and use it as necessary.** For the convenience of readers, compose an electronic file giving your name, title, physical address, telephone number and extension, fax number, and e-mail address. You can't always anticipate how a reader will need or want to reply to your message. Sometimes a telephone call or a fax might be needed, or a package of materials will have to be shipped. A signature file can be appended automatically, without typing in your contact information, and the ready availability of the data will make it easy for people to cooperate with you.

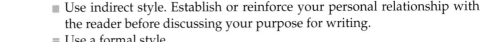

SPECIAL CONSIDERATIONS FOR INTERNATIONAL CORRESPONDENCE

In Chapter 6, Writing for International Readers, where we discuss the importance of understanding the culture of your readers, we advise adjusting your communication style to reflect awareness of the differences between American and other cultures:

- Use indirect style. Establish or reinforce your personal relationship with the reader before discussing your purpose for writing.
- Use a formal style.
- Avoid criticizing individuals or groups. Such criticism is often perceived as bringing disgrace both to the people criticized and to you (for being rude enough to express criticism). Instead of identifying problems, focus on solutions. Emphasize easier, quicker, and cheaper ways of achieving superior results.
- Address business issues from a wider human, social, and organizational perspective.
- Avoid rushing people to make decisions. Allow time for group consensus to build.
- Keep your language common and simple. Your message may need to be translated for some readers, and unusual words or idiomatic expressions could prove difficult for interpreters lacking full fluency in English.

WILD COMPUTERS www.wildcomputers.com
100 Water Street, Seattle, Washington 98194, USA, 1-800-555-WILD

February 19, 2005

Mr. Gu Bao-hui
General Manager
Suzhou Winmedia Co., Ltd.
25 Qinghua Road
Suzhou 215008
P. R. CHINA

Dear Mr. Gu:

During my visit to your company, you expressed interest in our products and asked for additional information.

I am enclosing a brochure that lists our products and services. If you will let me know your exact requirements, I will be happy to provide you with more details.

I look forward to your reply.

Sincerely,

Theresa Ricco

Theresa Ricco
Sales Representative

Enclosure

FIGURE 12-7 · THERESA'S LETTER USING A DIRECT STYLE

Consider the following situation. Theresa Ricco is a sales representative of Wild Computers. She is back in the office after a sales trip to the People's Republic of China, drafting a letter to a potential client. Using direct style, she writes a letter that is clear and to the point. The language is simple and the paragraphs brief. It is a crisp and efficient piece of correspondence and typical of the letters she usually writes. While this letter (see Figure 12-7) might be ideal for American readers, a Chinese audience might consider it discourteous— possibly thinking that Theresa didn't care enough to write a longer or personal letter.

WILD COMPUTERS www.wildcomputers.com
100 Water Street, Seattle, Washington 98194, USA, 1-800-555-WILD

February 19, 2005

Mr. Gu Bao-hui
General Manager
Suzhou Winmedia Co., Ltd.
25 Qinghua Road
Suzhou 215008
P. R. CHINA

Dear Mr. Gu:

I hope that you and your family enjoyed the happiest of New Year celebrations. I arrived home from my trip to China just in time to welcome the Year of the Rooster at a special dinner with my husband.

I wish to thank you for taking the time to meet with me during my visit to Suzhou. That meeting left me with a deep impression of your kindness and sincerity.

As I mentioned to you during my visit, my company is located in Seattle, a primary location for the computer industry in the United States. We have established a reputation for innovative products and friendly service to customers. We have been selling computer hardware for 15 years to 25 countries all over the world. We started selling computers to China in 1995.

I would like to know if you think Wild products could help your organization be more productive and profitable. I have enclosed a price list of our equipment. I would be pleased to provide more information.

I believe that our acquaintance could bring good fortune to us both.

Sincerely,

Theresa Ricco

Theresa Ricco
Sales Representative

FIGURE 12-8 · THERESA'S LETTER USING AN INDIRECT STYLE

A careful analysis of the audience inspires Theresa to try a different kind of letter. Choosing the indirect style, for her second draft, Ricco starts the letter with good wishes, shows familiarity with Chinese culture, and offers personal information about herself, such as that she is married. In the second paragraph, she proceeds to thank and compliment the reader. In the third paragraph, she describes her company, reminding the reader of matters that she no doubt covered orally during their meeting. Since, however, she knows that information that establishes her company's credibility is important in China, the points bear repeating. In the fourth paragraph, she arrives at the point of this letter, the offer of information about her company's products. Finally, she adopts the kind of formal closing that a Chinese writer might use.

This letter, while considerably longer, is more likely to be positively received. Like all effective correspondence, it is tailored to its audience (see Figure 12-8).

KEEPING COPIES OF CORRESPONDENCE

Keep a file—paper or electronic—of every substantive letter, memo, and e-mail message that you write or receive. By controlling this material, you have more control over the subjects discussed. And you will establish a reputation among your clients and colleagues for being organized, efficient, and indispensable.

For example, a dispute might arise about details in a letter, memo, or e-mail message written several weeks ago. If you have a copy of that correspondence, you are the person who can resolve that dispute. You don't have to trust your recollection of the message or rely on the recollection of others. You have the facts in a file at your fingertips.

By keeping copies of your correspondence, you will also protect yourself from being misrepresented, misquoted, or misinterpreted. If colleagues claim you never answered a request, you can present a copy of your response. If others forget what you wrote, you have copies to remind them. If clients claim you promised x and y, you have copies to prove you promised y and z. Keeping on top of your on-the-job correspondence gives you more control of your working environment.

✓ Planning and Revision Checklist

You will find the Planning and Revision Checklists that follow Chapter 2, Composing, and Chapter 8, Designing and Formatting Documents, valuable in planning and revising any presentation of technical information. The following questions specifically apply to correspondence. They summarize the key points in this chapter and provide a checklist for planning and revising.

Planning

- What is your subject and purpose?
- What do you want to have happen as a result of your correspondence? What are your plans for achieving your objective?
- Who are your primary readers? Secondary readers? Do your primary and secondary readers have different needs? How will you satisfy everyone?
- Why will people read your correspondence?
- What is the attitude of your readers toward you? Toward your subject? Toward your purpose?
- If you are addressing international readers, do you understand their cultural practices? What adjustments in your correspondence will their traditions require?
- Will you write a letter, memo, or e-mail message?
- Will you choose a direct or indirect style?

Revision

- Are your topic and purpose clearly identified?
- Will your reader's purpose be satisfied upon reading your document?
- Have you adopted a style suitable to your reader's culture?
- Have you avoided jargon and clichés?
- Does your correspondence demonstrate a you-attitude?
- Is your message clear, concise, complete, and courteous?

■ ■ ■ ■ ■ ■ ■ ■ Exercises ■ ■ ■ ■ ■ ■ ■ ■

Further exercises related to the content of this chapter and all documents employed within the following exercises can be found on the book's companion Web site.

1. Write an unsolicited inquiry to a company that manufactures a product you wish to know more about. Request sample materials or information. If the company has a home page on the Web, you probably can make your request by e-mail from the "Contact Us" link on the site. Otherwise, write a letter.

2. Identify a service or product that has recently caused you dissatisfaction. Find out the appropriate person or organization to address, and send your complaint to that person or organization—by letter or by e-mail.

3. Exchange letters of complaint written for Exercise 2 with a classmate. Write two different answers to your classmate's letter. In one letter, offer the adjustment requested. In the other letter, refuse the adjustment requested.

4. Compose the memo called for in the following situation.

Civil Engineering Associates of Purvis, Ohio, was established in 1990 by Robert B. Davidson and Walter F. Posey, both graduates of the University of Ohio. Business for the company was good, and CEA took on six additional partners between 1991 and 2003: Alvin T. Bennett, Wayne S. Cook, Frank G. Reynolds, John W. Castrop, George P. Santos, and Richard M. Burke—all graduates of the University of Ohio.

For the last five years, the partners of CEA have met every Friday for a working lunch at Coasters, a local restaurant and bar that features attractive young waitresses wearing provocative swimsuits. The customers are almost exclusively men, and the interaction between customers and waitresses is often flirtatious.

This year CEA hired Elizabeth P. Grider, a Missouri State University graduate, as a new partner in the firm. She has attended two of the working lunches at Coasters and is uncomfortable in this environment. She does not feel, however, that she can just skip the events, which are the only regular occasions on which all the partners gather. Projects and work assignments are often discussed and decided at these Friday meetings. In addition, the lunches offer the opportunity for her, as the new kid on the block, to try to establish a comfortable working relationship with her partners. She realizes that "Friday at Coasters" is a long-standing tradition at the firm and she is reluctant to upset the status quo, but she wishes another venue, acceptable to everyone, could be found.

As Elizabeth P. Grider, write a memo to the senior partner, Robert B. Davidson, explaining your problem and recommending one or more solutions. You would like to speak to Davidson directly, but you think that writing a memo allows you to organize your thoughts. After the meeting, you plan to leave the memo with Davidson as a written record of your position.

5. Robert Braxton, 1296 Sycamore Avenue, Idaho Falls, ID 83401, is disappointed with the color laser printer he purchased from your company, INK.com, 4307 88th Street, St. Louis, MO 65407. He has written a letter to you to complain.

Braxton is a freelance technical editor. Earlier this year Braxton ordered one of your Horizon 9900 color laser printers. Braxton paid $2,870, including taxes, for the 9900, which is your top-of-the-line printer. He says he had no trouble setting it up and that initially it worked fine. But today, the printer jammed while printing flyers. To clear the jam, Braxton had to remove the print cartridge. In the process, the cartridge cracked and dumped ink all over the inside of the printer. And as he removed the jammed sheet of paper, two of the guide rollers loosened. Braxton has enclosed the mangled flyer and photographs of the equipment, which display the damage clearly but also show that Braxton was using a refilled print cartridge instead of a new cartridge, as specified on page 9 of the user's manual. In addition, you find that the paper that jammed the printer was a heavy weight (60-lb) cotton bond instead of the light to medium weight (16- to 24-lb) paper recommended on page 33 of the manual.

Braxton has a $283 estimate for cleaning and repairs from his local computer store and wants you to agree to pay it. You are concerned because Braxton has recommended the Horizon 9900 to several of his business clients, who have written you asking for more information. The possible additional sales would be important to your new company. However, you know that Braxton bears much of the blame for the damage to his printer because he had not read the user's manual carefully.

You decide to offer to send a new print cartridge without charge, but no more. You will also instruct Braxton to refer to page 9 and 33 in the user's manual and you will emphasize as politely as possible the instructions to use new print cartridges and 16- to 24-lb paper.

Write the letter to Braxton. Determine what additional actions are required to avoid similar complaints. Write the letters, memos, or e-mail messages necessary to direct such additional actions.

6. Compose the letter called for in the following situation.

You are a member of the permissions department at Educational Books, Inc., 25 Astoria Avenue, New York, NY 10027. One of your responsibilities is to be sure that lengthy excerpts from books published by your company are not illegally copied for classroom use. One of your company's textbook authors, Linda Davalos, is a visiting professor at Guixin University. According to Professor Davalos, one of the professors at Guixin University, a Chinese national, is copying several chapters of one of your company's textbooks for distribution to his students. The book, *Communicating in Technology and Business*, by Jane Fisher, is available for worldwide sale. Such extensive copying clearly violates international copyright law, as covered by the Berne Convention for the Protection of Literary and Artistic Works. However, you also have heard from several sources that instructors in China often ignore the law chiefly because textbooks are expensive for their relatively poor students.

You learn from your informant that the Chinese professor, Li Kua-fan, teaches English. He is a senior and respected member of the faculty. You wish to persuade the professor that requiring students to buy the book would be fairer to your company and the book's author and also in the best interest of his students. You intend to write a letter to the professor to accomplish that objective.

First, however, you must familiarize yourself with the Berne Convention so that in your letter you may refer to specific clauses. Also find out what you can about aspects of Chinese society and the Chinese school system today that might lead a respected professor to engage in the extensive copying of copyrighted material. What might be the arguments he would use to justify such copying? To be fully persuasive, you will need to anticipate and address such arguments.

Scenario

Adrien Kasmiersky has a large stack of material on her desk. Her department has been trying to resolve a major issue regarding replacement of the existing computer system used to track all patients at the family practice medical clinic where Adrien has worked as senior office manager for the past three years. During the past year, the clinic managing staff have heard presentations from three different vendors and received material from two more, as well as information packets from a half-dozen other companies that specialize in systems for medical groups. What to do?

The managing clinic director, Dr. Sylvia Hawes, thinks a presentation should be made to the physicians. Several have shared complaints from patients about the mounting paperwork generated by a routine clinic visit. Adrien knows that the doctors are not happy with the current system either and would like the tracking to be done reliably and cost-effectively. What kind of material should she prepare for them? How long? What about an oral presentation? Or, perhaps both?

This chapter will examine a process to help you think creatively as you plan reports that often require unique approaches to ensure that the material effectively targets intended readers.

CHAPTER THIRTEEN

Creating Reports for Any Occasion

- **The Variable Nature of Reports**

- **Liability and Report Writing**

- **General Report Requirements**

- **Determining Internal Report Development**

- **Importance of the Introduction and Summary**

- **The Online Report**

- **The Slide/Visual Presentation Report**

  Exercises

In Chapter 10, we discussed the standard elements of reports, providing examples of each as they appear in actual reports and indicating their respective purposes. The report guidelines are just that—guidelines. As you saw from report examples on the book's companion Web site, report design varies. As a writer, you must choose the report design that best assures the effectiveness of your report. You use and shape the design elements to respond to a specific communication need.

For example, you may be told to write a report

"summarizing what we have done in developing solutions to design problems with our computer batteries"

"discussing our major problems in choosing the best toxic waste cleanup plan for this neighborhood of 2,500 single-family houses"

"assessing four space-reallocation plans for our company and recommending the best two"

"recommending the offices that should be moved into the new building"

"explaining the results of tests on pumps that will remove 300–500 gallons of flood water per hour from golf course fairways located along creeks"

Each of these situations incorporates characteristics of different report types:

- Information reports
- Analytical reports
- Feasibility studies
- Recommendation reports
- Empirical research reports

The foregoing classifications simply provide a way of describing reports of different kinds. To help you understand how to develop reports, we will discuss types generally used, although some reports may combine elements of several types. Remember, each report you write is unique: the effective report will respond to the situation that led to the report. In Chapter 14, we will explain the development of feasibility studies and empirical research reports. But first we want to help you understand how to design a report for any purpose.

In this chapter we provide six distinct kinds of reports; each develops reports for five different situations. Each example shows how a writer can develop a report to respond to the conditions that necessitated the document.

THE VARIABLE NATURE OF REPORTS

Any type of report can be presented as a formal report, an informal report, or a letter or memo report.

Formal reports include all or most of the elements discussed in Chapter 10—transmittal letter/memo, title page, table of contents, list of illustrations, abstracts, summary, glossary, body of report, conclusions, and appendices. Long reports use formal format to help readers access information efficiently. Reports that will have a wide range of readers will also benefit from many of the formal report elements. Most of the report examples accessible from the book's companion Web site are formal reports and use some or all of these elements.

Informal reports usually do not include prefatory matter. Rather, they focus on the report content: summary/abstract, introduction, discussion, conclusion, and any attachments. Reports initially designed to be read on-screen usually begin with a table of contents. Each item links directly to that report section.

Letter and memo reports are seldom more than five pages. They do not need the complex format of the long report. Memo reports remain within the organization; letter reports go outside the organization. In this chapter, Figure 13-1 is a memo report, Figure 13-2 is an informal report carefully designed for internal uses in a university, and Figure 13-3 is a letter report. Study the situations surrounding each report: each is unique. Nevertheless, the reports have similar elements, and in all cases the development of a report and its constituent parts emerges from a consideration of the purpose and context of the report, and its intended readers.

LIABILITY AND REPORT WRITING

Before discussing report design, we want to stress that all reports carry legal responsibility. Reports, like letters, e-mail messages, proposals, and instructions, can be used as legal instruments and can be introduced as evidence in court. They document your activities and your competence as an employee. Always design and write with care; you can never be sure that a document will be used for your intended purpose only. All documents should be accurate, complete, mechanically correct, effectively designed, and visually appealing. They should testify to your professionalism, your ability to think logically and precisely. When you write a report to a client, he or she will expect to be able to make financial and technical decisions based on the accuracy and clarity of what you say. If your work is unclear or inaccurate, a client may hold you legally responsible for any costs incurred as a result.

Producing the kind of document that achieves its purpose begins and ends with the composing process, which we emphasize throughout this book. Always look for ways in which your report could be misconstrued as you analyze

- Audiences
- Your relationship with these audiences
- The context that prompted your report
- The contexts in which it is being written and received

As you plan, organize, and edit, be on the lookout for copy that could be subject to multiple interpretations. Revise as necessary to prevent readers from wrongly interpreting your words. Care in developing every document you write is the best defense against having your writing used to discredit you or your organization.

Reports document recommended actions. For that reason, their content must be credible, that is, well argued, clear from the perspective of the readers, and believable (even if readers disagree with the ideas presented). Be sure that you give credit to sources you use in compiling your report to avoid charges of plagiarism. If content in your report is challenged, accurate documentation of sources strengthens your credibility as the writer. Planning, revising, and editing are critical, for reports stored on paper or electronically may be accessible indefinitely, and once submitted, a report is beyond the control of the writer.

GENERAL REPORT REQUIREMENTS

Credible reports begin with planning, which we introduced in Chapter 2 on composing. To review, planning entails the following:

- Understanding the situation that has led to the report
- Understanding what you want readers to take away from the report
- Understanding the readers' perspective on the information you will present
- Applying your knowledge of how people read and process information to the development and presentation of the message
- Choosing content, organization, style, and tone that are suitable for your audience, the message, and your relationship to both in the organizational context

In short, good report writing begins with good planning. Because employees are often overwhelmed with paperwork, your reports need to be designed to encourage audiences to read them. Therefore, you will want to apply carefully the principles covered in Chapter 8, Designing and Formatting Documents, and Chapter 10, Developing the Main Elements of Reports.

The following situation explains the background, rationale, and purpose of Figure 13-1, a good example of a document intended to report information.

Situation 1 Graham and Simpson is an architectural consulting firm that specializes in analyzing construction problems. The membership of a historic New England church has hired Graham and Simpson to determine why the chimneys of the church are leaking and to make recommendations. First Church is, of course, interested in the cost of correcting the problems. Tim Fong,

TO: Eben Graham   DATE: August 15, 2005

FROM: Tim Fong

SUBJECT: Status of Investigation of Masonry Deterioration and Leakage, The
           First Church, Nashua, NH

Jane Hazel, James Portales, and I have examined two of the chimneys of The
First Church. The following points should aid you in preliminary conversations
with First Church to inform them of the extent of damage. We will have a final
letter report for the Church by 10/31/05 that will include repair estimates for all
three chimneys and the roof.

Report purpose

Conclusions

The two chimneys we have examined are badly deteriorated and require
extensive repairs to stop the leakage and to eliminate an increasing danger
of falling debris.

Conclusions and
recommendations
up front

(1) Leakage through faulty flashing can be stopped easily and effectively by
repairing the flashings.

(2) Leakage through the stone masonry cannot be stopped effectively because
the chimney masonry is badly deteriorated. The stones in the chimneys have
shifted, thereby enlarging the stone joints and creating horizontal surfaces that
catch water. Water entering the masonry can bypass the lead counterflashing
internally and enter the building. Freeze–thaw action on the wet stone masonry
has slowly pushed the stones apart. The stones in the chimneys are not
mechanically tied together to prevent movement.

(3) Unless major repairs are made to the two chimneys and the bell tower,
significant deterioration will occur in wood decking and masonry. The damage
will cause dangerous conditions from falling debris. Eventually, the structural
safety of the church will be threatened.

Recommendations

(1) We recommend removal of any chimney not needed. The work should
include removal of the chimneys to below the roof level, repair of all deteriorated
wood about the chimney, installation of a new roof deck, and installation of new
roofing slate to match the existing roof.

FIGURE 13-1 · INTERNAL PROGRESS REPORT ON A CHURCH REPAIR PROJECT

Eben Graham　　　　　　　　　　-2-　　　　　　　　　　August 15, 2005

(2) Any chimney scheduled to remain should be demolished and reconstructed with the exterior stones. Reconstruction should include new clay flue liners, steel reinforcing in the horizontal joints of the stone masonry, and new stainless steel or copper through-wall flashings. The roofing around the chimney should be repaired with new metal base flashing of the same metal used for the through-wall flashing.

(3) The wood components of the deck should be thoroughly examined and replaced or repaired, as required by the rotted conditions. Repairs should include exploratory operations into hidden conditions to ensure all deteriorated wood is repaired.

(4) Reconstruction of the interior finishes on the church should be delayed until all repair work is complete and tested.

Field Observations

We have examined both the interior and the exterior of both chimneys.

Support for conclusions

Chimney No. 1

Chimney 1 has extensive problems with wood and masonry.

Interior Observations

* The chimney is constructed of brick below the roof deck.

* The underside of the roof deck and joists are water stained on four sides of the chimney. The heaviest staining occurs on the north side of the chimney, which is the low point of the surrounding roof.

* A gutter has been installed at the low side of the chimney to catch leakage water and direct it to trash barrels.

* The low header in the roof deck framing on the north side of the chimney opening is wet and extensively rotted, allowing easy insertion of a two-inch pocket knife. Large pieces of the rotted wood are easily removed by hand.

* The remaining structural components surrounding the chimney are sound, based on our visual examination from the attic space.

FIGURE 13-1 · *CONTINUED*

Eben Graham   -3-   August 15, 2005

Exterior Observations

* The chimney is capped with roofing cement and fabric membrane (see Photo 1).

* The mortar joints have been coated with a caulking material (see Photo 1 and Photo 2).

* A cricket made of copper sheet metal is located on the roof above the chimney (see Photo 3). There is one small crack in a soldered seam on the ridge of the cricket.

* After lifting the counterflashing on the cricket, we found a large hole in the membrane (see Photo 3). Water draining over the roof and onto the cricket enters through this hole into the building.

* Some of the granite blocks in the chimney have shifted, and the mortar joints are open slightly.

Chimney No. 2

Our examination shows that Chimney 2 is also in a highly deteriorated state.

Interior Observations

* The roof deck and structural components surrounding the chimney are stained from water leakage, but all components are sound, as far as we can tell without removing any decking.

Exterior Observations

* The chimney is capped with roofing cement and fabric membrane.

* The mortar joints have been coated with a caulking material.

* The flue lining is brick, which has deteriorated. The mortar joints of the lining are heavily eroded (see Photo 4).

* The granite blocks that form the exterior width of the chimney have shifted outward, leaving many of the horizontal and vertical stone joints open (see Photo 5).

Report design separates the discussions of the two chimneys

FIGURE 13-1 · *CONTINUED*

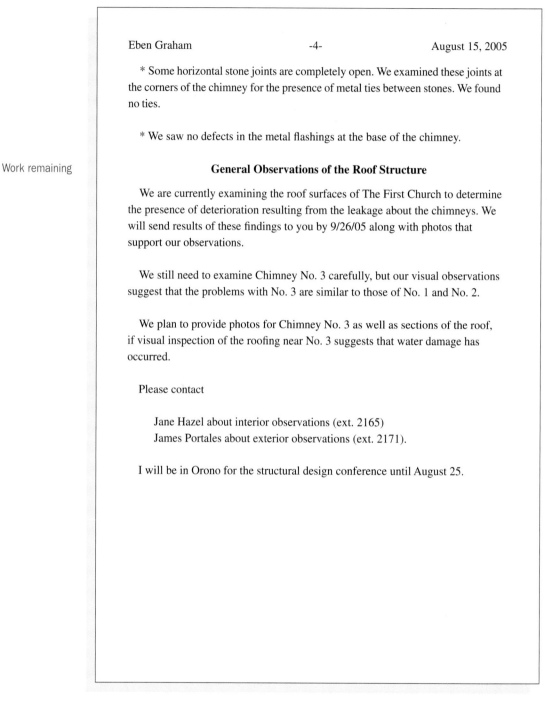

Eben Graham   -4-   August 15, 2005

* Some horizontal stone joints are completely open. We examined these joints at the corners of the chimney for the presence of metal ties between stones. We found no ties.

* We saw no defects in the metal flashings at the base of the chimney.

General Observations of the Roof Structure

We are currently examining the roof surfaces of The First Church to determine the presence of deterioration resulting from the leakage about the chimneys. We will send results of these findings to you by 9/26/05 along with photos that support our observations.

We still need to examine Chimney No. 3 carefully, but our visual observations suggest that the problems with No. 3 are similar to those of No. 1 and No. 2.

We plan to provide photos for Chimney No. 3 as well as sections of the roof, if visual inspection of the roofing near No. 3 suggests that water damage has occurred.

Please contact

    Jane Hazel about interior observations (ext. 2165)
    James Portales about exterior observations (ext. 2171).

I will be in Orono for the structural design conference until August 25.

Work remaining

FIGURE 13-1 · *CONTINUED*

the managing engineer on the First Church project, writes an internal report to Eben Graham, one of the principals of Graham and Simpson, who is dealing directly with the church committee on repairs.

Tim's main purpose is to let Eben Graham know what the team has done on the First Church project so that Graham can tell the chairperson of the church committee about the progress of the study. Eben Graham requires project reports—periodic and final—from all managing engineers on all projects so that he knows what's going on and can inform clients regularly. In a sense, this shows the characteristics of a good informal memo report prepared for internal use. Refer to Figure 13-1 as you review the six characteristics listed here.

1. State the subject of your informal report clearly. All reports need clear titles. Informal reports are no exception. A busy reader with a dozen items to read will often decide to read or not read an internal company report based on the subject line. For example,

SUBJECT: Description of First Church Masonry, Nashua, NH

is not as effective as

SUBJECT: Status of Investigation of Masonry Deterioration and Leakage, First Church, Nashua, NH

Similarly, the paragraph following the subject line tells the reader the purpose of the report, the date on which the report is due, and the meeting where the report will be discussed. Adding the date is also useful in aiding the reader's memory. If a report requires action by readers, you may want to add that immediately below the subject line:

SUBJECT: Recruiting Meeting—June 30, 2004—Change in Hiring Policies for Part-Time Research Technicians.

ACTION REQUIRED: Attached forms must be completed by all departments by June 13, 2004.

2. In a memo report, begin with the main information. That way, a reader who sees nothing but the subject line and the first page will get the point of your message. In Figure 13-1, the subject line is followed immediately by a short introduction. The conclusion and the recommendations based on those conclusions explain precisely what needs to be done. In many reports, conclusions and recommendations are of greater interest to the reader than the rationale for either.

In Figure 13-1, the introduction briefly alerts the reader to the reason for the report. Longer, more complex introductions, as you will see in our further examples, are appropriate when circumstances call for a fuller explanation.

3. Keep any additional paragraphs short to make the content easier to read, but be sure to provide all explanation needed to support the information in the opening paragraph.

4. Use design principles to highlight information. In Figure 13-1, the supporting part of the report is divided into two main headings—Chimney No. 1

and Chimney No. 2. Main problems with each chimney can be seen at a glance because of the listing arrangement.

5. Be sure to tell readers where to get additional information. Figure 13-1 includes the names and phone numbers of the engineers working on different aspects of the First Church project so that Graham can contact the appropriate individual if he has questions.

6. If you plan to e-mail the report as an attachment, make every effort to include the critical information on the first page. Many readers who open attachments will read only the first page unless that information suggests to them that they need to scroll through the rest of the report. Be sure to fill in the subject line of the e-mail.

Situation 2 The report presented in Figure 13-2 was written by an officer of a regional university that has been investigating its honor system and planning to revise it. The president of the university authorized a task force to develop a new honor system and an office to support its operation. The task force drafted a plan and submitted it to the president's staff for approval. The associate provost for academic operations writes the report to the university academic affairs committee—and the co-chairs of the task force—to launch the implementation of the new honor system. The task: to ensure that the new system is launched efficiently and effectively in terms of the university's constituent groups. The report is unique—it was designed to serve as an agenda for coming meetings. It is a problem-solving document that encourages thought, input, and answers to questions that the president's staff believe to be critical.

This organization scheme was selected for four reasons:

1. Readers can respond to questions and add others.
2. The introduction encourages input from the intended readers.
3. The bulk of the report, the list of questions, can be easily skimmed, a design feature planned to help readers during the forthcoming meetings. Note that the questions are numbered consecutively, across each section.
4. The report, sent as an e-mail attachment, allowed recipients to insert information before printing a copy of the report to bring to the first meeting.

▨ DETERMINING REPORT STRUCTURE

As mentioned earlier, many reports won't need all the design elements discussed in Chapter 10. Reports designed for internal communication can be set up in memo format (Figure 13-1) even if they are several pages long. Whatever the length of the report, you will want to organize it to achieve its intended goals, providing a logical system of headings to reveal the content to the reader.

Situation 3 John and Carol Hatfield have been married eighteen years. They have one child in high school, one in elementary school, and a toddler. The Hatfields have worked with Daniel Derringer for seven years. Dan tracks

MEMORANDUM

DATE: January 10, 2005

TO: All Members of the Academic Affairs Committee

FROM: Xavier Williams
                   Associate Provost for Academic Affairs

SUBJECT: **Establishing the New Honor System**

I've asked a number of questions in the attached memorandum to the Honor System Task Force. All need answers before we proceed with the launch of the new Honor System. With the concurrence of the Task Force, I want your input. Each of you represents a specific area of expertise and a perspective critical to our recommitment to integrity.

Please study as many of the questions as you can during the next week. You are invited to help answer these questions by participating in a series of meetings, beginning at 1:30 on Wednesday October 26, 2005, in Room 2092 Ranford Hall.

FIGURE 13-2 · TRANSMITTAL MEMO AND REPORT FROM AN ACADEMIC TASK FORCE ON THE ESTABLISHMENT OF THE NEW HONOR SYSTEM

TO: Ben Gonzales
                Miranda Mosier
                Peter Reichter, Co-chairs, Honor System Task Force

FROM: Xavier Williams

SUBJECT: **Launching the New Honor System**

The President and Provost have approved the plan submitted by the original task force on academic integrity. The university is now ready to proceed with launch of the plan. The body of this memo is a group of questions relating to implementation, operation, and resolution of resulting challenges from students, faculty, and parents (not to mention former students). The answer to many may be "too soon to tell," and perhaps other questions besides those below need to be addressed. As you peruse this list, look for areas I have omitted. I've arranged the question about the following areas: implementation, operation, and challenges. These questions have surfaced in the President's staff meeting.

The main purpose of this note is to try to determine the questions we must answer before we officially launch the office. The additional purpose is finding answers. That's why all the questions! We will focus on these questions at our meetings. Please come prepared to suggest answers and add additional questions. The better we are prepared, the smoother the new system will operate.

Implementation

1. What is the best target date for launching the new Honor System?

   Fall?

   Spring?

   Summer?

   What are the pros/cons of each date?

2. What's the best way to advertise the Honor System?

3. Who should be responsible for the announcement?

4. How do we let students know about the new system?

5. How can we ensure that students are informed about the effective date of the system?

6. How do we make clear that all students must comply?

7. What publicity/information strategies have been selected?

8. What about press conferences? Press releases? To what media? When?

FIGURE 13-2 · *CONTINUED*

Establishment of the New Honor System—2

9. How many people will be needed in the new honor system office by launch date?

10. Can we shift staff from existing positions? Or, will we have to hire all new staff?

11. What should the director's qualifications include? Should we hire an internal person or look for someone outside the university?

12. Is the budget realistic?

13. Can we retain the services of the Task Force to provide a resource for the director? Or, will the input from the Honor Council representatives be sufficient?

14. Have all colleges set election dates for their honor system representatives?

15. Do all deans and department heads have copies of the honor system manual (or the Web site address)?

16. What's the status of the Web site? Are the materials available sufficient? What else do we need on the Web site?

Operation

17. Who will maintain the honor system website?

18. If faculty report violations of the honor code via e-mail from the honor code Web site, will the database be complete by launch date?

19. Do we *need* a database? How far can we go without one?

20. Can we use our current student information system to track students with violations?

21. Can we collect reports of violations without spending $80,000 to develop a database?

22. What kind of record-keeping system will the honor office need? Have we defined what kind of system we will need?

23. Are we sure the registrar's office can provide the record keeping on F grades that students earn because of integrity violations?

Challenges

24 How do we encourage students to support the new system?

25. How do we impress them with the importance of academic integrity and that it affects the value of their degrees?

FIGURE 13-2 · *CONTINUED*

Establishment of the New Honor System—3

26. Has the Honor System manual been checked by Legal?

27. Have we made a list of "what ifs"?

28. What kinds of challenges can we expect to face

   from parents?

   from students?

   from former students? Do we need a campaign to inform the former students?

   from various advocate organizations and governmental entities?

29. Do we have a list of possible questions (with answers) to give faculty, deans, etc., who may be contacted by parents, the press, and the former students? Don't we want such a list of Q/A to ensure that everyone gives the same answers?

30. What kinds of incidents, from honor code violations during the past five years or so, has the university been required to resolve?

   Did we resolve them?

   Have we revised our student policy and procedures to deal with violations?

   Do our student policy/procedures mesh with the new system?

31. What are our biggest challenges in launching a new honor system and ensuring that it works from the beginning?

32. Do faculty support the establishment of an honor system office for academic integrity infractions? If not, why not?

Other Questions: Please list (and answer).

33.

34.

35.

36.

37.

FIGURE 13-2 · *CONTINUED*

changes in the tax laws and informs John and Carol of changes that affect them significantly. John, a psychologist who has a private practice, is fairly adroit about taking advantage of the tax system, as his father was an officer in the Well-Fargo Bank in San Francisco. John's sister, Jennifer, also a client of Dan's, is single and pursuing anthropology research in Europe. John is overseeing her financial affairs while she is out of the country.

Because of tax law changes, Dan writes letters informing his clients of issues they may want to consider. He expects that many of them will make appointments to discuss new financial planning strategies to take full advantage of tax law changes. Dan uses this communication strategy to keep a close working relationship with his clients. The letter tailored to the Hatfields' circumstances appears in Figure 13-3.

Note that Dan develops this letter report around headings. He begins with an introduction that states his purpose. He keeps paragraphs short and concise. His headings reveal the logic of the content he wants the Hatfields to know.

DETERMINING INTERNAL REPORT DEVELOPMENT

Reports can be arranged in various ways. As in Situation 2, the unique structure, organization, and presentation are determined by the purpose of the report, the intended readers, their expected attitudes, and the message to be delivered. The numbered lists *invite* rather than discourage perusal of the content by the average reader. We cannot overemphasize that the longer the report, the less likely any reader is to read all of it, and, the more important design is in encouraging readers to skim and surface read the content and find critical information.

IMPORTANCE OF THE INTRODUCTION AND SUMMARY

As Figures 13-1 through 13-3 illustrate, an introduction of some type is essential. The introduction may include the following components.

- the purpose of the report
- the rationale for the report
- the scope of the report—what will be covered
- perhaps a short summary of the main ideas covered in the main body of the report
- perhaps a combination of the summary and an introduction

Always, however, be clear about the report subject, its purpose, and the plan of the report.

As discussed in Chapter 10, the extensiveness (length) of the introduction depends on the readers' needs: Will they be expecting to receive the report (as in Situations 1 and 2)? Do they need to understand the circumstances surrounding the report and why it was written (as in Situation 3 and Figure 13-4)?

Daniel J. Derringer, CPA

961 Havner Street
Suite 300
San Carlos, CA 94070
(650) 592-1394

September 15, 1997

Mr. John and Carol Hatfield
1520 Hoover Street
Burlingame, CA 94010

Dear John and Carol:

On August 5, 1997, President Clinton signed the Balanced Budget Act of 1997 and the Taxpayer Relief Act of 1997 into law. The tax law makes more than 800 changes that affect nearly everyone. The effective dates for the changes vary. Some provisions are retroactive; some take effect immediately; and many go into effect in 1998 or later. As your CPA, I want to inform you about changes that affect your particular tax situation.

Please take a few minutes to read this synopsis of changes that affect you. Please give me a call if you wish to review your tax planning in light of this new law. I appreciate the opportunity to work with you on your financial planning. Please note that many of these changes will also affect Jennifer's tax situation. She may want to give me a call when she returns from London in early June.

Child Tax Credit

Beginning in 1998, each child under age 17 will be eligible for a $400 tax credit, which will increase to $500 per child in 1999. The credit is phased out at higher income levels beginning at $75,000 for singles and $110,000 for couples.

Capital Gains Tax Rates

The top capital gains tax rate, which had been 28%, has been lowered to 20%. People in the 15% income tax bracket will pay 10% on capital gains. The new rates apply to investments held for more than a year and sold after May 6, 1997, and before July 29, 1997.

For assets sold July 29th or later, the lower rate will apply only if the assets have been held 18 months or longer. Assets purchased in 2001 and later and held for at least five years will be taxed at

FIGURE 13-3 · LETTER REPORT

Mr. and Mrs. John Hatfield
September 15, 1997
Page 2

an 8% rate for lowest bracket taxpayers and at 18% for the higher bracket taxpayers. Depreciated real property is subject to special recapture provisions.

Home Sales

The law exempts from taxation profits on the sale of a personal residence of up to $500,000 for married couples filing jointly and $250,000 for singles. To qualify, sellers must have owned and used the home as their principal residence for at least two of the last five years before the sale. Effective for sales after May 6, 1997, this new provision replaces the prior rollover provision on home sales and the $125,000 exclusion of gain for those 55 and over. The new law does not change the rule prohibiting taxpayers from deducting losses on home sales.

Education Savings

Two new tax credits will be available for higher education expenses. The HOPE tax credit gives up to $1,500 of credit for each of the first two years of college (100% of the first $1,000 and 50% of the next $1,000). "Lifetime learning credits" are available for expenses paid after June 30, 1998. The maximum credit is $1,000 through 2002. After that date, the maximum increases to $2,000. Both the HOPE credit and the lifetime learning credit phase out for higher income taxpayers.

Interest paid on qualified education loans will be deductible even by those who don't itemize deductions on their tax returns, beginning in 1998. The maximum deduction allowed will be phased in over four years with a $1,000 maximum in 1998, increasing to $2,500 by 2001.

Beginning next year, taxpayers will be allowed to contribute annually up to $400 to an education IRA for each child under 18. The contributions are nondeductible, but withdrawals are tax-free if they are used to pay qualified higher education expenses. The annual contribution limit is phased out for married filers once their income reaches $150,000.

Retirement Accounts

Under prior law, couples with company pension plans could make the full $2,000 IRA deductible contribution if their income was $40,000 or less ($25,000 or less for singles). Beginning next year, these limits increase to $50,000 for married couples and $30,000 for singles. These amounts gradually increase until they reach $80,000 for married (2007) and $50,000 for singles (2005).

FIGURE 13-3 · *CONTINUED*

Mr. and Mrs. John Hatfield
September 15, 1997
Page 3

The new law allows penalty-free IRA withdrawals of up to $10,000 for the purchase of a first home and penalty-free withdrawals with no dollar limit if the money is used to pay for qualified higher education expenses.

The tax law creates a new type of individual retirement account called a Roth IRA (also called IRA Plus), which allows nondeductible contributions of up to $2,000 a year. Earnings within the account accumulate tax-free, and withdrawals are tax-free after five years if the money is used for retirement (that is, the distribution occurs after age 59 1/2) or for a first-time home purchase ($10,000 lifetime limit). The distribution can also be tax-free if attributable to the disability or death of the individual. No more than $2,000 per year may be contributed to the combined IRAs of an individual.

The law also eliminates the 15% excise tax on large distributions from 401(k) or other pension plans. (This tax has been under a three-year moratorium scheduled to expire at the end of 1999.)

Tax Planning—More Important Than Ever!

Tax planning becomes more challenging with every new tax legislation. The 1997 tax law, is, to quote the generally accepted observation of one analyst, "mind-numbing in its complexity." The law introduces additional complexity even for the average taxpayer because it offers such a variety of new and often bewildering options in many areas.

To take advantage of what the new law offers, you may need to decide whether to put your retirement savings in a regular IRA or a new Roth IRA, whether or not to keep long-term capital assets in retirement accounts at all because of the lower capital gains rates. Saving for college and paying college expenses now provide tax breaks, but finding the best option for you will take some analysis. Clearly, however, wills and estate plans, especially where small businesses are involved, must be reviewed and updated to take advantage of the new law.

I will continue to analyze this latest tax law and inform you of provisions that may affect your tax-paying decisions. If you have questions, please call my office. We can set up an appointment to discuss questions you may have and changes in your financial planning.

Sincerely,

Dan Derringer, CPA

FIGURE 13-3 · *CONTINUED*

**Texas Agricultural Extension Service
The Texas A&M University System**

**Common Concerns in West Texas Sunflower Production
and Ways to Solve Them**

Calvin Trostle, Extension Agronomy, TAEX-Lubbock, c-trostle@tamu.edu, (806) 746-6101
Pat Porter, Extension Entomology, TAEX-Lubbock, p-porter@tamu.edu, (806) 746-6101
Photographs courtesy of North Dakota State University

February 27, 2001

Sunflower acreage in West Texas may more than double in 2001 as growers facing high irrigation costs look for options to move away from high-irrigation crops, particularly corn. This will lead more farmers to consider sunflower production, and many to try it for the first time. Sunflower production in West Texas, both confectionary and oilseed, has been the subject of several recurring concerns from growers. These primary concerns are:

1. Crop losses due to sunflower (head) moth,
2. "Sunflowers were hard on my ground," as many growers feel their subsequent crop performs poorly,
3. Volunteer sunflowers the year after sunflower production.

These are legitimate concerns. The lack of answers or understanding of these three issues has soured many Texas sunflower growers after only one or two years. Our objective is to help both prospective and current growers anticipate potential problems (or equipment needs) before they occur and offer some options to compensate for and overcome these and other secondary concerns in Texas sunflower production. Comprehensive production and insect guides for sunflowers offer more detail, but for now we want to highlight these issues in advance of increased acreage and new sunflower growers *before* the 2001 crop is planted.

General information on sunflower production in West Texas is available from TAEX (Calvin Trostle; Brent Bean, TAEX, Amarillo Research & Extension Center), sunflower contractors, sunflower seedsmen, and others. Kansas State Univ. publishes an excellent resource for sunflower production useful for Texas growers, "High Plains Sunflower Production Handbook," publication MF-2384 (January 1999). It is available via the Internet at http://www.oznet.ksu.edu/library/crpsl2/samplers. MF2384.htm. Also, Carl Patrick, TAEX-Amarillo, has written "Managing Insect Pests of Texas Sunflowers" (B-1488, January 1998) available at http://texaserc.tamu.edu/catalog (click on "Extension Publications" then "Insects" and find it in the list to read or download).

FIGURE 13-4 · INFORMATION REPORT ON SUNFLOWER CONCERNS

Sunflower Concerns—2

Primary Concerns

Sunflower (head) moth.

We call this insect "the Boll Weevil of Sunflowers" to help communicate how important and devastating it can be. No, it is not a season-long pest like the boll weevil of cotton, but is quite a different insect altogether. This insect left unchecked when sunflowers are entering full bloom, however, can be a disaster on par with any heavy loss boll weevils ever inflicted on cotton.

Sunflower moth is a tan or gray adult up to 1 inch long (but usually smaller). The moths are attracted to sunflowers beginning to bloom. Their regional population is often influenced by the many alternate hosts, especially wild sunflowers. Sunflower moths are only fertile when pollen is available. Egg laying occurs on the head. After hatching, young larvae are relatively exposed for the first 4 to 6 days as they feed on pollen and floral parts. However, older larvae tunnel into seeds and seed heads and are somewhat protected from insecticides. These older larvae may destroy up to ten seeds per larva. Often *Rhizopus* head rot, a fungus, follows the tunneling sunflower moth larvae. If the insects don't do the major damage *Rhizopus* can, turning sunflower heads black and mushy. The incidence of this fungal disease, to which there is little chemical control available, is closely tied to sunflower moth infestations.

For sunflower moth control, <u>timing is everything</u>, both in scouting and in treatment. Sprays control moths and young larvae, but not larvae that have tunneled into the seed. The presence of moths triggers insecticide applications. If you wait long enough to find worms, you have probably waited too long. Scouting is best in the early morning or evening, and needs to begin at the bud swelling (R-3) vegetative stage of development. Two facts have caught some growers unprepared to move on sunflower moth treatments: 1) buds of newer hybrids often open faster than older hybrids, and fields often go from 5% to near 50% bloom in as little as two days, and 2) a sprayer may not be available at the critical time you need it. When a field of sunflowers is at 5% bloom, i.e., 5% of the plants are in bloom (that is, yellow ray flower petals are visible), that is a safer time to make a spraying decision. If there is a delay of a couple days in scheduling a spray, then we still might be near 50%. In the past, spraying decisions were more likely to be made at 50% bloom, but in today's quicker-opening hybrids this leaves too little room for error. OK, so you sprayed for sunflower moth. Now what? Keep scouting! Particularly for early plantings many experienced sunflower growers anticipate a second spraying about five days later to avoid the possibility of any egg laying. In years of the most severe pressure, even a third spraying may be justified.

More northern states suggest that 1 to 2 sunflower moth adults per 5 plants is an economic threshold, especially for confectionary sunflowers. However, Texas guidelines say to spray when any moths are found in a field at 20–25% bloom. Many farmers, however, rightly feel more secure scheduling a spray even if only a few scattered sunflower moths are present. If you are a new grower and don't have experience with this insect we would advise that you consider hiring a consultant or working with your

FIGURE 13-4 · *CONTINUED*

Sunflower Concerns—3

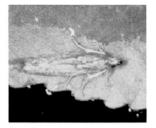

Sunflower (head) moth adult   Sunflower (head) moth larva   Sunflower (head) moth adult

local Extension IPM agent until you have learned how to respond appropriately to this insect. He or she can assist your scouting and spray decision-making, and offer suggestions on the several products that control this insect and how to apply them.

One agronomic practice that can affect our efforts to control sunflower moth is obtaining a uniform emergence of plants at once rather than spread out over several days or whenever the next rain occurs to germinate remaining seed. This is not always easily accomplished when planting in soils with poor moisture for good germination. But every effort should be made, plant 3 or even 2 inch deeper to moisture if you have to, to get a uniform stand. Else sunflower moths will be actively laying eggs in a field over a longer period of time than you have the patience or money to make additional sprayings.

Also, sunflower moth pressure is usually highest early in the season and declines with time. Early season sunflowers might require two sprayings, whereas a mid-June planting might require only one (or one early and none in June). This might steer growers to planting later, but keep in mind that sunflower yield potential declines with later planting dates.

Sunflower producers in the Dakotas have the nasty *Sclerotinia* disease. West Texas has the sunflower moth. We have practical treatment options, but they don't. As conditions allow we encourage growers to plant early and reach for the higher yield potential, provided they understand and are committed to staying on top of potential sunflower moth infestations.

While not useful for sunflower moth, crop rotation will help prevent some of the other potentially major insect pest problems, especially stem weevils (see Secondary Concerns section) and girdlers. However, sunflowers should not be planted in last year's soybean fields because girdler problems are usually worse in this rotation.

"Sunflowers were hard on my ground."

We believe there are two main contributing factors to the poor performance of crops following sunflower in West Texas. These factors are deep-water extraction and inadequate fertility programs for sunflower.

First, sunflowers are deeply rooted. It is common for sunflowers to extract soil moisture from a depth of six feet or more. Thus most of the time subsequent crops do not have as much stored soil moisture

FIGURE 13-4 · *CONTINUED*

Sunflower Concerns—4

to draw from, and there has been insufficient time for rainfall to recharge soils. Sunflower is a crop that can succeed in droughty conditions, but given the opportunity it will use amounts of water comparable to cotton, corn, and soybean. Sunflower is not a good candidate for the first crop in a double cropping system, as minimal moisture may be available for the next crop, particularly fall-planted small grains. On the other hand, the rainfall patterns in West Texas provide on average about 2.0 to 2.5 inches in September and 1.5 to 2.0 inches in October. Sunflowers planted by early May—this coincides with higher yield potential relative to the same hybrid planted in late May or June—even full season, will mature by about early September. Thus these two months of fall rains will help to replenish soil moisture in most years. This is a simple, cost-free means to alleviate soil moisture deficits after sunflower.

Secondly, sunflower soil fertility programs for much of West Texas are minimal or nonexistent. As a rule of thumb, each hundred pounds of sunflower production requires about 5 pounds of N, 1.5 pounds of P_2O_5, and 3.6 pounds of K_2O. As for micronutrients, for the same reason that deep moisture removal occurs due to deep rooting, we don't usually see micronutrient deficiencies in West Texas sunflowers. Kansas data suggest that a 2000 pound crop of sunflowers has about the same nutrient requirements as a 40 bushel per acre wheat crop. Our Panhandle wheat information suggests this level of wheat production would require about 60 pounds of N (soil and fertilizer). Using the above rule of thumb for sunflower would establish a requirement of 100 lbs. N per acre for a 2000 pound sunflower crop. This suggests that the fertility requirement for sunflower is higher than that of a 40 bu. per acre of wheat crop.

The nitrogen requirement for sunflower is a balance between ensuring that enough is present to reach yield potential, but don't add too much. Excess levels of N in sunflower are detrimental due to lodging and/or greater susceptibility to disease.

P and especially K are not normally a limitation in for West Texas sunflower unless soil test information warns that these nutrients might be short. In general, sunflowers will require some supplemental P if soil test is Very Low and Low. Fertilizing with some P if soil test P is medium might ensure better residual fertility for the subsequent crop.

Sunflowers are often seen as a low input crop fertility-wise, and this sets up your subsequent crop for a potential disappointment. For sunflower, if a certain nutrient is not readily available it will take all it can from the top two feet or so and go deep to get the rest. Other crops can't do this. In West Texas if you choose to not fertilize sunflowers we strongly encourage you to soil test after sunflower in advance of growing cotton, corn, or other crops the next year. In lieu of a soil test one might assume that a typical irrigated cotton crop the next year after unfertilized sunflower may require 30 to 40 additional pounds of N per acre above your normal application and up to 20 more pounds of P_2O_5 per acre. For dryland cotton, consider an additional 20 pounds of N per acre.

For further information on fertility, refer to the Kansas sunflower guide mentioned on the first page.

FIGURE 13-4 · *CONTINUED*

Sunflower Concerns—5

Volunteer sunflowers the next year.

This concern is less than it used to be with the advent of Round-Up-ready crops. Nevertheless, volunteer sunflower the following year can be a big nuisance. Significant amounts of sunflower seed may end up on the ground from shattering (particularly if harvested late), feeding birds, improper harvesting equipment, and not setting the harvester adjustments correctly. Experience at the Texas Agricultural Experiment Station, Halfway, TX, in 2000 underscored the need for timely harvest to reduce seed losses both to birds and losses due to over-dried crop. Round-Up is effective in controlling volunteer sunflower the next year, provided the plants don't get too big. Because Round-Up spraying on other crops could come several weeks after initial sunflower germination, we may need mechanical control in spite of Round-Up. Planting a Round-Up-ready crop the next year is not a requirement, but if you anticipate a volunteer sunflower problem it certainly is an option.

Year 2000 research reports several other herbicides may also be satisfactory for volunteer sunflower control. For further information please consult "Control of Volunteer Annual Sunflower" and "2001 Texas Panhandle Annual Sunflower Weed Control Guide" by Brent Bean and Matt Rowland of the Texas Agricultural Extension Service, Amarillo.

There are other actions we can take, however, to reduce the problem besides chemical treatments. Leaving the stubble and residues of harvested sunflowers on the surface in the late fall and over the winter allow ample time for birds to find the field and clean up a much of the seed. Given time, this can eliminate more than 80% of the seed remaining in heads or on the ground. Fields closer to town, bodies of water, or wooded areas are more likely to experience cleanup.

Sunflowers will germinate near 50°F (similar to corn). For many crops such as cotton or sorghum, which will be planted after these temperatures are reached, an early flush of volunteer sunflower can be controlled with tillage operations during field preparation.

Not having the proper harvesting equipment can contribute to volunteer sunflower, and we will discuss harvesting equipment tips below.

Secondary Concerns

Here are some lesser concerns we hear about sunflower production in West Texas. Some of these involve equipment, and it is not too soon to get the right planter plates, drums, etc., on hand to plant sunflowers the best you can. Also, experience in sunflower harvesting is a definite plus, and lest we get too busy and don't follow through we might leave too many dollars per acre of harvestable sunflowers in the field.

FIGURE 13-4 · *CONTINUED*

Sunflower Concerns—6

Achieving Desired Stand (Plant Population) and the Right Planter Equipment

These two issues go hand in hand. The best intention and understanding of why seeding rate is important is moot if you don't have the right planter equipment with any degree of seeding rate control. Achieving a targeted plant population is ESSENTIAL for sunflower production in West Texas. We need to limit overseeding. Any type of planter, even an air/vacuum planter, needs to be calibrated. From my own experience it is a real headache trying to calibrate a plate planter. Don't use one, as you probably won't or can't take the time to reliably and accurately calibrate it.

Seeding Rate Targets for West Texas

We encourage growers to think in terms of "seed drop" per acre rather than just plant population. This emphasizes our part in controlling the amount of seed we put out (and perhaps limiting the higher costs with too high seeding rates). Most sunflower guides talk about plant population and that is OK, but you have to have an understanding of how many seeds it takes to get that. Too many seed, hence plants per acre, is a potential economic pitfall for confectionary sunflowers due to the lower value, usually by half, of small seeds that don't pass a 20/64 screen.

Similar to sorghum in West Texas, we assert that we can manage drought risk with modest per acre sunflower seed drop rates. Particularly for dryland, growers should adjust typical seed drop down to poor moisture conditions, but don't increase the seeding rate above a targeted maximum just because you may have excellent soil moisture. A case in point: who would have thought with all the moisture Lubbock Co. had in June, 2000, that so much dryland cotton would fail? We believe we face this risk more than sunflower growers in Kansas and Colorado, and that is the primary reason that the sunflower seeding rates and resulting plant populations we suggest for West Texas are lower than you'll find in their recommendations.

Also, West Texas soils, especially as you move into the South Plains (Lubbock region), are sandier and have lower water holding capacity. We have higher evapotranspiration and irregular rainfall. Conditions are even more droughty. Although research suggests that sunflower seeding rates should not be adjusted for row width, the 40 inch row spacing is another reason to ensure that we don't go too high with seeding rates.

With that said, here are some assumptions underlying West Texas seeding rate targets. Assume 85% of seeds planted become productive plants (combination of germination and establishment). Adjust as needed for soil and moisture conditions (down but not up) and seed quality. The savior of many mistakes in sunflower production when rates were too low, and something we can take advantage of to manage risk, is that sunflower has great ability to flex upward (more seed, larger seed) at lower plant populations as conditions are favorable.

For West Texas confectionary (85% stand establishment), remembering that LESS (seed) is MORE ($) due to the approximate 2 × yield value of larger seeds, we suggest 16,000–18,000 seeds per acre irrigated, and 12,000–14,000 seeds per acre dryland.

**FIGURE 13-4 · ** *CONTINUED*

Sunflower Concerns—7

For <u>West Texas oilseed</u> (85% stand establishment), we suggest 20,000–23,000 seeds per acre irrigated and 14,000–18,000 seeds per acre dryland. For reasons given above, note that these oilseed rates are lower than often suggested for West Texas.

<u>Optimizing Your Sunflower Planting Equipment</u>

Foremost, <u>your sunflower seedsman should have the necessary technical information</u> to plant sunflowers at a targeted seed drop. We also have an old planter guide put out by Cargill that may have some useful information. In a nutshell, here is the planter equipment information for oilseed sunflower that a seedsman will probably hand you. For air/vacuum planters, medium corn plates work well for the large Size 2 oilseed, and OK with Size 3. Small corn plates help with plantability of the small Size 4 and perhaps Size 3.

For cyclo/drum planters always move the brush in the down position. In a normal corn seed drum, Size 2 works well and Size 3 is fair (but expect some unwanted doubles). A popcorn seed drum works OK with Sizes 3 and 4. Special sunflower drums work best with Size 4 rather than the larger seeds. Also, soybean drums can work fairly well if every other hole is plugged.

For finger pickup units, normal corn fingers work well with Size 2, OK with Size 3, but poor with Size 4. Special sunflower finger units are available and recommended for Size 4 and may also help with Size 3.

Alas, here is the plate planter information, just keep in mind you don't have that layer of "metering" protection you have for air/vacuum or drum units. Plate planters require a filler ring, and this ring varies with seed size (larger seed, thinner filler plate). For oilseed Size 2, use a yellow-colored BO-20 plate and a BFR-2 filler ring. For Size 3, use a BO-30 plate (red) and BFR-2 ring. For Size 4 use a BO-40 plate (blue) and BFR-40 ring.

Similar information for confectionary sunflowers may be obtained from your seed supplier.

Because we believe that achieving a targeted seed drop is so critical, we strongly suggest that if you can't get the right equipment together for sunflower planting we advise you to <u>lease an air/vacuum planter</u> or hire someone to plant your sunflowers. This is particularly so for confectionary sunflowers, where too many seeds hence plants per acre will yield more smaller seed, which is only worth half as much.

The Importance of Irrigation Timing

Sunflower can do about as much with the first five inches or so of timely irrigation water as any crop grown in West Texas. Data from Colorado suggest that after the first seven inches of water, each additional inch to sunflower can yield about 150 pounds of seed per acre. One or two timely rains or

FIGURE 13-4 · *CONTINUED*

Sunflower Concerns—8

irrigations can sometimes double yields as well as increase oil content for oilseeds (which increases your chance for a premium for oil content above 40%).

In general, the critical time for irrigation in sunflower is an approximately 40-day period from about 20 days (bud stage) before flowering to about 20 days after flowering, which roughly coincides with petal drop.

High water use begins at bud stage and peaks at flowering, which for most hybrids is about six weeks after emergence. Irrigation of confectionary sunflowers may be of slightly more value to ensure good seed size and quality.

Preplant irrigation for sunflower can have longer benefit than for other crops. Data from Kansas suggest that the yield return from one single large irrigation at either bud stage or full bloom (ray flowers fully extended outward) produced similar yields. In that scenario, if you could only water once (and there is enough soil moisture available during bud stage) then wait to water at full bloom.

In a limited irrigation scenario where you may apply two mid-season irrigations, apply the first irrigation at bud swelling, 0.75–1.0 inches in diameter (R-3) and the second watering at full bloom.

Sunflowers favor larger irrigations (2-4 inches per application) over frequent irrigation as deep percolation of water is still taken up by the deep-rooted sunflower. This also reduces the opportunity for disease development with fewer waterings and less humidity, particularly with sprinkler irrigation. Sandier soils of the Texas South Plains may not store all of the water in large irrigations, though if any crop can chase the water downward sunflower is it. Due to the limits of some pivot irrigations permitting more limited water per pass, the first watering may occur at bud with two subsequent irrigations 7 to 10 days apart (before early flower and full bloom).

Full irrigation of sunflower may be considered from early bud to petal drop. Research data suggests, however, that in many years optimal timely irrigation can perform just as well as full irrigation for sunflower.

Sunflower Stem Weevil

Sunflower stem weevil is an occasional pest, but can reach high numbers in early planted fields that have not been rotated. The small (3/16 inch) brown and white mottled adults emerge in early May and begin laying eggs 2 to 5 weeks later. The stem weevil is very difficult to detect. Larvae burrow into the stalk, feed, and eventually form a chamber near the base of the plant where they overwinter. This causes the plant to be highly susceptible to lodging, particularly when harvest is delayed. Many growers do not realize they have had damaging levels of stem weevil until plants actually lodge. The larvae also seem to predispose the plant to charcoal rot. Crop rotation and delayed planting can be used to help avoid the problem. For reasons unknown, Castro County usually has a greater problem with stem weevils than other counties in the Texas South Plains.

FIGURE 13-4 · *CONTINUED*

Sunflower Concerns—9

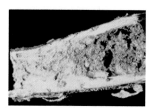

Sunflower stem weevil adult   Sunflower stem weevil larva   Sunflower stem weevil chamber

Bird Damage

Most sunflower growers deal with bird problems at one time or another. Just know that this nuisance can occur, but learn to recognize when it is truly an economic threat. A few turtledoves are not going to harm you—and you might even get some good hunting. But large numbers of birds such as crows and blackbirds have to be dealt with.

Being ready to harvest at maturity is probably the best thing we can do to keep bird damage in check. But left unchecked they will clean a field given enough time. In 2000 we had to hand harvest some of the early maturity hybrids at the Halfway, TX, plots while waiting about three weeks for the small plot combine to come.

Some of the following cultural practices may also help minimize potential for bird damage (from Kansas sunflower guide noted in introduction). Consider hybrid plant types with heads that turn down after flowering (note that the efficacy of any over-the-top insecticide may be reduced); plant early hybrids at early planting dates with early harvest (good for near town?); avoid planting near streams, playas, etc., where there are large numbers of birds and a late-summer water supply; leave a 100-yard buffer strip of crop not attractive to birds (small grains, cotton) adjacent to shelter belts or other wooded areas. On the other hand, after harvest don't be in a rush to plow harvested sunflower fields. Birds will help clean up shattered seed thus reducing volunteer sunflower potential.

Several mechanical noisemakers are possible to deter birds or drive them away, but this is difficult over large areas. Avitrol (4-aminopyridine) has been registered as a bait control, but may be only somewhat effective in the face of heavy bird pressure. This bait is really a last resort, and it may also kill game and other non-pest birds.

Proper Harvesting Equipment

Due to limited acreage the most advanced sunflower harvesting equipment may not be readily available in West Texas. All-crop headers (soybean headers, low-profile crop headers) are often the best choice if available. There is little need for modification, but this equipment can pose minor and

FIGURE 13-4 · *CONTINUED*

Sunflower Concerns—10

solvable problems as it runs more stalk through the combine. Small grain platform headers work satisfactorily only if "catch pans" are installed to cut losses due to seed shattering on the ground. In contrast to most equipment, the catch pans should be relatively inexpensive and worth the investment.

West Texas has more of the platform headers available and that would appear to be a favorable option as compared to an all-crop header. How many sunflower acres you may harvest over time will dictate whether you decide to purchase an additional combine attachment. Corn headers have also been used, and can work satisfactorily if you install stationary knives to cut the stalks. This, however, is not the best choice relative to the suitability of an all-crop header or a platform header with catch pans. Regardless what equipment you use, <u>check your losses</u> at the header.

If you have not harvested sunflowers before, we advise you to consult someone who has the experience. Talk to them beforehand and maybe have them check with you in the field at the beginning of harvest. Poor harvesting is the major reason why volunteer sunflower was cited earlier as a major concern.

Over-threshing is the most common problem with combine thresher settings. Preferably the sunflower head will get to the straw walker and pass on through the combine in one or maybe two pieces, but broken heads can overload the cleaning shoe with small pieces of head and trash. Experienced harvesters have learned to keep the cylinder speed slow (e.g., 250–450 rpm), but keep the concaves well open. Harvest at a reasonable ground speed when seed moisture is low (e.g., $< 15\%$), and use a minimum amount of air. As needed it is better to decrease concave clearance than increase cylinder speed to improve threshing.

All the right harvesting equipment won't stop significant harvest losses in a sunflower crop well past maturity. Eventual bird damage, the manifestation of lodging due to stem weevil, potential storm damage, etc., are good reasons to be ready to harvest once sunflowers are mature.

Disease Potential

We do not yet have much experience with diseases of sunflower in West Texas. We see downy mildew in the Lubbock region from time to time, but because this disease thrives more under wet or humid conditions there are limits in how much economic damage it can do. Foremost among sunflower diseases in the U.S. is *Sclerotinia* (white mold) which, like downy mildew, is much more common in wet, humid conditions. It is a major problem in the Dakotas, and there is no in-season treatment. It can be managed by rotating sunflower at most one year out of four. Industry sources say they have rarely seen *Sclerotinia* in the Texas South Plains. Also, phoma black stem and the rusts can appear on West Texas sunflower, but we have not yet learned much about them. *Rhizopus* head rot was mentioned earlier in conjunction with sunflower moth larvae activity in the head. The mushy appearance of a head so infected is not unlike the result of *Sclerotinia*.

According to Charlie Rush, plant pathologist, TAES-Amarillo, the big increase of sunflower acreage in the northern Panhandle in the early 1980s crashed in part because of disease problems. Unfortunately, we do not have a record of which diseases were present.

FIGURE 13-4 · *CONTINUED*

Figure 13-4, an information report, explains to ranchers the issues associated with planting sunflowers as a crop. The introduction explains the rationale and the plan of the report—a discussion of primary and secondary factors that will help growers decide whether they want to plant sunflowers. The headings articulate concerns expressed by ranchers. Remember: readers must be prepared for the information you intend to impart before they can process it. Take a few minutes to reread the introductions of the four example reports presented in this chapter. How do the introductions prepare readers for the content that follows?

If a reader may resist the report, then the introduction must encourage a consideration of the content (Figures 13-3 and Figure 13-4). Even readers who are not likely to object to the content will approach the document with several questions:

- What is this report about?
- What does it cover?
- Do I need to read it?
- How does it affect me?
- Will it be useful to me?

The introduction should answer those questions, and Figure 13-4 employs this approach. Thus, any report introduction should stand alone. It should make sense to a reader who is seeing the material for the first time. The length is determined by how much your reader needs to know before moving to the conclusions or the discussion.

THE ONLINE REPORT

Pursuant to our illustration of online reports in Chapter 10 (Figures 10-9 and 10-10), Figure 13-5 presents the table of contents of an online report that focuses on graphics. Note that the report is topically presented, with each section accessible as a link and the entire series of related graphic files listed separately.

The need to provide reports online, to comply with the U.S. government paperwork reduction effort, and the development of computer graphics, have led to innovative report design. For example, the U.S. Geological Survey reports (USGS) effectively combine visuals, such as maps, with short textual explanations. These reports are specifically designed for online presentation. As in most online reports, in a USGS report the content is demarcated by a table of contents available at the beginning of the Web page. Each section can be accessed from the link. Reports like the following may be read online or printed: http://pubs.usgs.gov/of/1999/of99-508/frames/display.html.

Another excerpt from the report whose contents appears in Figure 13-6 discusses beach profile of west-Central Florida Coastal Transect #4. Sections of the report can be viewed by clicking on the headings of interest, shown as links at

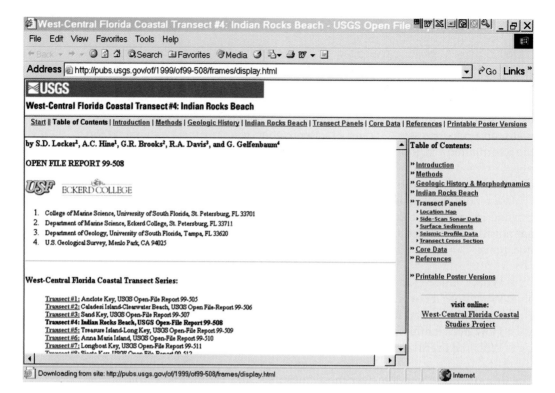

FIGURE 13-5 · OPENING PAGE OF ONLINE REPORT

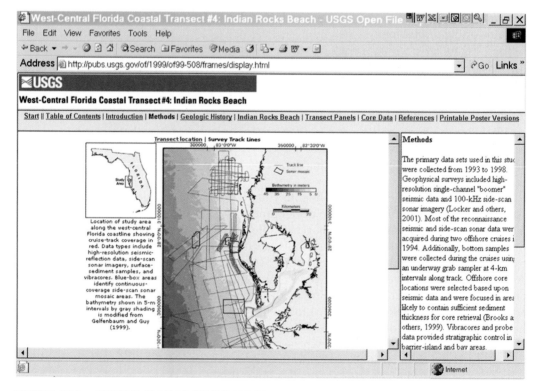

FIGURE 13-6 · OPENING PAGE OF ONLINE REPORT

the top of the screen. Individual segments within the page can be printed. Note, too, that the style is concise, almost succinct. The text focuses heavily on explication of the graphics and maps. In such reports, which incorporate more visual presentation than standard text, many of the standard elements will be used, but the text will be succinct. Most online reports are introduced by a table of contents composed of links to each report section.

THE SLIDE/VISUAL PRESENTATION REPORT

Some reports comprise slide presentations alone. The elements of the report are evident in the structure of these presentations, but the "report" itself appears in outline form only. Consulting firms often submit reports in a slide format only. Sensitive or proprietary details may be delivered during private meetings between the consultants.

Figure 13-7 shows a segment of a slide presentation report summarizing a 689-page environmental impact statement prepared by the Virginia Department of Transportation. The 28-page slide report summarizes the issues discussed in the study. Note that the slide report contains the elements of a

FIGURE 13-7 · SLIDE PRESENTATION

Timeline Recap
Major Investment Study & EIS

- **1994:** Develop new forecast model
- **1994:** Purpose and need approved
- **1995-96:** Corridor Study conducted
- **July 1997:** Selection of a locally preferred alternative by the MPO
- **Sept. 1997:** CTB endorsement of the MPO's LPA
- **Oct. 1999:** Draft EIS approved
- **Early 2000:** Public Hearings
- **July 2000:** CTB approved Alternative 9

HAMPTON
ROADS
CROSSING
STUDY

MIS Selection Criteria

- Important criteria for meeting purpose and need:
    - Reduce volumes at HRBT by 10% or more
    - Address existing and future regional O&D
    - Connect ports and major freight corridors
    - Connect to controlled access freeways
    - Relative cost
    - Relative ease of implementation

HAMPTON
ROADS
CROSSING
STUDY

FIGURE 13-7 · *CONTINUED*

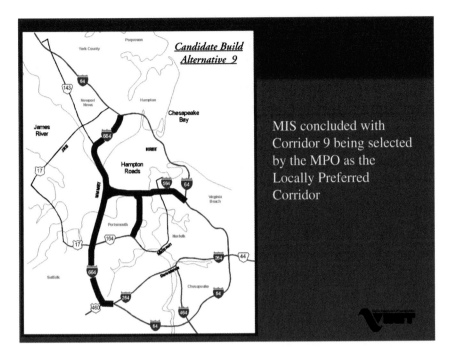

FIGURE 13-7 · *CONTINUED*

full-text report, but major segments are presented in bulleted format. Both reports may be examined under "Reports for Examination" at the book's companion Web site (select Chapter 13: Sample Documents).

Exercises

Further exercises related to the content of this chapter and all documents employed within the following exercises can be found on the book's companion Web site.

1. As you plan your reports throughout the semester, develop different plans for each report.

   · Design the plan that best meets the needs of the report context and the intended readers.
   · Pay attention to report design to improve the readability of your report. Teachers are no different from any other kind of reader you will encounter: they dislike unreadable reports.
   · Share a draft of your report and your general plan with one or two students in your class. Listen to their reaction. Redesign your report plan based on their input.

2. Choose one of the reports from the book's companion Web site, and adapt it for a slide presentation to the class.

Scenario

You are employed by a software company that develops building security systems. The owners wonder whether moving to a small town from the current metropolitan location would reduce operations cost and provide a better quality of life for employees. The operations vice president establishes a group to study the feasibility of a move and to find several locations for consideration. The feasibility study group, which includes representatives from each department in the company, has suggested several sites on the basis of criteria announced by the project director. Group members will visit each site, collect pertinent information, and prepare a report that assesses and then compares the merits and weaknesses of the potential relocation sites.

This chapter will explain the development of such reports, which must present conclusions and recommendations based on careful analysis.

CHAPTER FOURTEEN

Developing Analytical Reports: Recommendation Reports and Feasibility Studies

- **Analytical Reports**
- **Recommendation Reports**
- **Feasibility Studies**
- **Environmental Impact Statements**

  Exercises

Many routine reports are prepared solely to provide information, such as Figures 4-4 and 9-3, and the reports shown in Chapter 9 (pp. 203–230)—disease management of citrus fruits and the cultural control of boll weevils. The reports prepared for Congress by the Congressional Research Service (http://www.ncsonline.org/NLE/CRS) exemplify information reports: they summarize what is known, what work has been done on issues of importance to Congress as it considers legislation. On the book's companion Web site, www.oup.com/us/houp, we have selected several of these reports to illustrate basic, effective report design.

Reports in another category, which may be called extensive analytical reports, go beyond simply reporting information: they partition and then analyze or investigate information or data. From the analysis, the writer may evaluate information, draw conclusions, and perhaps recommend action based on those conclusions. Analytical reports often defy rigid classification, but for the purpose of learning to write them, we can list the following types:

- **Recommendation reports,** in which the analysis focuses on a recommendation.
- **Evaluation reports,** in which the analysis emphasizes evaluation of personnel, data, financial options, or possible solutions to problems or avenues for exploration.
- **Feasibility reports,** which analyze a problem, present possible solutions, determine criteria for assessing the solutions, assess the solutions against the criteria, and then show the best solution(s) based on the reported analysis; the recommendation is critical, but the analysis is as valuable as the conclusion and recommendation.

Many reports both inform and analyze. A progress or status report, which will be discussed in Chapter 16, Proposals and Progress Reports, describes and evaluates the work done on a project, the cost, and problems encountered. A trip report documents information gathered on a trip, evaluates this information, and may suggest action based findings. A personnel report describes an employee's performance, analyzes the effectiveness of the person's work, suggests methods of improvement, and estimates the employee's potential for promotion. An economic justification report explains the cost of a project or action and then argues for the cost-effectiveness of the project based on the cost.

In this chapter, we discuss the design of (1) the analytical report in general, (2) the recommendation report, and (3) the feasibility study.

The discussion of each report type is accompanied by illustrations that show how design works in specific situations. However, we continue to emphasize that report guidelines are just that—guidelines. Many organizations will have standard report templates, but information used in these templates should still reflect the writer's awareness of the reader, the report purpose, and the context in which the report will be received.

ANALYTICAL REPORTS

Because analytical reports focus on problem analysis, they usually contain more internal partitions. Analytical reports usually have a more complex structure than information reports, particularly in the introduction. The introduction to an analytical report can include any of the following:

- The purpose of the report
- The reason the report was written
- The history of the issue analyzed and rationale for the analysis
- The scope of the report—what issues will and will not be covered
- The procedure for investigating (analyzing) the topic of the report

The introduction should orient the reader to the problem and the approach described in determining the solution and recommendations.

After the introduction, the report can be developed in several ways. As always, the organization you choose should depend on your reader and the situation surrounding the report or the format used by your organization.

Plan 1:   Use this plan, which places the conclusion first, if readers will be more interested in the conclusions than the justification used to reach them.

- Introduction
- Conclusion—results of the analysis
- Recommendation(s)—if required by the investigation or if one or more has evolved from the analysis and conclusions
- Criteria for evaluation
- Presentation of information
- Discussion/evaluation of information

Plan 2:   Use this plan when readers will want analysis. In the second plan, you present the data or information, evaluate it, then present the conclusion and any recommendations you have. Here the focus is on analysis, rather than the conclusions, although you may decide to use the second plan if you believe readers will resist your conclusion, since careful analysis can be crucial in justifying a report's findings. This type of report is arranged as follows:

- Introduction
- Criteria for evaluation
- Discussion/evaluation of information
- Conclusion—results of the analysis
- Recommendation(s)—if required by the report purpose

Situation 1 The analytical report presented in Figure 14-1, prepared as part of a study to determine the feasibility of instituting brush control measures in a Texas reservoir, analyzes the economics of using these measures to enhance

APPENDIX C

ASSESSING THE ECONOMIC FEASIBILITY OF BRUSH CONTROL TO ENHANCE OFF-SITE WATER YIELD

Linda Dumke, Research Assistant; Brian Maxwell, Research Assistant; J. Richard Conner, Professor; Department of Agricultural Economics M.S. 2124, Texas A&M University, College Station, Texas 77843-2124 E-mail: JRC@tamu.edu

Abstract: A feasibility study of brush control for off-site water yield was undertaken in 1998 on the North Concho River near San Angelo, Texas. In 2000, feasibility studies were conducted on eight additional Texas watersheds. This year, studies of four additional Texas watersheds were completed and the results reported herein. Economic analysis was based on estimated control costs of the different options compared to the estimated landowner benefits from brush control. Control costs included initial and follow-up treatments required to reduce brush canopy to between 8% and 3% and maintain it at the reduced level for 10 years. The state cost share was estimated by subtracting the present value of landowner benefits from the present value of the total cost of the control program. The total cost of additional water was determined by dividing the total state cost share if all eligible acreage were enrolled by the total added water estimated to result from the brush control program. This procedure resulted in present values of total control costs per acre ranging from $35.57 to $203.17. Rancher benefits, based on the present value of the improved net returns to typical cattle, sheep, goat, and wildlife enterprises, ranged from $37.20 per acre to $17.09. Present values of the state cost share per acre ranged from $140.62 to $39.20. The cost of added water estimated for the four watersheds ranged from $14.83 to $35.41 per acre-foot averaged over each watershed.

INTRODUCTION

As was reported in Chapter 1 of this report, feasibility studies of brush control for water yield were previously conducted on the North Concho River near San Angelo, Texas (Bach and Conner, 1998) and in eight additional watersheds across Texas (Conner and Bach, 2000). These studies indicated that removing brush would produce cost effective increases in water yield for most of the watersheds studied. Subsequently, the Texas Legislature, in 2001, appropriated funds for feasibility studies on four additional watersheds. The watersheds (Lake Arrowhead, Lake Brownwood, Lake Fort Phantom Hill, and Lake Palo Pinto) are all located in North Central Texas, primarily in the Rolling Plains Land Resource Region. Detailed reports of the economic analysis results of the feasibility studies for each of the four watersheds are the subject of subsequent chapters.

Objectives

This chapter reports the assumptions and methods for estimating the economic feasibility of a program to encourage rangeland owners to engage in brush control for purposes of enhancing off-site (downstream) water availability. Vegetative cover determination and categorization through use of Landsat imagery and the estimation of increased water yield from control of the different brush type-density categories using the SWAT simulation model for the watersheds are described in Chapter 1. The data created by these efforts (along with primary data gathered from landowners and federal and state agency personnel) were used as the basis for the economic analysis.

FIGURE 14-1 · APPENDIX FROM AN ANALYTICAL REPORT

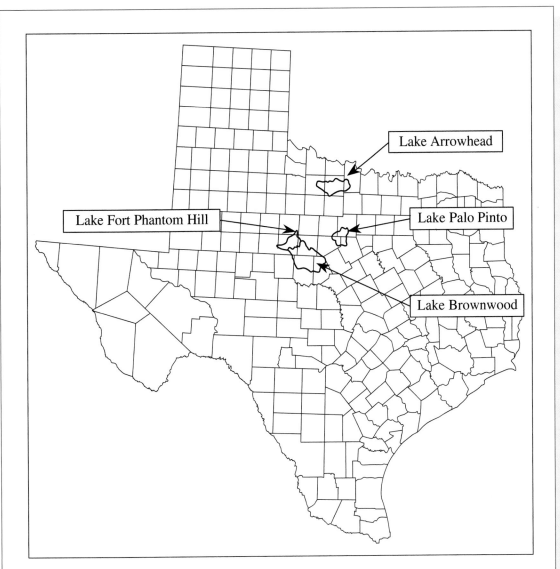

Figure 1.1. Watersheds included in the study area

This chapter provides details on how brush control costs and benefits were calculated for the different brush type-densities and illustrates their use in determining cost-share amounts for participating private landowners-ranchers and the State of Texas. SWAT model estimates of additional off-site water yield resulting from the brush control program are used with the cost estimates to obtain estimates of per acre-foot costs of added water gained through the program.

FIGURE 14-1 · *CONTINUED*

BRUSH CONTROL

It should be noted that public benefit in the form of additional water depends on landowner participation and proper implementation and maintenance of the appropriate brush control practices. It is also important to understand that rancher participation in a brush control program primarily depends on the rancher's expected economic consequences resulting from participation. With this in mind, the analyses described in this report are predicated on the objective of limiting rancher costs associated with participation in the program to no more than the benefits that would be expected to accrue to the rancher as a result of participation.

It is explicitly assumed that the difference between the total cost of the brush control practices and the value of the practice to the participating landowner would have to be contributed by the state in order to encourage landowner participation. Thus, the state (public) must determine whether the benefits, in the form of additional water for public use, are equal to or greater than the state's share of the costs of the brush control program. Administrative costs (state costs) which would be incurred in implementing, administering, and monitoring a brush control project or program are not included in this analysis.

Brush Type-Density Categories

Land cover categories identified and quantified for the four watersheds in Chapter 1 included four brush types: cedar (juniper), mesquite, oaks, and mixed brush. Landowners statewide indicated they were not interested in controlling oaks, so the type category was not considered eligible for inclusion in a brush control program. Two density categories, heavy and moderate, were used. These six type-density categories were used to estimate total costs, landowner benefits, and the amount of cost-share that would be required of the state.

Brush control practices include initial and follow-up treatments required to reduce the current canopies of all categories of brush types and densities to 3–8% percent and maintain it at the reduced level for at least 10 years. These practices, or brush control treatments, differed among watersheds due to differences in terrain, soils, amount, and distribution of cropland in close proximity to the rangeland, etc. An example of the alternative control practices, the time (year) of application and costs for the Lake Arrowhead/Watershed are outlined in Table 2-1. Year 0 in Table 2-1 is the year that the initial practice is applied while years 1–9 refer to follow-up treatments in specific years following the initial practice.

The appropriate brush control practices, or treatments, for each brush type-density category and their estimated costs were obtained from focus groups of landowners and NRCS and Extension personnel in each watershed.

Control Costs

Yearly costs for the brush control treatments and the present value of those costs (assuming a 6% discount rate as opportunity cost for rancher investment capital) are also displayed in Table 2-1. Present values of control programs are used for comparison since some of the treatments will be required in the first year to initiate the program, while others will not be needed until later years. Present values of total per acre control costs range from $35.57 for moderate mesquite that can be initially controlled with herbicide treatments to $175.57 for heavy mesquite that cannot be controlled with herbicide but must be initially controlled with mechanical tree bulldozing or rootplowing.

FIGURE 14-1 · *CONTINUED*

Landowner Benefits From Brush Control

As was mentioned earlier, one objective of the analysis is to equate rancher benefits with rancher costs. Therefore, the task of discovering the rancher cost (and thus, the rancher cost share) for brush control was reduced to estimating the 10 year stream of region-specific benefits that would be expected to accrue to any rancher participating in the program. These benefits are based on the present value of increased net returns made available to the ranching operation through increases or expansions of the typical livestock (cattle, sheep, or goats) and wildlife enterprises that would be reasonably expected to result from implementation of the brush control program.

Rancher benefits were calculated for changes in existing wildlife operations. Most of these operations were determined to be simple hunting leases with deer, turkeys, and quail being the most commonly hunted species. For control of heavy mesquite, mixed brush and cedar, wildlife revenues are expected to increase about $1.00 per acre due principally to the resulting improvement in quail habitat and hunter access to quail. Increased wildlife revenues were included only for the heavy brush categories because no changes in wildlife revenues were expected with control for the moderate brush type-density categories.

For the livestock enterprises, increased net returns would result from increased amounts of usable forage (grazing capacity) produced by removal of the brush and thus eliminating much of the competition for light, water, and nutrients within the plant communities on which the enterprise is based. For the wildlife enterprises, improvements in net returns are based on an increased ability to access wildlife for use by paying sportsmen.

As with the brush control methods and costs, estimates of vegetation (forage production/grazing capacity) responses used in the studies were obtained from landowner focus groups, Experiment Station and Extension Service scientists, and USDA-NRCS Range Specialists with brush control experience in the respective watersheds. Because of differences in soils and climate, livestock grazing capacities differ by location; in some cases significant differences were noted between sub-basins of a watershed. Grazing capacity estimates were collected for both pre- and post-control states of the brush type-density categories. The carrying capacities range from 45 acres per animal unit year (Ac/AUY) for land infested with heavy cedar to about 15 Ac/AUY for land on which mesquite is controlled to levels of brush less than 8% canopy cover (Table 2-2.).

Livestock production practices, revenues, and costs representative of the watersheds, or portions thereof, were also obtained from focus groups of local landowners. Estimates of the variable costs and returns associated with the livestock and wildlife enterprises typical of each area were then developed from this information into production-based investment analysis budgets.

For ranchers to benefit from the improved forage production resulting from brush control, livestock numbers must be changed as grazing capacity changes. In this study, it was assumed that ranchers would adjust livestock numbers to match grazing capacity changes on an annual basis. Annual benefits that result from brush control were measured as the net differences in annual revenue (added annual revenues minus added annualized costs) that would be expected with brush control as compared to without brush control. It is notable that many ranches preferred to maintain current levels of livestock, therefore realizing benefit in the form of reduced feeding and production risk. No change in perception of value was noted for either type of projected benefit.

FIGURE 14-1 · *CONTINUED*

The analysis of rancher benefits was done assuming a hypothetical 1,000 acre management unit for facilitating calculations. The investment analysis budget information, carrying capacity information, and brush control methods and costs comprised the data sets that were entered into the investment analysis model ECON (Conner, 1990). The ECON model yields net present values (NPV) for rancher benefits accruing to the management unit over the 10 year life of the projects being considered in the feasibility studies. An example of this process is shown in Table 2-3 for the control of heavy mesquite in the Lake Brownwood Watershed.

Since a 1,000 acre management unit was used, benefits needed to be converted to a per acre basis. To get per acre benefits, the accumulated net present value of $28,136 shown in Table 2-3 must be divided by 1,000, which results in $28.14 as the estimated present value of the per acre net benefit to a rancher. The resulting net benefit estimates for all of the type-density categories for all watersheds are shown in Table 2-4. Present values of landowner benefits differ by location within and across watersheds. They range from a low of $17.09 per acre for control of moderate mesquite in the Lake Palo Pinto Watershed to $37.20 per acre for control of heavy Shinnery Oak in the Lake Palo Pinto Watershed.

State Cost Share
The total benefits that are expected to accrue to the rancher from implementation of a brush control program are equal to the maximum amount that a profit maximizing rancher could be expected to spend on a brush control program (for a specific brush density category).

Using this logic, the state cost share is estimated as the difference between the present value of the total cost per acre of the control program and the present value of the rancher participation. Present values of the state cost share per acre of brush controlled are also shown in Table 2-4. The state's cost share ranges from a low of $42.53 for control of moderate mesquite in the Fort Phantom Hill Watershed to $131.61 for control of heavy cedar in the Lake Brownwood Watershed.

The costs to the state include only the cost for the state's cost share for brush control. Costs that are not accounted for, but which must be incurred, include costs for administering the program. Under current law, this task will be the responsibility of the Texas State Soil and Water Conservation Board.

COSTS OF ADDED WATER

The total cost of additional water is determined by dividing the total state cost share if all eligible acreage were enrolled in the program by the total added water estimated to result from the brush control program over the assumed ten-year life of the program. The brush control program water yields and the estimated acreage by brush type-density category by subbasin were supplied by the Blacklands Research Center, Texas Agricultural Experiment Station in Temple, Texas (see Chapter 1). The total state cost share for each subbasin is estimated by multiplying the per acre state cost share for each brush type-density category by the eligible acreage in each category for the subbasin. The cost of added water resulting from the control of the eligible brush in each subbasin is then determined by dividing the total state cost share by the added water yield (adjusted for the delay in time of availability over the 10-year period using a 6% discount rate). Table 2-5 provides a detailed example for the Lake Arrowhead Watershed. The cost of added water from brush control for the Lake Arrowhead Watershed is estimated to average

FIGURE 14-1 · *CONTINUED*

$14.83 per acre-foot for the entire watershed. Subbasin cost per added acre-foot within the watershed range from $6.84 to $26.38.

ADDITIONAL CONSIDERATIONS

Total state costs and total possible added water discussed above are based on the assumption that 100% of the eligible acres in each type-density category would enroll in the program. There are several reasons why this will not likely occur. Foremost, there are wildlife considerations. Most wildlife managers recommend maintaining more than 10% brush canopy cover for wildlife habitat, especially white-tailed deer. Since deer hunting is an important enterprise on almost all ranches in these four watersheds, it is expected that ranchers will want to leave varying, but significant amounts of brush in strategic locations to provide escape cover and travel lanes for wildlife. The program has consistently encouraged landowners to work with technical specialists from the NRCS and Texas Parks and Wildlife Department to determine how the program can be used with brush sculpting methods to create a balance of benefits.

Another reason that less than 100% of the brush will be enrolled is that many of the tracts where a particular type-density category are located will be so small that it will be infeasible to enroll them in the control program. An additional consideration is found in research work by Thurow et al. (2001) that indicated that only about 66% of ranchers surveyed were willing to enroll their land in a similarly characterized program. Also, some landowners will not be financially able to incur the costs expected of them in the beginning of the program due to current debt load.

Based on these considerations, it is reasonable to expect that less than 100% of the eligible land will be enrolled, and, therefore, less water will be added each year than is projected. However, it is likewise reasonable that participation can be encouraged by designing the project to include the concerns of the eligible landowners-ranchers.

LITERATURE CITED

Bach, Joel P. and J. Richard Conner. 1998. Economic Analysis of Brush Control Practices for Increased Water Yield: The North Concho River Example. In: Proceedings of the 25th Water for Texas Conference—Water Planning Strategies for Senate Bill 1. R. Jensen, editor. A Texas Water Resources Institute Conference held in Austin, Texas, December 1–2, 1998. Pgs. 209–217.

Conner, J.R. 1990. ECON: An Investment Analysis Procedure for Range Improvement Practices. Texas Agricultural Experiment Station Documentation Series MP-1717.

Conner, J.R. and J.P. Bach. 2000. Assessing the Economic Feasibility of Brush Control to Enhance Off-Site Water Yield. Chapter 2 in: *Brush Management/Water Yield Feasibility Studies for Eight Watersheds in Texas.* Final Report to the Texas State Soil & Water Conservation Board. Published by Texas Water Research Institute, TWRI TR-182.

Thurow, A., J.R. Conner, T. Thurow and M. Garriga. 2001. Modeling Texas ranchers' willingness to participate in a brush control cost-sharing program to improve off-site water yields. *Ecological Economics:* 37(Apr. 2001): 137–150.

FIGURE 14-1 · *CONTINUED*

the water reserves in the Fort Phantom Hill Reservoir. The report considers the brush control costs for ten years, discussing the cost share of landowners, the cost share of the state, and finally, the cost of the water that would be made available by controlling brush that consumes large amounts of water. The four tables of supporting data cited in the discussion are not reproduced here. To view the report in its entirety, go to the book's companion Web site.

Note that this report emphasizes the analysis itself; it does not call on readers to make any specific decisions related to the analysis. For that reason, the writers provide the analysis first and then the cost data. This design decision is immediately apparent from a table of contents for the report:

Abstract

Introduction—what is covered in this specific report; objectives of the report

Brush Control

Brush Type–Density Categories
Control Costs
Landowner–Benefits from Brush Control
State Cost Share
Costs of Added Water
Additional Considerations
Literature Cited

RECOMMENDATION REPORTS

A common type of analytical report is the recommendation report—you examined a short one in Chapter 13 (Figure 13-1). Here you analyze a problem or situation, present possible solutions, analyze each solution as it relates to the problem, and then recommend the one you think is best. In organizations, recommendation reports document strategy used to investigate and then report solution(s) to problems. The location of the recommendation, however, depends on the focus of the report. If readers want the recommendation, then place that first. If readers want recommendations based on analysis—if they want to see how the recommendations are supported—then place the recommendation at the end of the report. The following examples illustrate the difference in approach.

Situation 2 In 1998, the Comptroller of the State of Texas began to move state government offices toward electronic ways of doing business. The following report—from a report series authorized by the comptroller—recommends

Search

Limited Government, Unlimited Opportunity

January 2003

ED 7
Allow Four-year Colleges and Universities to Waive Tuition and Fees for Concurrently Enrolled Students

Summary

"Concurrent" enrollment allows high school students to earn college credit before graduation. State law allows community colleges to waive tuition and fees for students currently enrolled in high school. It does not, however, provide this option for four-year colleges and universities. Given the high success rate of concurrent enrollment and the expanded individual choice that extending the policy would provide, the state should allow four-year institutions to waive tuition and fees for concurrent enrollment courses.

Background

Concurrent enrollment allows high school students to earn college credit before graduation by enrolling in community college classes that award dual high school and college credit. About 38,000 Texas students concurrently enroll in more than 80,000 semester courses each year, and receive credit for 94 percent of them, a significantly higher pass rate than the state average for regular high school courses.[1]

Students typically take these courses on their own high school campuses, although in some cases they may attend class at the college. Either way, concurrently enrolled students become more familiar with the demands and instructional approach of higher education. This familiarity can increase their confidence in handling college-level work and facilitate their planning for college after high school.

State law allows community colleges to waive tuition and fees for concurrently enrolled students, and most large community colleges do so, a factor that can significantly reduce the cost of a college education.[2] Concurrent enrollment also allows students to earn a college degree more quickly.

Community colleges often provide concurrent courses in unused high school classrooms, making it more economical for them to forego tuition and fee income. Such courses also are an important recruiting tool that can encourage students to continue at the community college after high school graduation.

Schools and districts also receive recognition for concurrent enrollment courses in the Gold Performance Acknowledgment category of the Texas Education Agency's Accountability Rating System, as part of the Advanced Academic Courses indicator.

FIGURE 14-2 · SHORT ANALYTICAL REPORT FROM A STATE AGENCY

Finally, concurrent enrollment can benefit state finances as well. The state pays schools the full amount of per-student funding for each student taking four or more hours of high school-only instruction in a class day, and provides half-day funding for students taking at least two, but less than four hours of high school-only courses.[3] A student schedule with concurrent courses, then, can reduce or even eliminate state high school funding for that student.

At present, Texas law does not allow public four-year institutions of higher education to waive tuition and fees for concurrently enrolled students.

Recommendation

State law should be amended to allow Texas public four-year higher education institutions to waive tuition and fees for concurrently enrolled high school students.

This recommendation would increase educational options for students, provide a recruitment tool for four-year institutions and would save student and taxpayer dollars. The institutions would absorb the reduction in income from waived tuition and fees.

Fiscal Impact

To the extent that four-year higher education institutions opt to waive tuition and fees and students respond by enrolling in more concurrent classes, local school district taxpayers could benefit from reduced instructional costs. The state could save money due to reductions in educational funding for participating students. Savings also would accrue to students and parents from lower college costs. Universities that choose to waive tuition and fees for these courses will have to cover the tuition and fee costs of providing the courses. Four-year institutions would gain an important recruiting tool for attracting motivated students. Savings from this recommendation, however, would depend upon future events and cannot be estimated.

Endnotes

[1] Data compiled from the Texas Education Agency's PEIMS database, August 16, 2002.

[2] Tex. Educ. Code, §130.008.

[3] Texas Education Agency, *2002–2003 Student Attendance Accounting Handbook* (Austin, Texas, July 1, 2002), pp. 25–26.

Carole Keeton Strayhorn
Texas Comptroller of Public Accounts

Window on State Government
Contact Us
Privacy and Security Policy

FIGURE 14-2 · *CONTINUED*

specific changes in State of Texas education policy and state government. Note the documentation used to support the analysis.

Now refer again to Figure 10-9, which you examined when you were studying report structure. This report, designed as an online report, assesses the effectiveness of services for mentally retarded residents of intermediate care facilities in Texas. It can be read beginning with the executive summary, moving immediately to the committee recommendations; or study of the introduction and background can precede consideration of the recommendation. In this report, the writers are focusing on problem assessment. Because the report also discusses on effectiveness of the solution, the analysis is as critical as the conclusions and recommendations.

Another excellent example report discusses a U.S. government climate change technology initiative (go to the book's companion Web site). This report discusses each federal agency before assessing the efforts of each program's initiatives. Thus, this report places the conclusion last.

FEASIBILITY STUDIES

Another major type of analytical report, usually the longest and most complex of the analytical report types, is the feasibility study. A feasibility report assists its readers in determining whether an organization should take a specific action. The writers of this kind of study may consider several solutions to a problem, examine each one, and determine the best solution based on specific criteria.

Because of the complexity of feasibility reports, writers should first establish the purpose and scope of the report and the problem to be analyzed. Sometimes, however, investigators don't develop the report outline until a feasibility report has been commissioned, and they have additional information to use in deciding how to approach the actual study.

The central focus of the feasibility study is whether to do or not do something, or to determine which option among several is the best choice. A feasibility study is similar to a recommendation report but usually produces recommendations on a variety of issues. The status of the recommendations will provide the answer to the question: is X feasible or not? The following situations call for a feasibility study.

■ A university research group wants to know whether fire ants can be controlled effectively in open fields of one hundred acres or less by using a specific group of chemicals.

■ A state legislature wants to know whether partially treated sewage can be deposited on land in a semiarid part of the state to improve the soil without the leaching of harmful bacteria into the water table.

■ A company wants to know whether land it currently owns near a major highway should be developed as a shopping mall, a business park, or some combination of both.

■ A city council wants to know whether building an additional airport on property the city owns near a major waterway will cause flooding in residential areas downstream from the proposed airport.

■ A state water resources board wants to know whether a water district will improve conservation and effective usage of water in three counties of the state.

Usually, the feasibility study helps organizations make decisions based on the value of the proposed solutions as well as their cost. Those decisions must emerge from careful, complete, accurate, ethical presentation of the answers to questions that must be answered as well as the cost of the possible solutions. The study itself reports the problem at hand, the methods selected to study the problem, the subissues that comprise the problem, and the results of the analysis of each issue; the conclusion, recommendation(s), and data on costs of implementation emerge from the analysis of all aspects of the problem. Completing such a report can take months or years.

Situation 1 (continued) The report in Figure 14-1, which analyzes the cost of controlling brush to enhance water yield, served as an appendix to a larger report, a feasibility study on the value of brush control to save water in several Texas watersheds. By studying the table of contents of this report, *The Fort Phantom Hill Reservoir Watershed: Brush Control Assessment and Feasibility Study,* you can see how feasibility studies develop:

TABLE OF CONTENTS

Source: Texas State Board & Water Conservation Board, State Brush Control Program, Fort
Phantom Hill Reservoir Watershed, 2002–2003
http://www.tsswcb.state.tx.us/programs/brush2.html

Note, again, from the table of contents, the importance of clearly worded headings to reveal the report plan and content. The entire report can be found at
the book's companion Web site (select Feasibility Studies in Chapter 14: Sample
Documents). Several additional excerpts from this document are provided
shortly.

When you access the full report online, we encourage you to study
Appendix A, as a ready example of a short but carefully documented feasibility
study. The study opens with an abstract, a short introduction, and a discussion
of the methods used to study the hydrologic and economic implications of brush
removal in the selected watersheds. Following the discussion of methods, the
writers describe the model inputs and the calibration of the inputs. The next section provides the results of the study and a final factual summary of findings.
The report concludes with a Literature Cited section.

To fully appreciate the value of a well-written introduction, you may wish
to examine the following excerpts from the full study of the Fort Phantom Hill
Reservoir Watershed before reading Appendix A online. This way you will be
able to see how these report elements provide readers with the perspective they
need to access and understand a lengthy technical document.

Purpose

The Fort Phantom Hill feasibility study begins with a clear statement of the
topic (project), followed by the rationale for the study:

The Brazos River Authority is participating in a study coordinated by the
Texas State Soil and Water Conservation Board (TSSWCB) to assess the feasibility
of instituting brush control measures in the Fort Phantom Hill Reservoir watershed. In 1985, the Texas Legislature created the Texas Brush Control Program. The
goal of this legislation is to enhance the State's water resources through selective
control of brush species. The TSSWCB was given jurisdiction over the program.

Brush control, as defined in the legislation, means the selective control, removal, or reduction of noxious brush such as mesquite, prickly pear, salt cedar, or other deep-rooted plants that consume large amounts of water.

Water will likely be the most limiting natural resource in Texas in the future. The ability to meet future water needs will significantly impact growth and economic well being of this State. The United States Department of Agriculture—Natural Resources Conservation Service (NRCS) estimated that brush in Texas uses over 3.5 trillion gallons of water annually. Control of brush presents a viable option for increasing the availability of water allowing the State to meet its future needs.

Since the European settlement of Texas, improper livestock grazing practices, fire suppression and droughts have led to the increase and dominance of noxious brush species over the native grasses and trees. The improper livestock grazing of the watershed's rangeland in the late 19th century and early 20th century reduced the ability of grasses to suppress seedling tree establishment and led to the establishment of invasive woody species, such as juniper and mesquite. This noxious brush utilizes much of the available water resources with little return to the watershed and reduced production capabilities of the region.

The final paragraph in the introduction states both the project purpose and the report purpose:

This project aims to increase stream flow and water availability in the watershed that drains into Fort Phantom Hill Reservoir. This reservoir and three smaller reservoirs in the watershed, Lake Abilene, Lake Kirby, and Lake Lytle, are used as a water supply for industrial, agricultural, and municipal uses. This report will assess the feasibility of brush management to meet the project goals by developing a historical profile of the vegetation in the watershed, developing a hydrological profile of the watershed, and evaluating historical climatic data in the watershed.

The scope statement, embedded in the project purpose, reflects a number of questions:

- Is controlling brush to meet the project goals feasible? Can it be done?
- To what extent?
- How much will control cost?
- Who will pay? Is the amount tentatively assessed from each entity feasible?

The historical profile of the vegetation in the watershed that impacts these questions might be outlined as follows:

topography and drainage
geology

population
climate
land use
wildlife
vegetation

A hydrological profile of the watershed that impacts the questions might incorporate the following areas:

surface water
springs
groundwater levels
geology
existing surface water hydrology
existing groundwater hydrology
description of the hydrologic system

From such an initial list of questions, the investigation can proceed, although each question and its associated subtopics will likely be broadened and rephrased as research warrants. Determining scope—which guides the investigation—is crucial and should be performed in light of the information gathered in response to the initial questions. Look for holes, overlaps, irrelevant questions. Often, a person unacquainted with the study is in a far better position than you to spot shortcomings and illogical assumptions in the scope statement. Those responsible for major feasibility reports, such as statewide brush control/water conservation studies, often consult colleagues in the agency who are not involved in the project. The task of these outsiders is to ensure that all questions relevant to the study are addressed, and that the analysis is logical and thorough.

You may be able to grasp the main features of the report by reading the executive summary, the introduction, and the summary and conclusions. In an effectively written report, the discussion documents the conclusions and recommendations found in the summary and in the final segment of the report. Since you have read the introduction to this report, take a moment to read 1.0, the Executive Summary, and 5.0, Summary and Conclusions, which serves as a factual summary and differs from the executive summary. However, the factual summary in a bulleted format allows readers to rapidly locate the central conclusions. Note, too, how the bulleted items echo the topics of the table of contents (pp. 388–89):

1.0 Executive Summary

The Brazos River and its tributaries, along with reservoirs in the Brazos River basin, comprise a vast supply of surface water to Texas. Diversions and use of this surface water occur throughout the entire basin with over 1,500 water rights currently issued. The western part of the basin is heavily dependent on surface water sources. Fort Phantom Hill Reservoir, operated by the West Central Texas Municipal Water District, is one of the major water supply reservoirs in the western part of the region.

The western part of the Brazos River basin ranges from desert-like conditions to semi-arid with minimal rainfall. Water availability is a critical factor as the population in the urban areas of the region grows. In an effort to guarantee adequate water supply for the future of this region, a variety of options are being considered by the State. One of the options is brush control. Brush control is the selective control, removal, or reduction of noxious brush such as mesquite, prickly pear, salt cedar, or other deep-rooted plants, which consume large quantities of water. Brush control can have positive results in increasing stream flow, aquifer levels, and water availability. In watersheds where the vegetation is dominated by noxious brush, replacing the brush with native grasses that use less water may yield greater quantities of available water. The goal of this study is to evaluate the climate, vegetation, soil, topography, geology, and hydrology of the Fort Phantom Hill Reservoir Watershed with respect to the feasibility of implementing brush control programs in the watershed.

Climate data have been collected in the region since 1950. This data reveals no major changes in temperature or precipitation levels between 1950 and 2000. While the climate has not changed, it appears that various changes in stream flow, spring discharge, and vegetation have occurred since the first European settlers began to arrive in the area in the 19th century. The first-hand accounts of the early settlers document ample water supplied through perennial springs and streams and a lush grassland void of mesquite and juniper infestation. In contrast to historical accounts, today the area is dominated by mesquite and juniper brush, springs are intermittent, and the water supply in the watershed is inadequate to meet demand without inputs of water from other watersheds.

Brush removal simulations reveal that rates of evapotranspiration will be reduced as a result of brush control, grass cover will increase, and there will be higher runoff and groundwater flows in the Fort Phantom Hill Reservoir Watershed. Simulations of brush control implementation estimate average annual water yield increases in the Fort Phantom Hill Reservoir Watershed to be about 111,000 gallons per treated acre.

An assessment of the economic feasibility of brush control in the Palo Pinto Reservoir watershed revealed the following results: the total cost of added water was determined to average $29.45 per acre foot if all eligible acreage is treated; present value of total control costs per acre range from $35.57 for herbicide control of moderate mesquite to $143.17 for mechanical control of heavy mixed brush; benefits to landowners range from $21.37 per acre for control of moderate mesquite to $35.50 per acre for the control of heavy mixed brush; and state cost share per acre is estimated to be $14.20 for moderate mesquite control to $112.53 for heavy cedar control.

5.0 Summary and Conclusions

This evaluation of the hydrology of the Fort Phantom Hill Reservoir Watershed has included a review and analysis of available data on climate, vegetation, geology, surface hydrology, and groundwater hydrology. The following conclusions summarize the findings:

- No significant changes have occurred in the historical climate pattens within the watershed, including precipitation frequency, duration, and intensity.
- Changes in the historical vegetation of the watershed have been dramatic. Based on first-hand accounts of the vegetation during the 19th century, the area was predominantly mixed grass prairie, with little to no stands of juniper or mesquite. There is a great indication that brush cover in the watershed is significantly more extensive today than it was historically.
- Good quality data on stream flow in the watershed has not been collected for an extended period of time. USGS gauging stations have been operated at many locations in the watershed but gauging at any one location has been limited to a maximum duration of 16 years.
- The available stream flow data reveal no major changes have occurred in stream characteristics during the period of record; however, the current intermittent nature of the streams in the watershed is in direct opposition to the first-hand accounts of water availability during the 19th century.
- Water levels in aquifers in the watershed have historically risen and fallen in response to rainfall patterns and artificial withdrawals. No systematic declines in aquifer water levels are indicated, except for the alluvial aquifer in the watershed.
- The watershed is dependent on diversions of water from other stream systems. Without diversions from Deadman's Creek and the Clear Fork of the Brazos River, Fort Phantom Hill Reservoir would not be able to meet the current demands of the population of the watershed.
- Soils in the watershed are typically thin, formed of large particles, and conducive to groundwater recharge.
- Water supply shortages are a current problem for the area and these shortages are projected to increase as demand increases. Brush management could help offset supply deficits in the watershed by reducing water losses in both the streams and alluvial aquifer.
- While hydrological studies reveal that brush control in the Fort Phantom Hill Reservoir Watershed is estimated to increase annual average water yields by only 111,000 gallons per treated acre, the cost of control is moderate and the need for water in this region is immediate. An organized brush control program will provide great benefit to this water-poor region.

- It is recommended that the Texas Legislature commit to appropriate $10,189,417 to implement brush control practices in the Fort Phantom Hill Watershed. Implementation should occur as soon as funding is available, with maintenance occurring throughout the ten-year period following implementation.

Situation 3 State and federal transportation agencies often perform feasibility studies to determine whether and how to develop highways, ferries, access roads, park-and-ride facilities, and rail systems. The Department of Transportation of the State of Virginia (VDOT) has prepared numerous feasibility studies on these topics. See http://www.virginiadot.org/projects.

For example, VDOT prepared a feasibility study for a park-and-ride lot in northern Virginia. The purpose of this study was to determine the demand for park and ride spaces to facilitate carpooling in the existing and planned high-occupancy vehicle corridors. The authors of the study also had to recommend feasible sites for constructing future commuter lots to meet the demand. The main tasks determined by the study team were as follows:

- State the inventory of existing park and ride lots.
- Review requirements regarding new lots being planned.
- Estimate the future short-, intermediate-, and long-term demand.
- Identify the future needs for parking spaces.
- Identify feasible sites to meet the demand.
- Develop a proposed implementation plan.

One of the VDOT feasibility studies provides another example of how such a report develops. Because the study is lengthy, we will show here only the table of contents, which reveals the main content elements:

Note that the report discusses current and future evidence of demands (Sections 2 and 3), needs of a park and ride facility, and possible sites. Section 5.0 exemplifies effective data evaluation to determine which site is best for this facility. You can read the entire report at the book's companion Web site.

─────Northern Virginia Park & Ride Lot Feasibility Study

PARK & RIDE →

TABLE OF CONTENTS

ENVIRONMENTAL IMPACT STATEMENTS

Perhaps the most complex form of feasibility study is the environmental impact statement (EIS). The U.S. government requires an EIS to determine, for example, how a large area of wilderness will be affected by proposed commercial development. Transportation projects considered by the Virginia Department of Transportation often require environmental impact statements. In late 1993, VDOT investigated methods of relieving congestion at the existing Interstate

64 Hampton Roads Bridge Tunnel. A final EIS was prepared to evaluate the impact of building a new crossing at Hampton Roads. Earlier EISs had projected the impacts that would be incurred by a proposed transportation improvement during the preliminary engineering stage of development. Other prominent issues included impacts on historical and archaeological sites and impacts on flora and fauna.

Because of the length and complexity of environmental impact statements—the VDOT document has 700 pages—the essence of the report is often made available in a slide presentation, which then becomes the popular summary version of the report. Often officials of the government agencies that must approve the project are exposed first to the slides. Thus, the slide presentation serves as an executive summary and must accurately echo the findings of the complete environmental study.

The complete environmental impact study can be found at the book's companion Web site. Note the importance of effective report design to help readers access the report and target the areas of particular interest to them.

The 24-page slide presentation provides a concise statement of the final EIS, which also can be found under Feasibility Studies at our companion Web site.

☑ Planning and Revision Checklist

You will find the Planning and Revision Checklist inside the front cover valuable in planning and revising any presentation of technical information. The following questions specifically apply to recommendation reports and feasibility studies. They summarize the key points in this chapter and provide a checklist for planning and revising.

Planning

- What is the purpose of your report? Have you stated it in one sentence?
- What is the scope of your report?
- Who are your readers? What is their technical level?
- What will your readers do with the information?
- What information will you need to write the report?
- How long should the report be?
- What format should you use for the report?
- What report elements will you need?

▪ What graphics will you need to present information or data?

▪ What elements should appear in your introduction?

▪ What arrangement will you use in presenting your report?

Revision

▪ At a minimum, does your recommendation report do the following:
  Introduce the report?
  Present enough data in words and visuals to justify any conclusions drawn?
  Discuss and evaluate the data fairly?
  Summarize the data?
  Draw logical conclusions from the data?
  Present recommendations based on the data and the conclusions?

▪ Are your data accurate?

▪ Do your visuals immediately reveal what they are designed to show?

▪ Is your format suitable for your content, audience, and purpose?

▪ Have you properly documented all information sources?

▪ ▪ ▪ ▪ ▪ ▪ ▪ Exercises ▪ ▪ ▪ ▪ ▪ ▪ ▪

Further exercises related to the content of this chapter and all documents employed within the following exercises can be found on the book's companion Web site.

1. Find an example of an environmental impact statement. Prepare an information report explaining the context of the report, the subject, and the purpose. Describe the structure of the EIS, then summarize the main findings of the report.

2. Use the template that follows to prepare a feasibility study that focuses on a possible topic for your major technical report assignment. From the results of your feasibility study, prepare a proposal for the project if it is feasible. You may wish to consider two projects in the same subject area and let your feasibility study help you decide which is more workable for your major semester report project.

Factual Summary

Results of the study, conclusions, and recommendations regarding feasibility

Introduction

One paragraph—topic(s) under consideration, approach studied

Topic(s) under consideration, purpose of the project, scope of the project

Rationale for topic(s) in this area—why important

Historical background

Procedures and methods to be used in investigating the study

Approach

Criteria for analysis of the feasibility of the topic(s) and rationale for these criteria: for example,

Criterion 1: time constraint

Criterion 2: availability of research resources

Criterion 3: quality of these resources

Criterion 4: probability of completion as measured against objective of the study

Analysis of the Topic(s) According to Criteria

Criterion 1:

Criterion 2:

Criterion 3:

Criterion 4:

Conclusion

Is project feasible when analyzed against these criteria? (Is one more feasible than the other?)

Recommendations

Need for redirection? Modifications of the proposed research?

List of Sources Consulted

3. The student feasibility study, Improvements for the Texas A&M On-Campus Bus System, can be found on the book's companion Web site. Examine this document and consider proposing a feasibility study on a similar campus issue for your semester project.

4. A week ago, the manager of the appliance department at Cooper's Department Store notified you of his intention to resign, giving you two weeks' notice. As the vice president of personnel, you must determine a suitable replacement and make your recommendation

to the company president, Theresa Knight. Although you promptly advertised the position in the local newspaper, you received only three applications. Having interviewed all three candidates, you know that: each has strengths and weaknesses, but none strikes you as clearly superior to the other two. Cooper's appreciates loyalty to the company, but it doesn't promote according to seniority. Instead, Cooper's looks for managers who have leadership skills, business experience, a college education, and the ability to work with others. You need to make a decision soon and write your recommendation report for Knight. As you look over your notes from the applications and interviews, here is what you see:

Kevin Root is studying part-time for a master's degree in marketing at the local university, from which he obtained a bachelor's degree in psychology two years ago. In the interview he seemed quite congenial and easy to talk to. In college, he was president of his fraternity. He has held part-time or full-time jobs during the summers between academic years but has no other work experience. While in graduate school, he has worked at a local electronics store, selling televisions and computers. A letter of recommendation from the store manager was enthusiastic, applauding Root's agreeable personality and pleasant disposition, but noting that he often seemed fatigued.

Thomas Sadowski is retired from the police department. He has been employed by Cooper's for seven years, with three years in the appliance department as assistant department manager. His employee evaluations have all been positive, praising his high level of efficiency, loyalty to the company, and enthusiasm for the job, but noting that a number of employees consider him aggressive and dictatorial. As assistant department manager, he has been responsible for scheduling, and he fills in for the department manager during absences and vacations. He often leads the appliance department in sales and does a good job of promoting highly profitable service contracts on every appliance he sells. Sadowski, who is in his middle fifties, is a high school graduate.

Alice Crowley received a bachelor's degree in chemistry and taught high school for nine years. The principal of the high school offered a good recommendation, noting especially Crowley's coordination of the school's annual science fair. Crowley says that she stopped teaching because she was tired of interference from aggressive parents. Five years ago, she started working at Cooper's in the hardware department and a year ago switched to appliances. Her employee evaluations are consistently positive. She is highly regarded by colleagues in the appliance department and often is assigned to train new sales clerks on policies and procedures.

Scenario

As a newly graduated food science major, you were delighted to find work with Sheffield Farms, a medium-sized supermarket chain in the Midwest. Your first assignment was to participate in a small group that studies consumer food preferences and habits. The group leader is Shirley Gomez. One of your tasks was to research the readability of the new food labels required by the U.S. Food and Drug Administration.

You had sought volunteers among people shopping in Sheffield Farm stores. You had asked them to read labels and then questioned them on their understanding of the printed information. To a high degree they understood the information provided about fat, cholesterol, fiber, protein, and so forth. But as you were conducting your research, you wondered how many Sheffield Farms customers actually made decisions based on the labels.

You brought the idea up to Shirley. "Good thought," she said. "How would you go about researching it?"

"I've thought of three ways," you said. "Observation for one. Simply watch to see how many people read a food label when they take a product off a shelf. Another would be to ask people at checkout how often they read food labels. Third, we could select a common purchase, like cereal, in customers' baskets and, while concealing the food label, try to find out how much of the information on the label they're actually aware of."

"Has anyone done this?" Shirley asked.

"My preliminary search in the journals and on the Web hasn't turned up anything like what I'm proposing," you said. "There has been a lot of focus group research about readability and decision making based on labels, but I found nothing that tried to quantify their actual use by ordinary consumers. In any case, we haven't done anything like this with Sheffield customers. If we find that most our customers really pay no attention to the labels, we might want to start an awareness program of some sort."

"Well, check the lit some more," Shirley said. "If you still find that the research you propose hasn't been done, you might have a journal article. In any case, the research report could be useful for us in-house." As an afterthought, she added, "Check your methodology with me before you begin, though. You can't be too intrusive, or you'll annoy our customers. Maybe you can offer people some small reward for answering your questions."

And so empirical research reports are born. Someone sees a need for the research and checks the literature carefully to see if it really needs to be done. If the need is perceived, the methodology for carrying out the research is planned and executed. When the results are in and analyzed, it's time to write the report, which is what this chapter is all about.

CHAPTER FIFTEEN

Developing Empirical Research Reports

As a student and later as a specialist, you may need to design a device; test an idea, mechanism, product, or process; perform an experiment; and report your findings. This kind of analytical report is called an empirical research report because it:

■ Explores a solution to a problem based on extant knowledge
■ Proposes a new solution or process based on what is and is not known
■ Justifies the reasons for this proposed solution or process, tests that solution, and then concludes whether the solution is viable

Many scientific journals are basically collections of empirical research reports targeted to specialists in the journal's discipline. Many research organizations test their products and then report the results in studies archived on the company's Web site. Readers of research reports will be interested in the kinds of research reported, but the presentation must allow rapid reading and unencumbered access to methods, data, and results. Specialists working on a question related to the topic covered in an empirical research report will want to know that your research procedure is valid; your hypothesis and rationale, logical; and your analysis of your findings, accurate. They will read the reports that interest them carefully, critically, and evaluatively. These readers may want to replicate your findings and then further their own research on the basis of your results.

Examples of empirical research reports include the following:

■ A report on research conducted by a bioengineering student who was attempting to design a monitor for use with infants who may be at risk for sudden infant death syndrome. Like many such documents, this one describes the progress of the research until the point at which the report had to be submitted. Thus, its conclusions are not definitive, but they suggest what needs to be done to pursue this research further.
■ A report to determine the effectiveness of a weed killer on vegetable crops of different kinds. The conclusions suggest which chemical controls can be used to eradicate or reduce weeds without harming the quality of the crop or the safety of the food made from it.
■ A report on the ease of use of online voting software and the effectiveness of the software to minimize voting errors.
■ A report on the use of dried blood spots from HIV-positive patients as a means of determining subtype.

Empirical research reports, like any other type of technical writing, should be designed for the audience(s) the writer intends to address. The level of language used and the specificity of the research will depend on the target readers. While most journals that publish research reports have their own required format, research reports nevertheless have similar elements.

How each is developed and where it is placed will depend on:

■ The topic
■ The intended readers
■ The preferences of the publication in which the report will appear
■ The purpose of the report

To illustrate the development of the empirical research report, we will focus on the development of sections of three research reports. Several additional empirical research reports, accessible at the book's companion Web site, exemplify effective use of style, visual presentation (to report data), or effective review of extant research, presentation of materials and methods, and/or discussion of results. The reports vary in length depending on the complexity of the project. Keep in mind that because the empirical research report needs to be easy for your audience to follow and comprehend, a clear, direct style is important.

MAJOR SECTIONS OF EMPIRICAL RESEARCH REPORTS

Nearly all empirical research reports contain the following content sections, which can be combined or appear as self-identified headings.

Abstract
  • Introduction—statement of problem, importance of problem,
  • Literature Review—What is known about the topic, summary of relevant research with parenthetical citations,
  • Purpose of the current empirical research report
Materials and methods used in this research project
Results of the research
Discussion of results
Conclusion/Recommendation
References
Acknowledgments

Abstract

The abstract is perhaps the most important section in an empirical research report. Abstracting services often capture and sell abstracts to researchers in various disciplinary areas. Subscribers to these services often read only abstracts, seldom consulting the full text of an empirical research report. Thus, the purpose and results of the study must be clearly and concisely stated. The following abstract is effectively written because it can be understood apart from the entire report. We will color code the parts of the abstract we highlight. Note that the abstract begins with the <u>project purpose</u>, then focuses on the specifics of the methods used in the project, and concludes with the results. The writers combine active and passive voice, use moderate sentence length, and define the ingredients of the chemicals they are testing. They alert readers to the shift from procedure to results by using "results" as the subject of each sentence that announces the findings:

Abstract

This research evaluated the efficacy of using a chemical barrier applied to the soil area under stacked bales of hay to prevent the red imported fire ant, *Solenopsis invicta* Buren (Hymenoptera: Formicidae), from infesting stacked hay. Specifically, we were interested in determining if we could protect "clean" hay bales stored in fire ant infested fields for up to several weeks. Chemicals selected as barrier treatments were Lorsban® 4E, active ingre-dient chlorpyrifos, which kills ants on contact, and Astro™ Insecticide, active ingredient the pyrethroid permethrin, which can also act as a repellent to ants. We established a series of 12 ft × 12 ft plots, with a 10 ft buffer between plots along a fence row in a fire ant infested field. Plots were grouped into four blocks of three stacks each. Plots within blocks were randomly assigned to each treatment (four plots treated with Lorsban® 4E and four treated with Astro™ Insecticide, and four control plots). Treatments included spraying a 12 ft × 12 ft soil area with a 1-gal solution of each chemical and water formulation. After soil treatments, we placed four square bales of hay, stacked two a side and interlocking in two layers, in the center of each plot. Stacked bales were sampled for fire ant infestation using 2.5 cm × 2.5 cm olive oil–soaked index cards; one bait card was placed on each side of the top layer of hay in each stack. Results from ANOVA show a significant difference in mean infestation levels among treatments. Stacks of hay sitting in the chlorpyrifos plots had fewer ant infestations compared to the permethrin and control plots. Results after one week showed that only one stack in the permethrin, and two in the control plots were infested with ants, while none in the chlorpyrifos plots were infested. Results show that after three weeks all four control stacks, three stacks in the permethrin treatment, and two in the chlorpyrifos plots were infested. These results indicate that on a short-term basis, such as 1 to 7 days, chlorpyrifos may be an effective short-term treatment option for protecting stacked hay from fire ant infestations.

Ronald D. Weeks, Jr., Michael E. Heimer, and Bastiaan M. Drees, Chemical (Chlorpyrifos and Permethrin) Treatments Around Stacked Bales of Hay to Prevent Fire Ant Infestations, Texas Imported Fire Ant Research & Management Project, Red Imported Fire Ant Management Applied Research and Demonstration Reports, 2000–2002, Texas Cooperative Extension Service.

Source: http://fireant.tamu.edu/research/arr/year/00-02/2000-2002ResDemHbk.htm#stackedbales

The complete report can be found on the book's companion Web site (select Empirical Research Reports in Chapter 15: Sample Documents).

Introduction and Literature Review

Like any other introduction, an empirical research report introduction gives the subject, scope, significance, and objectives of the research. The first example incorporates all these features in three concise introductory paragraphs. As a rule, the literature review summarizes what is known about the problem and cites relevant published articles and reports; our example, however, lacks a literature review per se and instead reviews the results of various chemicals used to control fire ants. The introduction and literature review may be placed in one section or separate sections, depending on the complexity and length of the literature review. All this material should support the objectives of the research: why the research is needed, what gap this research will fill in resolving the problem at hand.

This first complete empirical report we discuss in this chapter, shows the value of the abstract (or summary) and its relationship to the introduction. Note that a boldface heading introduces the problem statement; the research objective, with a second heading. The **summary**, **problem statement**, and **objective** provide a clear view of the intent of the report. Note that in this example report, the writer uses the literature review to justify a choice of methods selected for conducting the research.

Summary

The summary begins with a <u>rationale</u> for the research topic, moves to the description of the research conducted, and concludes with the results.

Evaluation of Citrex® Fire Ant Killer as a Drench Treatment for Red Imported Fire Ant Mounds

Summary

The red imported fire ant, *Solenopsis invicta* Buren (herein referred to as the fire ant), has become an important economic threat in urban Texas. The fire ant affects recreational activities as well as agricultural operations. This trial evaluated a product that contains a botanically derived insecticide, D-limonene, as a single mound treatment fire ant mounds, at lower- than-labeled rates, on the premises of the Johnson Space Center (JSC) of the National Aeronautics & Space Administration (NASA) in Houston, TX. The data indicates that Citrex® at the 3, 4, and 5 oz/gal rate, when compared to the untreated check, reduced mound activity within 3 days after treatment (DAT). This reduction was still evident 14 DAT, with the 4 and 5 oz/gal rates having fewer active mounds than the 3 oz/gal rate. This trial was applied April 27, 2001, when temperatures were moderate, moisture was good, and fire ant activity was good. This trial demonstrates that the 4 and 5 oz/gal rates are effective in reducing fire ant mound activity as single mound treatments.

Problem Statement The problem statement expands on the economic rationale for the experiment—the damage caused by fire ants, the costs of various
treatments, which become significant because of the size of the fire ant problem,
and issues surrounding the use of various treatments. This report uses the
"literature review" feature (i.e., the results obtained from the chemicals studied) to justify the choice of methods for conducting the research. Thus, the
problem statement is combined with a description of effects of chemicals used
to deal with fire ants.

Problem

The red imported fire ant, *Solenopsis invicta* Buren (Hymenoptera: ormicidae),
has become an important economic threat in urban Texas. According to a 1998
study conducted by the Department of Agricultural Economics, Texas A&M
University, of fire ant related costs in Dallas, Fort Worth, Austin, San Antonio,
and Houston, fire ants have serious economic effects for these metro areas of
Texas. Households experienced the largest costs among sectors examined, with
a average of $151 per household spent annually, which included repairs to
property and equipment, first-aid, pesticides, baits, and professional services.
A full damage assessment for Texas, including additional sectors, is estimated
at over $1.2 billion per year. Treatment costs accounted for over 50% of the total
cost of $581 million in the five major metroplex areas (Dallas, Fort Worth,
Houston, Austin, and San Antonio). In Houston, the average medical treatment
cost per household was $25.46. The fire ant limits outdoor activities, and home-
owners and producers incur added costs in managing the fire ant. Citrex® Fire
Ant Killer, containing 78.20% D-limonene (an extract from oil from citrus peels)
plus an emulsifier inert ingredient (Surfonic N-95), by EnviroSafe Laboratories,
was introduced in August 1999. This product is considered to be an "organic"
treatment. In 2000, the label rate was 8 fl oz per gallon of water. At $15.49/32 fl oz
(2002 price), the per-mound treatment cost using 8 fl oz/gal per mound, the
per-mound treatment cost was $3.87. Furthermore, treatments were observed
to cause discoloration and death (phytotoxicity) of common turf grasses like
Bermuda and St. Augustine grass. In contrast, one of the least expensive individual mound treatments is acephate. For Ortho® Orthene® Fire Ant Killer
(50% acephate), applied at 1 tbsp/mound, 1 lb treats 80 mounds. At $13.77/lb,
the treatment cost is $0.17 per mound.

Objectives The research objective emerges from the problem section, thus
showing the logical rationale for the study. This research will attempt to find a
treatment effective in terms of cost and toxicity to grass.

Objectives

This trial was established to evaluate several lower rates of Citrex® Fire Ant
Killer as a single mound treatment for fire ants to reduce treatment cost and
phytotoxicity problems associated with treatments. The trial was designed to

observe the effectiveness of concentrations of product below the 8 fl oz/gal labeled rate in 1999–2000 in reducing fire ant activity and phytotoxicity over a two-week period. Furthermore, reduced volumes of the diluted product below the conventional 1-gallon-per-mound amount used in this trial offer further reductions in treatment cost. This effort could help lower the treatment cost for fire ant control in turfgrass areas statewide.

Materials and Methods

To allow other experienced researchers to duplicate the research, writers should explain clearly and accurately how the research was performed. The materials and methods section also helps build the credibility of the report by explaining how was the research conducted, what methods and important materials were used, and what procedures were followed. This section may include the following:

- Design of the investigation
- Materials used
- Procedure—how you conducted the research
- Methods used for observation, analysis, and interpretation

Methods sections will vary, depending on the type of research. However, descriptions should be clear, complete, and accurate to allow replication of the experiment (if necessary) and to assure readers that the research approach is sound:

Materials and Methods

On Thursday, April 26, 2001 on the premises of the JSC and NASA in Houston, TX, approximately 280 active fire ant mounds were identified and flagged in an area approximately 120 ft × 900 ft. The mounds were located by walking back and forth the length of the area in 15-ft widths. Mounds were considered active if more than 100 aggressive fire ants surfaced within 10 seconds of probing the mound with the surveying flag wire. On Friday, April 27, 2001, 240 active fire ant mounds were divided into 24 plots of 10 mounds each. The 24 plots, each containing 10 active fire ant mounds, were then marked with a second identifying flag to be either left untreated, treated with Surfonic N-95 (surfactant), treated with Ortho® Orthene® Fire Ant Killer, or treated with Citrex® Fire Ant Killer. Plots were randomly selected in oval groupings for 4 repetitions of each of the various treatments. All mounds were then treated between 12:00 P.M. and 4:00 P.M.

Treatments included the following:

1. Untreated
2. Surfonic N-95 surfactant at 4 oz/gal of water
3. Orthene Fire Ant Killer at 1 tbsp watered in with 2 qt water

4. Citrex® Fire Ant Killer at 3 oz/gal of water
5. Citrex® Fire Ant Killer at 4 oz/gal of water
6. Citrex® Fire Ant Killer at 5 oz/gal of water

The Surfonic N-95, the inert ingredient in Citrex®, and Citrex® Fire Ant Killer (EPA Reg. No. 72244-1) were mixed at the concentrations listed in 2.5-gal containers. Application was made to the mounds using 2-qt plastic pitchers. Two quarts was applied to each mound (exceptionally small mounds received 1 qt and unusually large mounds received either 3 qt or 1 gal) by starting from the outside edge of the mound and working toward the center in a circular motion, causing the mound to collapse in from the center. Only enough of the Surfonic and Citrex® mixture was used to saturate the mounds. The label rate (1 tbsp/fire ant mound) of Orthene Fire Ant Killer (EPA Reg. No. 239-2632) was sprinkled on the designated mounds and watered in with 2 qt of water.

Results

The results section explains what happened when the procedure was applied. Results should coordinate clearly and precisely with the methods section. Outcomes should be tied to procedure. Many empirical research reports separate the results and discussion. In this report, one section contains both.

Results

On April 28 and 30, and on May 4 and 11, 2001, all 240 mounds were inspected starting at 10:00 A.M. and finishing around 2:00 P.M. Mounds were probed with a wooden stick and closely inspected for fire ant activity. If after 15 seconds ants were seen coming out of the probed mound, the mound was considered active **(Table 1).** Inspection for new ("satellite") mounds occurring within 2 to 3 ft of treated mounds was made. Satellite mounds, mounds that appear within 5 ft of treated mound, were counted and data are presented in **Table 2.**

All concentrations of Citrex® significantly reduced the number of active mounds when compared to the untreated check at each of the evaluation dates **(Table 3).** Statistically, the 4 and 5 oz/gal Citrex® concentrations performed identically and produced significantly quicker elimination of ants in treated mounds than the 3 oz/gal concentration in reducing mound activity 3 days after treatment (DAT), and lower mound activity from these concentrations rates was still seen at 7 and 14 DAT. Satelite mounds were also found around more of the 3 oz/gal-treated mounds, than the 4 and 5 oz/gal-treated mounds **(Table 2).**

Phytotoxicity

All 240 mounds were also inspected on each evaluation date for signs of plant damage. The field was a mixture of Bermuda grass, St Augustine grass, wild flowers and other unknown grasses and weeds. On a scale of 0 to 10,

Table 1 ■ Active Fire Ant Mounds. Raw Data from Citrex® Fire Ant Killer Field Test Results, NASA, 2001

Treatment	Apr. 28	Apr. 30	May 4	May 11
Untreated	9, 10, 10, 10	9, 10, 10, 10	8, 10, 9, 10	8, 8, 8, 8
Surfonic N-95 4 oz/gal	7, 9, 10, 10	5, 7, 6, 6	4, 4, 6, 4	4, 4, 3, 3
Citrex® 3 oz/gal	3, 5, 5, 3	1, 3, 1, 4	3, 2, 1, 2	0, 4, 4, 2
Citrex® 4 oz/gal	1, 2, 2, 1	2, 4, 1, 0	1, 2, 0, 0	0, 2, 0, 0
Citrex® 5 oz/gal	1, 1, 0, 0	0, 1, 2, 0	1, 1, 0, 1	0, 2, 1, 0
Orthene® 2 tsp/gal	4, 5, 6, 2	1, 1, 4, 4	1, 4, 2, 1	0, 0, 1, 0

Table 2 ■ Satellite Mounds Found within 5 ft of Treated Mound

Treatment	Day 1	Day 3	Day 7	Day 14
Untreated	—	—	—	—
Surfonic N-95	0	0	0	4
Orthene®	0	0	0	1
Citrex® 3 ounce	0	4	4	6
Citrex® 4 ounce	0	0	0	3
Citrex® 5 ounce	0	0	0	1

Table 3 ■ Citrex® Fire Ant Killer Field Test Results, NASA, Galveston Co., Texas, Treated April 27, 2001

Treatment	Mean no. active ant mounds/10 (4 replications)*			
	Apr. 28	Apr. 30	May 4	May 11
Untreated	9.75a	9.75a	9.25a	8.00a
Surfonic N-95 4 oz/gal	9.00a	6.00a	4.50a	3.50b
Citrex® 3 oz/gal	4.25b	2.50b	2.00b	2.50b
Citrex® 4 oz/gal	1.50c	2.25b	0.75c	0.50c
Citrex® 5 oz/gal	0.50c	1.75b	0.75c	0.75c
Orthene® 2 tsp/gal	4.25b	2.50b	2.00b	0.25c
Mean square	57.97	46.27	42.54	34.67
F	66.038	28.917	45.177	44.571
P	0.0000	0.0000	0.0000	0.0000
d.f. = 5				

* Means in columns followed by the same letter are not significantly different using analysis of variance (ANOVA) at $P \leq 0.05$ and separated using Duncan's Multiple Range Test (Microstat).

with 0 being no damage and 10 being death of foliage, the following was noted:

In total, 120 mounds were treated with Citrex®. No sign of phytotoxicity was noted on either day 1 or day 3. Day 7 produced some yellowing/reddening of the vegetation on some of the mounds treated with Citrex®. All the mounds treated with Citrex® still showed phytotoxicity symptoms on day 7, but by day 14 these symptoms weres less evident. The use of Citrex® at any of the applied solutions may cause yellowing to residential lawns. The higher the rate, the more intense the symptoms.

Conclusion

Following the reporting of results and any analysis of those results (discussion), a writer may include a conclusion to summarize the results. Or, the conclusion may be incorporated into the discussion. Conclusions focus on accuracy, any limitations of findings, and any questions that need further investigation. In this section the writer assesses the experiment and suggests further research directions.

Conclusion

Results from this limited study showed that rates of Citrex® Fire Ant Killer as low as 3 oz/gal will reduce fire ant mound activity. The 4 and 5 oz/gal rates gave the highest reduction in activity with reduced phytotoxicity problems. The rate of 5 oz/gal of Citrex®, suggested on product labels in 2002 as a result of this study, was effective in reducing the activity of treated fire ant mounds. However, data suggested that the 4 oz/gal rate may be just as effective and could offer the user a slightly more economical means treating fire ant mounds. The lower concentration, which is being considered for a future revision of the product's label, would cut the cost of the product in half, to $1.94 per mound treatment.

Furthermore, although the volume of solution used to drench each mound with these low rates of Citrex® was not recorded for each mound treated in this trial, application of less than one gallon of dilute drench per mound could result in additional reductions in treatment cost. For instance, treating a "small" ant mound with a quart of material would cost $0.48, which is comparable to many other individual ant mound drench products currently on the market.

Acknowledgments and References

Most researchers note the help of individuals who worked on the project. Acknowledgments may occur in a footnote or in a section at the end of the report.

Any specific resources mentioned in the report should have full citations in a section entitled References (see, e.g., Example 1) or Literature Cited (e.g., Example 2).

Acknowledgment

The author would like to thank Mr. Craig Gant and EnviroSafe Labs, Conroe, TX, for the Citrex® product used in this study and help in establishing and evaluating this study.

Paul R. Nestor, Red Imported Fire Ant Management Applied Research and Demonstration Reports 2000–2002, Texas Imported Fire Ant Research and Management Plan, Texas Agriculture Extension Service
Source: http://fireant.tamu.edu/research/arr/year/00-02/index.html

OTHER EXAMPLES FOR ANALYSIS AND COMPARISON

Example 1

Let's now look at a second example to see how the elements of a report occur in another kind of research report. The report explains results of a research procedure to improve African violets. The problem statement appears in paragraph 1; pertinent studies are cited in paragraph 2, which ends with the purpose statement. Thus, the introduction states the problem, identifies related research, and gives the purpose of this specific research effort.

Use of a Protoplast Regeneration System for African Violet Improvement

Traud Winkelmann, Institute of Breeding of Ornamental Species
Ahrensburg, Germany

Introduction

Since African violet growing began in Germany in 1893, breeders have improved this species in many ways. Vegetative habit, time to flowering, and flower retention have been altered. In addition, a wide spectrum of flower colors, patterns, and shapes is available in the modern African violet. This was done mainly by making crosses and subsequently selecting the desirable seedlings. Traditional breeding methods are limited by the range of species which can be combined, and certain desirable features, particularly the introduction of true red and yellow flowering plants, has not been achieved (the Blansit violets appear to be an exception).

Research demonstrating that African violets could be propagated easily in vitro under sterile conditions has opened new ways for increasing genetic variability through biotechnology (Start and Cumming, 1976; Grunewaldt, 1977). Some of these techniques require that plants be regenerated from protoplasts (naked cells without cell walls) rather than from leaf tissue. Before being able to use methods like direct DNA transfer into protoplasts or fusion of protoplasts of otherwise incompatible genera (*Saintpaulia* and *Episcia*, for example)

it is necessary to develop a reliable method for obtaining whole plants from protoplasts. The aim of our research was to establish such a protoplast regeneration system for African violets.

Methods and Results In this report, the methods and results sections are combined under one heading. Other research is noted in the methods section to justify why this method was employed. This report was planned as an online report. Thus, the figure numbers appear as links in the text.

Isolation of Protoplasts, Protoplast Culture, and Plant Regeneration

Protoplasts can be released from plant tissue by one of two methods. The first method involves mechanically isolating the naked cells by dissection or rupture of the cell walls (Bilkey and Cocking, 1982). The more common method involves treating the plant tissue with enzymes that digest the cell wall material. For our work we used a combination of three enzymes: 0.5% macerozyme, a pectinase, to dissolve the tissue; and 2% cellulase R10 and 0.1% driselase, two cellulases, to dissolve the cell walls (Winkelmann and Grunewaldt, 1992).

The starting plant material employed for a source of protoplasts proved to be very important for successful regeneration. Only when young shoots from tissue culture were used as the starting material were plants able to be regenerated from protoplasts.

After the enzymes had been removed by centrifugation, the protoplasts were embedded in alginate. Protoplasts plated in liquid or in a medium solidified with agarose did not develop. The successful medium contained macro- and micronutrients, organic acids, vitamins, and high concentrations of different sugars to stabilize the naked protoplasts until the cell walls reformed. Cell walls were formed and the cells began to divide after 8 to 10 days of growth in the dark (see Figure 1). The medium also contained two plant growth regulators; 1 mg/liter naphthaleneacetic acid (an auxin); and 1 mg/liter benzyladenine (a cytokinin). The complete details of the protoplast culture procedure can be found in the reference by Winkelmann and Grunewaldt (1992).

After 14 days growth on the initial culture medium, the osmotic strength and the concentration of growth regulators was reduced. The osmotic strength was reduced again 10 days later. Then, after about 4 weeks of culture, small clumps of unorganized cells, or calli, could be removed and plated on a medium solidified with agarose (see Figure 2). These calli were grown in the dark until they reached a size of 3 to 4 mm in diameter, and then, they were transferred to a medium containing 2 mg/liter benzyladenine to induce plant formation.

As soon as the young plants were visible under a stereomicroscope, the cultures were moved to the light and placed on a shoot elongation medium (see Figure 3). The number of plants per callus varied, ranging between 5 and 50. These plants rooted easily and could be grown on in the greenhouse with few losses. The scheme presented in Figure 4 summarizes the process from proto-plast isolation to growth of the plants in the greenhouse.

The next section of this report provides a discussion of the results:

Applications for African Violet Improvement

Protoplasts are useful in genetic manipulations because they do not have cell walls. They are ideal targets for taking up naked DNA and for fusing with pro-toplasts of other related species. In the related genus, *Episcia,* some species and selections have true yellow and red flowers. We have applied successfully our procedure for plant regeneration from African violet protoplasts to protoplasts of *Episcia cupreata* 'Tropical Topaz' (Winkelmann and Grunewaldt, 1993). As was suggested by Bilkey and McCown (1978), protoplast fusion between African violet and *Episcia* may lead to the production of new flower colors which have not been possible because of genetic barriers. Research toward this goal is now in progress.

Following the discussion of applications, the report ends with the Acknowl-edgments and References sections.

Acknowledgments

The experiments reported here are part of the Ph.D. thesis of T. Winkelmann, which was supported by the Federal Ministry of Research and Technology (BMFT, Bonn) and Fischer Company (Hannover-Isernhagen). The author would like to thank Prof. R. D. Lineberger for his critical review of the manu-script.

References

Bilkey, P. C., and E. C. Cocking. 1982. A non-enzymatic method for isolation of protoplasts from callus of *Saintpaulia ionantha* (African violet). *Z. Pflanzenphysiol* 105:285–288.

Bilkey, P. C., and B. H. McCown. 1978. Towards true red, orange and yellow-flowering African violets—asexual hybridization of *Saintpaulia* and *Episcia. African Violet Magazine* 31:64–65.

Grunewaldt, J. 1977. Adventivknospenbildung und pflanzenregeneration bei Gesneriaceae in vitro. *Gartenbauwissenschaft* 42:171–175.

Start, N. D., and B. G. Cumming. 1976. In vitro propagation of *Saintpaulia ionantha* Wendl. *HortScience* 11:204–206.

Winkelmann, T., and J. Grunewaldt. 1992. Plant regeneration from protoplasts of *Saintpaulia ionan-tha* H. Wendl. *Gartenbauwissenschaft* 57:284–287.

Winkelmann, T., and J. Grunewaldt. 1993. Plant regeneration from protoplasts of *Saintpaulia ionan-tha* H. Wendl. XVIIth Eucarpia Symposium Ornamental Section, San Remo Italy, March 1–5, 1993. Abstract of presentation.

Example 2

Examine the following empirical report. How do visuals document results?

MICROPROPAGATION OF CHIMERAL AFRICAN VIOLETS

R. Daniel Lineberger* and Mark Druckenbrod
Department of Horticulture, Ohio State University

Abstract

The pinwheel flowering African violets are periclinal chimeras. Plantlets produced from tissue cultured leaf explants do not flower true-to-type. When intact inflorescences were cultured in vitro, plantlets arose in the axils of small bracts on the peduncles. These plantlets flowered between 80% and 95% true-to-type depending on the cultivar under consideration. It is hypothesized that these plantlets result from the growth of dormant axillary buds in the inflorescence. This hypothesis would account for the ability to propagate the periclinal chimeras in a true-to-type fashion, since the apical organization of axillary buds is identical to that of the apical meristem.

Introduction

African violets which have bicolor flowers with a banded arrangement of the colors are termed "pinwheel flowering." The lateral edge of each corolla segment is a different color than the central portion, giving the whole flower a "spoked" appearance, with the "spokes" being one color and the "spaces between the spokes" a different color (Fig. 1).

This patterned arrangement of the flower is not maintained by plants propagated by leaf cuttings, but can be maintained if the terminal portion of the crown is removed and the resulting "suckers" are separated and rooted (1). This technique of propagation gives rise to few propagules per plant, necessitates using large, well-established plants for crown removal, and exposes the stock plants to potential disease problems. The cost of these chimeral plants is therefore very high compared to other African violet types which can be propagated by leaf cuttings.

During the course of experiments designed to separate the component genotypes of several cultivars of pin-wheel flowering African violets, it was

Figure 1. The pinwheel-flowering African violet cultivar 'Valencia' is characterized by corolla segments with methyl violet margins and a white center stripe.

noted that some plants produced from inflorescence explants produced pin-wheeling flowering plants (2). The procedure reported herein is a refinement of this technique suitable for the high fidelity production of chimeral African violets through tissue culture.

Materials and Methods

Whole inflorescences of the African violet cultivars 'Valencia,' 'Dardevil,' 'Desert Dawn,' and 'Mauna Loa' served as tissue explants for these studies. Inflorescences were harvested several days prior to the opening of the first flower. Explants were washed in 0.1% Alconox for 5 to 10 min, disinfected in 0.5% sodium hypochlorite for 15 min, and rinsed twice in sterile distilled water. The peduncle was cut 5 to 10 mm below the attachment of the lowest flower buds, and the whole inflorescence was placed in 25 × 150 mm test tubes containing 12.5 ml of tissue culture medium. The medium used contained the Murashige and Skoog salt formulation and organics (3), with 100 mg/L myo-inositol, 200 mg/L casein hydrolysate, 3% sucrose, 1 mg/L naphthaleneacetic acid, 1 mg/L benzyladenine, and 0.6% Difco Bacto agar (pH 5.7). Cultures were grown in a culture room providing 16 hr per day of cool white fluorescent light (40 µeinsteins/m^2/s). The small plantlets which had formed by 5 weeks were removed from the peduncle and placed in plastic covered foil tins containing moistened Reddi Earth soilless medium (W. R. Grace Co., Cambridge, MA 02140) for rooting. Plantlets were well rooted within 3 to 4 weeks, at which time the plastic lids were loosened to allow the plants to acclimate to lower relative humidities. After approximately 2 to 3 weeks of acclimation, plants were potted into 8 cm plastic pots containing Metromix 350 soilless medium (W. R. Grace Co., Cambridge, MA 02140), placed on a capillary mat watering system in a shaded greenhouse (70% shade), and grown to flowering according to standard African violet culture. Plants were observed through at least one full flowering cycle to ascertain trueness-to-type.

Results and Discussion

Plants produced through in vitro culture of leaf tissue displayed a wide variety of flowering patterns, none of which was the characteristic pinwheel flower (Fig. 2A; compare to Figs. 2B–2L).

Similar variation was observed in plants produced from 'Dardevil' leaf tissue (Table 1). Only one type of variant was produced by leaf culture of 'Desert Dawn' (Table 1). In general, the plants produced through culture of leaf tissue most often displayed monochromatic (solid color) flowers of the same color as the margin of the corolla segments. Some bicolor, irregular combinations of both colors were produced, but in these studies pinwheel flowering plants were never obtained from leaf tissue (Table 1).

When whole inflorescences were placed in culture, plantlets grew from the axils of the bracts in a short time period (Fig. 3).

Figure 2. Various flower patterns produced on tissue cultured 'Valencia' African violets. (A) 'Valencia,' true flower type. (B–J) Various unstable off-type flower patterns. (K, L) Monochromatic flowers of the same color as the corolla segment margin.

Figure 3. Plantlets produced in the bract axil of 'Valencia' after 5 weeks in vitro. These plantlets were large enough to be removed for rooting at the end of 5 weeks.

Table 1 ■ **Flowering Pattern of Plants Produced by In Vitro Culture of Leaf Explants of Three Cultivars of Pinwheel Flowering African Violets**

		Plants with Stated Flowering Pattern			
Cultivar	Number of Plants Observed	Margin Color	Stripe Color	Bicolor	Pinwheel
'Valencia'	82	67%	0	33%	0
'Dardevil'	49	43%	35%	22%	0
'Desert Dawn'	36	100%	0	0	0

These plantlets were large enough to be removed for rooting at the end of 5 weeks.

Adventitious shoots which differentiated on leaf or peduncle tissue were just barely visible to the naked eye by 5 weeks, suggesting that these shoots arose from dormant vegetative buds in the inflorescence structure. Further evidence in support of this hypothesis was obtained when small plantlets were observed growing in the inflorescence of an intact 'Valencia' plant in the greenhouse (Fig. 4).

The occasional production of true-to-type flowering plants from rooted inflorescences also has been reported (1).

Plants produced through short-term culture of inflorescence tissue exhibit a high frequency of true-to-type flowering (Table 2). All of the 'Mauna Loa'

Figure 4. Expanded vegetative
plantlets produced on a flowering
plant of 'Valencia' in the greenhouse.

plants regenerated through tissue culture were pinwheel flowering, while
about 80% of the 'Dardevil' and 'Desert Dawn' plants flowered true-to-type.
The multiplication rate varied with cultivar, with 'Valencia' achieving the highest multiplication rate (Table 2).

These rates of multiplication appear low for a tissue culture system, but
they are quite acceptable, since (1) the system has high fidelity, (2) the explant
source (i.e., inflorescence) is produced in abundance on a mature plant, and
(3) the taking of explants does not reduce the vigor of the stock plant.

It should be emphasized that the period of in vitro culture should not
extend beyond 5 or 6 weeks. Adventitious shoots are produced on the peduncle in the vicinity of the plants believed to be produced from the axillary buds
and these adventitious shoots would not be pinwheel flowering types. This

Table 2 ■ Flowering Pattern of Plants Produced by Short-Term Culture of Inflorescence Tissue

Cultivar	Average No. of Plants per Explant after 5 Weeks	No. of Plants Observed	Same Color as Segment Margin	Bicolor	Pinwheel
			Plants with Stated Flowering Pattern		
'Valencia'	9.0	236	1.5%	3%	95.5%
'Dardevil'	3.2	62	8%	0	82%
'Desert Dawn'	3.7	65	20%	0	79%
'Mauna Loa'	2.3	42	0	0	100%

phenomenon likely accounts for the observed variation in fidelity of the plants produced by the different cultivars. For example, the 'Desert Dawn' cultures may have been "contaminated" by adventitious shoots to a greater degree than the cultures of 'Valencia.'

The inflorescence culture technique should allow true-to-type propagation of other African violet cultivars which are periclinal chimeras. Plants are produced rapidly on the explants and these plants show excellent rooting and survival. Care must be taken, however, to determine the extent of variation in the tissue cultured plants, since trueness-to-type was cultivar dependent and varied between 80% and 100%.

Literature Cited

1. Eyerdom, H. Sept. 1981. Flower color sports and variations in *Saintpaulia* hybrids. *Afr. Viol. Mag.*, pp. 33–37.

2. Lineberger, R. D., and M. Druckenbrod. 1985. Chimeral nature of the pinwheel flowering African violets. *Am. J. Bot.* 72:1204–1212.

3. Murashige, T., and F. Skoog. 1962. A revised medium for rapid growth and bioassays with tobacco tissue cultures. *Physiol. Plant.* 15:473–497.

Acknowledgments

Partial funding for this project provided by grants from the African Violet Society of America and the Honors Program of the College of Agriculture. The authors thank Joe Takayama for excellent technical assistance. This manuscript is dedicated to the memory of Dale Eyerdom, Grainger Gardens, Medina, Ohio, whose encouragement was invaluable.

*Present address: Department of Horticultural Sciences, Texas A&M University, College Station, TX 77843-2133

Source: http://aggie-horticulture.tamu.edu/tisscult/chimeras/valprop/val.html

Example 3

Our final example here (Figure 15-1) is an empirical research report written by a college senior Maritime Studies major at Texas A&M University–Galveston. This report published in the *TAMU Undergraduate Journal of Science* shows how the empirical research approach can be applied to related fields, such as archaeological artifacts. This report opens with an introduction, moves to the history of the nails and fasteners found, provides a description (with drawings) of the nails, followed by a discussion of the findings. The conclusion notes the importance of the artifacts. All empirical research reports have one feature in common: careful use of description of method, procedures, and results.

An Archaeological Analysis of Nail Artifacts Recovered from A'asu, American Samoa

Molly L. Cooper[a,b]

Prior to the late 1700s, nails were hand forged and drawn to a point by the rhythmic blows of the practiced blacksmith's hammer. The invention of the more easily made cut nail, tapered by machine as opposed to by hand, redefined the availability, demand, and use of the nail in America and around the world. Historic nails and fasteners recovered from field excavations at A'asu, American Samoa, during the summer of 2002 by Texas A&M University at Galveston archaeologists were conserved and studied. The nine fasteners in the collection are machine cut square nails and fasteners, dated no earlier than 1790, though perhaps much later. The nails provide further information on the use and evolution of machine-made cut nails, and give valuable insight into their function and importance in this society.

Introduction

Texas A&M University at Galveston archaeologists began assessments of A'asu, American Samoa, in the summer of 2002.[3] The A'asu site potentially contains an abundance of information on an important European contact event of 1781, as well as a broad range of historic and prehistoric information. The contact event between native Samoans and European explorers was the primary reason for the selection of A'asu as the site for the project. Among the research questions proposed for the project were: "Is there an aceramic population present?"; "Is there a ceramic population present?"; and "Is an artifact assemblage present associated with the 1781 massacre?" In addition to these questions, it was important to discover undisturbed, stratified sites at A'asu in order to help recognize significant issues in Samoan archaeology.

In December of 1781, A'asu Valley in American Samoa was the setting for a violent encounter between European voyagers and native Samoans.[1] French explorer Jean-François Galaup de La Pérouse lost 12 men in this conflict, which occurred during the era of European discovery in the South Pacific Islands.[1] During initial contact, La Pérouse had found the inhabitants of the island agreeable and tranquil. On the morning of December 11, a small landing party including the second in command of the expedition, Paul-Antoine-Marie Fleuriot de Langle, returned to the spring fed A'asu stream to replenish the crew's water supply in preparation for departure from the islands.[1] According to the expedition records, a crowd of Samoans began to gather at the site, and as de Langle was unable to effectively communicate with or appease them, the unreceptive group of Samoans attacked. Twelve of the 61 Frenchmen present were killed, 20 wounded, and an unknown number of Samoans died as well.[1] Though La Pérouse never returned to American Samoa following his expedition, he recorded the expedition in the journals he sent back to France from Botany Bay, in present-day Australia. These journals contain the only current written record of the "A'asu Massacre".[2] This incident brought about significant developments in the evolution and history of the Samoan islands. The site was later named Massacre Bay, and a small monument was constructed

[a] Senior Maritime Studies major, Texas A&M University-Galveston
[b] Dr. Frederic Pearl, advisor

8

FIGURE 15-1 · AN EMPIRICAL RESEARCH REPORT

Source: Cooper, Molly L. "An Archaeological Analysis of Nail Artifacts Recovered from A'asu, American Samoa." *TAMU Undergraduate Journal of Science* 3, no. 2 (Spring 2003): 8–13.

here after the remains of both French and Samoan victims were discovered in 1882.[2]

Among the historic artifacts recovered from A'asu were nails and fasteners of various types. In an attempt to piece together a small part of Samoan history and culture, the nails were individually studied. Research on nail manufacture and use in general was conducted in order to reach initial conclusions about a simple but common artifact used by generations of Samoans and their visitors.

History of Nails and Fasteners

The invention of cut nails, sliced by machine from sheet iron, made massive and lasting impressions on American society. As new machines were improved and made more efficient, the demand for nails increased dramatically.[4] Few other objects could compare with the nail in regard to its numerous necessary functions and profound involvement in every aspect of ordinary life during the time period of nail technological advancements.

The cut nail forced the age-old art of hand forging wrought nails, the only nails available before the late 1700s, into near disappearance.[5] The ease of manufacture of the cut nail and later of the completely machine-made nail caused the sheer volume of nail manufacture and availability to rise and peak beyond what was ever before imagined. In 1790, the first recorded cut nails were produced, without machine-made heads.[6] After 1800, mass production of the cut nail began, and nail making became a common industry in the U.S., as the manufactured nails were exported from America to other countries.[4] The nail later evolved in 1815 when the heads also became machine made. Previously the heads had been hammered out and formed by a blacksmith on a nail header used on the hardy hole of the anvil.[6] By 1825, the practice of machined nail production became even more widespread. After about 1830, the head of a completely machine made nail had evolved to appear more uniform, more symmetrical, less thin, and comparatively more square than from about 1825-1829, at which time heads appeared thinner and fairly lopsided and uneven.[4]

In a span of about 15 years from 1880-1895, the wire nail industry, first invented in France, eventually put square cut nails out of business.[4] Although cut nails are still manufactured today for specialized purposes, they were never able to regain the high consumer demand they had in the late 18th and early 19th centuries.

A'asu Nail Artifacts

The important attributes of nails and fasteners related to this study are material type, head shape and markings (stamped vs. hand-wrought), shank shape (cross-section) and markings, projections underneath head, fasteners with roves, rove shape and diagnostic features, fastener length, and width. The following is a brief description of the diagnostic features of the A'asu historic nail collection.

Figure 1: AS 32.006-0339

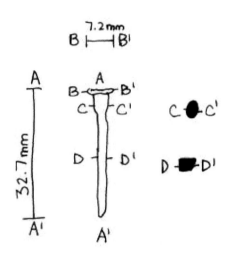

Figure 1: AS 32.006-0339, Copper Nail

The shank is rectangular, and underneath the fairly round head, there are two opposing ridge markings that protrude from the shank and meet the bottom of the head. These markings are "gripping marks," made during the manufacturing process when the machine grips the nail. Parallel striations can be seen on top of the head and along the shank, which are additional marks of the machine cutting mechanism.

Figure 2: AS 32.006-0192, Copper Nail

The shank is rectangular. The head is round and fairly lopsided. Underneath the head, there are two opposing gripping marks that protrude from the shank and meet the bottom of the head.

FIGURE 15-1 · *CONTINUED*

420

Figure 2: AS 32.006-0192

Figure 3: AS 32.006-0127

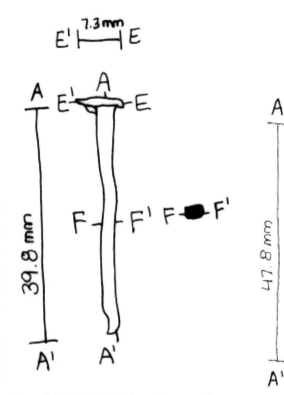

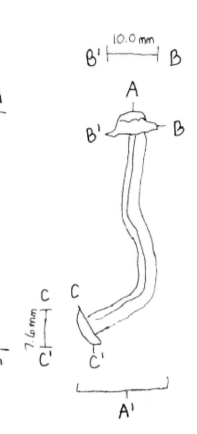

Figure 3: AS 32.006-0127, Copper Fastener- Rivet with Rove

The shank is rectangular, and the fastener appears to be heavily hammered on the rove end, almost to the point of disfigurement. The head is irregular and flawed because of intense hammer marks and resembles a square contour. Underneath the head, there are two opposing gripping marks that protrude from the shank and meet the bottom of the head. The head on the opposite end of the fastener is also uneven, but rounder than the first head. Two seams run the entire length of the rivet, on opposite sides of the shank. A rove is also found on this fastener. A rove looks similar to a conventional washer, and is used in boat building for clinching the end of a fastener.[7] The rove on this fastener has a jagged edge on the top portion where it meets the head.

Figure 4: AS 32.006-0135, Copper Fastener- Rivet with Rove

The shank is rectangular, and the head is square. There is a slight projection on the top of the head where the shank has met the bottom of the head. The head has a visible square protrusion when viewed from the top or side. Underneath the head, there are two opposing gripping marks that protrude from the shank and meet the bottom of the head. The opposite end of the fastener has an irregular shape, and looks almost like a smaller head, but has more of a curved L-shape, rather than a round or square head. Near the end of the L-shaped point, the profile begins to curve upward toward the head.

10

FIGURE 15-1 · *CONTINUED*

Figure 4: AS 32.006-0135

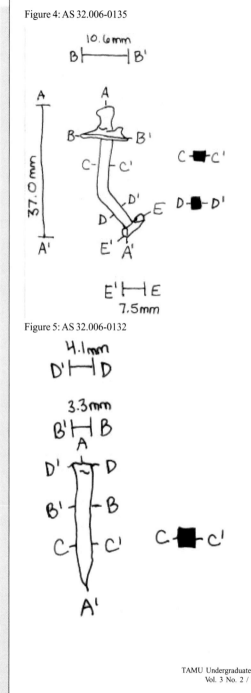

Figure 5: AS 32.006-0132

Figure 5: AS 32.006-0132, Headless Nail

The shank is rectangular and appears to have been clipped at the headless tip because of the fairly clean cut. The end of the nail, when compared to others in the collection, has a much more angular, pointed tip.

Figure 6: AS 32.006-0336

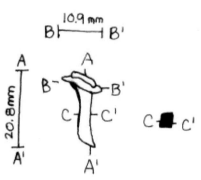

Figure 6: AS 32.006-0336, Copper Fastener with Rove

The shank is rectangular, and there is a possible seam running the length of the rivet, but it is impossible to verify due to the corrosion of the fastener. The shape of the head is round. This fastener has been cut, possibly broken, at one end so that only the head and rove remain. The rove is level and flush against the head of the rivet. The rove is firmly attached to the head and is immovable, unlike the other roves present in the A'asu collection.

Figure 7: AS 32.006-0050, Copper Nail

The shank is rectangular with hammer marks running along it, giving the appearance of heavy hammering along the middle and lower portion. The shank is not uniform. Near the head, it is more round, gradually becoming a square shape towards the mid-section. The head of the nail is severely hammered to a downward angle on one side, and is round in contour. Underneath the head, gripping marks protrude from the shank. The tip of the nail is at a blunt angle, not pointed.

FIGURE 15-1 · *CONTINUED*

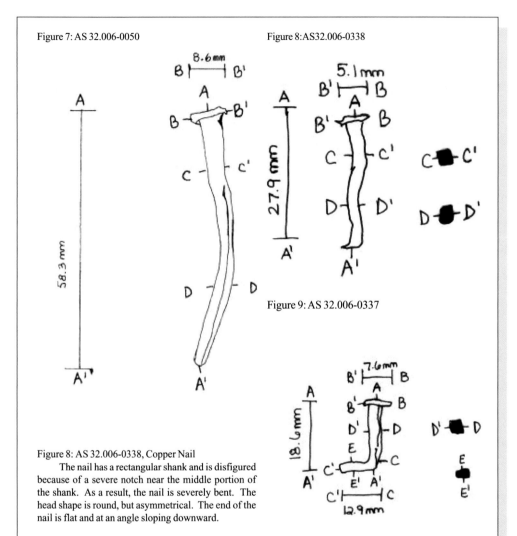

Figure 7: AS 32.006-0050

Figure 8:AS32.006-0338

Figure 9: AS 32.006-0337

Figure 8: AS 32.006-0338, Copper Nail

The nail has a rectangular shank and is disfigured because of a severe notch near the middle portion of the shank. As a result, the nail is severely bent. The head shape is round, but asymmetrical. The end of the nail is flat and at an angle sloping downward.

Figure 9: AS 32.006-0337, Copper Nail

The shank is rectangular, and has been bent into an L-shape. Along the shank, there are also intense hammer marks. The head is square, and the top is particularly hammered and deformed. The center of the head is offset, and has an L-shape when viewed from the side. There are no protrusion marks underneath the head, which were observed in the vast majority of other nails from the collection.

Discussion

The nine fasteners collected from A'asu are machine cut square nails and fasteners, dated no earlier than 1790, though probably much later. Of these, there are both manufactured heads and others with distinctly hand wrought heads. Head shape varies from a crude, hand-hammered appearance to later machine-made heads, which were more uniform and symmetrical.

12

FIGURE 15-1 · *CONTINUED*

The nails with protrusion marks underneath the head are certainly machine-made; the nail header pinching the head against the gripping mechanism of the machine could possibly cause these marks. Accordingly, these nails could not be from before the year 1815.

The rivets with roves were possibly used in boat manufacture. They are consistent with the designs of this time period for fasteners used in this type of construction. The Samoan people, since the time of their discovery, have been known as avid navigators. The Samoan islands were originally named by the explorer Louis Antoine de Bougainville as "The Navigator Islands" because of the large number of pirogues or canoes made by the hollowing out and shaping of large logs found there.[1] In boat building, certain material types are associated with particular types of fasteners. Copper was used frequently in boat construction because of the belief that it had a slower rate of corrosion when in contact with seawater. Copper was often used for tacks, nails, clench nails, and rivets, and iron was used for nails, clench nails, rivets, drifts, drift bolts, threaded bolts, and screws.[7] In addition to boat construction, the nails without roves could have been used for various purposes in the daily lives of Samoans. One only has to consider the importance of the nail in contemporary societies with recent advances in technology to imagine the impact of metal fasteners on a society without such technology during years of European contact and exploration into their lands.

A possible chronology for the fasteners, in order from most to least recent, is: AS 32.006-0339, AS32.006-0192, AS 32.006-0127, AS 32.006-0135, AS 32.006-0132, AS 32.006-0336, AS 32.006-0050, AS 32.006-0338, AS 32.006-0337. Artifact 0337, probably the oldest nail in the collection, can be dated between 1800-1825. Although it is a machine cut square nail, the head appears to be hand hammered, which would narrow its manufacture to this time period before machine stamped heads appeared. There is also an absence of the gripping marks, which provides further evidence of its antiquity and hand hammered head. The most recent fasteners in the collection with symmetrical and uniform attributes have a distinct appearance that sets them apart as being manufactured much later than more crude examples. The combination of gripping marks under the head and the appearance of seams running the length of the nail give the impression of a much later production date. Comparison of A'asu nails to a dated type collection would result in more conclusive results concerning the age and chronology of the fasteners.

Conclusions

The nail disbursement does not correlate with the European contact event, the focus of the project, because the collection can be dated only to 1790 and later. However, the assemblage does provide evidence of the use of nails among Samoans at various stages of nail development and in boatbuilding, which historians believe has been an integral part of the lives of these islanders for many years.

The nail itself is admittedly very small in size but carries great significance as far as historical influences on societies are concerned. From the time of its invention, the nail has been used for countless tasks both large and small and is widely used today. Few inventions can be traced back as far as the nail, which has withstood centuries of technological advancements and continues to be an essential part of civilization even today.

References

1. Brosse J. Great Voyages of Discovery. Paris: Bordas; 1983.
2. Dunmore J. The Journal of Jean-François Galaup de La Pérouse. London: The Hakluyt Society; 1994, 1785-1788.
3. Fontana BL. Johnny Ward's Ranch. The Kiva1962; 29(1-2): 1-115.
4. Fontana BL. The Tale of a Nail: On the Ethnological Interpretation of Historic Artifacts. Florida Anthropologist 1965; 18(3)(part 2): 85-111.
5. Noel-Hume I. A Guide to the Artifacts of Colonial America. New York: Alfred A. Knopf; 1972.
6. Davidson RB. The ABC's of Nails and Screws. The Magazine Antiques 1949; 55(3): 188-189.
7. [Lipke P.] Boats: A Manual for their Documentation. Nashville: Zenda, Inc. 1995.

Acknowledgements

Special thanks to Dr. Frederic Pearl for his guidance and encouragement, to Tom Oertling, Kimberly Hill, and Elizabeth Neyland for technical support, John Frederick, for his expertise in blacksmithing and enthusiastic support, to the American Samoa Office of Historic Preservation, and the Chiefs and people of A'asu for their support for this project.

FIGURE 15-1 · *CONTINUED*

▪ ▪ ▪ ▪ ▪ ▪ ▪ ▨ Exercises ▨ ▨ ▨ ▪ ▪ ▪ ▪ ▪ ▪

Further exercises related to the content of this chapter and all documents employed within the following exercises can be found on the book's companion Web site.

1. Find an example of an empirical research report in your field or a related field. In what ways is it similar to or different from the sample reports in this chapter?

2. What are the standard research methods in your field? List and explain each method in a slide presentation for the class.

3. Choose one of the sample reports in this chapter or on the book's companion Web site (select Empirical Research Reports in Chapter 15: Sample Documents). Use the strategies and techniques of Chapter 4 (Achieving a Readable Style) to make the methods section of this report both shorter and easier to read.

4. Using the same report as in Exercise 3, adapt the conclusions section as a one-page recommendation memo to a pertinent decision maker.

5. Study the report on bus routes given in the book's companion Web site. Can you think of a similar problem at your university that would provide a useful topic for an empirical research project?

Scenario

Glenville, a community of 70,000 people, has just received a tract of undeveloped land, approximately 70 acres, and a small development grant from the estate of a prominent citizen. The city council believes that the site will make a beautiful park. The council discusses several possible ways of using the land to provide recreation facilities while maintaining the natural beauty of the land.

The council decides to use a landscape design service but is not sure how to find and recognize the best company. The council decides to announce the project in the newspapers of Glenville and surrounding communities.

Helen Costillo notes: "Look, we don't know all that can be done with this land. We need input from several designers who can give us suggestions."

Paul Zetchen agrees but is concerned about the cost: "The estate has allotted about $40,000 toward development costs, but that won't begin to cover the work. We'll need to know the total amount for whatever we decide on."

Karen Schneider argues for credentials: "We want to know that the companies that we talk to are qualified. What other projects have they done like this? How much did those projects cost? How long did it take to do the job? Mason Valley had to fire one company that worked on their sports park because the company couldn't complete the work. Poor drainage was also a problem. What a mess!"

Laverne Roth then adds: "I don't want any trees removed that do not have to be removed. From my perspective, habitat preservation is critical. This land is too beautiful to be leveled by bulldozers!"

City Manager Catharine Cauthen wraps it up: "Let's mention these concerns in the notice we write for the newspapers. Let's also contact Felixville, which developed a new park when they began planning and find out who they talked to. We can then send letters to those companies.

"Let me draft a notice, then we can see if we need to include anything else. We want to target design firms that are environmentally conscious and experienced, as well as creative. What do we want firms to include in their proposals? We need to decide." The council members nod in agreement.

CHAPTER SIXTEEN

Writing Proposals and Progress Reports

Many times as an employee in an organization, you will generate a variety of documents relating to one particular problem or situation. You may send several e-mail memoranda to colleagues; you may write letters to individuals outside the organization; you may write memo reports "to file" that document your activities; you may also write a detailed formal report, such as a formal feasibility study (discussed in Chapter 14) at the conclusion of your work. In short, you will write various documents to different audiences about one project, problem, or topic. Proposals and progress reports are two additional types of document often required in connection with a project or problem.

THE RELATIONSHIP BETWEEN PROPOSALS AND PROGRESS REPORTS

The proposal describes work that is suggested, tells why it should be done, and discusses the methods proposed to accomplish the work. The progress report describes and evaluates a project as work is being done. Thus, if an individual or an organization decides to begin a work or research project, particularly one that requires several months or even several years to complete, it is usual to *propose* the project and then *report the progress* on that project at agreed intervals. The topic of each progress report emanates from the project that is proposed; the content and organization of the progress report are often directed by the written proposal.

In other instances, employees may be called on to report progress on the full range of projects or problems on which they are working. In situations like these, the employee writes a progress (or status) report to inform supervisors or others about what has been accomplished toward completing a job or solving a problem, stating what has been accomplished and by whom. The progress or status report thus becomes an official and legal record of work.

To help you understand how to design and write proposals and progress reports, we first discuss the development of proposals and use a student's research project proposal as an example. We then discuss progress reports in general. We illustrate the progress report by means of the student's progress report. Because progress or status reports are often written by employees to communicate and document their activities, we also present an example of a status report written by an employee to update his supervisor on an assignment.

In the world of work, proposals are most often used by organizations to solicit work contracts. Thus, we show you a situation in which a university proposes a distance education graduate program to a major oil company. The proposal is sent by e-mail.

After you study this chapter, you should be able to develop the sections typically included in a proposal as well as in progress or status reports. As the examples in the chapter will show you, progress reports and proposals—in fact, any kind of document we discuss in this book—can be submitted in a memo or letter format, as in Chapter 12, or as a longer, formal document, as in Chapter 10. The length of each document as well as the audience and the context in which the document is to be generated and received will determine which format you use.

PROPOSALS

All projects have to begin somewhere. In universities, in business, and in research organizations, the starting point is often a proposal. In simplest terms, a proposal is an offer to provide a service or a product to someone in exchange for money. Usually, when the organization—frequently a federal, state, or city government agency or a business enterprise—decides to have some sort of work done, it wants the best job for the best price. To announce its interest, the soliciting organization may advertise the work and invite contact from interested parties. In a university, the research and grants office may notify departments that money is available for research projects in a specific area. Faculty members are invited to submit project proposals that tell how much time they will need to complete the project; list any financial resources required for equipment, salaries, and release time from regular teaching duties; and describe the goals and benefits of the research to the researcher and the university. Thus, the proposal process usually begins when an organization that is interested in addressing a specific need or problem decides to hire outside help. The proposal is the written document that launches a possible solution to this need or problem by individuals or groups qualified to do the research or perform the work.

When an organization disseminates a description of the work it wants done, this document is usually called a request for a proposal (RFP) or a statement of work (SOW). The soliciting organization may send selected companies an RFP that includes complete specifications of the work desired. When, instead, an organization describes the needed work in general terms and invites interested firms to submit their qualifications, the document is usually called a request for qualifications (RFQ). A responding organization outlines its past accomplishments, giving the names of clients, describing the work it did, and supplying as references names of individuals who can substantiate the responder's claims. The soliciting organization will send full descriptions of the work to the groups it believes to be best qualified. Figure 16-1 shows the proposal cycle.

When the soliciting organization issues an RFQ, interested companies respond by describing briefly what they offer—their experience with similar projects, the qualifications of their personnel, their suggested approach to the project, and the approximate cost. Although RFQs are often published in newspapers (see Figure 16-2), the U.S. government posts theirs online at http://www.grants.gov/.

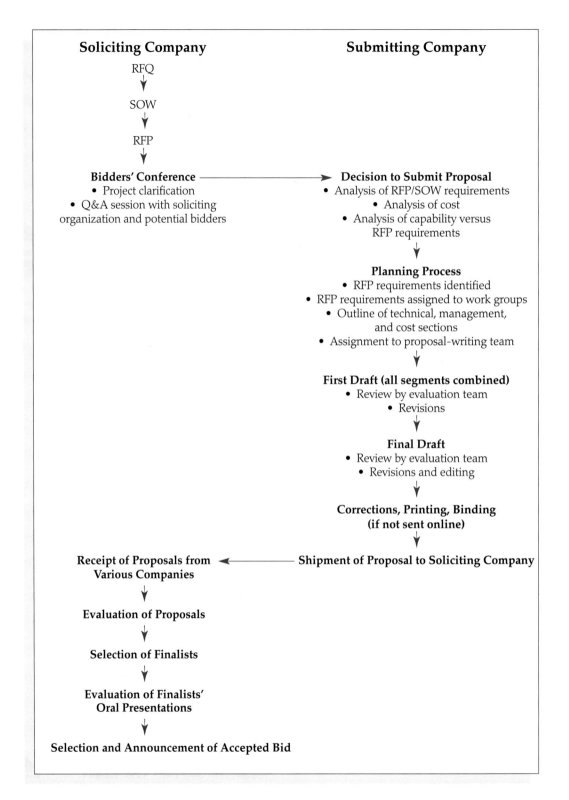

Soliciting Company

RFQ
↓
SOW
↓
RFP
↓

Bidders' Conference
- Project clarification
- Q&A session with soliciting organization and potential bidders

Submitting Company

Decision to Submit Proposal
- Analysis of RFP/SOW requirements
- Analysis of cost
- Analysis of capability versus RFP requirements
↓

Planning Process
- RFP requirements identified
- RFP requirements assigned to work groups
- Outline of technical, management, and cost sections
- Assignment to proposal-writing team
↓

First Draft (all segments combined)
- Review by evaluation team
- Revisions
↓

Final Draft
- Review by evaluation team
- Revisions and editing
↓

Corrections, Printing, Binding (if not sent online)
↓

Receipt of Proposals from Various Companies ← **Shipment of Proposal to Soliciting Company**
↓

Evaluation of Proposals
↓

Selection of Finalists
↓

Evaluation of Finalists' Oral Presentations
↓

Selection and Announcement of Accepted Bid

FIGURE 16-1 · THE PROPOSAL CYCLE

REQUEST FOR QUALIFICATIONS
For the construction of a
Clubhouse & Assorted Structures
for Universal City, Texas

The City of Universal City, Texas, through its agent, Granite Golf Management, Inc., seeks to identify interested and qualified companies to construct an 8,338-sf clubhouse with 5,834-sf underground cart storage, 5,000-sf maintenance building, and associated comfort stations and pump house/lift stations. Firms interested in being part of this process and having experience in construction of all the buildings herein described are invited to attend a prebid meeting March 3, 2001, 3:00 P.M. at Universal City, City Hall, 2150 Universal City Blvd, Universal City, TX 78148. Conference center and work plans may be picked up on February 26, 2001, between 9 A.M. and 12 P.M. and February 27, between 2 P.M. and 5 P.M. Limited special arrangements can be made by calling 210.659.6123.

If your firm is interested in being included in the bidding process and receiving more information for this project, please provide us with the following information:

1. Company name; if incorporated, give date and state of incorporation.
2. If your firm is not a corporation, specify type of entity.
3. If your firm is a partnership or individually owned, please give names of partners/owners and a résumé on each.
4. Number of years doing business under firm's current name, number of employees, and annual revenue. Any former name(s) under which firm has operated.
5. Jurisdictions and trade categories, registration numbers, and/or license numbers that indicate the categories in which your firm is legally qualified to do business, and the states of jurisdictions where those licenses or registrations allow you to do business.
6. Categories of work that your firm performs with its own labor.
7. Projects completed in the past five years where you provided similar services (scope of work). Enclose résumés for your key people to be utilized in planning or direct supervision of this project.
8. If you or any officer or member of your firm has failed to complete a project, or is involved in a litigation regarding a project, please give details.
9. Construction projects currently in process where you are providing similar services to those you wish to provide for this project, name of owner, percent complete and projected completion date.
10. Trade and bank references and name of surety bonding company, including agent's name and address.

Firms selected to be part of the bid process are required to provide financial statements prior to awarding contracts. Qualified firms interested in participating in the bid process may use AIA Form A305 for submission of qualifications which shall be submitted no later than the prebid meeting, March 2, 2001, to:

Granite Golf Management, Inc.
Jim Smith
9706 Gates Drive
Universal City, TX 78148
210.659.2267 phone
210.659.5546 fax

We expect construction to begin no later than September 1, 2001.

FIGURE 16-2 · EXAMPLE RFQ FROM A NEWSPAPER

Firms that respond with a price that best approximates what the soliciting organization wants to pay will be sent a complete description of the work needed and invited to submit detailed proposals.

To understand some of the many ways that proposals initiate projects, consider the following examples.

1. Professor X of the university's sociology department notes in the *Federal Register* that the U.S. Department of Health and Human Services (HHS) is soliciting studies of educational problems experienced by school-age children of single parents. Because Professor X has established a research record in this field and is looking for new projects, she decides to request a copy of the RFP. After studying it carefully and identifying a local school district willing to participate, she decides to submit a proposal. She describes her planned research and explains its benefits, stating her qualifications to conduct the research and detailing the costs of the project.

2. A county in Texas decides that it wants to repave a heavily used rural road and extend the paving another five miles beyond the existing pavement. The county public works office runs an advertisement in several county and state newspapers, briefly describing the work. Public works officials also send copies of the advertisement to road construction firms that have reputations for doing quality work at a fair price. The construction companies interested in submitting bids will notify the county officials and will be invited to attend a bidders' conference at which requirements of the job are discussed further. Public works officials may take potential contractors on a tour of the area. Those who decide to bid on the paving project will have four weeks to submit bids that meet the minimum specifications given in the published RFP and at the bidders' conference.

3. Alvin Cranston, a manager for a local telephone company, is charged with redesigning the operator service facilities for the company. Alvin knows that he will need to consider a number of issues (lighting, furniture, computers, and building layout), so he decides to publish a request for qualifications in telephone trade publications. He also asks the company's marketing department to help him locate a list of companies that specialize in ergonomic design. He writes each of the companies on the list and explains, in general terms, what his company wants to do and invites the design firms to submit their qualifications for performing such work.

4. Biotech Corporation is considering the development of a new organic dispersant for combating major oil spills in freshwater lakes. The company wants to know how much containers for transporting this new dispersant would cost, what kinds of container are currently available to transport the dispersant by rail or air, and whether chemical transport container companies would be interested in providing the containers and shipping the dispersant to purchasers.

5. The vice president for research at a large university has located funding for collaborative research grants, which will be available to faculty across various disciplines. He issues an RFP via e-mail to the university's research faculty.

December 5, 2003

MEMORANDUM

TO: Distribution A

FROM: Dr. Rolfe D. Auston
              Vice President for Research

SUBJECT: 2004 ABC University-Conacyt: Collaborative Research Grant
              Program

The Office of the Vice President for Research is pleased to issue a request for proposals for the ABCU-Conacyt: Collaborative Research Grant Program. This program annually awards one-year grants of up to $24,000 to faculty members to advance interinstitutional cooperation in science, technology, and scholarly activities that have a direct application in industry or government through the complementary efforts of scientists and scholars from ABCU and Mexican institutions. Two main objectives of the Collaborative Research Grant Program are to provide seed funding to support the completion of a 12-month interinstitutional project, and to support the development and submission of proposals for external funding of research from competitive granting agencies both domestic and international (e.g., NSF, NIH, DOE, World Bank, NATO, UNESCO) and industry.

ABCU and Conacyt have agreed on several research priority areas as noted in the request for proposals. The research proposed must be linked to the private sector and must have direct application to solving an industrial or governmental problem. All proposals must include research that directly improves security for the citizens of the region or explores issues relating to security challenges facing both countries.

A principal investigator (PI) is required from both ABCU and a Mexican institution. The PI from ABCU must be a tenured or tenure-track faculty member. The PI from Mexico must be a scientist or scholar from any Mexican institution of higher education and research that is registered with Conacyt. Other investigators may include faculty from branch campuses or Mexican faculty, postdoctoral students, graduate students, or research staff. A letter of intent must be received by 5:00 p.m. on Friday, March 4, 2005, to be eligible to submit a full proposal. Full proposals must be submitted, routed electronically for appropriate signatures, and received by 5:00 p.m. on Friday, April 29, 2005. The request for applications is available on the Web at http://conacyt.abcu.edu.

Additional information may be obtained by contacting Ms. Catharine J. Restivo (916-245-6517; cjr@abcu.edu).

In short, each aspect of the solicitation process, the RFP, the RFQ, and the SOW, has an appropriate use, but one or more of them is necessary to initiate action on a project.

The Context of Proposal Development

Because most proposals require substantial research and analysis, individuals and organizations wishing to respond to an RFP study the request and approach the decision to prepare a proposal carefully.

In deciding whether to respond to the invitation, the individual or company should study the RFP or RFQ carefully with a number of questions in mind:

- Can we do the work requested?
- Can we show that we can do this work, based on what we have already done?
- Can we do it within the time limit given in the RFP?

Businesses responding to RFPs are also interested in economic issues:

- How much will our proposed approach cost?
- How much money can we make?
- Who else will be submitting proposals?
- What price are they likely to be quoting for the same work?
- Can we be competitive?
- What other projects are we currently involved in?
- Could problems arise that would make us unable to complete the job on time and at the price we quote?
- Do we have personnel qualified to work on this project?

The e-mail RFP on page 433 would elicit other questions:

- Is my field applicable to the research opportunity described here?
- Can I develop a collaborative proposal by the deadline?
- What types of national security topic would be most likely to attract funding?

The university research office often has a person who answers these questions. Alternatively, like nonacademic organizations, the institution that has requested proposals will hold a bidders' conference at which companies interested in submitting a proposal can ask questions or seek clarification of the needs described in the RFP.

Most RFPs state a deadline for submitting proposals and require that submissions contain certain specific information. Proposals lacking that information may be omitted from consideration. Therefore, once an organization decides to submit a proposal, staff members carefully identify the information requirements and given each requirement is given to an individual or a group who will be responsible for furnishing necessary material and data.

Some proposals, such as university research proposals, may be written by one or two persons. For complex proposals, however, different sections may be written by individuals in different areas of the organization. An editor or proposal writer will then compile the final document. This writer/editor may be assisted by readers who help check the developing proposal to be sure that all requested information is included and that the information is correct. Once submitted, a proposal becomes legally binding. For that reason, the proposing organization carefully checks all information for accuracy. (Figure 16-1 will help you visualize the proposal process.)

When a large number of bidders submit proposals in response to an RFP, the soliciting organization may select several finalists and allow each to give an oral version of the proposal. During this oral presentation, the soliciting group asks questions; representatives of the proposing groups have one more opportunity to argue for the value of what they are proposing, present the merits of their organization, and then justify the cost attached to the proposed work.

Effective Argument in Proposal Development

All writing is persuasive—its intent is to convince readers that the writer has credibility and that his or her ideas have merit. The success of a proposal, however, rests totally on the effectiveness of the argument—how convincingly the writer argues for a plan, an idea, a product, or a service and how well the writer convinces readers that the proposing organization is the one best suited to do the work or research needed. In planning the content of the document, the writer must harmonize the soliciting company's needs with the proposer's capabilities. Writers must be acutely sensitive to what readers will be looking for, but mindful of the ethical responsibility to explain accurately and specifically what work can and cannot be done. To preclude any possibility of deceiving readers by means of promises that cannot be fulfilled, writers must be careful not to suggest actions outside the capability of the proposing individual or organization.

The following questions are useful in analyzing the effectiveness of the argument, whether in a written or an oral proposal:

■ What does the soliciting organization really want?
■ What is the problem that needs to be solved?
■ What approaches have people already tried?
■ What approaches to the solution are likely to be viewed most favorably?
■ What approaches are likely to be viewed unfavorably?
■ What objections will our plan elicit?
■ Can we accomplish the goals we propose?

To answer these questions, the proposer may be required to do research on the organization, its problems, its corporate culture, the perspective and attitudes this culture fosters, and the organization's financial status, goals, and problems.

As the different parts of the proposal are developed, each writer should examine it from the intended readers' perspective.

- What are the weaknesses of the plan, as we—the writers—perceive them?
- How can we counter any weaknesses and readers' potential objections?
- How can we make our plan appealing?
- How can we show that we understand their needs?
- How can we best present our capability to manage and complete this project?
- What are our strengths?
- From our own knowledge of our organization, what are our weaknesses— in personnel and in overall capability to complete this project as proposed?
- Do we need to modify our plan to avoid misleading readers about our ability to perform certain tasks on time, as proposed, and at cost?
- Can we sell our idea without compromising the accuracy of our statements about what we can actually do?

As a proposal writer, you should consider each question in determining what evidence you will need to support the merits of your idea and in developing the arguments needed to refute any objections. Although the proposal is designated a sales document, the writer is ethically obligated to make an honest presentation. In considering the ethical issues that confront proposal writers, you will want to review Chapter 5, Writing Ethically.

Standard Sections of Proposals

Proposals generally consist of a summary, a main body, and attachments. The main body focuses on the three main parts of the proposal:

- What the proposal's objectives are (technical proposal)
- How the objectives will be achieved (management proposal)
- How much the project will cost (cost proposal)

You may find it helpful to visualize the structure in this way:

Project summary
Project description (technical proposal)
    Introduction
    Rationale and significance
    Plan of the work
    Facilities and equipment
Personnel (management proposal)
Budget (cost proposal)
Appendixes (attachments)

Major proposals are submitted in complete report format, which requires a letter of transmittal, a title page, possibly a submission page, a table of contents, and a summary. Shorter proposals may be written in a memo or letter format. In most RFPs, the soliciting organization explains what should be included in the proposal (either specific information to be included or major elements), as shown in the RFP-based Exercise 1 at the end of this chapter. You will find the RFP on the companion Web site for this book. Often, RFPs indicate the maximum number of pages allowed. Writers are well advised to follow these instructions carefully to ensure that the proposal is not rejected during the initial screening process because it fails to follow the preparation guidelines stipulated.

Summary   The summary is by far the most important section of the proposal. Many proposal consultants believe that a project will be accepted or rejected based solely on the effectiveness of the summary, which is your readers' introduction to what you are proposing. The summary should concisely describe the project, particularly:

▮ How your work meets the requirements of the soliciting organization
▮ Your plan for doing the work
▮ Your or your company's main qualifications

The summary should be a concise version of the detailed plan, but it should be written to convince readers that you understand what the soliciting firm needs and wants, that what you are proposing can be done as you describe, and that your approach is solid because you have the required knowledge and expertise.

Project Description (Technical Proposal)   The technical proposal describes what you or your company proposes to do. The description must be as specific as possible. The technical proposal has a number of elements, described below.

Introduction   The proposal introduction should explain what you are proposing and why, and what you plan to accomplish. The introduction contains the same elements as any introduction. In short proposals, the summary and introduction can be combined.

Rationale and Significance   In this section you demonstrate that you understand your readers' needs—as stated in the summary or introduction—and show that your goals reflect careful identification and analysis of these needs. Although you will clearly be selling your idea, you should anticipate and answer any questions your readers might have as you present the merits of your project. Convincing your readers that you fully understand what they are looking for is critical in establishing your credibility. You have several approaches:

▮ You may want to define the problem, to show that you understand it.
▮ You may want to explain how the problem evolved by providing a historical review for background purposes.

- ▨ If you are proposing a research project, you will want to explain why undertaking your research will be worthwhile and what results can be expected.
- ▨ You may want to describe your solution and the benefits of your proposed solution.

Of greatest importance, however, is the *feasibility* of the work you propose:

- ▨ Is your proposed work doable?
- ▨ Is it suitable, appropriate, economical, and practicable?
- ▨ Have you given your readers an accurate view of what you can and will do?

Plan of the Work In this section, to prove that you understand the breadth of the work you are proposing, you will describe how you plan to achieve your stated goals. You will specify what you will do in what order, explaining and perhaps justifying your approach. A realistic approach is crucial in that if your plan omits major steps, a knowledgeable reader will realize this immediately. A flawed work plan can destroy your credibility and make irrelevant the merits of the solution you are proposing.

Scope The work plan section may need to describe the scope of the proposed work:

- ▨ What will and will you not do?
- ▨ What topics will and will not your study or work cover?
- ▨ What are the limits of what you are proposing?
- ▨ What topics will be outside the scope of your project?

As the writer of the proposal, you have both an ethical and legal obligation to make clear the limits of your employer's responsibility.

Methods A work plan may also require a statement of the methods you will use:

- ▨ If you are going to do on-site research, how will you proceed?
- ▨ If you plan to collect data, what method of analysis will you use?
- ▨ How will you guarantee the validity of the analysis?
- ▨ If you are going to conduct surveys, how will you develop them?
- ▨ If you plan to do historical research or a literature review, how will you ensure that your findings are representative of what is known about the subject area?

A precise, carefully detailed description of work methods can add to the credibility of the proposer as a person or organization competent to perform the work.

Task Breakdown Almost all proposals require you to divide your work into specific tasks and to state the amount of time to be allotted to each task. This information may be given in a milestone chart, as illustrated in the third main section of the student research report presented later (see Figure 16-3). A realistic time schedule also becomes an effective argument. It suggests to readers that you understand how much time your project will take and are not promising miracles just to win approval.

If a project must be completed by a deadline, the task breakdown and work schedule should indicate how you plan to fit every job into the allotted time. However, do not make time commitments that will be impossible to meet. Readers who sense that your work plan is artificial will immediately question your credibility. Remember, too, that a proposal is a binding commitment; failure to meet it can destroy your professional credibility and leave yourself or your organization open to litigation.

Problem Analysis Few projects are completed without problems. If you have carefully analyzed the situation you intend to address, you should anticipate where difficulties could arise. Problems that may be encountered can often be discussed in the rationale section. However, if you have reason to believe that major obstacles will occur during the course of the project, you may wish to isolate and discuss these in a separate section. Since organizations that request work or solicit research proposals tend to be aware of problem areas, anticipating and designing solutions to predictable difficulties can enhance your credibility. Readers will not be favorably impressed if you fail to diagnose points in your work plan that could be troublesome and even hinder your completion of the project.

Facilities The facilities section of the proposal is important if you need to convince the reader that your company has the requisite equipment, plant, and physical capability. Facilities descriptions are particularly crucial if hardware is to be built at a plant site owned by your organization. Even in study proposals, readers may want to know what research resources you will use. If the client's existing facilities are not adequate for a particular job, your company may have to purchase certain equipment. The facilities section enables you to explain how any such outlays will be included in the cost proposal.

Researchers may need to travel to visit specialized libraries or research sites. Moneys to pay for this travel will be included in the cost proposal. Thus, the nature of any extra research support, its importance, and its cost to the project should be explained here.

Personnel (Management Proposal) Any technical proposal or project is only as good as the management strategy that directs it. The management proposal should explain how you plan to manage the project, telling:

- Who will be in charge
- What qualifications that person or team has for the work at hand

Management procedures should harmonize with the methods of pursuing the work described in the technical proposal.

Readers should see the same kind of management applied to the proposed work as to the company and other projects it manages. Any testimony to or evidence of the effectiveness of the management approach will lend credibility to the proposal. Proposal reviewers must be convinced that you and your organization have a sound approach supported by good management.

In research proposals, the researcher who is soliciting funds will want to explain his or her expertise in the subject area proposed. This explanation may focus on:

educational background
projects successfully undertaken
published research on the topic
general experience

Cost (Cost Proposal) The cost proposal is usually the final item in the body of the proposal, and indeed, may ultimately be the most crucial factor. Information on cost appears as a budget for the length of the proposal period. The technical and management sections of the proposal, with their descriptions of methods, tasks, facilities, required travel, and personnel, should help justify the cost. The rationale for items that will produce the greatest cost should have been given earlier; however, any items not discussed in the technical and management sections (e.g., administrative expenses, additional insurance benefits costs, reserve for unexpected legal costs) should be explained. Often submitted as a separate document, an itemized budget includes items such as the proposing organization's liability for not meeting project deadlines, for cost overruns, and for unforeseen strikes and work stoppages. Many budget sections include standard statements such as descriptions of union contracts with labor costs, insurance benefits costs, nonstrike costs, and statements of existing corporate liability for other projects—any existing arrangements that could affect the cost of the proposed contract. Clearly, the goal is to estimate the cost of the project and explain how that figure was arrived at.

Conclusion The proposal concludes by restating your central argument: why you or your company should be selected to perform the work, and the benefits of the project, when completed, for the client.

Appendixes As in any report, the appendix section includes materials to support information you give in the main body of the proposal—in the technical, management, or cost proposal sections. For example, an appendix might contain résumés of principal investigators, managers, or researchers. These résumés should highlight the individuals' qualifications as they pertain to the specific project.

To help you understand proposals written by students, study the following situation and the corresponding student research proposal (Figure 16-3).

Situation 1 Anessa Jones is a senior agricultural engineering major. As part of her senior design project, Anessa has been asked by the Lombardy Irrigation

Proposal

for

Lombardy Irrigation Ditch Restoration Plan

DATE SUBMITTED: October 1, 2001

Summary and Introduction

This project proposes to develop a restoration plan for the Lombardy Irrigation Company in Rio Frio, Texas. In 1866, the company was founded, and an irrigation ditch was dug adjacent to the Frio River in Real County, Texas, to divert water from the Frio River to water crops and livestock. The community of Rio Frio grew up around the ditch and relied on the water it provided for more than 100 years.

Water has since stopped running the entire length of the ditch because of faulty design, frequent flooding, and lack of maintenance. The Lombardy Irrigation Company wants a restoration plan that will include a ditch design, soil volume estimates for any excavation, cost estimates for construction, and vegetation control and ditch maintenance recommendations.

As an engineering consultant, I am qualified to perform the research, calculation, and design required by the client. The completion date for the research will be December 3, 2001.

Introduction

Background

Description of the Problem

The original ditch had a bottom width of 6 feet; a water depth of 5 feet; a top width of 7 feet; a slope of 4 1/6 feet per 1000 feet; and a flow rate of 109.62 cubic of feet per second (cfs). This configuration translates to a cross section with 10:1 side slopes, which is almost vertical. The design, combined with the location of the ditch about 50 feet from the river, contributed to ditch failure.

The company currently maintains water rights to divert 3,460 acre-feet of water per year from the Frio River and rights to use 1600 acre-feet of water per year for irrigation purposes. Over the years, land uses adjacent to the ditch have become less agricultural and more residential. Restoring the Lombardy Irrigation Ditch is necessary for historical value and for the water value, although few farms are left along the ditch that would use the water for irrigation. This situation make cost analysis of the project difficult, since most of the economic gain from the ditch would come from increased tourism of the area, higher property values along the ditch, and other factors that cannot be predicted with certainty at this time.

A simple solution would be to redesign the canal and install an impermeable liner to prevent water loss and increase velocity. This solution is impractical because of economic considerations, land constraints, and the certain objections of adjacent landowners. Lombardy thus has the problem of finding the most economical way to restore the ditch to its original carrying capacity, taking into account socioeconomic factors and technical design factors.

Plan to Restore Lombardy Irrigation Ditch
October 1, 2001

1

FIGURE 16-3 · PROPOSAL SUBMITTED BY A BEGINNING ENGINEERING CONSULTANT

Project Goal

I propose to develop a restoration plan to bring the Lombardy Ditch Irrigation system in Rio Frio, Texas, back to its full carrying capacity. The plan will include a ditch design, soil volume estimates for any excavation, cost estimates for construction, and vegetation control and ditch maintenance recommendations.

Scope of the Proposed Work

The project will include an economic analysis of plan implementation; a ditch design including any excavation, side-slope stabilization, and line requirements; recommendations for vegetation control and ditch maintenance; and a suggested timeline for implementing the plan. The economic, social, and legal implications of ditch modification will be considered throughout all phases of the design and resulting recommendations.

Plan of the Proposed Work

The proposed plan has three specific phases, during which the following will be accomplished: a site inspection, a field survey, a literature review, a ditch design, and a final restoration plan. To conclude these tasks by the proposed December 14 date, I will work according to the following schedule:

Project Duration—Weeks: 10/8/01–12/14/01

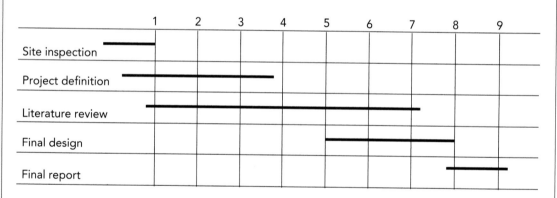

Site Inspection

Visiting the site will allow me to determine the cause of the problem and to help identify possible solutions. During a site inspection, I will collect soil and water samples, take photographs of unique elements in and around the ditch, and talk to community residents to learn about their concerns. Soils sample will identify soil characteristics in the area, including soil bulk density and permeability. A water sample will show turbidity of the water, or how much sediment the water is carrying that could be deposited in the ditch. A velocity meter will be used at the head gate of the ditch to estimate the flow rate from the river into the ditch. The flow rate will be used in the final ditch design.

2

Plan to Restore Lombardy Irrigation Ditch
October 1, 2001

FIGURE 16-3 · CONTINUED

Field Survey

The site inspection phase will include a survey of the ditch to determine the current slope and average cross section of the ditch. I will use this data to calculate soil excavation volumes and estimate construction costs. If the company decides to implement the plan, the construction contractor will use the survey to help with construction work.

Literature Review

A literature review will provide information about canal design, linear materials, soil survey information, historic Frio River flood dates, and other canal structures. This information will be critical to the final design and development of the recommendations for maintenance. A working bibliography of sources is included as Appendix A.

Ditch Design

The ditch design must address both land constraints and socioeconomic concerns of the area. The company owns the ditch and 10-foot easement on each side. The actual ditch cross section must fall within these boundaries because it is not feasible for the company to purchase more right-of-way from the adjacent landowners. Construction of the ditch is also an issue because there is not enough space to bring in heavy earthmoving equipment for ditch construction. The final design will include any excavation, side-slope stabilization, and linear requirements for the ditch.

Final Restoration Plan

The proposed final plan will include the ditch design, soil volume estimates for any excavation, cost estimates for construction, and vegetation control and ditch maintenance recommendations. I will provide an economic analysis of plan implementation, side-slope stabilization recommendations, and technical implications of plan implementation. A working outline for the final report is included as Appendix B.

Qualifications and Experience

I am completing the B.S. in agricultural engineering this fall. An integral part of my education has required me to develop expertise in design work, field surveying, and soil volume calculation. My résumé, included as Appendix C, describes my experience in these areas. During the three summers I worked as an engineering intern to the agricultural engineer group in Comal County, we diagnosed and developed designs for solving four large-scale irrigation problems in the county. After graduating from Texas A&M, I will be employed by Hydro Collins Engineering Group in San Antonio, Texas, as a consulting engineer. Collins has hired me because of my background and interest in solving problems such as the one faced by Lombardy and my knowledge of irrigation and terrain problems in the Rio Frio area.

Plan to Restore Lombardy Irrigation Ditch
October 1, 2001

3

FIGURE 16-3 · *CONTINUED*

Budget

The budget for this project includes all expenses for completing the project by December 14.

Table 1: Estimated Budget for Proposed Restoration Plan

Item	Time	Cost
Travel, $0.45/mile	250 miles	$112.50
Surveying equipment rental, $65/day	3 days	195.00
Surveying, $20/hour	24 hours	480.00
Engineering services, $50/hour	80 hours	$4,000.00
Total		$4,787.50

Anticipated Problems

Completing the project on time and at cost could be hindered by weather. Usually, the county is relatively dry during the fall. However, additional days have been figured into the project schedule to allow for adverse weather that would delay surveying. Therefore, I see no problem in completing the project by December 14, as proposed.

Progress Reports

At our initial meeting, you requested that progress reports be submitted to you by e-mail every two weeks. Thus, assuming initiation of the project by October 15, I would send e-mail progress reports on the following dates:

> November 9, 2001
> December 3, 2001
>
> Final report will be submitted December 14, 2001, to your office.
> Oral briefing. Date to be determined.

Conclusion

This project, as proposed, will furnish the design necessary to return the Lombardy Ditch to operative condition. Preliminary assessment of the problem and my initial research suggests that traditional methods can be used to redesign the ditch. With a minimum of expenditure for analysis and design, I can provide you a working plan for the ditch. All recommendations will focus on ensuring that reclamation is not only design-effective but also cost-effective.

Appendixes

Appendix A: Working Bibliography
Appendix B: Working Outline for Final Report
Appendix C: Résumé [not included here]

4

Plan to Restore Lombardy Irrigation Ditch
October 1, 2001

FIGURE 16-3 · *CONTINUED*

Appendix A

Working Bibliography

Aisenbrey, A. J., Jr., R. B. Hayes, H. J. Warren, D. L. Winsett, and R. B. Young. 1974. *Design of small canal structures.* Denver, CO: U.S. Department of the Interior, Bureau of Reclamation.

Certificate of Adjudication for Lombardy Irrigation Company. May 1984. No. 21-3158, Real and Uvalde Counties, Texas.

Deed of Trust of Lombardy Irrigation Company. March 1916. Real County, Texas.

Drainage and water table control. December 3–15, 1992. *Sixth International Drainage Symposium,* Nashville, TN: American Society of Agricultural Engineering.

Gunatillake, G. G. W. August 1987. Linings for irrigation canals. Student report. Department of Agricultural Engineering, Texas A&M University.

Haan C. T., B. J. Barfield, and J. C. Hayes. 1994. *Design hydrology and sedimentology for small catchments.* New York: Academic Press.

James, Larry G. 1988. *Principles of farm irrigation design.* Malabar, FL: Krieger.

Jones, C. W. August 1981. *Performance of granular soil covers on canals.* Denver, CO: U.S. Department of the Interior, Bureau of Reclamation.

Kraatz, D. B. 1977. *Irrigation canal lining.* Rome: Food and Agriculture Organization of the United Nations.

Morrison, W. R., E. W. Gray, Jr., D. B. Paul, and R. K. Frobel. September 1981. *Installation of flexible membrane lining on Mt. Elbert Forebay Reservoir.* Denver, CO: U.S. Department of the Interior, Bureau of Reclamation.

Sally, H. L. 1965. *Lining of earthen irrigation canals.* Los Angeles: Asia Publishing House.

Schwab, G. O. 1981. *Soil and water conservation engineering.* 3d ed. Baltimore: Durham-Hill.

Siddiqui, I. H. 1979. *Irrigation canals: planning, design, construction and maintenance.* Lahore, Pakistan: National Book Foundation.

Slagle, S. E. October 1992. Irrigation canal leakage in the Flathead Indian Reservation, northwestern Montana. U.S. Geological Survey, Water Resources Investigations Report No. 92-4066: Helena, MT: USGS.

Stevens, J. W., and D. D. Richmond. 1970. *Soil survey of Uvalde County, Texas.* Washington, DC: U.S. Department of Agriculture, Soil Conservation Service.

Plan to Restore Lombardy Irrigation Ditch
October 1, 2001

5

FIGURE 16-3 · *CONTINUED*

Appendix B

Proposed Report Outline

I. Executive Summary

II. Introduction
   A. Background
   B. Problem Definition
   C. Scope of Work

I. Literature Review
   A. Soil Information
   B. Channel Types
   C. Liner Types
   D. Channel Design

I. Design Alternatives

II. Final Ditch Design
   A. Slope and Cross Section
   B. Lined Areas
   C. Other Structures

I. Implementation Plan

II. Maintenance Plan

III. Vegetation Control Recommendations

IV. Technical Implications

V. Cost Analysis
   A. Tree Removal
   B. Construction
   C. Annual Maintenance

Plan to Restore Lombardy Irrigation Ditch
October 1, 2001

6

FIGURE 16-3 · *CONTINUED*

Company to design a plan to restore the Lombardy Irrigation Ditch, which was dug in 1866, the year the company was founded. Anessa presented a proposal (see Figure 16-3) to the president of Lombardy, Harold R. Cole, to explain how she proposes to design a plan to restore the ditch for irrigation use. To be able to finish the project by the end of the fall semester, Anessa describes her work plan, time line for completing the project by December 14, and project report schedule.

PROGRESS REPORTS

When a soliciting organization requests a proposal, it often states that a certain number of progress reports will be required, particularly if the project covers a long time period. Progress or status reports, usually submitted at agreed intervals, tell readers how a project is coming along. Their immediate purpose is to inform the authorizing person of the activities completed, but their long-range purpose should be to provide ongoing documentation of the competence of the proposing organization or the individual to pursue a task and complete it. A progress report has three main purposes, which provide *documentation* of work accomplished:

- To explain to the reader what has been accomplished and by whom, the status of the work performed, and problems that may have arisen that need attention
- To explain to the client how time and money have been spent, what work remains to be done, and how any problems encountered are being handled
- To enable the organization or individual to assess the work and plan future work

Writers use several different strategies in designing progress reports. The report should begin with an introduction and a project description to familiarize the reader with the project. A summary of work completed follows. The middle section explains what has been accomplished on specific tasks and what work remains, followed by a statement of work planned for the next progress report period. The final section assesses the work done thus far. Any problems that are encountered are also presented, and conclusions and recommendations outline proposed methods of coping. Cost can be dealt with in either the middle or final section.

Like reports of other types, progress reports can be structured by work performed or chronologically. Progress reports also can be organized by main project goals.

Structure by Work Performed

The structure of a progress report might follow one of the two basic plans given in this section. The beginning and the end have the same structure in both plans. The middle can be organized around work completed and work remaining, as shown in the left-hand column, or around tasks, as shown in the right-hand column.

Beginning
▪ Introduction/project description
▪ Summary

Middle

▪ Work completed	or	▪ Task 1
Task 1		Work completed
Task 2, etc.		Work remaining
▪ Work remaining		▪ Task 2
Task 3		Work completed
Task 4		Work remaining
▪ Cost		▪ Cost

End
▪ Overall appraisal of progress to date
▪ Conclusion and recommendations

In this general plan, you emphasize what has been done and what remains to be done and supply enough introduction to be sure that the reader knows what project is being discussed.

Situation 1 Continued Anessa Jones's first e-mail progress report (Figure 16-4) is organized by main tasks. Her report on the status of her restoration design for the Lombardy Irrigation Ditch follows the main elements of her work plan given in the proposal.

In her proposal, Anessa had said that she would send Harold Cole e-mail progress reports on two dates. These reports will be brief because each will cover only a few weeks, but they will tell Cole what Anessa has accomplished. An e-mail copy of the reports will also go to the faculty member who is monitoring the project, to allow him to ascertain whether the work is on schedule and can be completed by the end of the semester.

Beginning

Anessa begins the first report with a clearly worded subject line, followed by a brief introduction. Because Lombardy is funding the project, Anessa explains how much she

TO: hrc@pioneer.com

FROM: jones@aol.com

CC: d-canton@overland.com

RE: Progress Report #1-Lombardy Irrigation Ditch Restoration

This is the first of three e-mail progress reports on the status of my design plan for restoring the Lombardy Irrigation Ditch.

COST TO DATE

Travel

Two trips to Frio County	250 miles @ 0.45/mile	$225.00
Surveying Equipment	3 days	$95.00*
Engineering Services	50/hour × 20 hours**	$1000.00***

*I was able to rent surveying equipment for $100 less than anticipated.

**I was able to complete the survey in 20 hours rather than 24 hours, as projected.

***I have currently used 25% of the required engineering services that I proposed.

WORK COMPLETED

Site Inspection and Field Survey

I have completed two site evaluations. During the second site evaluation, I completed a partial field survey of the ditch and actual water velocity measurements through the headgate of the ditch. The survey data will be used to help calculate the amount of soil that will need to be removed when the new ditch is constructed and will tell us how much the soil removal will cost. The water velocity measurement will tell us if enough water is moving into the ditch from the river at the present time.

Ditch Design

I have also completed a new ditch design. The design is a trapezoidal channel with a 5-foot bottom width, 13-foot top width, and a 5-foot depth of flow.

FIGURE 16-4 · E-MAIL PROGRESS REPORT

Literature Review

In addition to the sources listed in my proposal, I have continued to search for additional information. The sources used in my literature review are dated, as canal design is not new technology. I have been unable to find any cutting-edge information regarding the flow of water through a channel. In this project, I am applying standard technology to design a new ditch.

WORK REMAINING

Data Analysis

I have to analyze the data collected during the field surveys and generate a topographic map of the site. I am having a problem locating the software that I need to perform this task. I am working with a computer specialist in the Department of Agricultural Engineering to determine how to resolve this problem. An older type of software is available, which I can use if current software I need cannot be located easily.

Final Design for the Restoration

Work remaining includes contacting contractors to get cost estimates for the different types of work that will need to be done to restore the ditch. I will use this information to complete a cost analysis of restoring the ditch.

A major remaining task is recommending maintenance and vegetation control. I have contacted the Extension Service in Real County about this issue and will discuss this topic by phone next week.

An additional trip to the ditch will be needed, which will add another $112.50 to the initial projected cost. I had not anticipated that two trips, rather than one, would be needed.

ASSESSMENT OF PROGRESS TO DATE

The project should be completed on time and within the cost proposed. The unexpected travel cost will be offset by the savings in surveying equipment and time allocated to that activity. Because I have not encountered any problem that will delay or hinder completion, the project will be complete by December 14.

FIGURE 16-4 · *CONTINUED*

has spent on the project and notes that she is under budget on the cost of surveying equipment and engineering services.

Middle

In the Work Completed and Work Remaining sections, Anessa says that she has completed the site inspection, field survey, ditch design, and the literature review but has not begun to develop the restoration plan.

End

She provides an overall appraisal of her work and mentions the one problem she has experienced—the need to make an extra trip to Real County. While this trip has increased the proposed cost of travel, the overall cost remains as projected because costs of surveying equipment and engineering services are under budget.

As we note early in the chapter, progress reports to document employee activities simply explain major accomplishments. Because business and technical organizations frequently stipulate a time frame for a project, status reports may emphasize completion dates as well as deadlines. Sometimes status reports indicate "action required" notices to keep the project on schedule. Figure 16-5, featured in Situation 2, illustrates a status report written by an employee in charge of a project that must be completed according to a schedule.

Situation 2 Dean Smith, a training manager for a software development company, develops training sessions for sales personnel. He decides what training should be conducted, develops the training courses, and arranges for all training of employees. Dean routinely writes progress reports to the director of personnel, Sharon Sanchez, to keep her informed of his activities—training he thinks will need to be offered, training programs he is developing or planning to develop, and training programs he and other training staff are currently teaching or directing.

Dean writes status reports like that in Figure 16-5 about once a month (or whenever he has something he wants Sharon to know about). Because the personnel director is familiar with what Dean does, he omits an elaborate introduction. He begins with a concise summary and then proceeds to describe pertinent activities. In this particular report, he is asking for increased funding for a training program. Thus he includes an "action required" statement in the heading.

For this routine progress report, Dean modifies the general plan as follows:

Beginning (introduction)

Purpose of the report—to report the status of a project
Purpose of the work being performed
Summary of current status

DATE: November 7, 2000

TO: Sharon Sanchez

FROM: Dean Smith

SUBJECT: TQI Workshop Plans for Spring 2001

     ACTION REQUIRED BY 12/8/00

Summary of Plans

 Planning for our TQI workshops, scheduled for April and May, is nearly complete. Our TQI facilitators have produced detailed training schedules for technical service and customer relations. I have approved these, and materials are being ordered and prepared. A TQI program package, developed by HI-TOP Sales Materials, specifically for computer sales personnel, can be purchased for $3,600. The format would be an excellent follow-up for all employees. As you and I had already discussed, quality off-the-shelf TQI packages should be considered.

TQI Workshop Activities as of 11/17/00

TQI Preparations for Technical Service Personnel

 Johnette Darden and her group have been preparing materials for the technical service employees. The workshop will help the staff to look at software purchases from the customer's perspective and to begin to look for ways of decreasing the time required to resolve customer complaints. The second part of the workshop will encourage TSP to examine faulty products and find ways of eliminating the problems by working with design. We will probably recommend a quality control unit be formed that uses people from both technical service and design.

TQI Preparations for Customer Relations

 Responding to customer needs and filling orders rapidly will be the main focus of the TQI workshop for CR employees. Robert Newmann will be working with the group (1) to reassess what we are doing in our current customer relations efforts, and then (2) to improve and even eliminate processes that do not help us serve customers quicker and better.

FIGURE 16-5 · ROUTINE PROGRESS REPORT

Sharon Sanchez
Page 2
11/17/00

HI-TOP Sales Program Available for Purchase by 12/15

HI-TOP developed a superb TQI program for sales personnel about two years ago. The program was so effective that HI-TOP is now selling the program, which has received rave reviews from a half-dozen companies that sell hardware and software. I reviewed the program two days ago and think it would be an excellent follow-up for any TQI work we do. I would like to purchase the program and use it with both Technical Services and Customer Relations. We can continue to use the package with other TQI training.

An outline of the program is attached. As you can see, it addresses the major issues we believe our own TQI programs need to emphasize.

Please Note

HI-TOP will sell us the program package priced at $4,500 for $3,600 if we act before December 8. They are currently redesigning the program to include extensive teaching aids, which are nice but unnecessary for our purposes. The new package will be available 1/1 for $6,000. **The original package will not be available after 12/15.**

Please give me a call so that we can discuss.

FIGURE 16-5 · *CONTINUED*

Middle (work completed on the project)

Task A

Task B

Task C

End (conclusion and perhaps recommendations)

For progress reports that cover more than one period, the basic design can be expanded as follows.

Beginning

■ Introduction

■ Project description

■ Summary of work to date

■ Summary of work in this period

Middle

■ Work accomplished by tasks (this period)

■ Work remaining on specific tasks

■ Work planned for the next reporting period

■ Work planned for periods thereafter

■ Cost to date

■ Cost in this period

End

■ Overall appraisal of work to date

■ Conclusions and recommendations concerning problems

Structure by Chronological Order

If your project or research is broken into time periods, progress reports can be structured to emphasize those periods.

Beginning

■ Introduction/project description

■ Summary of work completed

Middle

■ Work completed

■ Period 1 (beginning and ending dates)

    Description

    Cost

    Period 2 (beginning and ending dates)

    Description

    Cost

- Work remaining
- Period 3 (or remaining periods)
    Description of work to be done
    Expected cost

End
- Evaluation of work in this period
- Conclusions and recommendations

Structure by Main Project Goals

Many research projects are pursued by dividing the tasks into major groups. Then, the writer describes progress according to work done in each major group and perhaps the amount of time spent on that group. Alternatively, a researcher may decide to present a project by research goals. Thus, progress reports will explain activities performed to achieve those goals. In the template that follows, the left-hand column of the middle section is organized by work completed and work remaining, and the right-hand column by goals.

Beginning
- Introduction/project description
- Summary of progress to date

Middle

■ Work completed or	■ Goal 1	
Goal 1	Work completed	
Goal 2	Work remaining	
Goal 3, etc.	Cost	
■ Work remaining	■ Goal 2	
Goal 1	Work completed	
Goal 2	Work remaining	
Goal 3, etc.	Cost	
■ Cost		

End
- Evaluation of work to date
- Conclusions and recommendations

Situation 3 Linbeth Consulting Company was founded to help U.S. firms gain the knowledge and expertise necessary to do business abroad, either in specific countries or in specific market areas. When approached by potential clients, Linbeth provides letter proposals explaining its services and discussing how the client's situation might be approached. While working on a project, Linbeth consultants will frequently use a PowerPoint presentation to convey status information.

The segment of a PowerPoint progress report illustrated in Figure 16-6 was submitted to a company considering doing business in Mexico. We have included only eight slides from the presentation, but you can see how the narrative of the progress report surfaces in the presentation. Linbeth consultants provided copies of the presentation with space for notes. Note that the main segments of the progress report are clearly evident, even in this excerpt. Presentation software like PowerPoint allows you to develop graphics and then paste them into the presentation, as exemplified in Figure 16-6, Slides 7 and 8.

PHYSICAL APPEARANCE OF PROPOSALS AND PROGRESS REPORTS

The importance of the appearance of any proposal or progress report cannot be overstated. A printed report that is neat and effectively formatted suggests the competence of the proposing organization or individual and reflects interest and investment in securing the project. Proposals and progress reports longer than letter or memorandum length should have a protective cover. The title page should be tasteful. Colored paper and covers should convey a professional attitude. A professional appearance is the first argument for the merits of the proposal.

STYLE AND TONE OF PROPOSALS AND PROGRESS REPORTS

The proposal and its related report documents are, in effect, sales documents, but writers have an ethical commitment to present information about a project clearly and accurately. Once a proposal has been accepted, the representations it contains become legally binding. Thus, the style should be authoritative, vigorous, and positive, suggesting the competence of the proposer. Generalizations must be bolstered by detailed factual statements. Problems should be discussed honestly, but positive solutions should be stressed. Neither in the proposal nor in the progress report should writers resort to vague, obfuscatory language.

Doing Business in Mexico

A Status Report

Linbeth
Consulting, Inc.

March 15, 2001

Topics

▌ Project Description
▌ Summary of Progress to Date
▌ Work Remaining
▌ Analysis of NAFTA Economic Effects
▌ Problems Encountered
▌ Costs Incurred to Date

Project Description

▌ Describe the Mexican social, economic and political structure
▌ Detail the regulatory climate in Mexico
▌ Highlight profit opportunities for Texas businesses
▌ Clarify the implications of NAFTA on U.S./Texas/Mexico business operations
▌ Clarify effects of culture on business

FIGURE 16-6 · EXCERPT FROM A POWERPOINT PROGRESS REPORT

Summary of Progress

- Applicable demographics and statistics data are complete.
- Mexican regulations that affect business relationships have been compiled.
- Status of the Mexican economy through 1997 has been summarized.
- Status of the U.S. economy since passage of NAFTA has been summarized.

Work Remaining

- Analysis of assessments of NAFTA
  - From Mexican perspective
  - From congressional reports
  - From Department of Commerce
  - From border economic groups
  - From Canada

Economic Profile of NAFTA

- Effects of NAFTA are not clear.
- Problem areas:
  - Changes in environmental impact
  - Changes in U.S. economy
  - Real effects of NAFTA on the economy
  - Results in specific product sectors

FIGURE 16-6 · *CONTINUED*

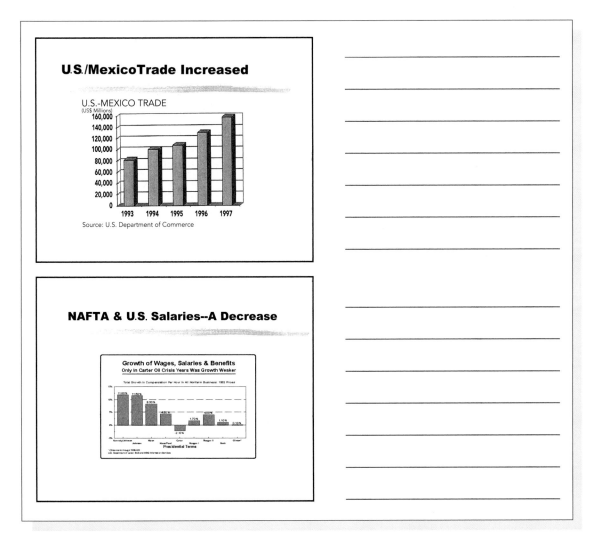

FIGURE 16-6 · *CONTINUED*

 OTHER FORMS OF PROPOSALS AND PROGRESS REPORTS

Proposals and progress reports can be prepared

    as memo reports
    as formal reports
    as letters

Yet, no matter what the format, proposals and progress reports will incorporate the same elements described earlier and illustrated in the reports written by Anessa Jones the senior engineering student, and Dean Smith, the software development company training manager. To see how proposals and progress reports might appear in another format and another context, examine Situation 4, which describes an internal proposal and a letter progress report.

Situation 4 ABC University, which has a highly ranked petroleum engineering department, has been discussing the possibility of offering graduate courses in petroleum engineering to Texaco, Inc. Some Texaco engineers would be interested in applying for the master's program in petroleum engineering, while others would like to take courses toward a certificate. Josh Jones, from ABC's Department of Petroleum Engineering, decides to submit a proposal to offer the certificate and the master's program—by distance learning (DL)—to enable engineers to access the program no matter where they work. Josh sends the initial proposal, shown below, within the e-mail conversation with Ron Harper:

Josh,

Thanks for the response. Let me forward this to our Denver group (Gary Gutierrez & Jeff Ferrante) and get their thoughts before moving forward.

We will get back to you quickly.

Ron

At 03:58 PM 5/14/2004 -0500, you wrote:

Ron,

I have completed a proposal for the Distance Learning Master of Engineering Degree to be offered in conjunction with the NSIP program in Denver, and I have outlined the major points below—all in response to your request for something concrete.

Curriculum

> PENG 603—Advanced Reservoir Engineering (Basic Simulation)
> PENG 611—Applications of Simulation
> PENG 608—Logging
> PENG 630—Geostatistics
> PENG 663—Formation Evaluation and Analysis of Reservoir Performance
> PENG 664—Petroleum Economics and Reservoir Management
> PENG 665—Petroleum Reservoir Engineering
> PENG 685—6 hours—credit for the NSIP reservoir study project
> PENG 692—3 hours—credit for a written report and oral presentation on the NSIP reservoir study project

The first seven courses represent the courses to be taken by distance learning. (There would be interaction with ABC faculty on the 685 and 692, but they would not be formal courses.) I recommend no more than two courses per semester; accordingly, a person would require four semesters (including summer) to complete this program at the rate of 2 + 2 + 2 + 1. The engineer could work half-time and still accomplish these objectives. If a person demonstrates the ability to handle 3 DL courses if allowed to spend full time on them in the final semester, the course work could be completed in one calendar year. Very few engineers have the motivation to work this hard at a distance, however, so be careful about adopting this idea.

The final two courses represent credit for work done in Tulsa on the project. I believe that a resident graduate student at ABC who had completed the seven courses I have listed could proceed directly to reservoir characterization project work without further classroom instruction. You may not agree, but I do think the need for additional classroom instruction will be minimal. I also think that a person could complete the project in Denver in six months or less after taking the suggested courses. Thus, total time for the degree would be 1.5 to 2.0 years.

I can provide additional detail to you as needed (e.g., course syllabi). We can modify the course list within limits if you want to negotiate on that. We can't go to fewer than 7 courses, though.

One final point to keep in mind: ABC would have to be in control of admissions to this program. A person qualified for the existing NSIP program would not necessarily be qualified for admission to PENG at ABCU, and you will need to be careful about what you promise.

We could start the program this summer (June 1) if you were to come up with candidates. We can certainly start in September—quite comfortably.

I look forward to your reaction.

Regards,

Josh

Following a phone call from Ron Harper, Josh sends a longer proposal, as an e-mail attachment (Figure 16-7), to Daniel Kotzner, his contact at Texaco. The following e-mail message to Kotzner serves as the letter of transmittal:

At our past two meetings in Dallas, you have mentioned that Texaco has engineers who need some or all of our master's program in petroleum engineering. As I mentioned, we can offer this degree by distance technology. We also have courses that could be taken for graduate certificate credit.

If you find the attached proposal acceptable, please forward it to Ben Zapista, whom I believe you said is in charge of education opportunities in Houston.

If you have any questions, please give me a call.

Josh

A Partnership between Texaco and ABC University to Expand Technology Training through Distance Learning

Executive Summary

We propose to develop a partnership between Texaco and ABC University to expand technology training. This partnership will expand and enhance the distance/continuing education program within the Harold Vance Department of Petroleum Engineering at ABC University to provide world-class, technically superior courses to Texaco professionals and to employees of other petroleum operators and service companies throughout the world. Texaco's commitment to underwrite this program will serve as a vehicle to advertise the role Texaco plays as a leader in exploring the boundaries of technology when providing solutions for maximizing client productivity and profit. This support, estimated at just over $200,000 per year, includes staff and student worker salaries, investment in equipment, and amounts provided for faculty members as incentives to take on these additional loads. This budget is outlined in Appendix 1.

Currently the petroleum engineering graduate program serves students from 32 countries with varied experiences and cultures. To broaden the reach of this program ABC University has made a significant investment in the distance learning programs by implementing the Blackboard system, one of the leading providers of e-learning software, as a gateway for distance learning courses. This system provides the petroleum engineering faculty with a secure environment in which to manage courses and the content material and enables students to communicate more efficiently with each other, the instructor, and the support staff.

Additionally, top-ranking faculty members have made personal and professional commitments to place the department at the forefront of distance learning. These efforts, when combined with the world wide reach of the Internet, make it possible to serve the greatest number of students and provide practicing professionals in the industry the ability to maintain their current employment and lifestyle while pursuing educational endeavors. When requested, the department can tailor the distance learning program to fit company/industry needs in the form of short courses, certificate programs in focused technical areas (see Appendix 2 for examples), a master's degree program, and, in the near future, a doctoral program—all delivered by means of distance education technologies.

The Goal: Learning and teaching are at a threshold of change that is revolutionary rather than evolutionary. Texaco recognizes the benefits of a more proficient employee with the ability to provide new, innovative solutions that maximize production and profit and the need to push the boundaries of technology. Texaco cannot afford to follow trends but must be leaders in every aspect of the industry. Leading in the expansion of the ABCU distance learning program would place Texaco at the forefront of educational technology development in this industry.

We propose to use Texaco funding to capitalize on the current momentum in the development of our graduate distance program and to develop additional training, certificate, or degree programs specifically to meet Texaco's needs using existing or added resources. Courses offered in cooperation with Texaco will include the company logo on materials made available to students enrolled in the program worldwide, including those resident on the College Station campus. (See Appendix 3 for current course offerings.) By taking courses jointly with resident ABC students and with professionals from other companies, Texaco professionals will benefit from the interaction with a diverse population of fellow students.

Our effort will be the delivery of the same quality instruction provided in graduate-level courses at ABC University in a way that meets the unique needs of remote students and makes accessible to other students their experiential knowledge, both on campus and in other remote locations. We will strive to enhance and develop effective and efficient courses and to employ the best support staff possible.

FIGURE 16-7 · PROPOSAL SUBMITTED AS AN E-MAIL ATTACHMENT

Proposal—2

The Proposal: This proposal advocates (1) continued employment of two expert staff members (i.e., the distance education team) in the designing and delivering of superior online material using the appropriate technology, (2) updating and maintaining computer software and hardware to ensure the best environment to facilitate student learning, (3) supporting additional staff to assist in development of course materials, and (4) providing faculty incentive funds.

The distance education team will work to facilitate learner-centered instructional design, use multiple delivery strategies, evaluate the program regularly, and research new techniques and technologies for the exchange of knowledge. This program will encourage direct student communication with professors by whatever media means is available. Learning must stretch beyond the classroom. E-learning is transforming traditional teaching systems. The efficiency and effectiveness of telecommunications and informatics will lead faculty and students to transform their curricula. E-learning employs distance technology and methods as intranet access as well as Internet access (campus networks and global networks). A focused line of inquiry into the development and use of E-learning is critical for teacher and learner adoption of this technology. These new technologies and methodologies provide access to global knowledge. Learner-centered instructional designs are crucial for engagement and meaning.

To exploit the lead the ABCU petroleum engineering program currently enjoys in the creation of a technically advanced product, investments must be made in additional equipment and personnel to guarantee that materials reach participants (students, staff, and faculty) in a timely and efficient manner. This effort can best be led by the department, highly ranked by *U.S. News & World Report* and now engaged in the development of the most technically advanced program.

Appendix 1

Budget Requirements

Estimated Salary and Expenses by Semester (3 courses per semester)

Item Description	Expense	Fringe	Total Expense
Kendra Jones (full-time staff)	($10,500.00)	($2,887.50)	($13,387.50)
Lloyd Richards (full-time staff)	($10,500.00)	($2,887.50)	($13,387.50)
Ted Binaldi (student worker)	($3,480.00)	$0.00	($3,480.00)
Additional student worker			
not yet hired	($3,480.00)	$0.00	($3,480.00)
Equipment	($5,000.00)	$0.00	($5,000.00)
Faculty Incentive Grants	($30,000.00)	$0.00	($30,000.00)
TOTAL EXPENSES FOR ONE SEMESTER			**($68,735.00)**
TOTAL EXPENSES FOR ENTIRE YEAR			**($206,205.00)**

This budget information is based on expected annual operations. Faculty salaries will come from state support (subvention) for courses taught on campus, since no distance courses have been offered solely to off-campus sites

FIGURE 16-7 · *CONTINUED*

Proposal—3

but are always offered concurrently to on-campus students. However, offering distance learning sections of the courses increases the faculty workload substantially, and the budget ($10,000 per course for three courses) reflects unrestricted accounts provided for faculty members as incentives to take on these additional duties. Since all students, on- and off-campus, are enrolled in exactly the same courses as local students, the only additional expenses of this program are the administrative, communications, and supplies costs.

Appendix 2

Certificate Programs

Certificate programs provide focused training in specialized areas of technology. Sequences of two or more courses comprise these programs, which can be of value to satisfy job skills requirements and professional governmental certification requirements. Where needed, ABC University can provide Continuing Education unit (CEU) certifications for these programs. This appendix suggests example certificate programs, but others can be readily developed when clients request alternatives.

<u>**Reservoir Simulation**</u>
PENG 603 – Advanced Reservoir Engineering I (Basic Simulation)
PENG 611 – Application of Petroleum Reservoir Simulation
PENG 610 – Numerical Simulation of Heat and Fluid Flow in Porous Media

<u>**Formation Evaluation**</u>
PENG 608 – Well Logging Methods
PENG 648 – Pressure Transient Testing
PENG 663 – Formation Evaluation and Analysis of Reservoir Performance

<u>**Asset Management and Economics**</u>
PENG 617 – Petroleum Reservoir Management
PENG 621 – Petroleum Development Strategy
PENG 664 – Petroleum Project Evaluation and Management

<u>**Natural Gas Engineering**</u>
PENG 613 – Natural Gas Engineering
PENG 648 – Pressure Transient Testing

<u>**Reservoir Fluids**</u>
PENG 605 – Phase Behavior of Petroleum Reservoir Fluids
PENG 616 – Engineering Near-Critical Reservoirs

<u>**Well Stimulation**</u>
PENG 602 – Well Stimulation
PENG 618 – Modern Petroleum Production
PENG 629 – Advanced Hydraulic Fracturing

FIGURE 16-7 · *CONTINUED*

Proposal—4

<u>**Production Engineering**</u>
PENG 662 – Production Engineering
PENG 618 – Modern Petroleum Production

<u>**Drilling**</u>
PENG 661 – Drilling Engineering
PENG 626 – Offshore Drilling
PENG 628 – Horizontal Drilling

<u>**Reservoir Description**</u>
PENG 630 – Geostatistics
PENG 633 – Data Integration for Petroleum Reservoirs
STAT 601 – Statistics

Appendix 3

Graduate Courses Offered by the Michael Vance Department of Petroleum Engineering.

Courses being offered through distance learning are indicated with an asterisk (*). Other courses can be developed for distance learning as requested.

*602. Well Stimulation. (3-0). Credit 3. Design and analysis of well stimulation methods, including acidizing and hydraulic fracturing; causes and solutions to low well productivity. Prerequisite: Approval of graduate adviser.

*603. Advanced Reservoir Engineering I. (3-0). Credit 3. Petroleum reservoir simulation basics including solution techniques for explicit problems. Prerequisite: Approval of graduate adviser.

604. Advanced Reservoir Engineering II. (3-0). Credit 3. Advanced petroleum reservoir simulation with generalized methods of solution for implicit problems. Prerequisites: PENG 603; approval of graduate adviser.

605. Phase Behavior of Petroleum Reservoir Fluids. (3-0). Credit 3. Pressure, volume, temperature, composition relationships of petroleum reservoir fluids. Prerequisite: Approval of graduate adviser.

606. EOR Methods-Thermal. (3-0). Credit 3. Fundamentals of enhanced oil recovery (EOR) methods and applications of thermal recovery methods. Prerequisites: PENG 323; approval of graduate adviser.

*608. Well Logging Methods. (3-0). Credit 3. Well logging methods for determining nature and fluid content of formations penetrated by drilling. Development of computer models for log analysis. Prerequisite: Approval of graduate adviser.

609. Enhanced Oil Recovery Processes. (3-0). Credit 3. Fundamentals and theory of enhanced oil recovery; polymer flooding, surfactant flooding, miscible gas flooding, and steam flooding; application of fractional flow theory; strategies and displacement performance calculations. Prerequisites: PENG 323; approval of graduate adviser.

610. Numerical Simulation of Heat and Fluid Flow in Porous Media. (3-0). Credit 3. Various schemes available for the numerical simulation of heat and fluid flow in porous media. Application to hot water and steam

FIGURE 16-7 · *CONTINUED*

Proposal—5

flooding of heavy oil reservoirs and to various geothermal problems. Prerequisites: PENG 604; approval of instructor or graduate adviser.

*611. Application of Petroleum Reservoir Simulation. (3-0). Credit 3. Use of simulators to solve reservoir engineering problems too complex for classical analytical techniques. Prerequisites: PENG 400 and 401; approval of graduate adviser.

613. Natural Gas Engineering. (3-0). Credit 3. Flow of natural gas in reservoirs and in well bores and gathering systems; deliverability testing; production forecasting and decline curves; flow measurement and compressor sizing. Prerequisites: PENG 323 and 324; approval of graduate adviser.

*616. Engineering Near-Critical Reservoirs. (3-0). Credit 3. Identification of reservoir fluid type; calculation of original gas in place, original oil in place, reserves and future performance of retrograde gas and volatile oil reservoirs. Prerequisites: PENG 323, 400, 401; approval of graduate adviser.

*617. Petroleum Reservoir Management. (3-0). Credit 3. The principles of reservoir management and application to specific reservoirs based on case studies presented in the petroleum literature. Prerequisite: Approval of graduate adviser.

*618. Modern Petroleum Production. (3-0). Credit 3. An advanced treatment of modern petroleum production engineering encompassing well deliverability from vertical, horizontal, and multilateral/multibranch wells; diagnosis of well performance includes elements of well testing and production logging; in this course the function of the production engineer is envisioned in the context of well design, stimulation, and artificial lift. Prerequisite: Approval of graduate adviser.

*620. Fluid Flow in Petroleum Reservoirs. (3-0). Credit 3. Analysis of fluid flow in bounded and unbounded reservoirs, wellbore storage, phase redistribution, finite and infinite conductivity fractures; dual-porosity systems. Prerequisites: PENG 323; approval of graduate adviser.

621. Petroleum Development Strategy. (2-3). Credit 3. Applications of the variables, models, and decision criteria used in modern petroleum development. The case approach will be used to study major projects such as offshore development and assisted recovery. Both commercial and student-prepared computer software will be used during the lab sessions to practice methods. Prerequisites: PENG 403; approval of graduate adviser.

*622. Exploration and Production Evaluation. (2-3). Credit 3. Selected topics in oil industry economic evaluation including offshore bidding, project ranking and selection, capital budgeting, long-term oil and gas field development projects, and incremental analysis for assisted recovery and acceleration. Prerequisites: PENG 403; approval of graduate adviser.

623. Waterflooding. (3-0). Credit 3. Design, surveillance, and project management of water floods in reservoirs. Prerequisites: PENG 323; approval of graduate adviser.

624. Rock Mechanic Aspects of Petroleum Reservoir Response. (3-0). Credit 3. Reservoir rocks and their physical behavior; porous media and fracture flow models; influence of rock deformability, stress, fluid pressure, and temperature. Prerequisites: PENG 604; approval of graduate adviser.

FIGURE 16-7 · *CONTINUED*

Proposal—6

625. Well Control. (3-0). Credit 3. Theory of pressure control in drilling operations and during well kicks; abnormal pressure detection and fracture gradient determination; casing setting depth selection and advanced casing design; theory supplemented on well control simulators. Prerequisites: PENG 411; approval of graduate adviser.

626. Offshore Drilling. (3-0). Credit 3. Offshore drilling from fixed and floating drilling structures; directional drilling including horizontal drilling; theory of deviation monitoring and control. Prerequisites: PENG 411; approval of graduate adviser.

628. Horizontal Drilling. (3-0). Credit 3. Changing a wellbore from vertical to horizontal; long- and short-radius horizontal wells; bottom hole assemblies for achieving and maintaining control of inclination and direction; drilling fluids; torque and drag calculations; transport of drilled solids. Prerequisites: PENG 411; approval of graduate adviser.

629. Advanced Hydraulic Fracturing. (3-0). Credit 3. Physical principles and engineering methods involved in hydraulic fracturing; an advanced treatise integrating the necessary fundamentals from elasticity theory, fracture mechanics, and fluid mechanics to understand designs and optimization and to evaluate hydraulic fracturing treatments including special topics such as high-permeability fracturing and deviated well fracturing. Prerequisite: Approval of graduate adviser.

*630. Geostatistics. (3-0). Credit 3. Introductory and advanced concepts in geostatistics for petroleum reservoir characterization by integrating static (cores/logs/seismic traces) and dynamic (flow/transport) data; variograms and spatial correlations; regionalized variables; intrinsic random functions; kriging/cokriging; conditional simulation; non-Gaussian approaches. Prerequisites: Introductory course in statistics or PENG 322; approval of graduate adviser.

*631. Petroleum Reservoir Description. (3-0). Credit 3. Engineering and geological evaluation techniques to define the extent and internal character of a petroleum reservoir; estimate depositional environment(s) during the formation of the sedimentary section and resulting effects on reservoir character. Prerequisites: PENG 324 and 620; approval of graduate adviser.

632. Physical and Engineering Properties of Rock. (3-3). Credits 4. Physical and engineering properties of rock and rock masses including strength, deformation, fluid flow, and thermal and electrical properties as a function of the subsurface temperature, in-situ stress, pore fluid pressure, and chemical environment; relationship of rock properties to logging, siting and design of wells, and structures in rock. Prerequisite: Approval of instructor or graduate adviser.

633. Data Integration for Petroleum Reservoirs. (3-0). Credit 3. Introduction and application of techniques that can be used to incorporate dynamic reservoir behavior into stochastic reservoir characterizations; dynamic data in the form of pressure transient tests, tracer tests, multiphase production histories or interpreted 4-D seismic information. Prerequisites: PENG 620 and STAT 601; approval of instructor or graduate adviser.

*648. Pressure Transient Testing. (3-0). Credit 3. Diffusivity equations and solutions for slightly compressible liquids; dimensionless variables; type curves; applications of solutions to buildup, drawdown, multirate, interference, pulse, and deliverability tests; extensions to multiphase flow; analysis of hydraulically fractured wells. Prerequisites: PENG 324 and 620; approval of graduate adviser.

FIGURE 16-7 · *CONTINUED*

Proposal—7

*661. Drilling Engineering. (3-0). Introduction to drilling systems: wellbore hydraulics; identification and solution of drilling problems; well cementing; drilling of directional and horizontal wells; wellbore surveying abnormal pore pressure; fracture gradients; well control; offshore drilling; and underbalanced drilling. Prerequisites: Approval of instructor of graduate adviser.

*662. Production Engineering (3-0). Development of fundamental skills for the design and evaluation of well completions; monitoring and management of the producing well; selection and design of artificial lift methods; and modeling and design of surface facilities. Prerequisites: Approval of instructor or graduate adviser.

*663. Formation Evaluation and the Analysis of Reservoir Performance. (3-0). Current methodologies used in geological description/analysis, formation evaluation (the analysis/interpretation of well log data), and the analysis of well performance data (the design/analysis/interpretation of well test and production data); specifically, the assessment of field performance data and the optimization of hydrocarbon recovery by analysis/interpretation/integration of geologic, well log, and well performance data. Prerequisites: Approval of instructor or graduate adviser.

*664. Petroleum Project Evaluation and Management. (3-0). Introduction to oil industry economics, including reserves estimation and classification, building and using reservoir models, developing and using reservoir management processes, managing new and mature fields, and investment ranking and selections. Prerequisites: Approval of instructor or graduate adviser.

*665. Petroleum Reservoir Description. (3-0). Reservoir description techniques using petrophysical and fluid properties; engineering methods to determine fluids in place, identify production-drive mechanisms, and forecast reservoir performance; implementation of pressure-maintenance schemes and secondary recovery. Prerequisites: Approval of instructor or graduate adviser.

681. Seminar. Credit 1 each semester. Study and presentation of papers on recent developments in petroleum technology. Prerequisite: Approval of graduate adviser.

685. Directed Studies. Credit 1 to 12 each semester. Offered to enable students to undertake and complete limited investigations not within their thesis research and not covered in established curricula. Prerequisites: Graduate classification; approval of instructor or graduate adviser.

689. Special Topics in . . . Credit 1 to 4. Special topics in an identified area of petroleum engineering. May be repeated for credit. Prerequisite: Approval of instructor or graduate adviser.

691. Research. Credit 1 or more each semester. Advanced work on some special problem within field of petroleum engineering. Thesis course. Prerequisite: Approval of committee or graduate adviser.

692. Professional Study. Credit 1 to 12. Approved professional study or project. May be taken more than once but not to exceed 6 hours of credit toward a degree. Prerequisite: Approval of graduate adviser.

FIGURE 16-7 · *CONTINUED*

☑ Planning and Revision Checklist

You will find the Planning and Revision Checklists that appear in Chapter 2, Composing, and Chapter 8, Designing and Formatting Documents, valuable in planning and revising any presentation of technical information. The following questions, which specifically apply to proposals and progress reports, summarize the key points in this chapter.

Proposals

Planning

■ Have you studied the RFP carefully?

■ Have you made a list of all requirements given in the RFP?

■ Who are your readers? Do they have technical competence in the field of the proposal? Is it a mixed audience, some technically educated, some not? To whom could the proposal be distributed?

■ What problem is the proposed work designed to remedy? What is the immediate background and history of the problem? Why does the problem need to be solved?

■ What is your proposed solution to the problem? What benefits will come from the solution? Is the solution feasible (both practical and applicable)?

■ How will you carry out the work proposed? Scope? Methods to be used? Task breakdown? Time and work schedule?

■ Do you want to make statements concerning the likelihood of success or failure and the products of the project? Who else has tried this solution? What was their success?

■ What facilities and equipment will you need to carry out the project?

■ Who will do the work? What are their qualifications? Can you obtain references?

■ How much do you expect the work to cost? Consider such things as materials, labor, test equipment, travel, administrative expenses, and fees. Who will pay for what? What will be the likely return on the investment?

■ Will the proposal need an appendix? Consider including biographical sketches, descriptions of earlier projects, and employment practices.

■ Will the proposal be better presented in a report format or in a letter or memo format?

■ Do you have a student report to propose? Consider including the following in your proposal:

    ■ Subject, purpose, and scope of report

    ■ Task and time breakdown

    ■ Resources available

    ■ Your qualifications for doing the report

Revision

- Does your proposal have a well-planned design and layout? Does its appearance suggest the high quality of the work you propose to do? Does your proposal both promote and inform?
- Does the project summary succinctly state the objectives and plan of the proposed work? Does it show how the proposed work is relevant to the readers' interest?
- Does the introduction make the subject and the purpose of the work clear? Does it briefly point out the so-whats of the proposed work?
- Have you defined the problem thoroughly?
- Is your solution well described? Have you made its benefits and feasibility clear?
- Will the target audience be able to follow your plan of work easily? Have you protected yourself by making clear what you will do and what you will not do? Have you been careful not to promise more than you can deliver?
- Have you carefully considered all the facilities and equipment you will need?
- Have you presented the qualifications of project personnel in a favorable but honest way? Have you asked permission from everyone you plan to use as a reference?
- Is your budget realistic? Will it be easy for the readers to follow and understand?
- Do all the items in the appendix lend credibility to the proposal?
- Have you included a few sentences that urge the readers to accept the proposal?
- Have you satisfied the needs of your readers? Do they have all the information they need to make a decision?

Progress Reports

Planning

- Is a clear description of your project available, perhaps in your proposal?
- Are all the project tasks clearly defined? Do all the tasks run in sequence, or do some run concurrently? In general, are the tasks going well or badly?
- What items need to be highlighted in your summary and appraisal?
- Are there any problems to be discussed?
- Can you suggest solutions for the problems?
- Is your work ahead of schedule, right on schedule, or behind schedule?
- Are costs running as expected?
- Do you have any unexpected good news to report?

Revision

■ Does your report have an attractive appearance?

■ Does the plan you have chosen show off your progress to its best advantage?

■ Is your tone authoritative, with an accent on the positive?

■ Have you supported your generalizations with facts?

■ Does your approach seem fresh?

■ Do you have a good balance between work accomplished and work to be done?

■ Can your summary and appraisal stand alone? Would they satisfy an executive reader?

■ ■ ■ ■ ■ ■ ■ ■ Exercises ■ ■ ■ ■ ■ ■ ■ ■

Further exercises related to the content of this chapter and all documents employed within the following exercises can be found on the book's companion Web site.

1. Examine the RFP from the Department of Health and Human Services. See the book's companion Web site. Answer the following questions:

· According to the RFP, what work does HHS want done, or what product does it want? What problem does the RFP present to be solved?
· Does the RFP specify a length for the proposal?
· Does the RFP make clear the information the proposal must contain?
· Does the RFP furnish an outline to follow? If so, what does the outline require?
· Does the RFP require a specific format for the proposal? What is it?
· Does the RFP make clear the criteria by which submitted proposals will be evaluated and who will do the evaluation?

2. Write an information report to your instructor based on this RFP. Summarize what content items the proposal should emphasize and what criteria will be used to evaluate the proposals.

3. You will probably be instructed to write a complete technical report as part of the requirements for your course in technical writing.

· Choose two or three topics you would consider to be suitable for a one-semester project.
· Write a feasibility report to your instructor examining each topic in terms of availability of information, suitability of the topic for the amount of time available during the semester, and the significance of each topic to your discipline or to your career goals. Decide which topic seems most feasible.

Forestry Research Associates
222 University Avenue
Madison, Wisconsin 53707
June 29, 2001
Mr. Lawrence Campbell, Director
Council for Peatlands Development
420 Duluth Street
Grand Forks, ND 58201
Dear Mr. Campbell:

Well, we have our Peatland Water-Table Depth Research Project under way. This is our first progress report. As you know, by ditching peatlands, foresters can control water-table depths for optimum growth of trees on those peatlands. Foresters, however, don't have good data on which water-table depths will encourage optimum tree growth. This study is an attempt to find out what those depths might be. We have to do several things to obtain the needed information. First, we have to measure tree growth on plots at varying distances from existing ditches on peatland. Then we have to establish what the average water-table depth is on the plots during the growing season of June, July, August, and September. To get meaningful growth and water-table depth figures, we have to gather these data for three growing seasons. Finally, we have to correlate average water-table depth with tree growth. Knowing that, foresters can recommend appropriate average water-table depths.

When the snow and ice went out in May, we were able to establish 14 tree plots on northern Minnesota peatlands. Each plot is one-fortieth of a hectare. The distances of the plots from a drainage ditch vary from 1 to 100 meters. The plots have mixed stands of black spruce and tamarack. During June, we measured height and diameter at breast height (DBH) of a random selection of trees on each plot. We marked the measured trees so that we can return to them for future measurements. We will measure them again in September of this year and in June and September of the next two years.

While we were measuring the trees, we began placing two wells on each plot. The wells consist of perforated plastic pipes driven eight feet into the mineral soil that underlies the plot. We should have all the wells in by the end of next week. We'll measure water-table depths once a month in July, August, and September. We will also measure water-table depths any time there is a rainfall of one inch or more on the plots.

We have our research well under way, and we're right on schedule. We have made all our initial tree measurements and will soon obtain our first water-table depth readings.

By the way, the entire test area seems to be composed of raw peat to a depth of about 20 cm with a layer of well-decomposed peat about a meter thick beneath that. However, to be sure there are no soil differences that would introduce an unaccounted-for variable into our calculations, we'll do a soil analysis on each of the 14 plots next summer. We will do this additional work at no extra cost to you. During the cold-weather months, between growing seasons, we'll prepare water-table profiles that will cover each plot for each month of measurement. At the completion of our measurements in the third year, we'll correlate these profiles with the growth measurements on the plots. This correlation should enable us to recommend a water-table depth for optimum growth of black spruce and tamarack. We have promised two progress reports per growing season and one each December. Therefore, we will submit our next report on September 28.

Sincerely,
Robert Weaver
Principal Investigator

FIGURE 16-8 · INEFFECTIVE PROGRESS REPORT FOR REVISION

- Once your instructor has approved your choice of topic, write a proposal to your instructor, using memo format. In your proposal, include all elements commonly found in proposals.
- Write a progress report to describe the status of your semester report project. Design the progress report to reflect the tasks or project goals you used in developing your project proposal.

4. The progress report in Figure 16-8 is poorly organized and formatted. Reorganize it and rewrite it. Use a letter format, but furnish a subject line and headings to help readers find their way through the report.

Oral Exercises

Before doing these exercises, see Chapter 19 on oral reports.

5. Prepare an oral version of the proposal you wrote for Exercise 3, and plan on delivering it to your class. You will be allowed eight minutes maximum. Enhance your presentation with computer graphics or overhead transparencies to show anticipated costs, your project schedule, and any visuals that will help explain the significance of your project or the methods you propose to use.

6. Prepare an oral progress report to deliver to your class. You will be allowed five minutes maximum. Enhance your presentation with graphics to show work completed, work remaining, project costs to date, and status of your project.

Video examples of oral reports can be found on the book's companion Web site (select Chapter 19).

Scenario

Your organization has recently purchased a new financial information system (FIS). As assistant director of finance, you know that the new FIS has excellent security and manageability for users at all levels. For example, financial records personnel in each office will be able to access parts of the system, while data entry personnel in the main fiscal office will enter information daily. The personnel office can access only specific payroll information. Financial officers will have access to the entire system, along with the audit group. The system has full documentation—both in a large manual and on several CD-ROMs. The problem: too much information, not organized in a way that people can easily find what they need. You decide to write specific instructions on how to use the new FIS for each group of employees who will need to access the system.

CHAPTER SEVENTEEN

Formulating Instructions, Procedures, and Policies

- **Planning Instructions and Procedures**

- **Structure and Organization**

- **Theory Governing the Procedure or Instruction**

- **Conditions under Which the Task Is to Be Performed**

- **Steps in Performing the Task**

- **Procedures**
  Format Considerations for Instructions and Procedures

- **Policies**

- *Exercises*

In addition to memoranda, letters, routine reports, and e-mail, documents that explain how to perform operations will likely be a major kind of job-related writing. Procedures provide general guidelines for performing a task, while instructions provide specific, detailed steps. Procedures—or general guidelines—will be appropriate for some situations, while instructions may be necessary for readers who need detailed directions to perform a job task.

Instructions and procedures can appear:

- In letters and memoranda
- In reports
- In technical papers or notes
- In stand-alone documents to accompany mechanisms
- In complete manuals

Complex procedures and instructions usually appear as manuals, which use many of the report elements described in Chapter 10. They will have a title page, a table of contents, perhaps a glossary, and a list of illustrations used through-out the procedures. E-mail is not a suitable method of sending such materials, unless you send them as an attachments (in a standard rtf, Word, or pdf file) that can be printed or read on the receiver's computer screen. Differences in e-mail platforms can produce text that is awkwardly formatted and difficult to read.

In this chapter, we will use examples of instructions, procedures, and policies used by real organizations. We will first illustrate instructions from excerpts taken from Cisco Corporation's online instruction site to show you how the guidelines discussed may look in practice.

PLANNING INSTRUCTIONS AND PROCEDURES

Instructions and procedures should enable those who need them to perform the tasks covered. They are successful if the intended users can *easily* read, *easily* understand, and *correctly* follow them. Understanding your target readers and, in some cases, the context in which the documents will be used, is paramount. If readers are confused or misled by statements, they may sue the company for injury or financial damage incurred, citing poorly written instructions or procedures.

Who are the readers? What do your readers know about the subject? These questions will determine how much information you provide and the type of language you use. The audience for instructions should be specified if necessary. Remember that many people tend to avoid reading general instructions and routinely attempt to perform a task or use new equipment without them. When people do read the instructions, they want to understand them immediately without having to reread them. Design features (layout), format, choice of language, and use of visuals can be critical to encouraging readers to focus on the content of instructions or procedures.

What is your purpose? What should your readers know and be able to do after they have read these instructions/procedures? Any information that isn't applicable should be omitted.

What is the context in which the instructions/procedures will be used? Knowing how documents will be read can help you design effective documents. Some people will be focusing on the documents while at their desks, away from the task at hand. Others will read the entire document before trying to perform the task; still others will perform the task as they read. Do your readers need to read the instructions/procedures before they begin? You may want to draft detailed instructions for first-time readers, as well as a summary of steps for readers already familiar with the longer instruction set. Context-sensitive instructions can be found in numerous settings:

- Some manufacturers place instructions for how to right a sailboard that has turned over in the water on the bottom of the boat, where they will be most useful.
- "Quick start" instructions for many handheld devices can be found inside the top or lid of the equipment.
- Instructions used in machine shops on manufacturing floors are often printed in large type (14–16 pt) on eye-saver yellow or green paper and laminated to resist grease. These instructions may be kept in metal loose-leaf notebooks from which pages cannot be torn out.
- Basic operation instructions and troubleshooting information for large clocks, found inside the clock mechanism access compartment, are designed to accompany the booklet that contains detailed instructions.
- Some instructions are placed on large signs at construction sites (see Figure 4-3).

You will be interested in seeing more examples of instructions, located on the book's companion Web site.

Will the instructions be available online or on paper? If instructions will be available online, you will want to focus on clarity and readability of the material. You will want to avoid instructions that are "text heavy" and lengthy. You will want to partition your instructions into distinct parts. You may want to tell readers immediately that they should print a copy of instructions that are too long and complex for online reading.

Before you begin your instruction or procedures, complete the document analysis presented in the Planning and Revision Checklist at the end of this chapter. Note that how you answer each question will help you determine:

- Whether you need instructions or procedures
- The delivery medium (online or paper)
- The content you will include
- Any design strategies that could help you produce an effective document

Effective instructions require sensitive attention to readers' information needs, their knowledge of the topic, and what you want them to know as a result of reading the instructions.

STRUCTURE AND ORGANIZATION

After you have analyzed the audience, purpose, and context, you can begin to structure your instructional document. The following plan provides a general guide for what elements to include; but these will depend, in each case, on your topic, your audience, your purpose, and the reader's context. Manuals of instructions and procedures usually contain most of these elements, while shorter instructions or procedures may have fewer. To illustrate instructions, we have selected a series of examples from Cisco Systems, whose products and support services are designed to improve the overall efficiency of network operations and network performance. Excerpts from the vendor's instructional material are preceded by our comments.

Introduction

The introduction to an instruction set should familiarize readers with the task to be performed: what the instructions will allow readers to do; what skill level the user should have to perform the task successfully. This section of the vendor's instructions for installing the Cisco 800 Router prepares users for what follows and also describes the audience for which the instructions are designed.

Audience

Source: http://www.cisco.com/en/US/products/hw/routers/ps380/products_installation_guide_chapter09186a008007e675.html#xtocid1

This guide is intended for service technicians with all levels of experience in installing routers. The goal of all technicians is to connect the router to the network as quickly as possible. Where relevant, this guide explains how the router is implemented and why. Conceptual information is usually in a separate section or appendix so that technicians who are not interested can skip this information.

Organization

This guide contains the following information:

- About This Guide—Describes audience, organization, conventions used in this guide, and how to access related documentation.
- Overview—Contains router features and a description of router LEDs, ports, and other components.

- Installation—Provides information on safety, preventing damage, unpacking, and preparing for installation as well as installing, mounting, and verifying the connections to your router.
- Troubleshooting—Describes how to identify and solve problems with your router.
- ISDN and IDSL Concepts—Describes how ISDN is implemented on the router.
- Specifications and Cables—Provides router, port, and cable specifications.
- Glossary—Defines technical terms frequently used in this guide.

Conventions

This section describes the conventions used in this guide.

 Note Means *reader take note*. Notes contain helpful suggestions or references to additional information and material.

nice to know

 Caution This symbol means *reader be careful*. In this situation, you might do something that could result in equipment damage or loss of data.

Warning This warning symbol means danger. You are in a situation that could cause bodily injury. Before you work on any equipment, be aware of the hazards involved with electrical circuitry and be familiar with the standard practices for preventing accidents.

Table of Contents

Overview
Feature Summary
Router Ports Summary
Front Panels
Back Panels
LEDs

Overview

The Cisco 800 series routers connect small professional offices or telecommuters over Integrated Services Digital Network (ISDN) Basic Rate Interface (BRI) lines to the corporate LANs and the Internet. The routers offer bridging and multiprotocol routing capability between LAN and WAN ports.

This chapter contains the following topics:

- Feature Summary
- Router Ports Summary
- Front Panels
- Back Panels
- LEDs

THEORY GOVERNING THE PROCEDURE OR INSTRUCTION

Readers will perform better at tasks associated with some procedures if they have a general overview of the process. Such knowledge may help them to avoid errors by enhancing their ability to recognize their own mistakes at the earliest possible moment. If you are going to give instructions for operating a mechanism, a brief explanation of the value and purpose of the mechanism can prepare readers for the process. The preceding excerpt from the Cisco manual, and the following one, which gives the text accessible from the ISDN and IDSL Concepts link, provide background information for readers. The use of links in online instructions enables writers to separate units of information: readers can select what they need and want to read. Note that the Cisco instructions are "layered" to allow readers to move from general to detailed information, depending on their knowledge needs.

ISDN and IDSL Concepts

Source: http://www.cisco.com/en/US/products/hw/routers/ps380/products_installation_guide_chapter09186a008007e698.html

This appendix provides further explanation of ISDN and IDSL concepts.

The Cisco 800 series routers provide one basic rate interface (BRI). The ISDN BRI service provided by your telephone service provider offers two bearer channels (B channels) and one data channel (D channel). The B channel operates at 64 kbps and carries user data. The D channel operates at 16 kbps and carries control and signaling information although it can support user data transmission under certain circumstances.

Cisco 801 and Cisco 803 routers have an ISDN S/T port. Cisco 802 and Cisco 804 routers have an ISDN U port, and Cisco 802 IDSL and Cisco 804 IDSL routers have an IDSL port.

Outside North America, telephone service providers typically provide an S/T interface. The S/T interfaces are four-wire (two pairs of wires) interfaces from the phone switch that supports full-duplex data transfer over two pairs of wires.

In North America, telephone service providers typically provide a U interface. The U interface is a two-wire (single pair) interface from the phone switch that supports full-duplex data transfer over a single pair of wires.

Cisco 803 and Cisco 804 routers support data and voice applications. The data applications on these routers are implemented through the ISDN port on these routers. The voice applications on these routers are implemented with ISDN BRI and through the telephone ports.

Warnings, Cautions, Hazards, and Notes Regarding Safety or Quality

In today's litigious workplace environment, you cannot "overwarn" readers. Give the warning, followed by the reason for it or the result of ignoring it, and be sure that all warnings, cautions, and notes are clearly visible. Warnings may also be repeated if they are associated with a particular step or direction. Warnings given at the beginning alert readers to possible problems they may encounter. Be sure that warnings stand out so that readers' eyes are drawn to them. The American National Standards Institute articulates standards for hazard statements:

http://www.safetylabel.com/safetylabelstandards/iso-ansi-faq.php

A good literature review of hazard message can be found at

http://www.osha.gov/SLTC/hazardcommunications/hc2inf2.html

Cisco's instructions, shown next, contain ample warnings with explanations of the warning icons.

Safety

Source: http://www.cisco.com/en/US/products/hw/routers/ps380/products_installation_guide_
chapter09186a008007e69e.html#xtocid1

Before installing the router, read the following warnings:

 Warning Only trained and qualified personnel should be allowed to install or replace this equipment.

 Warning Read the installation instructions before you connect the system to its power source.

 Warning Before working on a system that has a standby/off switch, turn the power to standby and unplug the power cord.

 Warning Before working on equipment that is connected to power lines, remove jewelry (including rings, necklaces, and watches). Metal objects will heat up when connected to power and ground and can cause serious burns or weld the metal object to the terminals.

 Warning The ISDN connection is regarded as a source of voltage that should be inaccessible to user contact. Do not attempt to tamper with or open any public telephone operator (PTO)-provided equipment or connection hardware. Any hardwired connection (other than by a nonremovable, connect-one-time-only plug) must be made only by PTO staff or suitably trained engineers.

 Warning To avoid electric shock, do not connect safety extra-low voltage (SELV) circuits to telephone-network voltage (TNV) circuits. LAN ports contain SELV circuits, and WAN ports contain TNV circuits. Some LAN and WAN ports both use RJ-45 connectors. Use caution when connecting cables.

Warning Ultimate disposal of this product should be handled according to all national laws and regulations.

 Warning If the symbol of suitability with an overlaid cross (⊗) appears above a port, you must not connect the port to a public network that follows the European Union standards. Connecting the port to this type of public network can cause severe injury or damage your router.

CONDITIONS UNDER WHICH THE TASK IS TO BE PERFORMED

Some instructions, particularly laboratory instructions, may require you to describe the physical conditions required to perform a task:

■ Room size
■ Temperature
■ Time required for the entire process and for individual steps. If time is a constraint, readers need to know how much time will be needed before they begin the task.
■ Specific safety processes. As for safety, the Cisco instructions, for example, explain how to avoid electrostatic discharge. They also provide drawings and names of items required to install the router—items that should be included in the packaging. As the following excerpt illustrates, warnings are inserted at the point the reader needs to be alerted.

Preventing Electrostatic Discharge Damage

Source: http://www.cisco.com/en/US/products/hw/routers/ps380/products_installation_guide_chapter09186a008007e69e.html#xtocid1

Electrostatic discharge (ESD) is a transfer of electrostatic charge between bodies of different electrostatic potentials, such as a person and a piece of electrical equipment. It occurs when electronic components are improperly handled, and it can damage equipment and impair electrical circuitry. Electrostatic discharge is more likely to occur with the combination of synthetic fibers and dry atmosphere.

Always use the following ESD-prevention procedures when removing and replacing components:

1. Connect the chassis to earth ground with a wire that you provide.
2. Wear an ESD-preventive wrist strap that you provide, ensuring that it makes good skin contact.

> Connect the clip to an unpainted surface of the chassis frame to safely channel unwanted ESD voltages to ground. To properly guard against ESD damage and shocks, the wrist strap and cord must operate effectively. If no wrist strap is available, ground yourself by

touching the metal part of the chassis. Always follow the guidelines in the preceding section, "Safety."

3. Do not touch any exposed contact pins or connector shells of interface ports that do not have a cable attached.

If cables are connected at one end only, do not touch the exposed pins at the unconnected end of the cable.

 Note This device is intended for use in residential and commercial environments only.

 Caution Periodically check the resistance value of the antistatic strap, which should be between 1 and 10 megohms (Mohms).

Preventing Router Damage

Use the following guidelines when connecting devices to your router:

- Connect the color-coded cables supplied by Cisco Systems to the color-coded ports on the back panel.
- If you must supply your own cable, see the "Cabling Specifications" section in Appendix B, "Specifications and Cables." If this appendix does not provide specifications for a particular cable, we strongly recommend ordering the cable from Cisco Systems.
- If the symbol of suitability (⊗) appears above a port, you can connect the port directly to a public network that follows the European Union standards.

 Warning If the symbol of suitability with an overlaid cross (⊗) appears above a port, you must not connect the port to a public network that follows the European Union standards. Connecting the port to this type of public network can cause severe injury or damage your router.

Unpacking Your Router

Figure 2-1 shows the items included with your router. If any of the items is missing or damaged, contact your customer service representative.

Preinstallation Activities

Before you begin installing your Cisco 800 series router, perform the following steps:

Step 1 Order an ISDN BRI line from your telephone service provider. For more information, refer to the *Cisco 800 Series Routers Software Configuration Guide*.

Step 2 If you have a Cisco 801 or Cisco 803 router, do the following:

- If you are outside of North America, ask your telephone service provider if you must provide an external Network Termination 1 (NT1) and the ISDN U cable that connects the NT1 to the ISDN wall jack. Ask for NT1 vendors if necessary.

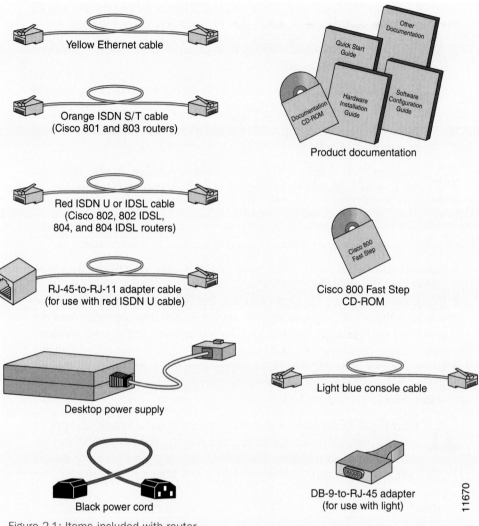

Figure 2-1: Items included with router

- If you are in North America, ask your telephone service provider for external NT1 vendors. Provide the ISDN U cable that connects the NT1 to the ISDN wall jack.

Step 3 Be aware of Ethernet, ISDN, and IDSL cable distance limitations. For more information, see the "Maximum Cable Distances" section in Appendix B, "Specifications and Cables."

Step 4 Gather the Ethernet devices to be connected to the router: hub, server, workstation, or PC with 10- or 10/100-Mbps network interface card (NIC).

Step 5 If you have a Cisco 801 or Cisco 803 router and plan to connect a digital telephone, you must provide an NT1 with two S/T interfaces and one U interface, a telephone cable to connect the telephone (usually this cable is provided with the device), and an ISDN U cable that connects the NT1 to the ISDN wall jack.

Step 6 If you have a Cisco 803 or Cisco 804 router, gather the devices (such as an analog telephone, fax machine, or modem) that you plan on connecting to the router. You must also provide the telephone cable to connect each device (usually this cable is provided with the device).

Step 7 If you plan to configure the software using a terminal or PC connected to the router, provide the terminal or PC.

Step 8 If you plan to mount your router on a wall or vertical surface, you need to provide two number-six, 3/4-in. (M3.5 $\times$ 20 mm) screws. If the wall on which you mount your router is drywall, you instead need to provide two hollow wall-anchors (1/8-in. with 5/16-in. drill bit or M3 with 8-mm drill bit) to secure the screws.

Step 9 If you plan to use the cable lock feature, you need to provide a Kensington or equivalent locking cable.

STEPS IN PERFORMING THE TASK

As exemplified in the preceding Cisco excerpt, you should name and number each step in the procedure. Or, in online instructions, provide an outline of the process with a link to each. Then, readers may proceed to the appropriate link or continue to scroll through the instructions, from beginning to end. Readers need to grasp the complexity of the process, particularly if there are a number of steps. For example, in the troubleshooting section of the Cisco 800 Series Instructions, readers learn that problems will occur at one of three stages of the installation. A table gives the problem, cause, and solution. Each section follows the same plan, as illustrated for a portion of the troubleshooting section of the manual.

Table of Contents

Troubleshooting
    Problems During First Startup
    Problems After First Startup
    Problems After Router Is Running
When Contacting Your Cisco Reseller

Troubleshooting

This chapter describes problems that could occur with the Cisco 800 series router hardware, reasons for the problems, and steps to solve the problems. The problems are grouped as follows:

- Problems during first startup
- Problems after first startup
- Problems after router is running

For information on problems that could occur with the software, refer to the *Cisco 800 Series Routers Software Configuration Guide.*

Problems During First Startup

Table 3-1 lists problems that could occur after you turn on the power switch for the first time.

Table 3-1 ■ **Problems During First Startup**

Symptom	Problem	Solutions
All LEDs, including OK LED, are off.	No power to router.	Perform the following steps in the following order: • Make sure that the power switch is ON. • Make sure that all connections to and from the power supply are securely connected. • Make sure that the power outlet has power. • If the problem continues, the router might have a faulty power supply. Contact your Cisco reseller.

Name of Each Step

For each step, you may wish to give the following: purpose of the step; warnings, cautions, and notes; any conditions necessary for performing the step; time required to perform the step; and materials needed for the step.

- ▨ Place instructions in chronological order.
- ▨ Number each step.
- ▨ Limit each instruction to one action.
- ▨ If you believe that readers might not follow an instruction as written without an explanation, state, in a separate sentence, the reason for the instruction. Include warnings whenever necessary.
- ▨ Be sure to explain all warnings if you believe that people will not understand why caution is necessary. For an example, we present a final excerpt from the Cisco instructions.

Installing Your Router

To install the Cisco 800 series routers, you need to perform the following tasks in the following order:

1. Connect the Ethernet devices to the router.
2. Connect the ISDN or IDSL line to the router.
3. If you have a Cisco 801 or Cisco 803 router, connect an optional digital telephone.
4. If you have a Cisco 803 or Cisco 804 router, connect an optional analog telephone, fax, or modem.
5. Connect a terminal or PC to the router (for software configuration using the command-line interface [CLI] or for troubleshooting).

6. Connect the router to the power source.

7. Mount your router.

8. Verify the router installation.

Connecting Ethernet Devices

Table 2-1 lists the Ethernet devices you can connect to the router, connections for each device, and the settings of the router HUB/NO HUB or TO HUB/TO PC button (the default setting is IN).

......

To connect an ISDN line to Cisco 802 and Cisco 804 routers, follow the steps in Figure 2-6.

 Warning Network hazardous voltages are present in the ISDN cable. If you detach the ISDN cable, detach the end away from the router first to avoid possible electric shock. Network hazardous voltages also are present on the system card in the area of the ISDN port (RJ-45 connector), regardless of when power is turned to standby.

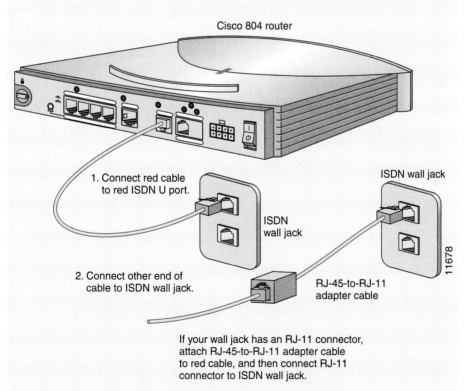

Cisco 804 router

1. Connect red cable to red ISDN U port.

ISDN wall jack

ISDN wall jack

2. Connect other end of cable to ISDN wall jack.

RJ-45-to-RJ-11 adapter cable

11678

If your wall jack has an RJ-11 connector, attach RJ-45-to-RJ-11 adapter cable to red cable, and then connect RJ-11 connector to ISDN wall jack.

Figure 2-6: Connecting ISDN to Cisco 802 or Cisco 804 Routers

Warning Do not work on the system or connect or disconnect cables during periods of lightning activity.

 Warning To reduce the risk of fire, use only No. 26 AWG or larger telecommunications line cord.

Caution Always connect the red cable to the red ISDN U port on the router. Do not connect the cable to a yellow Ethernet port. This will damage your router.

Caution The Cisco 800 series routers do not support the Australian IUT requirement, which specifies that the routers must communicate for 1/2 hour after a power failure. If a power failure occurs, a Cisco 800 series router stops communicating with other devices.

PROCEDURES

Procedures and instructions differ in specificity. Guidelines usually require less detail, but their development parallels development of instructions in the general content required.

Introduction

Introduction: purpose of the procedure—why it is important, what it should achieve
Scope: who/what activities are involved; who/what is affected
Definitions: specialized vocabulary readers need to know to follow the procedure

Description of the operation to be performed: steps in the process

References: other documents readers will need or wish to consult before beginning the procedure

Attachments: sample forms and explanatory material alluded to in the procedures

Figure 17-1 illustrates procedures for applying pesticides to orchards. Note that the discussion is much less detailed than the instructions for the Cisco router installation. The procedures explain calibration, maintenance, and safety as these apply to the process of spraying *in general*. Users can apply these guidelines to their particular situation and equipment.

Format Considerations for Instructions and Procedures

Figure 17-2, illustrates a technical note, really a set of instructions. These instructions, available on the Web, are provided by the company that makes the chemicals needed by criminalists involved in crime scene investigations (http://www.redwop.com/technotes.asp). An internationally recognized

Orchard Spraying
Thomas D. Valco and Henry O'Neal
Extension agricultural engineers
The Texas A&M University System

Air carrier sprayers are used in citrus orchards to discharge a high velocity of air for propelling and distributing pesticides throughout the trees. These sprayers, sometimes called air-blast sprayers or mist blowers, produce air speeds ranging from 80 to 150 miles per hour (mph) and air volumes of 5,000 to 60,000 cubic feet per minute (cfm). The air stream breaks up liquid into droplets by airshear or centrifugal energy and distributes droplets to all leaf, bark and fruit surfaces for effective pest control. Good pesticide distribution requires low ground speeds with small droplets or large volumes of pesticide.

The required volume of spray material per acre will vary with the sprayer type used and the target pest. Effective pest control has been obtained with from 5 gallons per acre to 500 gallons per acre. A low-volume application in small droplet size readily adheres to tree parts with virtually no drip or runoff. This generally occurs with high volume sprays. By decreasing application rate, application costs may be lowered through improved sprayer field capacity, achieved by reducing filling and mixing time. **Pesticide label rates should still be used with low-volume application rates.**

Selecting a Sprayer

Different air carrier sprayers operate on different principles to form spray droplets and propel them into the trees. The volume of solution applied per acre dictates the type of sprayer needed. Spray coverage is a result of the air horsepower, a relationship between air velocity, capacity, outlet area and fan pressure—not just air velocity or capacity alone.

A sprayer is a combination of many design features, and it cannot be stated that one machine is better than another because one feature is different. The proof of the sprayer's overall worth comes from how well it controls pests in a crop and not from a higher outlet velocity, a bigger capacity, a mechanical agitator or some other single factor. The most expensive sprayer a grower can buy is one that is not designed for the job.

Some of the low-volume sprayers on the market are much less expensive than most medium- to high-volume machines. Some growers who cannot justify the expenditure necessary for a larger machine may find low-volume sprayers to be the most satisfactory machine.

Nozzles and Nozzle Arrangement

The orifice size of the nozzle and nozzle operating pressure determine the rate of discharge for any given nozzle type. To determine operating pressure at the nozzle, place a pressure gage in the discharge manifold or

FIGURE 17-1 · PROCEDURES FOR APPLYING PESTICIDES TO ORCHARDS

replace a nozzle with a pressure gage. Check the pressure on both sides of the sprayer and set it to conform with the manufacturer's specifications. As the nozzle orifice wears, the spray rate increases. This causes most growers to apply more spray than intended. Once a nozzle output exceeds 10 percent of its designed rate, it should be replaced.

To determine nozzle arrangement on the sprayer, determine the gallons-per-minute discharge for each nozzle. This rate may be found in the nozzle manufacturer's catalog. Arrange the nozzles to apply 70 percent of the spray to the top half of the tree and 30 percent to the bottom half (Figure 1). Use larger nozzle sizes rather than more nozzles to obtain the additional output in the upper zone. Larger nozzles produce larger droplets, which are desirable for tree tops.

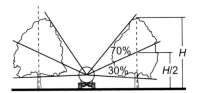

Figure 1. Recommended proportioning of spray material.

Spray distribution in the orchard should then be checked with water and dye. The addition of hydrated lime to the spray water leaves a traceable residue to help determine coverage. (Mix approximately 10 pounds of hydrated lime per 100 gallons of water. Do not allow material to settle in tank; maintain constant agitation.) Check to see that spray material reaches the tops of the trees but is not excessive. Then examine the lower tree skirts and inside the tree canopy for coverage. If coverage is not as desired, rearrange the nozzles and nozzle orifice sizes to get the desired coverage.

Sprayer Calibration

Proper maintenance and calibration are critical for achieving good pest control. Air carrier sprayers are equipped with centrifugal pumps, piston pumps or roller pumps. Occasionally check engine speed and pump drive belts for slippage because these two factors affect pump operating pressure. Check pump pressure at the discharge manifold rather than at the pump. Clogged strainers, air leaks into pump suction line, worn pump impellers and faulty pressure regulators also affect pump operating pressure.

Other factors which can cause variation in the rate of application are: (1) number of nozzles on the manifold, (2) ground speed and (3) nozzle orifice size. All of these factors must be considered to calibrate a sprayer. Sprayers should be calibrated prior to initial use and periodically during the season.

Step 1: Determine sprayer ground speed.

Load the sprayer tank with water and make a test run to find a ground speed at which desired coverage is obtained and record the gear and throttle setting.

Then make a test run at the desired ground speed and count the number of trees passed in 1 minute. Determine the ground speed in miles per hour (mph) with this formula:

$$\text{mph} = \frac{\text{tree spacing (ft)} \times \text{trees passed/minute}}{88}$$

FIGURE 17-1 · *CONTINUED*

Example: If trees are on 15-foot spacings and six trees are passed in 1 minute,

$$mph = \frac{15 \times 6}{88}$$

Table 1 may be used to determine sprayer speed based on tree spacing. Select the column which represents the tree spacing in your orchard and the speed you desire to travel. From that you can get the number of trees to pass in 1 minute.

Table 1 ■ Tractor Speed Based on Tree Spacing

	Tree Spacing (ft)					
	10	12.5	15	18	20	22
Speed of Tractor (mph):	(Number of trees passed per minute)					
1.0	9.0	7.0	6.0	5.0	4.5	4.0
1.5	13.0	10.5	9.0	7.0	6.5	6.0
2.0	18.0	14.0	12.0	10.0	9.0	8.0
3.0	26.5	21.0	17.0	15.0	13.0	12.0
5.0	44.0	35.0	29.5	24.5	22.0	20.0

Step 2: Determine the desired sprayer discharge rate with these formulas:

One-sided delivery

$$gpa = \frac{gpa \times mph \times d}{990}$$

Two-sided delivery

$$gpa = \frac{gpa \times mph \times d}{495}$$

gpm = sprayer discharge rate in gallons per minute
gpa = application rate in gallons per acre
mph = sprayer speed in miles per hour
 d = tree spacing between rows in feet

Example: Orchard requires 100-gallons-per-acre application rate with tree rows spaced 25 feet apart. Ground speed is 1 mph for a two-sided delivery sprayer.

$$gpm = \frac{gpa \times mph \times d}{495}$$

$$gpm = \frac{100 \times 1 \times 25}{495}$$

$$gpm = 5.1$$

FIGURE 17-1 · *CONTINUED*

Step 3: Determine the actual sprayer discharge rate.

The discharge rate in gallons per minute for a specific nozzle and pressure may be obtained from the nozzle manufacturer's catalog. To verify sprayer output, make a trial run with the sprayer in operation.

a. Fill the tank with water to a known level.
b. Set gear and throttle setting for desired ground speed.
c. Select the pressure setting recommended by sprayer manufacturer.
d. Spray for a measured interval of time.
e. Refill the tank to the known level, measuring the number of gallons of water required.
f. Determine the actual discharge rate with this formula:

$$\text{gpm} = \frac{\text{spray discharged in gallons}}{\text{time interval in minutes}}$$

Example: 15 gallons sprayed in 3 minutes.

$$\text{gpm} = \frac{\text{spray discharged in gallons}}{\text{time interval in minutes}} \quad \frac{15}{3}$$

$$\text{gpm} = 5$$

Compare the desired spray discharge with the actual discharge from the sprayer. If the actual output is greater than desired, increase sprayer ground speed or reduce nozzle sizes or do both. (**Caution:** Because the volume of air discharged is an integral part of the coverage, increasing ground speed beyond 2 miles per hour can cause poor spray distribution.) When the output is less than desired, reduce tractor speed, select larger nozzles or do both.

Sprayer Maintenance

The owner's manual supplied by the manufacturer is the best source of information on routine sprayer maintenance. Maintenance of the engine is the same as for any engine. Lubricate, check filters and service cooling systems daily.

More pumps are ruined by improper maintenance than are worn out. Pump wear and deterioration occur with ordinary use but these are accelerated by misuse. These suggestions will help minimize problems and prolong the useful life of the pump and sprayer:

- Put clean chemicals, solutions and water in the sprayer. Small amounts of silt or sand particles rapidly wear pumps and parts of the sprayer system.
- Use chemicals that the sprayer and pump were designed to use. For example, liquid fertilizers are corrosive to copper, bronze, ordinary steel and galvanized surfaces. If the pump is made of these materials, it can be ruined by just one application of liquid fertilizer. Use stainless steel pumps for fertilizer application.

FIGURE 17-1 · *CONTINUED*

- Before using a new sprayer, clean the screens and nozzles of all metal chips and other foreign material.
- Flush the spray system with clean water after each day of spraying.
- Inspect screens and nozzle tips after each day of spraying. Clean by soaking and using a soft brush such as a toothbrush. Tips and screens may be damaged if cleaned with metal objects such as a knife or wire.
- Clean the sprayer thoroughly after each use or when chemicals are changed. Many chemicals cause metal corrosion and a chemical residue sometimes will react with succeeding chemicals, causing a loss of effectiveness.

Safety

Pesticides are potentially dangerous when not used properly. Sprayer operators should be thoroughly familiar with both chemicals and equipment before spraying. Read the pesticide label for all safety instructions. Wear protective clothing and equipment such as goggles, respirator, gloves, hat, boots, long-sleeved shirt and long trousers during mixing and application of pesticides. Do not smoke or eat while handling chemicals.

Always store and dispose of chemicals and containers properly. Triple rinse empty containers and dispose of them in accordance with the pesticide label instructions.

Make sure that sprayer operators and chemical handlers have the telephone numbers of poison control centers and physicians and that they know which chemical is being applied. Know the signs and symptoms of pesticide poisoning and what to do if it occurs.

Equipment and machinery also can be dangerous. Some important safety practices in machine operation include:

- Stop the engine, disconnect the power source and wait for all machine movement to stop before servicing, adjusting or cleaning equipment.
- Make sure all persons are clear of machinery before starting the engine, engaging power or operating a machine.
- Keep all guards in place when machines are in operation.
- When operating a tractor with a sprayer, watch for dangers and obstacles, especially at row ends, on roads and around trees.
- When possible, avoid operating the sprayer near ditches, embankments and holes.
- Allow no riders on equipment.

References

1. *Hidalgo County Demonstration Handbook,* 1982, "Citrus Spray Volume Demonstration," Texas Agricultural Extension Service.
2. *Citrus Growers Guide to Air Spraying,* Florida Cooperative Extension Service, Circular 351.

Educational programs conducted by the Texas Agricultural Extension Service serve people of all ages regardless of socioeconomic level, race, color, sex, religion, handicap or national origin.

FIGURE 17-1 · *CONTINUED*

Technical Note—Amido Black

(http://www.redwop.com/technotes.asp?ID=70)

Introduction

Amido Black, also known as Naphthalene Blue-Black or Naphthalene Black 12B, is a protein dye, sensitive to the protein in blood. It will stain the protein residue in a blood-contaminated latent print and turn a blue-black color. It will not stain the normal constituents found in latent print residue so it should only be used in the case of blood-contaminated latent prints to be successful.

Safety

As with all chemicals, always read the MSDS (material safety data sheet) to learn about the safe handling and health hazards of each chemical. Gloves and protective clothing should be worn when using Amido Black. It should be mixed and used in a fume hood or with an appropriate respirator. Amido Black is not a carcinogen. Some of the solvents used to mix it are hazardous and/or flammable. Use the proper safety precautions in handling and disposal.

Mixing Instructions

Working Solution

- 1 gram Amido Black
- 50 ml glacial acetic acid
- 450 ml methanol

1. Weigh out 1 gram of Amido Black and place it in a one-liter glass beaker.
2. Place it on a magnetic stirrer.
3. Add 50 ml of glacial acetic acid and stir.
4. When completely mixed, add 450 ml of methanol and stir for at least 30 minutes.
5. When completely mixed, store the solution in a glass bottle.
6. Properly label the bottle.

Glacial Acetic Acid–Methanol Solution

- 100 ml glacial acetic acid
- 900 ml methanol

1. Pour 100 ml of glacial acetic acid into a glass storage bottle.
2. Add 900 ml of methanol. Stir the solution until mixed.
3. Properly label the bottle.

FIGURE 17-2 · TECHNICAL NOTE

Glacial Acetic Acid–Distilled Water Solution

- 50 ml glacial acetic acid
- 950 ml distilled water

1. Pour 50 ml of glacial acetic acid into a glass storage bottle.
2. Add 950 ml of distilled water.
3. Stir the solution until mixed.
4. Properly label the bottle.

Processing Instructions

Step One: Fix the blood on the surface of the evidence

- Pour enough methanol to cover the item of evidence into a clean glass tray.
- Immerse the evidence in the methanol for about one hour.
- Cover the tray to prevent evaporation. Replenish the methanol in the tray, if necessary.
- Discard the methanol after use.
- If the evidence cannot be immersed in methanol, heat the surface with a lamp, heater or oven for at least one hour. Be careful of the risk of fire.

Step Two: Using the working solution

- Pour enough working solution to cover the item of evidence into a clean glass tray.
- Soak the evidence in the working solution for about two to three minutes or until the latent prints become a blue-black color.
- If the solution in the tray becomes heavily contaminated, it should be replaced with fresh solution.
- If the solution is not badly contaminated, it can be poured back into the bottle and used again.

Step Three: First rinse

- Pour enough glacial acetic acid–methanol solution to cover the item of evidence into a clean, glass tray.
- Immerse the evidence into the solution and rock the tray gently.
- When excess dye has been removed from the background, take the evidence out of the rinse.
- If the solution in the tray becomes heavily contaminated, it should be replaced with fresh solution.
- Discard after use.

FIGURE 17-2 · *CONTINUED*

Step Four: Second rinse

- Pour enough glacial acetic acid–distilled water solution to cover the item of evidence into a clean glass tray.
- Immerse the evidence into the solution and rock the tray gently for about 30 seconds.
- If the solution in the tray becomes heavily contaminated, it should be replaced with fresh solution.
- Discard after use.

Step Five: Drying and photographing

- Allow the evidence to dry at room temperature.
- Photograph any useful latent prints.

Sequential Processing

Treatment with physical developer may be done after Amido Black to try to improve the developed latent prints. It is suggested to photograph any latent prints developed with Amido Black before treating the evidence with physical developer.

Photography

Photography of latent prints developed with Amido Black should not pose any problems if the surface background is a light color. If the surface is a dark color but will fluoresce, it may be beneficial to use fluorescence examination to enhance the photographic contrast.

Additional Reading

Advances in Fingerprint Technology, edited by Dr. Henry Lee and Dr. R. E. Gaensslen.

Friction Ridge Skin: Comparison and Identification of Fingerprints, by James F. Cowger.

Manual of Fingerprint Development Techniques, by the British Home Office, 2nd edition.

FIGURE 17-2 · *CONTINUED*

distributor and manufacturer of crime scene investigation equipment, Lightning Powder Company provides a wide selection of products, services, and technical expertise. The product line includes thousands of items, ranging from fingerprint cards to forensic light sources, specialty magnifiers, and books and videotapes. The company provides free technical notes. These can be downloaded and used by crime scene professionals.

The instructions for mixing and using Amido Black, like the segments we selected from the Cisco router installation instructions, illustrate a number of format techniques discussed in Chapter 8 as they are applied to instructions and procedures. These techniques enhance readability by making the instructions easy to follow.

1. Use clearly worded headings and subheadings to describe and highlight the content of each section.
2. Leave plenty of white space around headings. Most margins should be at least one inch on each side. If the instructions are online, be sure that headings are separate from their content.
3. Use white space to make items in lists easy to read.
4. Highlight safety information and warnings.
5. Leave space between major sections and between main steps.
6. Place visuals at the point where the reader needs them. Don't ask readers to locate a visual several pages away from the instruction to which it pertains.
7. Label each visual, and at the appropriate point in the related text, say "See Figure x."
8. Make visuals as clear and simple as possible.
9. Label all visuals to correspond to the accompanying text.
10. Do not begin an instruction at the bottom of one page and complete it at the top of the next page. Insert a "page break": move the entire instruction to the next page.

POLICIES

Policies define an organization's position on subjects important to the daily operation of the company. Policies exist to document required actions and to help achieve uniform practice in adhering to them. Because policies describe the stance or action considered appropriate, they may also describe the actions that should be taken to execute the necessary procedures. Most organizations have policies on sick leave, employee development, travel, and a host of other topics. Most organizations develop and revise policies as they are needed. Most policies include the number of the current revision, the date it took effect, and all the information any employee would need to be able to comply.

Policies are structured like procedures—they begin with an introduction that states the policy, tells who is covered by it, and announces the policy's purpose. Then follows a general description, and perhaps a list of forms, reports, and correspondence examples required to execute the policy. This section is usually followed by additional sections that give detailed information about main issues covered in the general statement. Policies frequently contain procedures to explain how a policy should be implemented. Organizations usually have an institutionalized plan for writing policies. The following general pattern, in some form, is usually followed:

- Name of the policy—clearly stated, either in a subject line or a title line
- Statement of the policy—brief summary of the policy
- List of forms needed to execute the policy
- General applications—the situations to which the policy applies
- Specific applications—procedures for applying the policy in the specific case(s) described under general applications.
- Attachments—examples of forms, reports, and letters that will need to be completed in executing the policy.

Organizations often have their legal departments prepare policies for final dissemination; but managers and operations personnel may produce drafts based on the perceived need for a policy to deal with a specific problem. Legal specialists then check the policy for compliance with applicable laws, union contracts, and federal regulations. Policies are carefully screened for clarity and accuracy because they become the "rules" by which an organization operates.

Figure 17-3 illustrates a policy used by a major corporation. The policy explains the responsibilities of each person in the organization for tracking employees who miss work. While many policies are heavily procedure oriented, others appear in a narrative format. We include additional policy examples on the book's companion Web site (select Policies, in Chapter 17: Sample Documents).

Procedures and Policy Manuals

Both procedures and policy manuals are common documents in organizations. Policies may be grouped according to areas of an organization—hiring, safety, transportation, and communication policies, for example. Each policy type will usually follow the same plan. Organizations usually develop a common plan for policies and require that all new policies follow the plan. The plan used in Figure 17-3 exemplifies a common policy format.

Procedure manuals are collections of procedures that pertain to operating a specific system or process. Both procedure and policy manuals are developed as formal documents. They have tables of contents, lists of illustrations, glossaries— working definitions of terms used throughout the manual, and pictorial descriptions. Examples of procedure manuals can be found on the book's companion Web site. Procedure manuals may be prepared for paper distribution in binders, stored as pdf files for on-demand printing, or prepared for online access only. Online policy/procedures often use an electronic link from the policy/procedure in the table of contents to the actual policy or procedure. Policy and procedure manuals offer another example of how report elements can be applied.

Number: 21
Subject: Absence from Illnesses or Disability
Effective Date: January 1, 2004
Version: 1.0

POLICY

This policy ensures that eligible employees promptly receive correct payment of accident and sickness benefits. The policy also controls excessive absenteeism and the costs of disability insurance.

FORMS REQUIRED

All forms are in pdf format and are available on the company intranet under Human Resources, along with this policy.

1. Form 224-812, Absence Form—Disability (See Attachment A.)
2. FAS Form (See Attachment B.)
3. Insurance Claim Form 223 and instructions (See Attachment C.)
4. Physician's Statement, Form 225 (See Attachment D.)
5. Employee Return Notice, Form 111 (See Attachment D.)

POLICY DESCRIPTION

Group Manager's Responsibilities

1. Complete an Absence Form—Disability (Attachment A) immediately when one of the following events occurs.

    The day an employee tells you about impending hospitalization or home confinement.
    The first day an employee is hospitalized.
    The first and last days an employee is injured.
    The third consecutive day of absence due to illness. Notify the Health Benefits Administrator the morning of the third day.

2. Send the form to the Health Benefits Administrator, and notify the employee and/or the employee's physician to check the status of the Claim Form.

3. Help the Health Benefits Administrator obtain information about an employee's absence.

4. Visit the employee at the hospital at least once every two weeks. Encourage the employee to return to work as soon as possible. Let the employee know you care.

5. Get clearance from the Health Benefits Administrator before you allow the employee to return to work (Attachment B).

Personnel Department Responsibilities

1. Health Benefits Administrator—After receiving the Disability Absence Report from the Project Manager, give Insurance Claim Form 220 (Attachment C) with instructions to the employee.

FIGURE 17-3 · CORPORATE POLICY

2. During the first week the employee is absent, notify the employee and/or the employee's physician to check the status of the Claim Form.

3. See that the Claim Form is properly completed and includes the physician's definite or estimated date of return to work.

4. Forward the completed form to the insurance company.

5. See that the employee receives payments as quickly as possible. Correct any problems causing delayed payment. Check for accuracy of the payment.

6. Request interim supplemental information (by e-mailed form or phone) when the physician has not indicated a return to work date or when the absence may exceed the normal expected absence for this disability or illness (see Attachment D).

7. Periodically advise the Project Manager about the employee's status and the date the employee is expected to return to work.

8. Call or visit the employee to secure any supplementary information.

9. When the employee returns to work, see that a prompt and correct final Accident and Sickness Insurance benefit payment is made.

10. Send notification (Attachment E) to the payroll office.

11. Tell the employee the following:

    • Report to the Medical Department before returning to work.
    • Obtain a "Return to Work Advice Form" (Attachment B).
    • Submit the form to the Project Manager.

FIGURE 17-3 · *CONTINUED*

☑ Planning and Revision Checklist

▪ Who are your readers? Describe them in terms of their knowledge of the subject: educational level, technical level, responsibilities in the organization.

▪ What do you want them to be able to do as a result of these instructions/ procedures? What equipment or materials do they need?

▪ What is the situation that has led to the need for these instructions/ procedures to be written?

▪ How will the materials be used? Will readers need to read an entire document before they begin the task? In what context will readers be using these instructions?

▪ What problems could readers encounter in attempting to execute these procedures?

▪ What types of safety and/or quality control problems must be emphasized? What warnings or notes will you need to include?

▪ What topics do you want to be sure to include/exclude?

▪ What format will you use: online—to be printed? Paper? Manual? Poster?

▪ Given the context in which the instructions will be read, what format strategies would enhance accessibility?

▪ What types of visual will you need to include?

▪ What is your basic outline of your instructions? Does this outline meet the needs of your readers? Is it likely to achieve your purpose?

▪ Might any material be so misconstrued that readers who believed they were following the instructions could perform actions resulting in legal liability to you or your organization?

▪ ▪ ▪ ▪ ▪ ▪ ▪   Exercises   ▪ ▪ ▪ ▪ ▪ ▪ ▪

Further exercises related to the content of this chapter and all documents employed within the following exercises can be found on the book's companion Web site.

1. Develop a set of instructions or procedures. Or bring to class a poorly written set of instructions or procedures. Revise your selection. Submit both the original and your revision. Explain, in a memorandum to your instructor, why you find the original set of instructions ineffective.

2. If you are a member of a campus student organization, write or revise a policy for the organization. Or, for a semester project, you may wish to write or revise an existing policy manual. Develop an organization plan that you will use for the entire policy manual.

3. Your company is a small manufacturer of organic ready-to-eat breakfast cereals. The company would like to expand its market overseas, especially to China. Ready-to-eat cereals, however, are not a traditional Chinese breakfast. To minimize confusion for potential consumers, your package will need to include instructions on how to prepare and eat a bowl of cereal. Compose the instructions, including rough sketches of necessary illustrations, to fit on the side of the box. Keep in mind the advice from Chapter 6 on writing for international audiences.

Scenario

You have been asked to chair a committee to develop a recommendation report on changes that need to be made to the organization's health insurance program. The fourteen-person committee includes six members who work in branch offices located in three other cities. The challenge is to produce a collaborative document that reflects the concerns of all offices. Committee members from the branch offices may be able to come to corporate headquarters for occasional face-to-face meetings, but you realize that you need to use technology to develop much of the report. Trying to schedule frequent meetings that accommodate the work schedules of more than a dozen people will not be productive use of your time. The vice president for human resources wants the report in two months. Efficient, effective collaboration will be essential. Therefore, you must develop a plan that will allow maximum interaction among busy committee members.

Writing Collaboratively

Throughout this book we discuss methods of planning, developing, and revising standard types of documents prepared in the workplace. Our goal has been to help you become an effective writer prepared for the communication demands of the workplace. To reiterate: effectively written documents don't "just happen": they result from a writer's ability to plan a document and then write or revise it to best serve its purpose. Being able to write effectively means that you can do a good job when you prepare documents as the sole author. However, in the workplace, documents prepared by committee are common. Knowing how to work collaboratively is another step in your developing communication skills.

ISSUES IN COLLABORATION

Collaboration does not mean that several people form a group, whereupon one person does most of the writing. In a work environment, collaboration requires contributions from those involved in the project and equal legal responsibility and accountability to the organization; the same is true, usually without legal accountability, of academic collaborative projects undertaken in classes. Many collaborative reports will have a title page followed by a submission page signed by all participants. Or, the names of all writers may appear on the title page. In some cases, the name of the individual who manages the group developing the document may go on the title page. While that person is legally responsible for the content of the document, others who participated in preparing the document will also be held responsible for the content if issues and questions arise. In short, while all individuals may not always contribute equal amounts of information, the group—individually and wholly—is responsible for the document.

VALUE OF COLLABORATION

Collaborative projects have the potential to generate powerful documents. The knowledge and expertise of individuals working together, as a team, can create high-quality, highly informative documents that could not be produced without the combined expertise of those individuals. Several individuals working collaboratively can enhance the main planning stages in the writing process. Having several people consider the audience and purpose of a project ensures that the planning process covers essential perspectives. In addition, individuals with specific, needed expertise often can produce a collaborative report more quickly than a single person who is responsible for the entire process. Long reports, such as feasibility studies, environmental impact statements, and many empirical research reports, are usually collaborative. The more involved and complex the project, the more collaboration becomes necessary to the project and to the report that emerges.

The secret to effective collaboration is planning:

- Deciding who will do what
- Deciding the time line for completion
- Learning to work together to plan, develop, and write the document

TECHNIQUES FOR DEVELOPING COLLABORATIVE DOCUMENTS

The On-site Collaborative Group

Assume that five individuals are needed to prepare a feasibility study and that a supervisor has appointed a leader who is responsible for convening the group and tracking project development and performance, with each group member responsible for one area of the study. During the initial discussion, the group might decide to proceed in this way:

■ A project leader is named who will be responsible for convening meetings and tracking all written portions of the document; he or she is the facilitator who keeps the project moving. The group will need to decide, based on the report deadline, how often the group should meet.

■ The group divides the work—who will do what.

■ The structure of the report is developed—what sections will be needed (or required) based on any existing requirements of the organization's style sheet or the client's needs. For example, documents prepared for the U.S. government often require specific sections in a stated order and will not be accepted if the announced structure is not followed.

■ The group decides time lines. Once the plan has been determined, the following must be set:
- What section can be done by what date
- How long the drafting will take
- How much time should be allowed for discussion and revision of the initial draft

Once a work schedule has been developed, the group leader will serve as the liaison among members and collect segments as they are prepared. The project then proceeds as follows:

■ Members transmit their sections to the group leader by means of e-mail, whereupon the leader will assimilate the sections into a draft, which all members of the team receive, also via e-mail.

■ Alternatively, the group leader may wish to display the working segments on a Web page. Documents may be in Word files, which group members can open and annotate by means of comment tags. The group may then want to meet to discuss the draft, with the understanding that any additional changes of their sections will be e-mailed to the group leader. Similarly, two people who are developing a section may wish to proceed by e-mailing versions of the draft

Project name:	
Project goals:	
Team members:	
Duties:	
Deliverables:	
Timeline:	
Obstacles:	
Initial outline:	

FIGURE 18-1 · GROUP PLANNING SHEET

back and forth. Word processing software allows individuals to add comments and queries to items and to track insertions and deletions. Collaborative software is also available.

■ A group project is challenging because it is desirable to produce a seamless document that does not reflect the individual writing styles of multiple writers. Once the group leader has compiled the document and the members have approved content, one person should be charged with making the document seamless by

creating a single style,

ensuring that format is consistent,

ensuring the report includes all needed content and required sections.

■ The person who styles the document should submit it via e-mail, to all group members who helped with the final revisions and editing. In long, complex documents, collaborative revision and editing can improve the final document, as individual contributors read not only their work but the work of others.

Figure 18-1 shows a group planning sheet that can be used for this type of group collaboration.

The Distributed Collaborative Work Group

In national and multinational organizations, writers who do not know each other may be selected to work on a collaborative project because of their expertise. Such a group may work online via one of many collaborative software tools that allow individuals to "post" their materials and their comments. They may also communicate via online interactive audio-video software, which also allows them to discuss the document. Even so, the group will have a leader, perhaps a trained technical writer, or someone who is an effective project manager. Rather than having face-to-face discussions on site, the group will "meet" via the collaborative software or via a listserv, so that participants may pose questions and answers. In complex, lengthy report situations, the group may arrange a single face-to-face meeting or conduct numerous meetings via interactive conferencing technology. You can find exercises using this method on the book's companion Web site.

When people from different countries and cultures collaborate, the need for sensitivity to cultural differences is critical. The group leader will do well to provide his or her agenda to all collaborators well before each online meeting. The agenda may be sent as an e-mail attachment. Focusing the discussion on the business issue can help the collaboration proceed smoothly. Nevertheless, collaborators should not underestimate the importance of understanding others' perspectives. Development of an effective document will require positive contributions for all participants.

The Lead Author Work Group

Collaboratively written documents may also begin as a draft that is developed by a lead writer. This method is particularly effective for short documents that require input and approval from a wide range of contributors. The lead writer will begin with a broad outline of the document that also includes a draft of most sections. The lead writer then distributes the document via e-mail to other contributors, who submit any corrections and revisions that prove to be necessary. Microsoft's revision tool on Word exemplifies an effective means of dealing with revisions, although a number of powerful collaborative tools are available for groups that wish to collaborate online. Examples of these documents can be found on the book's companion Web site.

MAKING COLLABORATIVE PROJECTS WORK

A key ingredient of a successful project is the ability of collaborators to work together, showing mutual respect to other members of the group, meeting deadlines, and providing high-quality content. Face-to-face collaboration requires people to exhibit tact and respect; online collaboration is no different. As discussed in Chapter 4, Achieving a Readable Style, the "tone" or sound of an e-mail message can generate a hostile environment. The ability to accept constructive criticism is critical to any successful collaborative project, but each collaborator is responsible for communicating, either face to face or online, in a professional manner.

It may be helpful to assign one person to take notes during face-to-face, interactive audio-video and computer conferencing. After the meeting, notes can be distributed to ensure that everyone is clear about responsibilities. Many organizations also require that meeting notes be archived in case accountability issues arise.

Working face to face is often easier than working via e-mail, but either method requires a few fundamental strategies:

- ▨ Maintain a positive approach.
- ▨ Watch your tone: be aware of how you sound.
- ▨ Be respectful of the views of others.
- ▨ Solicit input.
- ▨ Encourage timely response.
- ▨ Be sure that everyone knows the time line for the project.
- ▨ Send "reminder" messages by e-mail, if necessary.
- ▨ Keep records of all messages in an e-mail folder.

COLLABORATIVE PROJECTS IN ACTION

Situation 1 John Stewart, a distinguished professor at a major university, has been named chair of the task force on student housing established by the president of the university. The committee, composed of ten people, will meet for the first time in two weeks. From the president's charge to the committee, John develops an agenda for the meeting and distributes it electronically to the other nine individuals. Having asked for other agenda items, John receives additional recommendations. He revises the initial agenda, resends it to the committee via e-mail, and then asks for volunteers to discuss four of the topics on the agenda. Each volunteer should write a short summary of his or her position and send it to John, who will add it to the agenda.

Three days before the meeting, John sends out the agenda for the third time. This revision includes the summaries by those who have agreed to lead the discussion of the topics. The result? By the first meeting, the committee has had opportunity to think about the charge of the committee and some of the critical issues that will be discussed.

Situation 2 Three members of an engineering firm with offices throughout the United States have been asked to draft a response for a business that is seeking an "RFQ" (request for quote) on a job project. The three are experts in pricing and are often called upon the develop responses to RFQs. The project manager in Atlanta e-mails the necessary information from the business to his three price experts, all of whom reply to him and to each other. With their comments in hand, the project manager drafts a two-page RFQ and sends a copy to each contributor for notes and comments. When the document looks good to all three pricing experts, the project manager completes and submits the response to the RFQ to the company.

Situation 3 A professional organization decides to develop a procedures manual that describes the duties of each officer. Charged by the Executive Committee to develop the manual, the vice president of the organization invites several individuals who have served as officers to draft parts of the manual. The VP provides a general plan for each section to guide the development of the drafts. The writers e-mail their drafts to the VP, who then begins to write the procedures manual by compiling information from the individual contributors.

■ ■ ■ ■ ■ ■ ■   Exercises   ■ ■ ■ ■ ■ ■ ■

Further collaborative exercises can be found on the book's companion Web site.

1. Issues in staying healthy now require that people understand as much as possible about various physical and mental ailments and diseases. Work in groups of three to research what is currently known about a particular illness or medication that interests all members of the working group. Focus on resources that can be understood by nonmedical readers. Assess the validity of the material you find. Write a collaborative summary, about three pages, recording your findings. Be sure to note the sources of your information and state why you believe that information is valid. Explain how your team worked together to prepare this assignment.

2. The president of your university has asked student groups to outline what they believe to be the major problems students confront on campus and to describe each problem briefly. Solutions can also be suggested. Work as a team (a fraternity, service organization, or student activity group), and draft a response.

3. Assume that the president of the university has taken a stand on an issue and your group finds it objectionable. Write the president, stating the group's reasons for disagreeing with what he has said and presenting your collective position.

4. Prepare a collaborative report assignment to any report assignment given by your instructor. Have one person on the team keep a record of who is doing what. As a team, prepare a cover memo to the report that explains the collaborative process your team used; in addition, provide individual narratives on the success of the effort. Describe difficult issues that arose and explain how these issues were resolved collaboratively.

As you and your team members plan your response.in each situation, discuss each of the questions presented in Chapter 2 as your consider your readers and how you want them to respond to your message.

Scenario

As a newly hired engineer with a large engineering firm, you believe that you can handle report writing fairly well. However, your supervisor has told you that you need to improve your oral presentation skills. "You talk too fast and use too many PowerPoint slides," she said. "Your listeners can't keep up with you. You need to work on a more effective presentation of project summaries and project reports. To make partner in this firm, an ability to present proposals is critical."

CHAPTER NINETEEN

Preparing Oral Reports: The Basics

Workplace communication studies show that the ability to speak effectively is as crucial as the ability to write effectively. If you doubt that claim, go to Google and type "importance of communication skills." That phrase will return more than two millions hits from a variety of sources—universities, government agencies, and consulting firms, for example. You will see that writing and speaking well are critical skills for sustained employment and professional progress. Employees actually spend less time writing than speaking—for example, when:

> using the phone
> conversing informally with colleagues, subordinates, and superiors on routine work issues
> conducting/attending meetings
> working in problem-solving groups
> conducting employee evaluation sessions
> participating in teleconferences and sales presentations
> being tapped to deliver formal speeches inside and outside the organization

Communication research also reveals that the higher an employee moves in an organization, the more important speaking skills become.

This chapter describes basic strategies for orally presenting technical and business information in formal presentations. You will see that developing the effective written document and making a good oral presentation use many of the same strategies and techniques. Understanding similarities between preparing written and oral reports can be helpful for several reasons:

1. You may be asked to document an oral presentation you have given; that is, you must submit what you said in written form.
2. You may be asked to make an oral summary of a written document.
3. Many consulting reports consist only of the oral presentation articulated from PowerPoint slides; no written report accompanies the oral presentation. Particularly in consulting reports containing highly sensitive information, details are presented during confidential meetings. The slide presentation outlines the background of the problem and makes general recommendations, but specific marketing or financial recommendations are not textualized. Thus, for many in the audience, the presentation report is the entire report.

UNDERSTANDING THE SPEAKING/WRITING RELATIONSHIP

Effective oral and written presentations share the following requirements. As a writer/presenter, you must

> ■ Analyze your audience
> ■ Understand the context in which your presentation will be received

- Understand and articulate your purpose clearly
- Develop sufficient and appropriate supporting material
- Organize and arrange content so that it is easy for the audience to follow
- Choose a speaking style—level of language, approach to the subject, and tone—suitable to your role as well as to your audience and purpose
- Select the presentation format and visuals that will enhance your audience's understanding of your message

Many of the concepts covered in Chapters 2, 3, and 4 can be applied directly to the oral presentation. Although you will need to know how to adapt guidelines for organization, style, and graphics to fit the speaking situation, since listening differs from reading, you will see that writing and speaking are similar communication activities.

ANALYZING THE AUDIENCE

Just as the reader determines the success of the written communication, the audience determines the success of the oral presentation. Both have succeeded if the audiences respond appropriately: that is, if they are informed, persuaded, or instructed according to your plan and then respond as you had wished.

You cannot speak or write effectively to anyone without first understanding the other party's perspective. You must determine, as accurately as possible, how a business audience will likely respond based on their:

educational and cultural background
knowledge of the subject
technical expertise
position in the organization

But you must also consider your audience as human beings who react emotionally to you, your topic, your message, and even the context in which the communication occurs. Because the results of this analysis suggest what you should and shouldn't say or write and the tone you should use, you will need to answer each of the following questions carefully:

- How much does my audience know about the subject?
- How much do they know about me?
- What do they expect from me?
- How interested are they likely to be in what I say?
- What is their attitude toward me and my organization?
- What is their attitude toward my subject?
- What is their age group?
- What positions do they occupy in the organization?

www.oup.com/us/houp

CHECKLIST

- What is their educational background?
- What is their cultural/ethnic background?
- What is their economic background?
- What are their political and religious views?
- What kinds of cultural and social biases will they likely have toward me and my topic?
- How busy are they?
- Home much time do I have?

In reviewing this list, note the prevalence of questions on *attitude*—the audience's *attitude* toward *you* as well as the *subject*. Will your audience consider you trustworthy and credible? Can you reasonably expect them to accept what you say, or will you have to establish empathy with them before they can really "listen" and take your points? How will you build empathy? For example, your clothing, your demeanor, and the tone of your language can help or hinder your efforts to build empathy. When you speak to people from other countries, you should plan to do research on the culture of that country. Be aware that some of the hand gestures people use routinely with U.S. audiences have different meanings in other cultures. The clothing you choose and the color of clothing should also be selected with the audience culture in mind.

ANALYZING THE CONTEXT

Analysis of context is often difficult to separate from analysis of audience: context is one facet of the larger issue of audience. In analyzing context, you need to know why your presentation—oral or written—is required.

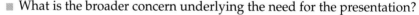

- What is the broader concern underlying the need for the presentation?
- What primary issues underlie the presentation?
- How does your presentation relate to these issues?
- What will be happening in the organization when you make your presentation?
- How does your presentation fit into the organizational situation?
- If you are one of several speakers, what kinds of presentation will the others be making? What has already been covered? Where do your remarks fit in?
- In what surroundings will you be making the presentation?
- If your report contains confidential or proprietary information, what can you include in your presentation that will not compromise the company's security?

For example, delivering a presentation at a regular meeting of project directors differs from a presentation to stockholders or the board of directors.

Speaking at a company Christmas dinner differs from making a presentation at the annual meeting of a professional or trade society. Knowing the context in which you will deliver your speech is as important as knowing your audience and your purpose. Context will often be inextricably bound to questions of audience attitude and the way you shape your purpose. Audience attitude frequently reflects current organizational issues. For example, if you work for a consulting firm that specializes in helping business organizations that must downsize, your report to the management group anticipating the elimination of managerial positions may be received with concern, even hostility. What you can or should say in your presentation—your purpose and the content you choose to present—may be dictated by the context surrounding your presentation and the perspective that your audience brings. You may present general strategies for downsizing to the management group, but you may reserve specific percentages of employees to be terminated for discussion during a private conversation with the CEO of the firm.

DETERMINING THE GOAL OF YOUR PRESENTATION

Whatever the business or professional goal of your presentation, you will always want to achieve the following communication goals. You want any presentation to be:

clear
truthful
easy to follow
memorable

With these goals in mind, and based on your analysis of your audience, state your purpose(s) in terms suited to your audience's perspective. Oral presentations may also have multiple purposes: to report the status of a project, to summarize a problem, to describe a plan, or to propose an action. But the long-range objectives of highlighting or documenting important specific issues and further establishing your credibility within the organization may be as important as making your purpose and your message clear.

Like the report or letter, the oral presentation must make its purpose clearly evident at the beginning so that the audience knows how to listen to what the speaker is saying. As you plan your presentation, state your goal in one sentence in terms of your audience's background and presumed attitude: announce your purpose early in the presentation to prepare your audience for the main ideas to come.

Speeches, like written documents, can enhance the presenter's reputation in an organization. Therefore, consider every speaking opportunity an opportunity to sell not only your ideas but also your competence, your value to the organization.

CHOOSING AND SHAPING CONTENT

Preparing the oral presentation often requires the same kind of research needed for the written report. You must determine what information you will need and then shape it to appeal to your audience—bearing in mind particularly their attitudes, interests, and prejudices. In selecting content, consider a variety of information types: statistics, testimony, cases, illustrations, history, and particularly narratives and examples of interest to the audience that help convey the goal you have for your presentation. While statistics and data are often necessary in building your argument, narratives interspersed with data provide an important change of pace needed to keep your listeners attentive. In short, vary your content, but be sure that every item you include furthers the achievement of your goal.

DECIDING HOW TO ARRANGE AND ORGANIZE THE CONTENT

Like written communications, oral presentations must be *organized* with your audience's needs and perspective in mind. Ask yourself the following:

- Is your audience interested in what you will say?
- What are the main ideas you want to convey?
- Based on your purpose and the needs and expectations of audience, in what order should you present these ideas?

Answers to these questions will help you decide how to go about organizing your presentation. Generally, oral presentations have an introduction, a main body, and a conclusion. The introduction should clearly tell the audience what the presentation will cover to prepare the audience for what is to come. The body should develop each point stated in the introduction, and the conclusion should repeat and reinforce the critical ideas.

While you are doing your research and collecting content, you should begin to organize your information. Divide your content into main categories of ideas:

I.   Idea 1

II. Idea 2

III. Idea 3, etc.

Then subdivide each main idea. Order the subdivisions so that the information moves in a logical sequence, one that will be easy for listeners to follow:

I. Idea 1

    A, B, C . . .

II. Idea 2

    A, B, C . . .

III. Idea 3

    A, B, C . . .

DESIGNING EACH SEGMENT: GUIDELINES

The structure of the oral presentation is crucial for one main reason: once you have articulated a statement, the audience cannot "hit Rewind" to "rehear" what you just said. A reader who does not understand a sentence or paragraph can stop and reread the passage. A speaker's audience, however, must be able to follow the ideas presented and understand them without having to stop and consider a particular point the speaker has made, thereby missing later statements in the presentation. To help your audience follow what you say, you must design your presentation with your audience's listening limitations in mind.

Audiences generally do not enjoy long presentations. Listening attentively is difficult, and audiences will tire even when a presentation is utterly smashing. For that reason, as you design your presentation and select content, look for ways to keep your message as concise as possible. Don't omit information your audience needs, but look for ways to eliminate nonessential material. Again, without carefully analyzing your audience—their attitude toward the subject, their background, knowledge of the topic, their perspective toward you—you will be ill equipped to make accurate decisions regarding content, design, or structure.

Choose an Interesting Title

A good way to grab your audience's attention is to develop a title that reflects the content in an interesting way. Like the title of a formal report or the subject line in a letter, memo, or informal report, the title of an oral presentation should prepare your audience for what is to come.

Develop Your Presentation about Three Main Divisions

To help your audience follow your message, build into your structure a certain amount of redundancy, that is, repeat main points. When you divide your presentation into introduction, main body, and conclusion, you are building in this necessary redundancy. In the introduction, you "tell them what you are going to tell them"; in the main body, you "tell them"; and in the conclusion, you "tell them what you told them." This kind of deliberate repetition helps your audience follow and remember your main points. Use verbatim repetition sparingly, however; it is preferable, as a rule, to reword statements made a second or third time.

Plan the Introduction Carefully

An effective introduction focuses your audience's attention on your theme and reveals how you plan to present it. Unless the introduction interests the audience, you will have difficulty keeping their attention. You must convince them that your presentation is worthwhile.

An effective introduction thus tells your audience:

how to listen
what to expect
the path you will follow in presenting your message

To prepare your audience for what you will say, the introduction states the topic, the purpose of the presentation, and the main points that will be covered.

You may also wish to introduce your topic with an attention-getting device: a startling fact, a relevant anecdote, a rhetorical question, a quotation or a statement designed to arouse your audience's interest. Again, the device you choose will depend on the audience, the occasion, the purpose of the presentation.

If your audience is not readily familiar with the subject, you may want to include background material to help them grasp your main points. Then, having introduced your topic, you can establish your purpose and tell the audience how your presentation will be structured. In introducing the topic, you have two other concerns: (1) motivating your audience to listen to you and (2) establishing your credibility.

If your analysis of your audience and the situation in which you will be speaking indicates a need to establish credibility, then you may need to include in the design of your introduction a number of strategies:

1. Acknowledge your perception of the audience's perspective.
2. Establish a common ground with the audience, points of agreement.
3. Attempt to refute (if you can do so efficiently) erroneous assumptions that your research indicates may be harbored by the audience.
4. Ask the audience to allow you the opportunity to present your information as objectively as possible.

Remember that the success of the presentation rests on your ability to make your audience want to listen and to prepare them to follow and understand what you say.

Design the Body to Help People Comprehend Your Ideas

In designing the body of the presentation, you develop what you want to say about each of your main points or ideas. You may want to present your ideas in a chronological sequence, in a logical sequence, or as a simple sequence organized by topic. The method that will best help your audience follow your ideas may depend on whether you are giving an informative speech, an analytical speech, or a persuasive speech. It is important, however, to let your audience know when you have completed one point and begun another.

Design the Conclusion to Reinforce Your Main Ideas

The conclusion reinforces the main ideas or perspective you wish your audience to retain. In a presentation that has covered numerous points, you should be sure to reemphasize the main points. But the conclusion also allows you to emphasize the importance of specific ideas, or you can reemphasize the value of the ideas you have presented. How you design the conclusion will depend on your initial purpose. A strong conclusion is nearly as important as a strong introduction, as the beginning and the end are the parts most likely to be remembered.

Strict organization may not be necessary in short, informal presentations. However, in oral proposals, progress reports, final project reports, financial reports, feasibility studies, and similar analytical reports, the audience will benefit from a clear, precise organization that allows them to follow your content and remember what they have heard.

CHOOSING AN APPROPRIATE SPEAKING STYLE

How you sound when you speak is crucial to the success of your presentation. You may have effective content, excellent ideas, and accurate supporting statistics. But if the style you use in speaking is inappropriate to the occasion, to the audience (as individuals and as employees in a particular part of the organization), and to your purpose, your presentation will more than likely be ineffective. Choosing the appropriate style will help you achieve your purpose, but tone and degree of formality will be dictated by your organizational role and your relationship to your audience.

- Do the people know you?
- Is your rank in the organization above or below them?
- Are you speaking to an audience of individuals from all levels within the organization?
- What demeanor, approach, and level of formality does the organization usually expect from those giving oral presentations?
- Is the audience composed of people who understand English well?

If you are speaking before a group consisting largely of people from another country, you need to determine beforehand their fluency in English. If they are uncomfortable with English, be sure that you speak slowly; avoid idiomatic expressions; choose concrete words; and speak in relatively short sentences, limiting each sentence to one idea.

As we discussed in Chapter 4, style in writing refers basically to:

your choice of words
the length and structure of your sentences
the tone, or attitude, you express toward your audience

Style in delivering oral presentations is also defined by the same characteristics, plus many nonverbal cues that can either enhance or detract from your presentation. While the style you use will vary with the audience, topic, and context, the following guidelines can enhance your delivery:

- Avoid speaking in a "written" style.
- Use phrases, and vary the length of your sentences.
- Speak as if you were talking with your audience. If you concentrate on getting your point across by having a conversation with the audience, you will likely use a natural, conversational style.

Many suggestions for clarity in writing also apply to clarity in speaking:

- Avoid long, cumbersome sentences. Long sentences can be as hard to attend to in a speech as they are to read.
- Avoid overuse of abstract, polysyllabic words. Instead, use concrete language that your audience can visualize.
- Avoid overuse of jargon, unless you are sure that your audience will be readily familiar with specialized terms.
- Use sentences that follow natural speech patterns.

SPEAKING TO MULTICULTURAL AUDIENCES

As organizations become more international, you may find yourself giving presentations to groups in other countries or to audiences composed of individuals from all over the world. Because you will want any audience to respond positively to your presentation, you will need to do research to understand how people from other cultures will likely interpret what you say, how you say it, how you dress, how you act in your dealings with them. The graphics and visuals you use may also have to be changed, as symbols in one culture may have an entirely different meaning in another culture (see Chapter 6). As you consider your audience and the content you want to present, remember that your understanding the ethnic profile of your listeners is part of your audience assessment homework.

USING TECHNIQUES TO ENHANCE AUDIENCE COMPREHENSION

Because your audience cannot "hit Rewind" to "rehear" ideas that have been stated aloud, look for ways to help people easily follow your ideas:

1. Be sure you clearly demarcate the beginning and end of each point and segment of your presentation:

- Announce each main topic as you come to it. Your audience should know when you have completed one topic and are beginning the next one.

▪ Pause slightly after you have completed your introduction; then announce your first topic.

▪ After completing your final topic in the main body of your presentation, pause slightly before beginning your conclusion.

2. Speak slowly and enthusiastically. Be sure you enunciate your words carefully, particularly if you are addressing a large group.

3. Use gestures to accentuate points. Move your body deliberately to aid you in announcing major transition points. Avoid standing transfixed before your audience.

4. Maintain eye contact with your audience. Doing so helps you keep your listeners involved in what you are saying. If you look at the ceiling, the floor, the corners of the room, your audience may sense a lack of self-confidence. Lack of eye contact also tends to lessen your credibility. In contrast, consistent eye contact enhances the importance of the message. By looking at your audience, you can often sense their reaction to what you are saying and make adjustments in your presentation if necessary.

5. Do not memorize or write out your entire presentation. Otherwise, your speech will sound as if you are reading it. With PowerPoint, use a laser highlighter to call attention to individual points. Avoid note cards and voluminous notes. If you suddenly forget what you are trying to say, you can easily lose track of your place in several pages of notes. If possible, have an outline of the presentation on one sheet of paper. If you do forget where you are, a quick glance will usually refresh your memory.

6. Rehearse your presentation until you are comfortable. Try walking around, speaking each segment, and then speaking aloud the entire presentation. Rephrase ideas that are difficult for you to say—these will likely be hard for your audience to follow. Be sure to time your presentation so that it does not exceed the limit given. To keep the speech as short as possible, avoid adding information as you rehearse.

7. If possible, record or videotape your speech. Listen to what you have said as objectively as possible. As you listen, consider the main issues of audience, purpose, organization, context, content, and style.

8. Listen for tone, attitude, and clarity. Is the tone you project appropriate for your audience and your purpose? Is each sentence easy to understand? Are you speaking too rapidly? Are the major divisions in your presentation easy to hear? Are any sentences difficult to understand? If you have a videotape to study, watch for body language, fidgets, eye contact, and so on.

9. If you develop a PowerPoint presentation, use a notes format as exemplified in Figure 19-1 and distribute the handouts before you begin your presentation. Note that the opening page begins with the title. On each subsequent slide, the main point is emphasized, and the most important subpoints briefly listed. The slide presentation should form a summary that will be meaningful to listeners who consult their notes later. The notes should also be meaningful to those who receive an e-mail version of your presentation.

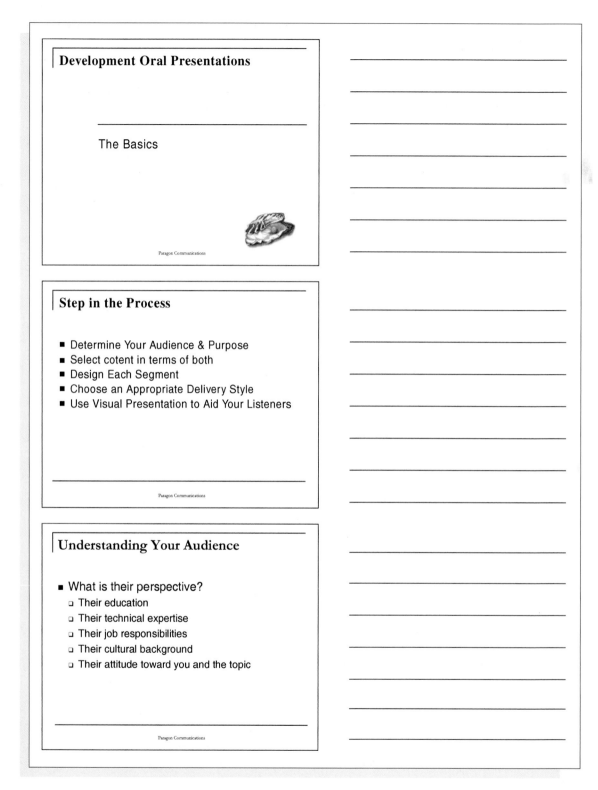

FIGURE 19-1 · NOTES FORMAT FOR POWERPOINT PRESENTATION

Analyzing the Context

- What primary issues underlie th presentation?
- How does your presentation release to these issues?
- What will be happening in the organization when you make your presentation?
- How does your presentation fit into the organizational situation?

Paragon Communications

Designing Each Segment

- Choose an interesting title
- Develop the presentation about three main divisions.
- Plan the introduction carefully
- Design the body to reveal each main point
- Design the conclusion to reinforce main points

Paragon Communications

Choosing an Effetive Delivery Style

- Avoid long sentences.
- Use language your audience will easily understand.
- Use pauses to help listeners follow the arrangement of your ideas.
- Maintain eye contact with your audience.
- Use gestures to accentuate points.

Paragon Communications

FIGURE 19-1 · *CONTINUED*

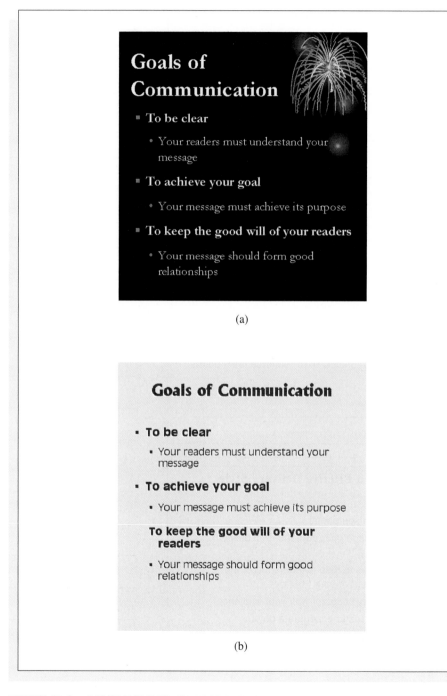

FIGURE 19-2 · SAMPLE SLIDES: (A) POORLY CHOSEN TEXT–BACKGROUND
COMBINATION; (B) AND (C) MORE READABLE COMBINATIONS

(c)

FIGURE 19-2 · *CONTINUED*

10. Make your slides clear but simple. PowerPoint can help you avoid the use of excessive data; but PowerPoint allows you almost unrestricted use of color and a wide choice of slide backgrounds, which in turn can be overdone, to the detriment of readability. Figure 19-2A uses a background that makes the text more difficult to read than it would be on a plain background (Figure 19-2B). If you are preparing slides or transparencies for video conferencing, use the plain background and a color—such as yellow or light green—and black text. Color can enhance a visual, but it can also reduce the effectiveness of the message. Figure 19-2C shows how sans serif type can be even more readable on some templates than serif type (Figure 19-2A).

11. When you are planning your presentation, determine how you will handle questions. Although in many presentations questions will not be an issue, during oral reports, such as status reports, project reports, and proposals, you can expect your audience to have questions.

You must decide whether you want to allow questions during your presentation or afterward. Asking your audience at the outset to refrain from questions until you have completed the presentation avoids interruptions that may cause listeners to lose track of what you are saying. Couple this request with the assurance that you will be happy to answer questions later.

12. As you plan your presentation, prepare for questions your audience may ask and decide how you will answer each one. If you have properly analyzed your audience and the reason for your presentation, you should be able to anticipate likely questions.

13. Keep the question-and-answer time moving briskly. Answer each question as concisely as possible; then move to the next one. Reword a difficult question, or break it into several parts; then answer each part. If any listeners are unable to hear questions asked by persons not using a microphone, be sure to restate the question.

PLANNING VISUALS TO ENHANCE YOUR PURPOSE AND YOUR MEANING

Because we live in a time when communication is both visual and verbal, visual enhancements of oral communication are important. Visuals help your audience understand your ideas, follow your arguments and remain attentive, and remember what you said. Visuals used in oral presentations need to fit the needs of the audience. Even for a highly specialized audience, however, they should be as simple, clear, and easy to understand as possible.

How many visuals? Avoid an excessive number! Listeners resist an unbroken barrage of words and visuals just as readers become numb when they encounter unbroken pages of prose and long sections of visual clutter. Thus, many of the guidelines for visuals discussed in Chapter 11 also apply to oral presentations.

Formal presentations should use PowerPoint or similar software or associated visual aids to help listeners follow your ideas. PowerPoint allows you to give your listeners the outline of your presentation and insert pictures and provide graphs, tables, drawings, photographs, diagrams, flowcharts as well as sound, video clips, and animations. Keep the following guidelines in mind.

- Avoid too much information on any single slide. Use a font size that can be easily read.
- Use sans serif type because it produces a sharper image for slides and transparencies.
- Limit the fonts you use to two per visual.
- Avoid all caps. Use type that contrasts distinctly with the background.
- If possible, before you begin, provide the audience a copy of your slides with an area for notes, as shown earlier (Figure 19-1). This arrangement allows listeners to track what you are saying.
- Avoid visuals that are "busy" or crowded with data (Figure 19-3 and Figure 19-4). Be sure that your slides, as a group, present the concepts your want to convey in your presentation.

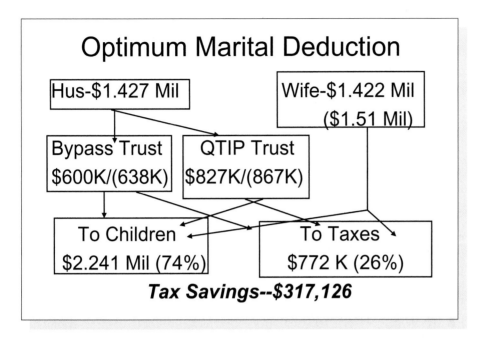

FIGURE 19-3 · SAMPLE SLIDE: HARD TO READ, TOO COMPLEX

HHSC LAR Summary Request

GOAL		FY 2004-2005 Expend/Budgeted	FY 2006-2007 Base Request	FY 2006-2007 Exceptional Items	FY 2006-2007 Total Request
Goal A	HHS Enterprise Oversight and Policy	1,156,491,420	1,125,138,046	42,468,432	1,167,606,478
Goal B	Medicaid	23,471,668,180	23,332,396,442	8,422,660,942	31,755,057,384
Goal C	CHIP Services	992,065,427	928,111,026	497,720,007	1,425,831,033
Goal D	Encourage Self Sufficiency	934,980,796	916,383,704	14,941,490	931,325,194
Goal E	Program Support	80,253,130	76,907,054	3,806,910	80,713,964
Goal F	Information Technology Projects	143,844,506	60,721,687	73,692	60,795,379
	Total, Agency Request	$ 26,779,303,459	$ 26,439,657,959	$ 8,981,671,473	$ 35,421,329,432
Method of Financing:					
	General Revenue	$ 9,362,087,538	$ 9,758,885,600	$ 3,636,164,845	$ 13,395,050,445
	General Revenue - Dedicated	66,328,051	33,559,464	10,921,685	44,481,149
	Earned Federal Funds	7,278,695	6,956,612	264,586	7,221,198
	Federal Funds	16,601,429,277	15,945,771,302	5,331,446,091	21,277,217,393
	Other Funds	742,179,898	694,484,981	2,874,266	697,359,247
	Total, Method of Financing	$ 26,779,303,459	$ 26,439,657,959	$ 8,981,671,473	$ 35,421,329,432

Page 11

FIGURE 19-4 · SAMPLE SLIDE: HARD TO READ, CROWDED WITH DATA

■ Do not talk and present a slide at the same time. What you say should complement the information on the slide. Present the slide, wait a second or two, then add your commentary. People who are busy trying to absorb the content of your slides will not be listening to you.

■ Avoid talking too fast and showing too many slides. Many PowerPoint presentations are ineffective because the speaker proceeds at a pace that prevents the audience from attending to the commentary, viewing the slide content, and taking notes. Remember your audience must understand the plan of your presentation, process your message, and then internalize the content. Use visuals that help listeners follow, understand, and remember what you have said.

Computer graphics and programs such as PowerPoint and Excel, in combination with color printers and slide projection equipment, give you the opportunity to experiment with graphic design to develop visuals that are pleasing as well as clear. For example, Figure 19-5 was generated with Excel and transported into PowerPoint. Figure 19-6 has transported information from the Web into a slide. Note that the source of the information is included. Use color carefully: choose white letters on a dark background, or dark letters on a light background.

Slides, prepared as handouts for the audience, may be printed in grayscale with an area included for notes. Be sure that handouts of your slides convey a coherent content message. That is, the slides should make sense independently of the presentation. In addition, the handouts should be meaningful to people who were unable to attend the presentation itself. If you believe that your presentation would benefit from more explanation, you may wish to add notes after each slide. See Figure 19-7, as an example.

Although the room in which you are to give a presentation may not have Ethernet connections, you can print Web materials via a color copier onto paper or transparency masters. People who give many presentations carry transparencies of their slide presentations, along with a portable projector, in case of Internet access problems or computer or projector malfunctions.

▨ DESIGNING AND PRESENTING THE WRITTEN PAPER

While most business presentations can be developed about a PowerPoint presentation, some occasions may require you to read from a manuscript. Papers presented at professional meetings that cover complex material are frequently read from manuscripts. These papers may then be published in the official proceedings of the professional society.

However, presentations may need to be read for other reasons:

1. A carefully written presentation that discusses company policy, a sensitive issue, or a topic that has been approved in a specific form by someone

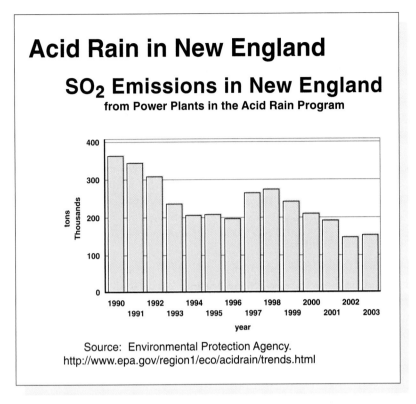

FIGURE 19-5 · SAMPLE SLIDE TRANSPORTED FROM EXCEL TO POWERPOINT

FIGURE 19-6 · SAMPLE SLIDE TRANSPORTED FROM EPA WEB SITE TO POWERPOINT

Understanding Your Audience

- ■ What is their perspective?
  - ❑ Their education
  - ❑ Their technical expertise
  - ❑ Their job responsibilities
  - ❑ Their cultural background
  - ❑ Their attitude toward you and the topic

When you analyze your audience, you must focus on their professional profile; but you must also focus on your audience as human beings who react emotionally to you, your topic, your message, and even the context in which the communication occurs. The results of this analysis then suggest what you should say or write, what you should not say, and the tone you should use.

Therefore, as you consider the audience you will address, you will need to answer each of the following questions as carefully as possible:

- How much does my audience know about the subject?
- How much do they know about me?
- What do they expect from me?
- How interested are they likely to be in what I say?
- What is their attitude toward me?
- What is their attitude toward my subject?
- What is their age group?
- What positions do their occupy in the organization?
- What is their educational background?
- What is their cultural/ethnic background?
- What is their economic background?

FIGURE 19-7 · ANNOTATED POWERPOINT SLIDE GIVING DETAILS OF PRESENTATION

· What are their political and religious views?

· What kinds of cultural biases will they likely have toward me and my topic?

In viewing this list, you will note the prevalence of questions on attitude—the audience's attitude to ward you as well as the subject.

These questions are particularly crucial ones, as you need to know, before you begin planning your presentation, if your audience will consider you trustworthy and credible. Can you reasonably expect them accept what you say, or will you have to establish an empathy between you and the audience before they can really "listen" and take your point? To be an effective speaker, you must know your audience, establish an effective relationship with them by being sincere and knowledgeable about the subject, then conform to their standards about dress, demeanor, choice of language, and attitude toward them and the topic.

When you speak to people from other countries, you should plan to do research on the culture of that country. Be aware that hand gestures you use routinely with U.S. audiences may have different meanings in other cultures. Also, the clothing you choose to wear should be selected with the culture of the audience in mind.

FIGURE 19-7 · *CONTINUED*

other than the presenter may be read from the approved, typed manuscript to ensure accuracy;

2. A presentation that will be circulated or filed as documentation may be read by a spokesperson, particularly if a possibility exists that material may be misconstrued.

3. Inexperienced speakers who have been assigned a difficult topic may be more comfortable reading from a prepared manuscript, as insurance against losing their train of thought or forgetting important details.

Obviously, from the listener's perspective, the extemporaneous presentation will likely be more effective and interesting than the presentation that is read from a manuscript. With an extemporaneous presentation, the speaker can work more actively to involve the audience in the content. Effective use of eye contact and body movement enables the speaker to judge audience reaction to each idea as it is presented and to make adjustments if the presentation does not seem to be achieving its purpose. On the other hand, written presentations can be effective if the speaker plans and writes the presentation carefully and then utilizes a number of delivery techniques to enhance the effectiveness of the reading.

Structuring the Written Speech

The structure of the written speech is the same as that of the extemporaneous speech. That is, there are three main parts: the introduction, the main body, and the conclusion. Each section should be structured according to the principles set forth for extemporaneous speeches. However, you will need to write each section completely. If you know that the speech will be published, you may wish to write it like an article for publication, or a report, designating content and organization by means of headings and subheadings.

Writing the Speech

After you have designed the content of your paper and made final revisions in your ideas, you will need to give close attention to your sentences and paragraphs, since you will be reading these directly from the page.

1. Be sure that each section is clearly demarcated from other sections. This means that each section should have an overview that clearly announces that a new topic has begun. Each paragraph should also begin with a topic sentence that summarizes the content of the paragraph. In short, in writing a speech to be read, you are following the advice given in Chapter 3; but since your audience cannot stop and "hit Rewind," you are making a concerted effort to accentuate every device for revealing organization.

2. Limit each section and each paragraph to one idea. Keep units short so that your audience will not lose track of the main ideas.

3. Avoid excessive detail.

4. Use enumeration to help your audience follow your main points and to know when one point has ended and the next point is beginning.

5. Avoid long sentences. Long sentences are as difficult (or more difficult) to follow in spoken form as they are to read.

6. Prune every sentence to make it as concise as possible.

7. Use active voice whenever possible so that your sentences will preserve the natural quality of spoken language. Use passive voice when you want to hide *who* is doing what ("The plan was executed") or *who* is responsible for specific actions ("Mistakes were made").

8. Type your presentation in a large type—12 points or larger. Triple space, and leave wide margins on both sides of the page.

9. With a marker, draw a "break" line after the introduction, between main points in the body, and before the conclusion.

10. Underline or highlight important phrases or sentences.

11. Consider using visual aids (e.g., as slides or on an easel or whiteboard), even though these may not be published separately. Again, use visual aids that will clarify any difficult or important points.

Practicing the Presentation

1. Read each sentence aloud. Rewrite sentences that are difficult for you to say.

2. As you practice reading the presentation, look directly at an imaginary audience and speak important phrases or sentences to that audience.

3. Use overviews and topic sentences to announce each major topic as you come to it. However, to further alert your audience to the beginning of a new point, pause briefly; look at your audience; then read your overview statement or topic sentence. If possible, plan to speak these to your audience.

4. As you practice reading your presentation, continue to listen for any sentences or words that are difficult to articulate. Recast sentences and paragraphs that do not sound organized, logical, and clear. If possible, replace difficult words with others that are easier to speak.

5. Speak slowly, and enunciate clearly and distinctly.

6. When you can read each sentence with ease and without haste, time your presentation to be sure that it does not exceed a time limit if you have been given one.

7. Read your speech into a recorder. Allow a few days or hours between recording and listening so that you can gain some objectivity. As you listen, check for sentences that are hard to follow. Listen for breaks between major sections and major points.

The policy speech that illustrates Situation 1 was written to be read at an employee–management meeting and then published in the company magazine.

Situation 1: Clarifying Continuing Education Policy Enbico, Inc., a security company, retains a staff of three hundred guards who work in banks, public buildings, grocery stores—any building that requires armed security guards to protect individuals or property. Ralph Ernst, owner of Enbico, has recently established an organizational career ladder. One prerequisite for promotion is continuing education, particularly in criminology and law enforcement. Ernst wants to encourage staff to enter these programs and is offering pay increases and promotional opportunities to employees who complete a series of prescribed courses. Ernst has a general staff meeting every two weeks. Everyone is required to attend these meetings, which are scheduled between 7:00 A.M. and 8:00 A.M. Whenever Ernst announces policy changes, he always reads a prepared statement, prints it in the Enbico newsletter, published every two weeks, then solicits questions at the next general staff meeting. Here is Ernst's written policy on continuing education.

Continuing Education Policy at Enbico

Last week, the board of advisers approved a continuing education policy for all employees. The board will release complete details of the policy in three weeks; this statement announces the purpose of the policy and general guidelines.

Purpose of the Policy

The board designed this continuing education policy to achieve three goals: to strengthen the qualifications of security personnel, to prepare personnel for advancement as the company continues to grow, and to emphasize our belief that knowledge of law and law enforcement improve the quality of service Enbico provides its clients. To achieve these goals, Enbico will reimburse employees for educational expenses they incur in approved institutions and in approved areas of study.

Approved Areas of Study

Embico personnel will receive reimbursement if they enroll in the following approved areas of study:

- in programs of study leading to a certificate in criminology and law enforcement at Groveton County College;
- in degree programs leading to a B.A. or B.S. degree in criminology and law enforcement at Groveton College;
- in programs leading to completion of law enforcement and criminology courses on the master's level at Fillmore State University;
- in programs that combine the study of law enforcement and criminology with studies in accounting, general business, and mechanical technology.

The board will consider adding approved programs of study at a later date, most likely after the current program has been operating for a year.

Reimbursement for Expenses

- Employees will receive reimbursement for the following expenses: tuition, fees, and texts.
- Employees will not be reimbursed for travel expenses to and from school or for supplies other than books.

Procedures for Reimbursement

1. Employees will complete a form (now in preparation) that states the institution the employee plans to enter and his or her educational goal.

2. Upon admission to a given program, the employee will receive the letter of acceptance.

3. After registration, the employee will submit invoices from the billing office of the institution and the bookstore from which books are purchased. The employee will receive reimbursement within seven working days.

4. At the completion of the term, the employee will submit the grade report to the company. An employee who drops courses during the semester must reimburse the company for the money lost. The employee will receive no further educational reimbursement until the company has recovered all costs from coursework not completed. Unrecovered education costs will be subtracted from that employee's paycheck.

Scheduling Continuing Education Courses

The board will appoint a director to manage the continuing education program and to ensure that reimbursement is handled accurately and efficiently. The director will help employees harmonize work schedules and class hours. However, employees must remember that their first obligation is to the company. They are not to schedule classes that would interfere with their primary work assignments.

Complete guidelines and application forms will be available from Henry Stouffer in three weeks. Please be ready to ask any questions at our next staff meeting. The name of the director of continuing education will be announced at that time.

SPEAKING EFFECTIVELY: PRACTICE, PRACTICE, PRACTICE

No matter what type of presentation you are giving, your ultimate success as a speaker and the success of the presentation depend on the credibility you have been able to establish with your audience. Guidelines on planning, structuring, and delivering the presentation are important because they are designed to build your credibility with your audience. However, no amount of planning and organization will substitute for practice, which builds confidence. Practice also enhances and displays your planning and the value of your ideas and helps define the timing of the presentation.

You audience expects you to be knowledgeable, prepared, organized, and trustworthy. To meet these standards, you must use and then practice the preceding guidelines. Some examples of oral presentations can be found on the book's companion Web site (select Chapter 19).

☑ Planning and Revision Checklist

www.oup.com/us/houp

CHECKLIST

The first step in preparing for the presentation is to analyze each point you list by answering the following questions, just as you did in planning your written communication. Then you will be ready to design, structure, and organize your presentation so that it will effectively satisfy the constraints that arise from your consideration of each point.

Audience
■ Who is my audience?

■ What do I know about my audience (background, knowledge, position in the organization, attitudes toward me and my subject)?

Purpose

■ What is my purpose in giving this oral presentation?

■ Is there (should there be?) a long-range purpose?

■ What is the situation that led to this presentation?

■ Given my audience's background and attitudes, do I need to reshape my purpose to make my presentation more acceptable to my audience?

Context

■ Where and when will I be speaking?

■ What events will be transpiring in the organization (theirs or mine) that may affect how my audience perceives what I say?

Content

■ What ideas do I want to include and not include?

■ Based on the audience and the context, what difficulties do I need to anticipate in choosing content?

■ Can any ideas be misconstrued, with potential harmful consequences to me or my organization?

Graphics

■ What kinds of visual aids will I need to enhance the ideas I will present? Where should I use these in my presentation?

■ Are my graphics immediately readable and understandable?

Style

■ What kind of tone do I want to use in addressing my audience?

■ What kind of image—of myself and my organization—do I want to project?

■ What level of language do I need to use, based on my audience's background and knowledge of my subject?

■ What approach will my audience expect from me?

■ How formal should I be?

Rethinking Your Plan: Other Questions to Consider

■ What is the relationship between you and your audience?

■ What is the attitude of your audience toward you and your presentation likely to be?

■ Is your audience from a culture markedly different from yours? What adjustments to your persona and your presentation will any such difference require?

▨ What are the conditions under which you will speak?

▨ What equipment is available to you?

▨ Which delivery technique will be more appropriate: extemporaneous? manuscript?

▨ If you are speaking extemporaneously, have you prepared a speech outline to guide you?

▨ If you will be reading from a manuscript, have you introduced a conversational tone into your talk? Is your typed manuscript easy to read from?

▨ Do you have a good opening that will interest your audience and create a friendly atmosphere?

▨ Have you limited your major points to fit your allotted time?

▨ Does your talk contain sufficient examples, analogies, narratives, and data to support your generalizations? Have you repeated key points?

▨ Can you relate your subject matter to some vital interest of your audience?

▨ Which visual aids do you plan to use?

Graphs?

Tables?

Representational art?

Photographs?

Words and phrases?

Cartoons?

Hardware?

▨ Which presentation tools will you use?

Computer programs?

Overhead projection?

Slides?

Charts?

Movies and videos?

Whiteboard?

▨ Have you prepared your graphics? Do they successfully focus the listeners' attention and augment and clarify your message? Do they meet the four criteria that govern good graphics?

Visibility

Clarity

Simplicity

Controllability

▨ Do you have a good ending ready, perhaps a summary of key points or an anecdote that supports your purpose? If you began with a story, do you want to go back to it now?

▨ Have you rehearsed your talk several times?

▪ ▪ ▪ ▪ ▪ ▪ ▪ ▪ Exercises ▪ ▪ ▪ ▪ ▪ ▪ ▪ ▪

Further exercises related to the content of this chapter and all documents employed within the following exercises can be found on the book's companion Web site.

1. Examine the graphics problems at the end of Chapter 11. Give an oral presentation explaining the results of your analysis to your class. Use either PowerPoint or transparencies to present the graphs that support your recommendations. Write a version of your report to be published in the company newsletter.

2. Adapt Situation 2 to fit your own college. Assume that your dean has asked you to evaluate your experience thus far. Give a five- to eight-minute presentation. Provide an outline one or two pages long.

Situation 2: Describing Problems to the Board of Regents

Assume that you are an active member of a campus student organization or professional society. Each student organization is asked to write a letter to the student government to describe what that organization considers to be the biggest problems students face.

The student government committee has invited you and representatives from two other student organizations to present your description to the university's board of regents, which has expressed interest in hearing the perspective of several student groups. The board may take action to correct the problems it learns of.

3. Develop your response to Situation 2. Assume you are giving a ten-minute presentation to the board of regents of your college. Provide a one- or two-page outline.

4. Modify Situation 3 by selecting another country.

Situation 3: Speaking to a Group of Students from Argentina

Your university has a study-abroad program in which you have enrolled for a summer. As one of your assignments, you will give a presentation to students in Buenos Aires who are interested in applying for graduate school at your university. These students speak English fairly well, but, you discover, they have never been outside Argentina. Plan a presentation in which you attempt to interest them in coming to your university.

5. If you have been assigned a semester project, prepare a ten-minute oral version of your written project proposal. Then, be prepared to give a five-minute status report to your classmates.

6. If you prepared a formal report defining, analyzing, and defining your field, prepare an oral presentation of this project. Prepare a handout summarizing your findings, and use it as a guide during your presentation. Or, develop a PowerPoint presentation to accompany one of the first five exercises.

7. Select an article from a journal in your field. Prepare an oral presentation summarizing this article for the class.

8. Prepare an oral briefing for freshman entering your college. Explain one of the following:

· Your university's rules on academic integrity
· Tips on how to survive the first year of college
· One or more of the traditions of your university

Video and audio files of oral presentation examples can be found on the companion Web site.

www.oup.com/us/houp

Scenario

You and Ms. Mansouri, the counselor at your college's career center, are discussing your upcoming job search. "Have you looked at the CareerOneStop Web site?" she asks. "It's a great site to start at, with information about industries, companies, job opportunities, and wages." She jots down the organization's URL, and you begin to investigate.

In a preliminary exploration of the site, you find a link for posting your résumé online. You also choose the link to the Career Resource Library and find links to information about specific occupations, such as the fastest growing and the highest paying. You also find links to detailed descriptions of potential employers and to lists of job openings.

Following a link to Miscellaneous Opportunities and to Nonprofit Opportunities, you find a link to OpportunityKnocks.org, a site identifying available jobs at nonprofit organizations. Searching by state, you find a job in California that sounds especially interesting: public relations and marketing specialist for the Asian Art Museum of San Francisco. The job is full time, and it offers a good starting wage and exceptional benefits, including medical insurance, a generous holiday and vacation policy, and a retirement program. The job description stipulates "Excellent oral and written communication skills; outstanding administrative and organizational skills; high level of research and analytical skills"—all of which you believe you have, thanks especially to your technical writing class. You will definitely apply for this job.

CHAPTER TWENTY

Understanding the Strategies and Communications of the Job Search

As a college student looking for a good job after graduation, you will likely experience both excitement and disappointment. It's exciting to imagine yourself working for different companies and living your life in different cities. If you don't immediately get your dream job, however, you will be disappointed. The search process is often a roller coaster of emotional highs and lows. You will need to keep in mind that a variety of factors entirely aside from your credentials go into every hiring decision. To help you in this emotionally exhausting environment, we cover in this chapter the major steps of the job search: preparation, letters of application, résumés, and interviews.

PREPARATION

The job search is itself a job. If you are still in college, it has to be at least a part-time job (most professional jobs occupy 40 to 50 hours a week). Job searching also requires that you schedule your time around various activities and that you dedicate at least 15 to 20 hours a week to the search. You start by equipping yourself for the search through self-assessment, gathering information about possible jobs, and networking.

Self-Assessment

The goal of self-assessment is twofold. First and most important, you want to avoid pounding a square peg (you) into a round hole (the wrong job). You want to determine what jobs among those that are available would suit you the best: What kind of work can you do well? What kind of work satisfies you? Second, in the job search, you'll be creating résumés, completing applications, and answering interview questions. Self-assessment will ensure that you list details that you will need about past work and educational experiences—such as dates, names, and job responsibilities.

In your self-assessment you should ask questions such as the following:

- What are my strengths?
- What are my weaknesses?
- How well have I performed in past jobs?
- Have I shown initiative?
- Have I improved procedures?
- Have I accepted responsibility?
- Have I demonstrated leadership?
- Have I been promoted or been given a merit raise?
- How can I present myself most attractively?
- What skills do I possess that relate directly to what the employer seems to need?
- How and where have I obtained those skills?

To go about your self-assessment in a serious and systematic way, use the questionnaire provided in Figure 20-1.

Work Experience (Use a sheet like this for each position you have held, including military service.)

Company: _____

Address: _____

Supervisor's Name and Title: _____

Dates of Employment: _____

Position(s)/Title(s)/Military Rank: _____

Duties and Responsibilities: _____

_____

_____

Accomplishments (including awards or commendations): _____

_____

_____

Skills, Knowledge, and Abilities Used: _____

_____

_____

Duties Liked and Disliked: _____

_____

_____

Education and Training

School, College, or University	Dates of Enrollment	Degree or Major Certificate	Date	GPA

Career-Related Courses: _____

_____

_____

Scholastic Honors, Awards, and Scholarships: _____

_____

_____

College Extracurricular Activities: _____

_____

_____

FIGURE 20-1 · SELF-ASSESSMENT QUESTIONNAIRE

Source: U.S. Department of Labor. *Job Search Guide: Strategies for Professionals.* Washington, DC: Government Printing Office, 1993.

Other Training (include courses sponsored by the military, employers, or professional associations, etc.): _____

_____

_____

Courses, Activities Liked and Disliked: _____

_____

_____

Skills, Knowledge, and Abilities Learned: _____

_____

_____

Professional Licenses: _____

_____

_____

Personal Characteristics (e.g., organizational ability, study habits, social skills, like to work alone or on a team, like or dislike public speaking, detail work): _____

_____

_____

Personal Activities
Professional (association memberships, positions held, committees served on activities, honors, publications, patents, etc.): _____

_____

_____

Community (civic, cultural, religious, political organization memberships, offices or positions held, activities, etc.): _____

_____

_____

Other (hobbies, recreational activities, and other personal abilities and accomplishments):

_____

_____

Overall Assessment
Tale a look at all the work sheets you have completed: Work Experience, Education, and Personal Activities. Considering all you have done, list your strengths and positive attributes in each of the areas below.
Skills, Knowledge, and Abilities: _____

_____

_____

Accomplishments: _____

_____

_____

FIGURE 20-1 · *CONTINUED*

Personal Characteristics: _____
_____
_____

Activities Performed Well: _____
_____
_____

Activities Liked: _____
_____
_____

FIGURE 20-1 · *CONTINUED*

When you have completed your self-assessment, you will have a good record of past job and educational experiences. You should have a good idea of what skills you have and what it is you really like to do. If you are uncertain about how your skills and experience match to jobs available, two publications from the U.S. Department of Labor can help you make the match:

- *Career Guide to Industries.* Describes over forty industries, including available occupations, working conditions, training and promotion opportunities, wages, and benefits. Available online at http://stats.bls.gov/oco/cg/home.htm.

- *Occupational Outlook Handbook.* Describes the educational requirements, duties, job prospects, and earnings for a wide variety of occupations. Avaliable online at http://stats.bls.gov/oco/home.htm.

What if, at the end of your self-assessment, you either can't decide what you want to do or decide that your college major is in a field that no longer interests or satisfies you? In either case you might seek professional job counseling. School or college career centers often offer such individualized assistance.

If you are fairly certain of your career direction, your next step is to find the jobs that will lead you in that direction.

Information Gathering

The easiest and quickest way to find information about companies and organizations with available jobs is to go online. The CareerOneStop Web site (www.careeronestop.org) offers information on every issue regarding the job search, either directly or through links to related sites. Its career resource library, for example, includes:

Occupational information: career and industry guides for exploring specific occupations

Job search aids: research on employers and tips for application letters, résumés, and interviewing

Job and résumé banks: lists and clearinghouses for a wide variety of jobs and resources for posting your résumé online

Relocation information: city and state guides, cost-of-living calculators, moving estimates

CareerOneStop also includes America's Job Bank, which allows you to search for a job by industry, location, or keyword as well as post a résumé, compose a cover letter to accompany your résumé, or create a search scout to search job lists automatically and e-mail you the results.

The U.S. Government Office of Personnel Management, at www.usajobs. opm.gov, has information for those interested in a job with the federal government. You can find here a list of jobs at all levels for which the government is taking applications.

MonsterTrak, at www.monstertrak.com, is designed specifically to assist college students and new graduates to locate full-time and part-time jobs and internships. MonsterTrak has an excellent career index that describes career employment outlooks and provides salary ranges, as well as cost-of-living calculators for different cities. MonsterTrak also offers good job-searching tips, including instruction on designing scannable electronic résumés.

Similar resources for finding jobs at all levels of experience are available at such sites as

EmploymentGuide.com	www.cweb.com
Hot Jobs	hotjobs.yahoo.com
JobFind	www.jobfind.com
Monster	www.monster.com
Vault	www.vault.com

CareerMag.com, at www.careermag.com, also offers job lists, a venue for posting résumés online, and extensive information resources, including free subscriptions to a wide variety of industry magazines.

The career center at your college will likely have additional materials, especially regarding companies and organizations that recruit at your school.

To supplement such sources, continuously check magazines and newspapers that carry up-to-date business news, such as *Forbes, Fortune, Business Week,* and the *Wall Street Journal,* as well as journals specific to your field.

Your information-gathering activities are crucial to the success of your job search. To find the right job, you must know as much as possible about the companies to which you apply, the jobs typical of your industry, and the usual earnings and working conditions. With such information, you will be equipped to make informed and careful decisions about your future employment.

Networking

Networking is a way of finding both job information and job opportunities. Networking starts with broadcasting the news that you are looking for a job. Tell your family and friends, including grandparents, aunts, uncles, cousins, in-laws, and neighbors. Mention your search to professors, clergy, favorite high school teachers, and former employers. If possible, join the professional associations in your field. If there are local chapters of such organizations, attend their meetings and introduce yourself to the chapter leaders. Often such associations have special membership rates and services for students.

Contact local businesses or organizations that have jobs you might be interested in and ask for the names of people who have or manage jobs. Contact these people and ask if you can come by for a visit. Make it clear that you are not suggesting a job interview but rather an informational discussion—to seek advice about looking for work or how to prepare for a certain kind of job. If you have a name you can use in making the initial contact, such as a relative or teacher, strangers may be more receptive to your request. But don't let not having a name deter you. Try to make the contact anyway. A lot of people simply won't have time to meet, but you may be surprised at how many people will be pleased to visit with you. People like to give advice in general, and in particular, they like to talk about their occupations. Furthermore, organizations are on the lookout for enterprising self-starters. In calling on an organization, you are showing yourself to be such a person.

In your contacts with local businesses and professional organizations, you are the interviewer. To help people remember you, have business cards made with your name and contact information, including your physical address, e-mail address, telephone number, and fax number. If you have a profession, such as computer programmer, list it on the card under your name. To help yourself remember the people you have seen and everything you have learned, keep good records as you proceed. Write down names (accurately spelled), addresses, and telephone numbers. Record what seems to be important to your contacts about work in general and the specific work that interests you.

All this may strike you as a little irregular, and it is. But research shows that most people find work in precisely this way because jobs are often filled, not through a methodical search, but through networking, as just described. Also, the majority of all jobs are in companies with fewer than twenty-five employees. Small companies often don't publish job notices or send recruiters to college campuses. You have to find such jobs yourself, and networking is an effective way to do it.

For a more systematic approach to networking, contact organizations dedicated to helping people find jobs. If you are a college student, your college career center will be the leading site for such networking. Many large firms regularly call on college career centers when seeking new employees. Also, the career center schedules campus interviews for graduating students, and many offices have job fairs once or twice a year, to which companies send representatives. Job fairs are good places to gather information about potential employers and to meet with recruiters. Regional and national conferences of professional associations also offer good networking opportunities: companies and organizations active in the industry will often use such conferences for recruiting and interviewing of job candidates.

We cannot overemphasize that finding a job is itself a job. Information and advice are available, but you can't be timid or passive. You have to search actively and assertively for support and opportunities.

THE CORRESPONDENCE OF THE JOB SEARCH

In some cases, prospective employers will first hear of you from the letter of application and résumé that you send to them. A persuasive letter of application, sometimes called a *cover letter,* and a clear, well-organized résumé won't guarantee that you get a job, but ineffective ones will usually guarantee that you don't. In this section we describe how to prepare the letter of application and both paper and electronic résumés. We also tell you about several follow-up letters you will need to write during the job search.

Letter of Application

Plan the mechanics of your letter of application carefully. Buy the best quality white bond paper. This is no time to skimp. Use a standard typeface. Don't use italics or bold. Make sure your letter is mechanically perfect, free of errors in grammar and spelling. Be brief, but not telegraphic. Keep the letter to a single page unless you have extensive pertinent experience to emphasize and explain. Don't send a letter that has been duplicated: each copy of your letter must be individually addressed, printed, and signed. Accompany each letter with a résumé. We discuss résumés later.

Pay attention to the style of the letter and the résumé that accompanies it. The tone you want in your letter is one of self-confidence. You must avoid both arrogance and timidity. You must sound interested and eager, but not obsequious or desperate. Don't give the impression that you *must* have the job, but, on the other hand, don't seem indifferent about getting it.

When describing your accomplishments in the letter and résumé, use action verbs. They help to give your writing specificity, and, by their brevity, they make you seem dynamic and energetic. For example, don't just say that you worked as a sales clerk. Rather, tell how you maintained inventories, promoted merchandise, prepared displays, implemented new procedures, and supervised and trained new clerks. Here's a sampling of action words and phrases:

administer	design	exhibit
analyze	develop	expand
conduct	direct	improve
create	edit	manage
cut	evaluate	operate

organize	produce	support
oversee	reduce costs	was promoted
plan	reorganize	write

You cannot avoid the use of *I* in a letter of application. But take the you-attitude as much as possible. Emphasize what you can do for the prospective employer—how your getting this job will benefit the company, how your experiences and abilities can assist the company in achieving its objectives. The letter of application is not the place to be worried about salary and benefits. Above all, be mature and dignified. Avoid tricky and flashy approaches. Write a well-organized, informative letter that highlights skills your analysis of the company shows that it most desires in its employees. The following discussion of the beginning, the body, and the ending of an application letter illustrates these general pointers.

The Beginning Start by explaining that you are applying for a job, and identify the specific job you want. Don't be aggressive or inventive here. A beginning such as "WANTED: An alert, aggressive employer who will recognize an alert, aggressive young programmer" will usually direct your letter to the reject pile. A bit of name-dropping is a good beginning, but only if you have permission and if the name you drop will mean something to the prospective employer. If you qualify on both counts, begin with an opening sentence such as this:

Dear Ms. Marchand:

Professor John J. Jones of Missouri State University's Food Science faculty has suggested that I apply for the post of food supervisor that you have open. In June I will receive my Bachelor of Science degree in Food Science from MSU. I have also spent the last two summers working in food preparation for Memorial Hospital in Kirkwood.

Remember that you are trying to arouse immediate interest about yourself in the potential employer. Another way to do this is to refer to something about the company that interests you. Doing so establishes that you have done your homework. Then try to show how some preparation on your part relates to this special interest. See Figure 20-2 for an example of such an opener.

Sometimes the best approach is a simple statement about the job you seek, accompanied by a description of something in your experience that fits you for the job, as in this example:

Your opening for a food supervisor has come to my attention. In June of this year, I will graduate from Missouri State University with a bachelor of science degree in food science. I have spent the last two summers working in food preparation for Memorial Hospital in Kirkwood. I believe that both my education and my work experience qualify me to be a food supervisor on your staff.

Be specific about the job you want. Although if this particular job is not open, a firm may offer you an alternative one, employers are not impressed

635 Shuflin Road
Watertown, CA 90233
March 23, 2005

Mr. Morell R. Solem
Director of Research
Price Industries, Inc.
2163 Airport Drive
St. Louis, MO 63136

Dear. Mr. Solem:

I read in the January issue of *Metal Age* that Dr. Charles E. Gore of your company is conducting extensive research into the application of X-ray diffraction to problems in physical metallurgy. I have conducted experiments at Watertown Polytechnic Institute in the same area under the guidance of Professor John J. O'Brien. I would like to become a research assistant with your firm and, if possible, to work for Dr. Gore.

In June, I will graduate from WPI with a bachelor of science degree in Metallurgical Engineering. At present, I am in the upper 25 percent of my class. In addition to my work with Professor O'Brien, I have taken as many courses relating to metal inspection problems as I could.

For the past two summers, I have worked for Watertown Concrete Test Services, where I have qualified as a laboratory technician for hardened concrete testing. I know how to find and apply the specifications of the American Society for Testing and Materials. This experience has taught me a good deal about modern inspection techniques. Because this practical experience supplements the theory learned at school, I believe I could fit into a research laboratory with a minimum of training.

You will find more detailed information about my education and work experience in the résumé enclosed with this letter. I can supply job descriptions concerning past employment and the report of my X-ray diffraction research.

In April, I will attend the annual meeting of the American Institute of Mining, Metallurgical, and Petroleum Engineers in Detroit. Would it be possible for me to talk with some member of Price Industries at that time?

Sincerely yours,

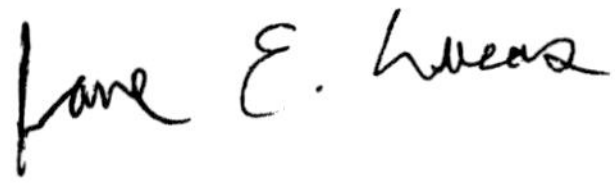

Jane E. Lucas

Jane E. Lucas

Enclosure

Margin annotations:

Opener showing knowledge of company

Specific job mentioned

Highlights of education

Highlights of work experience

Reference to résumé

Request for interview

FIGURE 20-2 · LETTER OF APPLICATION

with general statements such as, "I'm willing and able to work at any job you may have open in research, production, or sales." Instead of indicating flexibility, such a claim usually implies that your skills and interests are unfocused—that you would be adequate at several things but truly exceptional at nothing.

The Body In the body of your letter you highlight selected items from your education and experience that show your qualifications for the job you seek. Remember that you are trying to show the employer how well you would fit into the job and the organization.

In selecting your items, it pays to know what employers value the most. In evaluating recent college graduates, employers look closely at the major, academic performance, work experience, awards and honors, and extracurricular activities. They also consider recommendations, standardized test scores, military experience, and community service.

Try to include information from the areas that employers value the most, but emphasize those areas in which you are noteworthy. If your grades are good, mention them prominently. If you are in the lowest quarter of your class, maintain a discreet silence. Speak to the employer's interests, and at the same time highlight your own accomplishments. Show how it would be to the employer's advantage to hire you. The following paragraph, an excellent example of the you-attitude in action, does all these things:

I understand that the research team of which I might be a part works as a single unit in the measurement and collection of data. Because of this, team members need a general knowledge and ability in fishery techniques as well as a specific job skill. Therefore, I would like to point out that last summer I worked for the Department of Natural Resources on a fish population study. On that job I gained electro-fishing and seining experience and also learned how to collect and identify aquatic invertebrates.

Be specific about your accomplishments: otherwise, it will sound like bragging. It is much better to write, "I was president of my senior class" instead of "I am a natural leader." "I was president" is a piece of evidence; "I am a natural leader" is a conclusion. Give employers the evidence that will lead them to the right conclusions about you.

One tip about job experience: the best experience relates to the job you seek, but recent college graduates should mention any job experience, even if it does not relate to the work desired. Employers believe that a student who has worked is more likely to be mature than one who has not. You may also have transferable skills that would be useful in another job. If you have worked at a job, you know what it is like to:

- Come in and leave at a certain time
- Take directions from a boss
- Do a job that you are assigned to do
- Have effective work habits and a real work ethic

If you have worked at a job, it indicates that another employer once judged you worthy of hiring.

Don't forget hobbies that relate to the job. You're trying to establish that you are interested in, as well as qualified for, the job.

Don't mention salary unless you're answering an advertisement that specifically requests this information. Keep the you-attitude. Don't worry about insurance benefits, vacations, and coffee breaks at this stage of the game. Keep the prospective employer's interests in the foreground. Your self-interest is taken for granted.

If you are already working and not a student, construct the body of your letter of application much as we've described. The significant difference is that you will emphasize your work experience more than your college grades and activities. Identify your responsibilities and achievements on the job. Do not complain about your present employer. Such remarks will lead the prospective employer to mistrust you.

In the last paragraph of the body of the letter, refer the employer to your enclosed résumé. Mention your willingness to supply additional information such as references, letters concerning your work, research reports, and college transcripts.

The Ending The goal of the letter of application is to get you an interview with the prospective employer. In your final paragraph, you request this interview. Your request should be neither apologetic nor aggressive. Simply indicate that you are available for an interview at the employer's convenience, and give any special instructions needed for reaching you. If the prospective employer is in a distant city, indicate (if possible) a time and location that you might conveniently meet with a representative of the company, such as the upcoming convention of a professional association. If the employer is really interested, you may be invited to visit the company at its expense.

The Complete Letter The beginning of the complete letter of application in Figure 20-2 shows that the writer is interested enough in the company to investigate it. The desired job is specifically mentioned. The middle portion highlights the writer's course work and work experience that relate directly to the job she is seeking. The close makes an interview convenient for the director of research to arrange.

Keep in mind that a personnel officer skims your letter and résumé in about thirty seconds. If you have not grabbed his or her interest in that time, you are probably finished with that organization.

The Résumé

A résumé provides prospective employers with a convenient summary of your education and experience. As in the letter of application, good grammar, correct spelling, neatness, and brevity are of major importance. Although the traditional paper résumé is still commonplace, more and more organizations are using electronic media for screening their job candidates. Therefore, it's important to have versions of your résumé available in the following formats:

- Traditional paper
- Scannable format
- E-mail format
- Web format

We discuss content and organization first because résumés of different kinds are similar in those respects regardless of format.

Content and Organization The three most widely used résumés are chronological, functional, and targeted résumés. All have advantages and disadvantages, and we begin with these; brief discussions of the paper, scannable, and online modes of presentation follow.

Chronological Résumés A chronological résumé (Figure 20-3) is traditional and acceptable. If your education and experience show a steady progression toward the career you seek, the chronological résumé portrays that progression well. Its major disadvantage is that your special capabilities or accomplishments may get lost in the chronological detail. Also, any holes in your employment or educational history will show up clearly.

Put your address at the top. Include your e-mail address and your telephone number (including the area code). If you have a fax number, include it also.

For most students, educational information should be placed before work experience. People with extensive work experience, however, may choose to position this information most prominently.

List the colleges or universities you have attended in reverse chronological order—in other words, list the school you attended most recently first; the one before, second; and so on. Do not list your high school. Give your major and date, or expected date, of graduation. Do not list courses, but list anything that is out of the ordinary, such as honors, special projects, and emphases in addition to the major. Extracurricular activities also go here.

As with your educational experience, put your work experience in reverse chronological order. To save space and to avoid the repetition of *I* throughout the résumé, use phrases rather than complete sentences. Emphasize the experiences that show you in the best light for the kinds of job you seek. Use nouns and active verbs in your descriptions. Do not neglect less important jobs of the sort you may have had in high school, but summarize them even more briefly. You would probably put college internships and work–study programs here, though you might choose to put them under education. If you have military experience, put it here. Give the highest rank you held, list service schools you attended, and describe your duties. Make a special effort to show how your military experience relates to the civilian work you seek.

RÉSUMÉ OF JANE E. LUCAS

635 Shuflin Road
Watertown, California 90233
(213) 596-4236
jlucas@bga.com

<table>
<tr><td colspan="2">Education</td></tr>
<tr><td>2003–2005</td><td>Watertown Polytechnic Institute
Watertown, California</td></tr>
</table>

Education

2003–2005 Watertown Polytechnic Institute
Watertown, California

Candidate for Bachelor of Science degree in Metallurgical Engineering in June 2005. In upper 25% of class with GPA of 3.2 on 4.0 scale. Have been yearbook photographer for two years. Member of Outing Club, elected president in senior year. Elected to Student Intermediary Board in senior year. Oversaw promoting and allocating funds for student activities. Wrote a report on peer advising that resulted in a change in college policy. Earned 75% of college expenses.

2001–2003 San Diego Community College
San Diego, California

Received associate of arts degree in general studies in June 2003. Made dean's list three of four semesters. Member of debate team. Participated in dramatics and intramural athletics.

Business Experience

2003–2004 Summers Watertown Concrete Test Services
Watertown, California

Qualified as laboratory technician for hardened concrete testing under specification E329 of American Society for Testing and Materials (ASTM). Conducted following ASTM tests: Load Test in Core Samples (ASTM C39), Penetration Probe (ASTM C803-75T), and the Transverse Resonant Frequency Determination (ASTM C666-73). Implemented new reporting system for laboratory results.

2003–2005 Watertown Ice Skating Arena
Watertown, California

During academic year, work 15 hours a week as ice monitor. Supervise skating and administer first aid.

2001–2003 Summer and part-time jobs included newspaper carrier, supermarket stock clerk, and salesperson for large department store.

Personal Background Grew up in San Diego, California. Travels include Mexico and the eastern United States. Can converse in Spanish. Interests include reading, backpacking, photography, and sports (tennis, skiing, and running). Willing to relocate.

References Personal references available upon request.

February 2005

Fragmentary sentences used

Highlights of education

Special activities

Honors and activities

Special work skills

Summary of early employment

FIGURE 20-3 · CHRONOLOGICAL RÉSUMÉ

You may wish to provide personal information. This can be a subtle way to point out your desirable qualities. Recent trips indicate a broadening of knowledge and probably a willingness to travel. Hobbies listed may relate to the work sought. Participation in sports, drama, or community activities indicates that you enjoy working with people. Cultural activities indicate you are not a person of narrow interests. If you are proficient in a language other than English, mention this also: more and more organizations are looking for employees with the potential to support international business activity.

If you choose to indicate that you are married or have a family, you might also emphasize that you are willing to relocate. Don't discuss your health unless it is to describe it as excellent.

You have a choice with references. You can list several references with addresses and phone numbers or simply say "References available upon request." Both methods have an advantage and a disadvantage. If you provide references, a potential employer can contact them immediately, but you have used up precious space that might be better allotted to more information about yourself. Conversely, if you don't provide the reference information, you save the space but put an additional step between potential employers and information they may want. It's a judgment call, but, on balance, we favor saving space by omitting the reference information. Your first goal is to interest the potential employer in you. If that happens, there is still plenty of time later to provide the reference information.

In any case, do have at least three references available. Choose from among your college teachers and past employers—people who know you well and will say positive things about you. Get their permission, of course. Also, it's a smart idea to send them a copy of your résumé. If you can't call on them personally, send them a letter that requests permission to use them as a reference, reminds them of their association with you, and sets a time for their reply.

Dear Ms. Zamora:

In June of this year, I'll graduate from Watertown Polytechnic Institute with a B.S. in metallurgical engineering. I'm getting ready to look for work. If you believe that you know enough about my abilities to give me a good recommendation, I would like your permission to use you as a reference.

As a reminder, during the summers of 2004 and 2005, I worked as a laboratory technician in your testing lab at Watertown. These were highly instructive summers for me, and I qualified, with your help, to carry out several ASTM tests.

I want to start sending my résumé out to some companies by mid-March and would appreciate having your reply by that time. I enclose a copy of my résumé for you, so that you can see in detail what I've been doing.

Thanks for all your help in the past.

Best regards,

At the bottom of the traditional paper résumé, place a dateline—the month and year in which you completed the résumé. Place the date in the heading of scannable and e-mail résumés.

Functional Résumés The functional résumé (Figure 20-4) allows you to highlight the experiences that show you to your best advantage. Extracurricular experiences are displayed to particular advantage in a functional résumé. The major disadvantage of this format is the difficulty, for the first-time reader, of discerning a steady progression of work and education.

The address portion of the functional résumé is the same as that of the chronological. After the address, you may include a job objective line if you like. A job objective entry specifies the kind of work you want to do and sometimes the industry or service area in which you want to do it, like this:

Work in food service management.

or like this:

Work in food service management in a metropolitan hospital.

If you use a job objective entry in a functional résumé, place it immediately after the address and align it with the rest of the entries (as shown later in Figure 20-5).

For education, simply give the institution from which you received your degree, your major, your degree, and your date of graduation. The body of the résumé is essentially a classification. You sort your experiences—educational, business, extracurricular—into categories that reveal capabilities related to the jobs you seek. Remember that in addition to professional skills, employers want good communication skills and good interpersonal skills. Possible categories are *technical, professional, team building, communication, research, sales, production, administration,* and *consulting.*

The best way to prepare a functional résumé is to brainstorm it. Begin by listing some categories that you think might display your experiences well. Brainstorm further by listing your experiences in those categories. When you have good listings, select the categories and experiences that show you in the best light. Remember, you don't have to display everything you've ever done, just the activities that might strike a potential employer as valuable. Finish off the functional résumé with a brief reverse chronological work history and a dateline, as in the chronological résumé.

Targeted Résumés The main advantage of the targeted résumé (Figure 20-5) is also its main disadvantage: you zero in on one goal. If you can achieve that goal, fine—but the narrowness of the approach may block you from being considered for other possibilities. The targeted résumé displays your capabilities and achievements well, but, like the functional résumé, it's a format that is sometimes difficult for potential employers to interpret quickly.

The address and education portions of the targeted résumé are the same as those in the functional résumé. The whole point of a targeted résumé is

RÉSUMÉ OF JANE E. LUCAS

635 Shuflin Road
Watertown, California 90233
(213) 596-4236
jlucas@bga.com

Education	Candidate for degree in metallurgical engineering from Watertown Polytechnic Institute in June 2005.
Technical	· Qualified as laboratory technician for hardened concrete testing under specification E329 of American Society for Testing and Materials (ASTM). · Conducted following ASTM tests: Load Test in Core Samples (ASTM C39), Penetration Probe (ASTM C803-75T), and the Transverse Resonant Frequency Determination (ASTM C666-73). · Will graduate in upper 25 percent of class with a GPA of 3.2 on a 4.0 scale.
People	· Elected president of Outing Club. · Elected to Student Intermediary Board. · Oversaw promoting and allocating funds for student activities. · Participated in dramatics and intramural athletics.
Communication	· Wrote a report on peer advising that resulted in a change in institute policy. · Participated in intercollegiate debate. · Worked two years as yearbook photographer. · Completed courses in advanced speaking, small group discussion, and technical writing.

Work Experience

2003–2004 Summers	· Watertown Concrete Test Services, Watertown, California: laboratory technician.
2003–2005	· Watertown Ice Skating Arena, Watertown, California: ice monitor.
2001–2003	· Summer and part-time jobs included newspaper carrier, supermarket stock clerk, and salesperson for large department store. · Earned 75 percent of college expenses.
References	References available upon request.

February 2005

Academic, work, and extracurricular activities categorized by capabilities

Summary of work experience in reverse chronological order

FIGURE 20-4 · FUNCTIONAL RÉSUMÉ

RÉSUMÉ OF JANE E. LUCAS

635 Shuflin Road
Watertown, California 90233
(213) 596-4236
jlucas@bga.com

Job Objective	Research assistant in a testing or research laboratory.
Education	Candidate for degree in metallurgical engineering from Watertown Polytechnic Institute in June 2005.

Capabilities
- Find and apply specifications of the American Society for Testing and Materials (ASTM).
- Conduct X-ray diffraction tests.
- Work individually or as a team member in a laboratory setting.
- Take responsibility and think about a task in terms of objectives and time to complete.
- Report research results in both written and oral form.
- Communicate persuasively with nontechnical audiences.

Achievements
- Will graduate in upper 25 percent of class with a GPA of 3.2 on a 4.0 scale.
- Earned 75 percent of college expenses.
- Qualified as laboratory technician for hardened concrete testing under specification E329 of ASTM.
- Conducted major ASTM tests and implemented new reporting system for laboratory results.
- Completed courses in advanced speaking, small group discussion, and technical writing.
- Wrote a report on peer advising that resulted in a change in institute policy.

Work experience

2003–2004 Summers	• Watertown Concrete Test Services, Watertown, California: laboratory technician.
2003–2005	• Watertown Ice Skating Arena, Watertown, California: ice monitor.
2001–2003	• Summer and part-time jobs included newspaper carrier, supermarket stock clerk, and salesperson for large department store.
References	References available upon request.

February 2005

FIGURE 20-5 · TARGETED RÉSUMÉ

to focus on a specific job objective. Therefore, you express your job objective very precisely.

Next list your capabilities that match the job objective. Obviously, you have to know quite a bit about the job you are seeking to make the proper match. To be credible, the capabilities you list must be supported by achievements or accomplishments, which are given next in your résumé. You finish off the targeted résumé with a reverse chronological work history and a dateline, as in the functional résumé.

As with the functional résumé, brainstorming is a good way to discover the material you need for your targeted résumé. Under the headings Capabilities and Achievements, make as many statements about yourself as you can. When you are finished, select and use only the statements that best relate to the job you are seeking.

Paper Résumés As Figures 20-3, 20-4, and 20-5 illustrate, in a paper résumé, you use variations in type and spacing to emphasize and organize information. Make the résumé look easy to read—leave generous margins and white space. Use distinctive headings and subheadings. The use of a two-column spread is common, as is the use of boldface in headings. You might use 12-point sans serif type for headings and 10-point serif type for the text (see Chapter 11). Be careful, though, not to overdo the typographical variation, which can quickly change a sensible arrangement that showcases your qualifications into a chaotic mix of competing decorations that pulls the reader's attention away from your abilities.

If you are printing your résumé, use a high-quality bond paper in a standard color such as white or off-white. Ideally, your letter and résumé are on matching paper. If you are attaching the résumé as an electronic file to an e-mail message, make sure you submit it in one or more widely used formats (e.g., as a .doc file, .pdf file, or .rtf file) for easy access by the recipient. Make the file read-only so that nothing can be later inserted or deleted.

Who is the recipient for your letters and résumés? When answering an advertisement, you should follow the instructions given. When operating on your own, send the materials, if at all possible, to the person in the organization for whom you would be working—that is, the person who directly supervises the position. This person normally has the power to hire for the position. Your research into the company may turn up the name you need. If not, don't hesitate to call the company switchboard and ask directly for a name and title. If need be, write to human resources directors. Whatever you do, write to *a specific individual* by name. Don't send "To Whom It May Concern" letters on your job search. It's wasted effort.

Sometimes, of course, you may gain an interview without having submitted a letter of application—for example, when recruiters come to your campus. Bring a résumé with you and hand it to the interviewer immediately. The résumé gives the interviewer a starting point for questions and often helps to focus the interview on your qualifications for the job.

Scannable Résumés A scannable résumé is a paper résumé that has been modified so that it can be electronically scanned (see Figure 20-6). Many

JANE E. LUCAS
635 Shuflin Road
Watertown, CA 90233
(213) 596-4236
jlucas@bga.com
February 2005

JOB OBJECTIVE
Research assistant in a testing or research laboratory

EDUCATION
Watertown Polytechnic Institute, Watertown, CA
B.S., metallurgical engineering, June 2005
GPA: 3.2 on 4.0 scale

TECHNICAL
Qualified as laboratory technician for hardened concrete testing under specification E329 of American Society for Testing and Materials (ASTM).
Conducted following ASTM tests: Load Test in Core Samples (ASTM C39), Penetration Probe (ASTM C803-75T), and the Transverse Resonant Frequency Determination (ASTM C666-73).

PEOPLE
Elected President of Outing Club.
Elected to Student Intermediary Board.
Oversaw promoting and allocating funds for student activities.
Participated in dramatics and intramural athletics.

COMMUNICATION
Wrote a report on peer advising that resulted in a change in Institute policy.
Participated in intercollegiate debate.
Worked two years as yearbook photographer.
Completed courses in advanced speaking, small group discussion, and technical writing.

WORK EXPERIENCE
Watertown Concrete Test Services, Watertown, CA; laboratory technician; summers, 2003, 2004
Watertown Ice Skating Arena, Watertown, CA; ice monitor; 2003–2005.
Summer and part-time jobs included newspaper carrier, supermarket clerk, and salesperson for large department store; 2001–2003.
Earned 75 percent of college expenses.

REFERENCES
References available upon request.

FIGURE 20-6 · FUNCTIONAL RÉSUMÉ IN SCANNABLE FORMAT

organizations now scan the paper résumés they receive and enter the information into a special database for quick retrieval by keywords. For example, if the company needs an environmental specialist, the hiring manager scours the database for words such as *environment* and *ecology*. Only job candidates with the keywords on their résumé will be considered. In a scannable résumé, therefore, you must make sure that the keywords of your occupation, often nouns, are present in abundance. You must get through the computer to be considered by a human reader.

For your résumé to be scannable, you must also modify its format. Ordinarily, a scanner reads résumés from left to right and often makes mistakes if it encounters such features as italics, underlining, changes of typeface, and small sizes of type. In designing your scannable résumé, therefore, adopt the following guidelines:

▨ Display all information in a single column.
▨ Align all information on the left margin.
▨ Use spaces instead of tabs to separate headings from text, or place headings on a separate line.
▨ Use a single typeface.
▨ Use all capital letters for your name and major headings.
▨ Use 12-point type.
▨ Do not use italics and underlining.
▨ Do not use rules and borders.
▨ Submit a clean and crisp copy.
▨ Do not fold or staple the résumé.

Online Résumés As discussed earlier, the Internet offers a variety of job-finding sites that allow you to post résumés and cover letters. You submit specific biographical information electronically, usually by filling in the fields on a series of pages at the site. The fields solicit the details of your address, credentials, and objectives. This information is automatically compiled in a standardized résumé format. Several services give you a choice of formats. (If you have a choice, match the online format to that of your paper résumé for a consistent appearance in case you later choose to carry a newly updated paper copy to your job interview.) The service then adds your résumé to its international database of available employees, thus making your résumé readily accessible through the Internet to prospective employers all over the world.

Follow-up Letters

Write *follow-up letters:*

▨ If after approximately two weeks you have received no answer to your letter of application
▨ After an interview
▨ If a company fails to offer you a job
▨ To accept or refuse a job

No Answer If a company doesn't acknowledge receipt of your original letter of application within two weeks, write again with a gracious inquiry such as the following:

Dear Mr. Petrosian:

On 12 April I applied for a position with your company. I have not heard from you, so perhaps my original letter and résumé have been misplaced. I enclose copies of them.

If you have already reached a decision regarding my application or if there has been a delay in filling the position, please let me know at your earliest opportunity.

I look forward to hearing from you.

Sincerely yours,

After an Interview Within a day's time, follow up your interview with a letter. Such a letter draws favorable attention to you as a person who understands business courtesy and good communication practice. Express your appreciation for the interview. Emphasize any of your qualifications that seemed to be especially important to the interviewer. Express your willingness to live with any special conditions of employment, such as relocation. Make clear that you want the job and feel qualified to do it. If you include a specific question in your letter, it may hasten a reply. Your letter might look like this one:

Dear Ms. Kuriyama:

Thank you for speaking with me last Tuesday about the food supervisor position you have open.

Working in a hospital food service relates well to my experience and interests. The job you have available is one I am qualified to do. A feasibility study I am currently writing as a senior project deals with a food service's ability to provide more varied meals to people with restricted dietary requirements. May I send you a copy next week when it is completed?

I understand that the work you described would include alternating weekly night shifts with weekly day shifts. This requirement presents no difficulty for me.

I look forward to hearing from you soon.

Sincerely yours,

After Being Refused a Job When a company refuses you a job, it is good tactics to acknowledge the refusal. Express thanks for the time spent with you and state your regret that no opening exists at the present time. If you like, express the hope of being considered in the future. You never know; it might happen. In any case, you want to maintain a good reputation with this employer and its representatives. These are people working in the same industry as you: you may encounter them at professional conferences or meetings of the local chapter of your professional association. They may later come to work at your company as colleagues, supervisors, or subordinates. The few minutes you devote now to a courteous reply will create a lasting impression of your professionalism.

Accepting or Refusing a Job Writing an acceptance letter presents few problems. Be brief. Thank the employer for the job offer, and accept the job. Determine the day and time you will report for work, and express pleasure at the prospect of joining the organization and working with your new colleagues. A good letter of acceptance might read as follows:

Dear Mr. Gafaiti:

Thank you for offering me a job as research assistant with your firm. I happily accept. I will report to work on 1 July as you have requested.

I look forward to working with Price Industries and particularly to the opportunity of doing research with Dr. Ertas.

Sincerely yours,

Writing a letter of refusal can be difficult. Be as gracious as possible. Be brief but not so brief as to suggest rudeness or indifference. Make it clear that you appreciate the offer. If you can, give a reason for your refusal. The employer who has spent time and money in interviewing and corresponding with you deserves these courtesies. And, of course, your own self-interest is involved. Some day you may wish to reapply to an organization that for the moment you must turn down. A good letter of refusal might look like this one:

Dear Ms. White:

I enjoyed my visit to the research department of your company. I would very much have liked to work with the people I met there. I thank you for offering me the opportunity to do so.

After much serious thought, I have decided that the research opportunities offered me in another job are closer to the interests I developed at the university. Therefore, I have accepted that other job and regret that I cannot accept yours.

I appreciate the great courtesy and thoughtfulness that you and your associates extended to me during the application and interview process.

Sincerely yours,

INTERVIEWING

The immediate goal of all your preparation and letter and résumé writing is an interview with a potential employer. Interviews come about in various ways. If you network successfully, you will obtain interviews that allow you to ask questions of people already on the job. If you impress the person you are talking to, such information-seeking interviews may turn into interviews that assess your potential as an employee. Your letters and résumés may obtain interviews for you. Recruiters may come to your campus and schedule screening interviews with graduating students. As the name suggests, screening interviews are preliminary dialogs that enable recruiters to choose people to go further in the hiring process.

"Going further" often means multiple interviews at the organization's headquarters. In one day, you might have interviews with a human resources staff person, the person who would be your boss if you were employed, and, perhaps, his or her boss. If you make it to the point of being offered a job, you will likely have an interview in which you negotiate the details of your job, salary, and benefits. All this can be quite stressful. The better prepared you are, the easier it will go.

The Interview

If you have prepared properly, you should show up at the interview knowing almost everything about the organization, including its basic history, chief locations, key products and services, financial situation, and mission.

You should have the following with you:

- Your résumé, in both regular and scannable formats
- A portfolio of your work, if appropriate
- Pen and notebook
- A list of your references
- Your business card

For interviews, you will want to be groomed and dressed conservatively—usually in a suit. Arrive at the place of the interview early enough to be relaxed. Shake hands firmly but not aggressively, and make eye contact. Give the interviewer a copy of your résumé, and, when the interviewer sits, sit comfortably yourself. Body language is important—you want to appear relaxed but attentive. Don't be stiff, but don't slouch either.

Most interviews follow a three-part pattern. To begin, the interviewer may generate small talk designed to set you at ease. Particularly at a screening interview, the interviewer may give some information about the company. This is a good chance for you to ask some questions that demonstrate that you have been doing your research on the company. But don't impose your questions on the interviewer or try to steer the interview.

Most of the middle portion of the interview will be taken up with questions aimed at assessing your skills and interests and how you might be of value to the organization. A thorough self-assessment obviously is a necessity in answering such questions. Figure 20-7 lists commonly asked questions. If you prepare answers for them, you should be able to handle most of the questions

- What can you tell me about yourself?
- What are your strengths and weaknesses?
- What do you want to be doing five years from now?
- Do you know much about us? Why do you want to work for us?
- We're interviewing ten people for this job. Why should you be the one we hire?
- What in your life are you most proud of?
- Here is a problem we had (interviewer describes problem). How would you have solved it?
- If you won $10 million in a lottery, how would you spend the rest of your life?
- Why do you want a career in your chosen field?
- Which school subjects interested you the most (least)? Why?

FIGURE 20-7 · FREQUENTLY ASKED INTERVIEW QUESTIONS

you are likely to receive. Certainly, have well-thought-out answers to questions concerning both your short-range and long-range goals. In your answers, relate always to the organization. To answer the question, "What do you want to be doing five years from now?" by saying "Running my own consulting firm" might bring the interview to an early close.

An interviewer who asks "What can you tell me about yourself?" really doesn't expect an extended life history. This question is simply an opening for you to talk about your work, educational experiences, and skills. Try to relate your skills and experience to the needs of the organization. Don't overlook the people and communication skills essential to nearly every professional job. In your answer to this and other questions, be specific in your examples. If you say something like "I have good managerial skills," immediately back it up with an occasion or experience that supports your statement. Focus on offering the specific evidence that will lead the interviewer to the right conclusions about you. Such specifics are more memorable than unsupported claims and will help you stick in the interviewer's mind.

The question "Why do you want to work for us?" allows you to display what you have learned about the organization. In answering this question, you should again show that what you have to offer meshes with what the company needs.

The question about how you would spend your life if you won $10 million is an interesting one. "Lolling around the beach" is obviously the wrong answer, but "I'd continue to work for your corporation," might create a bad impression, as well. The question is intended to elicit a glimpse of the worthwhile things in your life that you really enjoy. Promoting community programs and services or supporting local charities might qualify as good answers. So might investing in research projects in your industry.

In answering questions about your strengths and weaknesses, be honest, but don't betray weaknesses that could eliminate you from consideration.

"I can't stand criticism," would likely disqualify you from further considera-tion. "Sometimes I don't know when to quit when I'm trying to solve a problem," given as a weakness could be perceived as a strength. Or admit to weaknesses that would have no impact on the specific job for which you are interviewing: for example, if the job has no managerial dtuies, it would be okay to acknowledge that your managerial skills are undeveloped. When mention-ing weakness, end with a positive statement of working toward improvement.

In the last part of the interview you will likely be given a chance to ask some questions of your own. It's a good time to get more details about the job or jobs that may be open. Ask about the organization's goals. "What is the company most proud of?" is a good question. Or "What are you looking for in the can-didate for this position? What will you want him or her to accomplish in the first year on the job?" Don't ask naive questions—questions you ought to know the answers to from your research—like the size of the company or the number of employees. Don't ask questions just to ask questions. The interview is a good time for you to find out whether you really want to work for an organization. Not every organization is going to be a good fit for what you have to offer and what you want to do. Unless the interviewer has raised the matter of salary and benefits, don't ask about them.

If you really want to go to work for the organization, make that clear before the interview ends. But don't allow your willingness to come across as desper-ation. Companies don't hire desperate people. At some point in the interview, be sure to get the interviewer's name (spelled correctly!), title, address, phone number, e-mail address, and fax number. You'll need them for later correspon-dence. When the interviewer thanks you for coming, thank him or her for seeing you and leave. Don't drag the interview out or linger at the door when the occasion is clearly over.

Negotiation

Interviewers seldom bring up salary and benefits until they either see you as a good prospect or are sure they want to hire you. If they offer you the job, the negotiation is sometimes done in a separate interview. For example, your future boss may offer you the job and then send you to negotiate with the human resources staff.

Sometimes, the negotiator may offer you a salary. At other times, you may be asked to name a salary. Now is the time to put to good use any information you have received through your networking activities. Or check the online job find-ing services or specialized services such as Salary.Com (at www.salary.com) for their estimates of earnings for different positions in different industries.

Your research in these sources will give you not a specific salary but a salary range. If asked to name a salary, do not give a figure at the bottom of the range. Ask for as near the top as you believe is reasonable given your education and experience. The negotiator will respect you the more for knowing what you are worth. However, balance the compensation package—vacations, pension plans, health care, educational opportunities, and so forth—against the salary. Some compensation packages are worth a good deal of money and may make it feasible for you to accept a lower salary.

The location of the job is also a critical factor to consider. Online services such as Salary.Com offer cost-of-living calculators that determine the comparative worth of salaries by location.

Before and after the Interview

If you have not participated in job interviews before, you would be wise to practice with friends. Using the information that we have given and that you have gathered for yourself, do several mock interviews, with an interviewer, an interviewee, and observers. Ask the observers to look for strengths and weaknesses in your answers, diction, grammar, and body language and to give you a candid appraisal. Practice until you feel comfortable with the process.

After you finish each interview, write down your impressions as soon as possible:

- Were there unexpected questions?
- How good were your answers?
- What did you learn about the organization?
- What did you learn about a specific job or jobs?
- Did anything make you uncomfortable about the organization?
- Do you think you would fit in there?

By the next day, get a thank you note (letter, e-mail, or fax) off to the interviewer.

☑ Planning and Revision Checklist

The following questions specifically apply to the job search. They summarize the key points in this chapter and provide a checklist for planning and revising.

Preparation

Planning and Revision

- Have you completed the self-assessment questionnaire in Figure 20-1?
- Do you have a complete record of your past job and educational experiences?
- Do you know your strengths and weaknesses, your skills, and your qualifications for the jobs you seek? Do you have clear career objectives?
- Do you need professional career counseling?

▪ Have you researched the Internet and sources to find job information?

▪ Have you found organizations that fit your needs?

▪ Have you started your networking?

▪ Have you had business cards made?

▪ Have you called on businesses and set up informational interviews?

▪ Have you joined a professional association relevant to your career field?

▪ Have you located agencies and individuals who can help you in your job search?

▪ Have you kept good records of your networking?

The Correspondence of the Job Search

Planning

▪ For your letter of application:

Do you have the needed names and addresses?

What position do you seek?

How did you learn of this position?

Why are you qualified for this position?

What interests you about the company?

What can you do for the organization to meet its needs?

Can you do anything to make an interview more convenient for the employer?

How can the employer reach you?

▪ For your résumé:

Do you have all the necessary details of your past educational and work experiences (dates, job descriptions, schools, majors, degrees, extracurricular activities, etc.)?

Which résumé format will suit your experience and capabilities best: Chronological? Functional? Targeted? Why?

Do you need to prepare your résumé in a scannable format? What keywords will you use?

In a functional résumé, which categories would best suit your experience and capabilities?

Do you know enough about the job you are seeking to use a targeted résumé?

Do you have permission to use the names of three people as references?

▪ What follow-up letters do you need? To follow up on no answer to your application? To follow up on an interview? To respond to a job refusal? To accept or refuse a job?

Revision

■ Do your letter of application and résumé reflect adequate preparation and self-assessment?

■ Are your letter of application and résumé completely free of grammatical and spelling errors? Are they designed for easy reading and skimming?

■ For the letter of application:

Have you the right tone—self-confidence without arrogance?

Does your letter show how you could be valuable to the employer? Will it raise the employer's interest?

Does your letter reflect interest in a specific job?

Have you highlighted the courses and work experience that best suit you for the job you seek?

Is it clear that you are seeking an interview and convenient for the employer to arrange one?

■ For the résumé:

Have you chosen the résumé type that best suits your experiences and qualifications?

Have you limited your résumé to one page?

Have you put your educational and work experience in reverse-chronological order?

Have you given your information in phrases rather than complete sentences?

Have you used active verbs to describe your experience? Appropriate keywords?

Does the personal information presented enhance your job potential?

Do you have permission to use the names of those you list as references?

If you are using a functional résumé, do the categories reflect appropriately your capabilities and experience?

Has your targeted résumé zeroed in on an easily recognizable career objective?

Do the listed achievements support the capabilities listed?

■ Have you followed up every interview with a letter? Are the follow-up letters you have written gracious in tone? Do the letters invite further communication in some way?

■ Does your letter of acceptance of a job show an understanding of the necessary details, such as when you report to work?

■ Does your letter of refusal thank the employer for time spent with you and make clear that you appreciate the offer?

The Interview

Planning

■ Have you found out as much about the organization as you can—its products, goals, locations, and so forth?

■ Have you practiced interviewing using the questions in Figure 20-7?

■ Do you have the proper clothes for an interview?

■ Do you have good questions to ask the interviewer?

■ Do you know the salary range for the jobs you seek?

Assessing

■ Did you answer questions well? Which answers need improving?

■ Did the interviewer ask any unexpected questions?

■ How did the interviewer respond to your questions? Did he or she seem to think they were relevant?

■ How well do you think you did? Why?

■ Do you think your career goals and this organization are a good fit?

■ ■ ■ ■ ■ ■ ■ ■ Exercises ■ ■ ■ ■ ■ ■ ■

Further exercises related to the content of this chapter and all documents employed within the following exercises can be found on the book's companion Web site.

1. Compose a schedule for your job search. Allocate time and determine deadlines for completing the following stages:

· Self-assessment
· Gathering job information
· Networking
· Preparing a letter of application
· Preparing a résumé
· Practicing the interview

2. Complete the self-assessment in Figure 20-1. What are your particular strengths and weaknesses? Which of your strengths will you emphasize in your letter of application, résumé, and interview? Which weakness will you acknowledge in your interview if you are asked to name one?

3. Write a letter of application to an organization that employs majors in your field. Brainstorm and develop the three kinds of résumé: chronological, functional, and targeted. Design each in both traditional paper and scannable form to accompany your letter of application (If you are actually seeking work, write your letter with a specific organization in mind.) In groups of four students, review your letters and résumés. As a group, determine which kind of résumé suits the candidate and the job for which he or she is applying.

4. Prepare a fact sheet for yourself that contains the following information that will be useful to you in your job search:

- Current and future job prospects in your field; the qualifications needed for the jobs available
- Salary ranges for jobs for which you are qualified
- Three places where you might post your résumé
- Five companies that hire people in your field
- Useful information about a city that interests you as a place to live
- Information about the job-finding assistance offered by a state in which you might wish to live
- Jobs available in the federal government in your discipline

5. Using both this chapter and Chapter 6, Writing for International Readers, as your guides, prepare a letter of application to an employer in a country in Latin America or Asia.

6. Develop a list of questions that you might ask about a company during a job interview. Share your favorite question with the class.

7. In groups of four, plan and carry out practice interviews. Everyone in the group should get a chance to serve as interviewer and as interviewee.

8. Investigate a potential employer of majors in your field. Locate information available from online job-finding services. Are their profiles of this employer basically similar? Do you notice important differences? Locate additional information available in newspapers, magazines, and government documents about this employer. What is the public perception of this employer? What is its reputation in the industry? Interview current employees or managers for their perspectives on the advantages and disadvantages of working for this employer. Compile the findings of your research in a report for your college's career center on the strengths and weaknesses of this employer.

Handbook

Any living language is a growing, flexible instrument with rules that are constantly changing by virtue of the way it is used by its live, independent speakers and writers. Only the rules of a dead language are unalterably fixed.

Nevertheless, at any point in a language's development, certain conventions of usage are in force. Certain constructions are considered to be errors that mark the person who uses them as uneducated. It is with these conventions and errors that this handbook primarily deals. We also include a section on sexist usage.

To make the handbook easy to use as a reference, we have arranged the topics covered in alphabetical order. Each convention and error dealt with has an abbreviated reference tag. The tags are reproduced on the back endpapers, along with some of the more important proofreading symbols. If you are in a college writing course, your instructor may use some combination of these tags and symbols to indicate revisions needed in your reports.

Abbreviations
Acronyms
Apostrophe
Brackets
Capitalization
Colon
Comma
Dangling Modifier

Dash
Diction
Ellipsis
Exclamation Point
Fragmentary Sentence
Hyphen
Italicization
Misplaced Modifier
Numbers
Parallelism
Parentheses
Period
Pronoun–Antecedent Agreement
Pronoun Form
Question Mark
Quotation Marks
Run-On Sentence
Semicolon
Sexist Usage
Spelling Error
Verb Form
Verb–Subject Agreement

ab ABBREVIATIONS

Every scientific and professional field generates hundreds of specialized terms, and many of these terms are abbreviated for the sake of conciseness and simplicity. A few principles for the use of abbreviations follow, the first probably being the most important:

- As a general principle, the style manuals that govern the various disciplines, discourage the indiscriminate use of abbreviations. The *Publication Manual of the American Psychological Society* (83) states this principle well:

  In general, use an abbreviation only (a) if it's conventional and if the reader is more familiar with the abbreviation than with the complete form or (b) if considerable space can be saved and cumbersome repetition avoided. In short, use only those abbreviations that will help you communicate with your readers. Remember, they have not had the same experience with your abbreviations as you have.

- You must decide if your audience is familiar enough with the term to allow you to use it without definition. Second, you must decide if your audience is familiar enough with the abbreviation for you to use it without spelling it out. For example, here is how one writer introduces the term *electric and magnetic fields* to a lay audience:

Electric and magnetic fields (EMFs) are produced by power lines, electrical wiring, and electrical equipment. There are many other sources of EMFs. The focus of this booklet is on EMFs associated with the generation, transmission, and use of electric power. EMFs are invisible lines of force that surround any electrical device. Electric fields are produced by voltage and increase in strength as the voltage increases. The electric field strength is measured in units of volts per meter (V/m). Magnetic fields result from the flow of current through wires or electrical devices and increase in strength as the current increases. Magnetic fields are measured in units of gauss (G) or tesla (T). Most electrical equipment has to be turned on, that is, current must be flowing, for a magnetic field to be produced. Electric fields, on the other hand, are present even when the equipment is switched off, as long as it remains connected to the source of electric power (Department of Energy 5).

Placing the abbreviation *EMF* in parentheses following the first use of the term allows the use of the abbreviation throughout the rest of the article. Following the parentheses, the writer provides an extended definition. The amount of provided detail depends as always on purpose and audience. For a more expert audience it would probably be enough to put the abbreviation in parentheses without a definition.

- In some cases the abbreviation is so well known to a technical audience and the term itself is so cumbersome that the abbreviation without the term will suffice. For example, an article for a computer-knowledge-able audience would probably use *ACSII* without providing or defining the term, *American Standard Code for Information Exchange*. A term such as *DNA* is so well known even if not thoroughly understood, that it can be used by itself, even in articles for lay people.
- Even when not used in the text, abbreviations are widely used to save space in tables and figures. When the abbreviations are standard abbreviations known to the audience, you don't need to spell them out or explain them. If they are not known, spell them out in the captions to the figure or table or in a note. If necessary, define them.
- Use Latin abbreviations like *cf.* (compare), *i.e.* (that is), and *e.g.* (for example) sparingly. In your text use them only in parenthetical explanations. Use them also in tables and figures where space constraints make their use practical and acceptable. Elsewhere, use the English equivalents.
- Do not begin a sentence with an abbreviation. If practical, spell the abbreviation out. If not, recast the sentence to move the abbreviation.

The formation of standard abbreviations follow a brief set of rules, illustrated by the following set of abbreviations and the rules that follow it.

absolute	abs
acre or acres	acre or acres
atomic weight	at. wt
barometer	bar.
Brinell hardness number	Bhn
British thermal units	Btu
meter	m
square meter	m²
miwatt or miwatts	mW
miles per hour	mph
National Electric Code	NEC
per	per
revolutions per minute	rpm
rod	rod
ton	ton

1. Use the singular form of abbreviation for both singular and plural terms:

cu ft	either cubic foot or cubic feet
cm	either centimeter or centimeters

   But there are some common exceptions:

no.	number
nos.	numbers
p.	page
pp.	pages
ms.	manuscript
mss.	manuscripts

2. In abbreviating units measurement, use lowercase letters except for letters that stand for proper nouns or proper adjectives:

at. wt	*but*	Btu or B
mph	*but*	mW

3. For technical terms, use periods only after abbreviations that spell complete words. For example, *in* is a word that could be confused with the abbreviation for inches. Therefore, use a period:

ft	*but*	in.
abs	*but*	bar.
cu ft	*but*	at. wt

5. Spell out many short and common words:

acre	rod	per	ton

6. In compound abbreviations, use internal spacing only if the first word is represented by more than its first letter:

rpm	*but*	cu ft
mph	*but*	at. wt

7. With few exceptions, form the abbreviations of organization names without periods or spacing:

NEC   ASA

8. Abbreviate terms of measurement only if they are preceded by an arabic expression of exact quantity:

55 mph   *and*   20-lb anchor

but

We will need an engine of greater horsepower.

The principles and rules we have provided here will cover most general situations. But in preparing a manuscript for a specific discipline, be sure to consult the applicable style or publication manual.

ACRONYMS

acro

Acronyms are formed in two ways. In one way, the initial letters of each word in some phrase are combined. An example would be WYSIWYG, an acronym for the computer phrase "What you see is what you get." In a second way, some combination of initial letters or several letters of the words in the phrase are combined. An example would be *radar* for *r*adio *d*etection *a*nd *r*anging.

Technical writing uses acronyms freely, as in this example from a description of a computer program that performs statistical analysis:

It has good procedural capabilities, including some time-series-related plots and ARIMA forecasting, but it doesn't have depth in any one area. Although it has commands to create EDA displays, these graphics are static and are printed with characters rather than with lines. (Levine 128)

Use acronyms without explanation only when you are absolutely sure your readers know them. If you have any doubts at all, at least provide the words from which the acronym stems. If you're unsure whether the words are enough, provide a definition of the complete phrase. In the case of the paragraph just quoted, the computer magazine in which it was printed provided a glossary giving both the complete phrases and definitions:

ARIMA (auto-regressive integrated moving-average): a model that characterizes changes in one variable over time. It is used in time-series analysis.

EDA (exploratory data analysis): The use of graphically based tools, particularly in initial states of data analysis, to inspect data properties and to discover relationships among variables. (Levine 120)

See also the Entry for Abbreviations. In general follow the same principles for using acronyms as for using abbreviations. Acronyms can be daunting to those unfamiliar with them. Even for an audience that knows their meaning, a too heavy use of acronyms can make your writing seem lumpish and uninviting.

apos APOSTROPHE

The apostrophe has three chief uses: (1) to form the possessive, (2) to stand for missing letters or numbers, and (3) to form the plural of certain expressions.

Possessives

Add an apostrophe and an *s* to form the possessive of most singular nouns, including proper nouns, even when they already end in an *s* or another sibilant such as *x:*

> man's
> spectator's
> jazz's
> Marx's
> Charles's

Exceptions to this rule occur when adding an apostrophe plus an *s* would result in an *s* or *z* sound that is difficult to pronounce. In such cases, usually just the apostrophe is added:

> Xerxes'
> Moses'
> conscience'
> appearance'

To understand this exception, pronounce *Marx's* and then a word like *Moses's* or *conscience's.*

To form plurals into the possessive case, add an apostrophe plus *s* to words that do not end in an *s* or other sibilant and an apostrophe only to those that do:

> men's
> data's
> spectators'
> agents'
> witnesses'

To show joint possession, add the apostrophe and *s* to the last member of a compound or group; to show separate possession, add an apostrophe and *s* to each member:

> Gregg and Klymer's experiment astounded the class.
> Gregg's and Klymer's experiments were very similar.

Of the several classes of pronouns, only the indefinite pronouns use an apostrophe to form the possessive.

Possessive of Indefinite Pronouns	Possessive of Other Pronouns
anyone's	my (mine)
everyone's	your (yours)
everybody's	his, her (hers), its
nobody's	our (ours)
no one's	their (theirs)
other's	whose
neither's	

Missing Letters or Numbers

Use an apostrophe to stand for the missing letters in contractions and to stand for the missing letter or number in any word or set of numbers from which a letter or number is omitted for one reason or another:

> can't, don't, o'clock, it's (it is), and similar constructions
> We were movin' downriver, listenin' to the birds singin'.
> The class of '49 was Colgate's best class in years.

Plural Forms

An apostrophe is sometimes used to form the plural of letters and numbers, but this style is gradually dying, particularly with numbers.

> 6's and 7's (but more commonly, 6s and 7s)
> a's and b's

BRACKETS

brackets

Brackets are chiefly used when a clarifying word or comment is inserted into a quotation:

> "The result of this [disregard by the propulsion engineer] has been the neglect of the theoretical and mathematical mastery of the engine inlet problem."
> "An ideal outlet require [sic] a frictionless flow."
> "Last year [2000] saw a partial solution to the problem."

Sic, by the way, is Latin for *thus*. Inserted in a quotation, it means that the mistake it follows is the original writer's, not yours. Use it with discretion.

cap CAPITALIZATION

The following are the more important rules of capitalization. For a complete rundown, see your college dictionary.

Proper Nouns

Capitalize all proper nouns and their derivatives:

Places

America   American   Americanize   Americanism

Days of the Week and Months

Monday   Tuesday   January   February

But not the seasons:

winter   spring   summer   fall

Organizations and Their Abbreviations

American Kennel Club (AKC)
United States Air Force (USAF)

Capitalize *geographic areas* when you refer to them as areas:

The Andersons toured the Southwest.

But do not capitalize words that merely indicate direction:

We flew west over the Pacific.

Capitalize the names of *studies* in a curriculum only if the names are already proper nouns or derivatives of proper nouns or if they are part of the official title of a department or course:

Department of Geology
English Literature 25
the study of literature
the study of English literature

Note: Many nouns (and their derivatives) that were originally proper have been so broadened in application and have become so familiar that they are no longer capitalized: *boycott, macadam, italicize, platonic, chinaware, quixotic.*

Literary Titles

Capitalize the first word, the last word, and every important word in literary titles:

But What's a Dictionary For
The Meaning of Ethics
How to Write and Be Read

Rank, Position, Family Relationships

Capitalize the titles of rank, position, and family relationship unless they are preceded by *my*, *his*, *their*, or similar possessive pronouns:

> Professor J. E. Higgins
> I visited Uncle Timothy.
> I visited my uncle Timothy.
> Dr. Milton Weller, Head, Department of Entomology

COLON

The colon is chiefly used to introduce quotations, lists, or supporting statements. It is also used between clauses when the second clause is an example or amplification of the first and in certain conventional ways with numbers, correspondence, and bibliographical entries.

Introduction

Place a colon before a quotation, a list, or supporting statements and examples that are formally introduced:

> Mr. Smith says the following of wave generation:
> > The wind waves that are generated in the ocean and which later become swells as they pass out of the generating area are products of storms. The low pressure regions that occur during the polar winters of the Arctic and Antarctic produce many of these wave-generating storms.

> The various forms of engine that might be used would operate within the following ranges of Mach number:

M-0 to M-1.5	Turbojet with or without precooling
M-1.5 to M-7	Reheated turbojet, possibly with precooling
M-7 to M-10+	Ramjet with supersonic combustion

> Engineers are developing three new engines: turbojet, reheated turbojets, and ramjets.

Do not place a colon between a verb and its objects or a linking verb and the predicate nouns.

Objects
The engineers designed turbojets, reheated turbojets, and ramjets.

Predicate Nouns
The three engines the engineers are developing are turbojets, reheated turbojets, and ramjets.

Do not place a colon between a preposition and its objects:

The plane landed at Detroit, Chicago, and Rochester.

Between Clauses

If the second of two clauses is an example or an amplification of the first clause, then the colon may replace the comma, semicolon, or period:

The docking phase involves the actual "soft" contact: the securing of lines, latches, and air locks.

Figure 2 illustrates the difference between these two guidance systems: The paths of the two vehicles are shown to the left and the motion of the ferry as viewed from the target station is shown to the right.

Generally, a complete sentence after a colon begins with a capital letter, whereas a simple list begins lowercase.

Styling Conventions

Place a colon after a formal salutation in a letter, between numerals representing hours and minutes, between a title and a subtitle, and between chapter and verse of the Bible:

Dear Ms. Jones:
at 7:15 P.M.
Working Women: A Chartbook
I Samuel 7:14–18

c COMMA

The most used—and misused—mark of punctuation is the comma. Writers use commas to separate words, phrases, and clauses. Generally, commas correspond to the pauses we use in our speech to separate ideas and to avoid ambiguity. You will use the comma often: About two out of every three marks of punctuation you use will be commas. Sometimes your use of the comma will be essential for clarity; at other times you will be honoring grammatical conventions. (See also the entry for Run-On Sentences.)

Main Clauses

Place a comma before a coordinating conjunction (*and, but, or, nor, for, yet*) that joins two main (independent) clauses:

During the first few weeks we felt a great deal of confusion, but as time passed we gradually fell into a routine.

We could not be sure that the plumbing would escape frost damage, nor were we at all confident that the house could withstand the winds of almost hurricane force.

If the clauses are short, have little or no internal punctuation, or are closely related in meaning, then you may omit the comma before the coordinating conjunction:

The wave becomes steeper but it does not tumble yet.

In much published writing there is a growing tendency to place two very short and closely related independent clauses (called contact clauses) side by side with only a comma between:

The wind starts to blow, the waves begin to develop.

Sentences consisting of *three* or more equal main clauses should be punctuated uniformly:

We explained how urgent the problem was, we outlined preliminary plans, and we arranged a time for discussion.

In general, identical marks are used to separate equal main clauses. If the equal clauses are short and uncomplicated, commas usually suffice. If the equal clauses are long or internally punctuated, or if their separateness is to be emphasized, semicolons are preferable and sometimes essential.

Clarification

Place a comma after an introductory word, phrase, or clause that might be over-read or that abnormally delays the main clause:

As soon as you have finished polishing, the car should be moved into the garage. (Comma to prevent over-reading)

Soon after, the winds began to moderate somewhat, and we were permitted to return to our rooms. (First comma to prevent over-reading)

If the polar ice caps should someday mount in thickness and weight to the point that their combined weight exceeded the equatorial bulge, the earth might suddenly flop ninety degrees. (Introductory clause abnormally long)

After a short introductory element (word, phrase, or clause) where there is no possibility for ambiguity, the use of the comma is optional. Generally, let the emphasis you desire guide you. A short introductory element set off by a comma will be more emphatic than one that is not.

Nonrestrictive Modifiers

Enclose or set off from the rest of the sentence every nonrestrictive modifier, whether a word, a phrase, or a clause. How can you tell a nonrestrictive modifier from a restrictive one? Look at these two examples:

Restrictive

A runway that is not oriented with the prevailing wind endangers the aircraft using it.

Nonrestrictive

The safety of any aircraft, whether heavy or light, is put in jeopardy when it is forced to take off or land in a crosswind.

The restrictive modifier is necessary to the meaning of the sentence. Not just any runway but "a runway that is not oriented with the prevailing wind" endangers aircraft. The writer has *restricted* the many kinds of runway he or she could talk about to one particular kind. In the nonrestrictive example, the modifier merely adds descriptive details. The writer doesn't restrict *aircraft* with the modifier but simply makes the meaning a little clearer.

Restrictive modifiers cannot be left out of the sentence if it is to have the meaning the writer intends; nonrestrictive modifiers can be left out.

Nonrestrictive Appositives

Set off or enclose every nonrestrictive appositive. As used here, the term *appositive* means any element (word, phrase, or clause) that parallels and repeats the thought of a preceding element. According to this view, a verb may be coupled appositively with another verb, an adjective with another adjective, and so on. An appositive is usually more specific or more vivid than the element for which it is an appositive; an appositive makes explicit and precise something that has not been clearly implied.

Some appositives are restrictive and, therefore, are not set off or enclosed.

Nonrestrictive

A crosswind, a wind perpendicular to the runway, causes the pilot to make potentially dangerous corrections just before landing.

Restrictive

In some ways, Mr. Bush the president has to behave differently from Mr. Bush the governor.

In the nonrestrictive example, the appositive merely adds a clarifying definition. The sentence makes sense without it. The appositives in the restrictive example are essential to the meaning. Without them we would have, "In some ways, Mr. Bush has to behave differently from Mr. Bush."

Series

Use commas to separate members of a coordinate series of words, phrases, or clauses if *all* the elements are not joined by coordinating conjunctions:

Instructions on the label state clearly how to prepare the surfaces, how to apply the contents, and how to clean and polish the mended article.

To mold these lead figures you will need a hot flame, a two-pound block of lead, the molds themselves, a file or a rasp, and an awl.

Under the microscope the sensitive, filigree-like mold appeared luminous and transparent and faintly green.

Other Conventional Uses

Date
On August 24, A.D. 79, Mount Vesuvius erupted, covering Pompeii with 50 feet of ash and pumice.

Note: When you write the month and the year without the day, it is common practice to omit the comma between them—as in June 1993.

Geographical Expression
During World War II, Middletown, Pennsylvania, was the site of a huge military airport and supply depot.

Title after Proper Name
A card in yesterday's mail informed us that Penny Hutchinson, M.D., would soon open an office in Hinsdale.

Noun of Direct Address
Lewis, do you suppose that we can find our way back to the cabin before nightfall?

Informal Salutation
Dear Jane,

DANGLING MODIFIER

dm

Many curious sentences result from the failure to provide a modifier with something to modify:

Having finished the job, the tarpaulins were removed.

In this example it seems as though the tarpaulins have finished the job. As is so often the case, a passive voice construction has caused the problem (see pages 69–70). If we recast the sentence in active voice, we remove the problem:

Having finished the job, the workers removed the tarpaulins.

dash DASH

In technical writing, you will use the dash almost exclusively to set off parenthetical statements. You may, of course, use commas or parentheses for the same function, but the dash is the most emphatic separator of the three. You may also use the dash to indicate a sharp transition.

> The target must emit or reflect light the pilot can see—but how bright must this light be?

d DICTION

For good diction, choose words that are accurate, effective, and appropriate to the situation. Many different kinds of linguistic sins can cause faulty diction. Poor diction can involve a choice of words that are too heavy or pretentious: *utilize* for *use, finalize* for *finish, at this point in time* for *now,* and so forth. Tired old clichés are poor diction: *with respect to, with your permission, with reference to,* and many others. We talk about such language in Chapter 4.

Sometimes the words chosen are simply too vague to be accurate: *inclement weather* for *rain, too hot* for *600°C.*

Poor diction can mean an overly casual use of language when some degree of formality is expected. One of the many synonyms for *intoxicated,* such as *bombed, stoned,* or *smashed,* might be appropriate in casual conversation but totally wrong in a police or laboratory report.

Poor diction can reflect a lack of sensitivity to language—to the way one group of words relates to another group. Someone who writes that "The airlines are beginning a crash program to solve their financial difficulties" is not paying attention to relationships. The person who writes that the "Steelworkers' Union representatives are getting down to brass tacks in the strike negotiations" has a tin ear, to say the least. Make your language work for you, and make it appropriate to the situation.

ell ELLIPSIS

Use three spaced periods to indicate words omitted within a quoted sentence; use four spaced periods if the omission occurs at the end of the sentence:

> "As depth decreases, the circular orbits become elliptical and the orbital velocity . . . increases as the wave height increases."

"As the ground swells move across the ocean, they are subject to head-winds or crosswinds. . . ."

You need not show an ellipsis if the context of the quotation makes it clear that it is not complete:

Wright said the accident had to be considered a "freak of nature."

EXCLAMATION POINT

Place an exclamation point at the end of a startling or exclamatory sentence.

According to the Centers for Disease Control and Prevention, every cigarette smoked shortens the smoker's life by seven minutes!

With the emphasis in technical writing on objectivity, you will seldom use the exclamation point.

FRAGMENTARY SENTENCE

Most fragmentary sentences are either verbal phrases or subordinate clauses that the writer mistakes for a complete sentence.

A verbal phrase has a participle, a gerund, or an infinitive in the predicate position, none of which functions as a complete verb:

Norton, depicting the electromagnetic heart. (participle)
The timing of this announcement about Triptycene. (gerund)
Braun, in order to understand tumor cell growth. (infinitive)

When your fragment is a verbal phrase, either change the participle, gerund, or infinitive to a complete verb or repunctuate so that the phrase becomes part of a complete sentence.

Fragment
Norton, depicting the electromagnetic heart. She made a mockup of it.

Rewritten
Norton depicted the electromagnetic heart. She made a mockup of it.
Norton, depicting the electromagnetic heart, made a mockup of it.

Subordinate clauses are distinguishable from phrases in that they have complete subjects and complete verbs (rather than verbals) and are introduced by relative pronouns (*who, which, that*) or by subordinating conjunctions (*because, although, since, after, while*).

The presence of the relative pronoun or the subordinating conjunction is a signal that the clause is not independent but is part of a more complex sentence unit. Any independent clause can become a subordinate clause with the addition of a relative pronoun or subordinating conjunction.

Independent Clause
Women's unemployment rates were higher than men's.

Subordinate Clause
Although women's unemployment rates were higher than men's.

Repunctuate a subordinate clause so that it is joined to the complex sentence of which it is a part.

Fragment
Although women's unemployment rates were higher than men's. Now the rates are similar.

Rewritten
Although women's unemployment rates were higher than men's, now the rates are similar.

Various kinds of elliptical sentence without a subject or a verb do exist in English, for example, "No!" "Oh?" "Good shot." "Ouch!" "Well, now." These constructions may occasionally be used for stylistic reasons, particularly to represent conversation, but they are seldom needed in technical writing. If you do use such constructions, use them sparingly. Remember that major deviations from normal sentence patterns will probably jar your readers and break their concentration on your report, the last thing that any writer wants.

---

hyphen HYPHEN

Hyphens are used to form various compound words and in breaking up a word that must be carried over to the next line.

Compound Numbers

See Numbers.

Common Compound Words

Observe dictionary usage in using or omitting the hyphen in compound words.

governor-elect	court-martial
ex-treasurer	Croesus-like

Russo-Japanese	drill-like
pro-American	self-interest

But:

neophyte	sweet corn
newspaper	weather map
radioactive	radio beam
bloodless	blood pressure

Compound Words as Modifiers

Use the hyphen between words joined together to modify other words:

a half-empty fuel tank
an eight-cylinder engine
their too-little-and-too-late methods

Be particularly careful to hyphenate when omitting the hyphen may cause ambiguity:

two-hundred-gallon drums
two hundred-gallon drums
a pink-skinned hamster

Sometimes you have to carry a modifier over to a later word, creating what is called a *suspended hyphen:*

GM cars come with a choice of four-, six-, or eight-cylinder engines.

ITALICIZATION

ital

Italic print is a distinctive typeface, like this sample: *Scientific American.* When you use a word processor, you can use an italic typeface or represent italics by underlining, like this:

Scientific American

Foreign Words

Italicize foreign words that have not yet become a part of the English language:

We suspected him always of holding some *arrière pensée.*
Karl's everlasting *Weltschmerz* makes him a depressing companion.

Also italicize Latin words for genus and species.

Cichorium endivia (endive)
Percopsis omiscomaycus (trout-perch)

But do not italicize Latin abbreviations or foreign words that have become a part of the English language:

etc.   bourgeois
vs.   status quo

A good collegiate dictionary should indicate which foreign words are still italicized and which are not.

Words, Letters, and Numbers Used as Such

The words *entrance* and *admission* are not perfectly interchangeable.
Don't forget the *k* in *picnicking*.
His *9s* and *7s* descended below the line of writing.

Titles

In general, italicize most titles, including the titles of books, plays, pamphlets, periodicals, movies, radio and television programs, operas, ballets, and record albums. Also italicize the names of works of art such as sculptures and paintings and the names of ships, airplanes, and spacecraft. Some examples follow:

The Chicago Manual of Style   *Sesame Street*
Othello   *Swan Lake*
Scientific American   *Mona Lisa*
Star Wars   *Sputnik II*

mm MISPLACED MODIFIER

As in the case of dangling modifiers, curious sentences result from a modifier's not being placed next to the element modified:

An engine may crack when cold water is poured in unless it is running.

Probably, with a little effort, no one will misread this example, but, undeniably, it says that the engine will crack unless the water is running. Move the modifier to make the sentence clear:

Unless it is running, an engine may crack when cold water is poured in.

It should be apparent from the preceding examples that a modifier can be in the wrong position to convey one meaning but in the perfect position to convey a different meaning. In the next example, the placement of *for three years* is either right or wrong. It is in the right position to modify *to work* but in the wrong position to modify *have been trying*.

I have been trying to place him under contract to work here for three years. (three-year contract)

As the examples suggest, correct placement of modifiers sometimes amounts to more than mere nicety of expression. It can mean the difference between stating falsehood and truth, between saying what you mean and saying something else.

NUMBERS

There is a good deal of inconsistency in the rules for handling numbers. Often the question is whether you should write the number as a word or as a figure. We will give you the general rules. Your instructor or your organization may give you others. As in all matters of format, you must satisfy whomever you are working for at the moment. Do, however, be internally consistent within your reports. Do not handle numbers differently from page to page of a report.

Numbers as Words

Generally, in technical and scientific writing, you write out all numbers from zero to nine and rounded-off large numbers, as words:

> six generators
> about a million dollars

However, when you are writing a series of numbers, do not mix up figures and words. Let the larger numbers determine the form used:

> five boys and six girls

But:

> It took us 6 months and 25 days to complete the experiment.

Numbers as Sentence Openers

Do not begin a sentence with a figure. If you can, write the number as a word. If this would be cumbersome, recast the sentence to get the figure out of the beginning position:

> Fifteen months ago, we saw the new wheat for the first time.
> We found 350 deficient steering systems.

Compound Number Adjectives

When you write two numbers together in a compound number adjective, spell out the first one or the shorter one to avoid confusing the reader:

> Twenty 10-inch trout
> 100 twelve-volt batteries

Hyphens

Two-word numbers are hyphenated on the rare occasions when they are written out:

> Eighty-five boxes

or:

> Eighty-five should be enough.

Numbers as Figures

The general rule for technical and scientific writing is to write all exact numbers over nine as figures. However, as we noted, rounded-off numbers are commonly written as words. The precise figure could give the reader an impression of exactness that might not be called for.

Certain conventional uses call for figures at all times.

Dates, Exact Sums of Money, Time, Address
1 January or January 1, 2002
$3,422.67 but about three thousand dollars
1:57 P.M. but two o'clock
660 Fuller Road

Technical Units of Measurement
6 cu ft
4,000 rpm

Cross-References
See page 22.
Refer to Figure 2.

Fractions

When a fraction stands alone, write it as an unhyphenated compound:

> two thirds
> fifteen thousandths

When a fraction is used as an adjective, you may write it as a hyphenated compound. But if either the numerator or the denominator is hyphenated, do not hyphenate the compound. More commonly, fractions used as adjectives are written as figures.

> two-thirds engine speed
> twenty-five thousandths
> 3/4 rpm

PARALLELISM

When you link elements in a series, they must all be in the same grammatical form. Link an adjective with an adjective, a noun with a noun, a clause with a clause, and so forth. Look at the boldface portion of the sentence below:

A good test would **use small amounts of plant material, require little time, simple to run**, and **accurate.**

The series begins with the verbs *use* and *require* and then abruptly switches to the adjectives *simple* and *accurate*. All four elements must be based on the same part of speech. In this case, it's easy to change the last two elements:

A good test would use small amounts of plant material, require little time, **be simple to run**, and **be accurate.**

Always be careful when you are listing to keep all the elements of the list parallel. In the following example, the third item in the list is not parallel to the first two:

The process has three stages: (1) the specimen is dried, (2) all potential pollutants are removed, and (3) atomization.

The error is easily corrected:

The process has three stages: (1) the specimen is dried, (2) all potential pollutants are removed, and (3) the specimen is atomized.

When you start a series, keep track of what you are doing, and finish the series the same way you started it. Nonparallel sentences are at best awkward and off-key. At worst, they can lead to serious misunderstandings.

PARENTHESES

Parentheses are used to enclose supplementary details inserted into a sentence. Commas and dashes may also be used for this purpose, but with some restrictions. You may enclose a complete sentence or several complete sentences within parentheses. But such an extensive enclosure would confuse the reader if only commas or dashes were used to enclose it.

The violence of these storms can scarcely be exaggerated. (Typhoons and hurricanes generate winds over 75 miles an hour and waves 50 feet high.) The study . . .

Lists

Parentheses are also used to enclose numbers or letters used in listing:

This general analysis consists of sections on (1) wave generation, (2) wave propagation, (3) wave action near a shoreline, and (4) wave energy.

Punctuation of Parentheses in Sentences

Within a sentence, place no mark of punctuation before the opening parenthesis. Place any marks needed in the sentence after the closing parenthesis:

> A runway that is regularly exposed to crosswinds of over 10 knots (11.6 mph) is considered to be unsafe.

Do not use any punctuation around parentheses when they come between sentences. Give the statement *inside* the parentheses any punctuation it needs.

per   PERIOD

Periods have several conventional uses.

End Stop

Place a period at the end of any sentence that is not a question or an exclamation:

> Find maximum average daily temperature and maximum pressure altitude.

Abbreviations

Place a period after certain abbreviations:

M.D.   etc.
Ph.D.   Jr.

See also the entry for Abbreviations.

Decimal Point

Use the period with decimal fractions and as a decimal point between dollars and cents:

0.4 $5.60
0.05% $450.23

p/ag   PRONOUN–ANTECEDENT AGREEMENT

Pronoun–antecedent agreement is closely related to verb–subject agreement. For example, the problem area concerning the use of collective nouns explained under Verb–Subject Agreement is closely related to the proper use of pronouns.

When a collective noun is considered singular, it takes a singular pronoun as well as a singular verb. Also, such antecedents as *each, everyone, either, neither, anybody, somebody, everybody,* and *no one* take singular pronouns as well as singular verbs:

Everyone had his assignment ready.

However, our sensitivity about using male pronouns exclusively when the reference may be to both men and women makes the choice of a suitable pronoun in this construction difficult. Many people object to the use of *his* as the pronoun in the preceding example. Do not choose to solve the problem by introducing a grammatical error, as in this example of incorrect usage:

Everyone had their assignment ready.

The use of male and female pronouns together is grammatically correct, if a bit awkward at times:

Everyone had his or her assignment ready.

Perhaps the best solution, one that is often applicable, is to use a plural antecedent that allows the use of a neutral plural pronoun, as in this example:

All the students had their assignments ready.

The same problem presents itself when we use such nouns as *student* or *human being* in their generic sense; that is, when we use them to stand for all students or all human beings. If used in the singular, such nouns must be followed by singular pronouns:

The student seeking a loan must have his or her application in by 3 September.

Again, the best solution is to use a plural antecedent:

Students seeking loans must have their applications in by 3 September.

See also the entry for Sexist Usage.

PRONOUN FORM

pron

Almost every adult can remember being constantly corrected by parents and elementary school teachers in regard to pronoun form. The common sequence is for the child to say, "Me and Johnny are going swimming," and for the teacher or parent to say patiently, "No, dear, 'Johnny and I are going swimming.'" As a result of this conditioning, many adults automatically regard all objective forms with suspicion, and the most common pronoun error is for the speaker or writer to use a subjective case pronoun such as *I, he,* or *she* when an objective case pronoun such as *me, him,* or *her* is called for.

Whenever a pronoun is the object of a verb or the object of a preposition, it must be in the objective case:

It occurred to my colleagues and me to check the velocity data on the earthquake waves.

Just between you and me, the news shook Mary and him.

However, use a subjective case pronoun in the predicate nominative position. This rule slightly complicates the use of pronouns after the verb. Normally, the pronoun position after the verb is thought of as objective pronoun territory, but when the verb is a linking verb (chiefly the verb *to be*), the pronoun is called a *predicate noun* rather than an object and is in the subjective case.

It is she.
It was he who discovered the mutated fruit fly.

ques QUESTION MARK

Place a question mark at the end of every sentence that asks a direct question:

What is the purpose of this report?

A request that you politely phrase as a question may be followed by either a period or a question mark:

Will you be sure to return the experimental results as soon as possible.
Will you be sure to return the experimental results as soon as possible?

When you have a question mark within quotation marks, you need no other mark of punctuation:

"Where am I?" he asked.

quot QUOTATION MARKS

Use quotation marks to set off short quotations and certain titles.

Short Quotations

Use quotation marks to enclose quotations that are short enough to work into your own text (normally, fewer than three lines):

According to Dr. Stockdale, "Ants, wonderful as they are, have many enemies."

Quotations longer than three lines should be set off by single spacing and indenting. See the entry for Colon for an example of this style. Do not use quotation marks when quotations are set off and indented.

Titles

Place quotation marks around titles of articles from journals and periodicals:

Nihei's article "The Color of the Future" appeared in *PC World*.

Single Quotes

When you must use quotation marks within other quotation marks, use single marks (the apostrophe on your keyboard):

"Do you have the same trouble with the distinction between 'venal' and 'venial' that I do?" asked the copy editor.

Punctuation Conventions

The following are the conventions in the United States for using punctuation with quotation marks:

Commas and Periods   Always place commas and periods inside the quotation marks. There are no exceptions to this rule:

G. D. Brewer wrote "Manned Hypersonic Vehicles."

Semicolons and Colons   Always place semicolons and colons outside the quotation marks. There are no exceptions to this rule:

As Dr. Damron points out, "New technology has made photographs easy to fake"; therefore, they are no longer reliable as courtroom evidence.

Question Marks, Exclamation Points, and Dashes   Place question marks, exclamation points, and dashes inside the quotation marks when they apply *to the quote only or to the quote and the entire sentence at the same time*. Place them outside the quotation marks when they apply to the entire sentence only.

Inside
When are we going to find the answer to the question, "What causes clear air turbulence?"

Outside
Did you read Minna Levine's "Business Statistics"?

RUN-ON SENTENCE

run-on

A run-on sentence is two independent clauses (that is, two complete sentences) put together with only a comma or no punctuation at all between them. Punctuate two independent clauses placed together with a period, a semicolon, or a comma and a coordinating conjunction (*and, but, for, nor,* or *yet*). Infrequently, the colon or dash is used also. (There are some exceptions to these rules. See the entry for Comma.) The following three examples are punctuated correctly, the first with a period, the second with a semicolon, the third with a comma and a coordinating conjunction:

> Check the hydraulic pressure. If it reads below normal, do not turn on the aileron boost.

> We will describe the new technology in greater detail; however, first we will say a few words about the principal devices found in electronic circuits.

> Ground contact with wood is particularly likely to cause decay, but wood buried far below the ground line will not decay because of a lack of sufficient oxygen.

If the example sentences had only commas or no punctuation at all between the independent clauses, they would be run-on sentences.

Writers most frequently write run-on sentences when they mistake conjunctive adverbs for coordinating conjunctions. The most common conjunctive adverbs are *also, anyhow, besides, consequently, furthermore, hence, however, moreover, nevertheless, therefore,* and *too.*

When a conjunctive adverb is used to join two independent clauses, the mark of punctuation most often used is a semicolon (a period is used infrequently), as in this correctly punctuated sentence:

> Ice fish are nearly invisible; however, they do have a few dark spots on their bodies.

Often the sentence will be more effective if it is rewritten completely, making one of the independent clauses a subordinate clause or a phrase.

Run-On Sentence
The students at the university are mostly young Californians, most of them are between the ages of 18 and 24.

Rewritten
The students at the university are mostly young Californians between the ages of 18 and 24.

---

semi   **SEMICOLON**

The semicolon lies between the comma and the period in force. Its use is quite restricted. (See also the entry for Run-On Sentences.)

Independent Clauses

Place a semicolon between two closely connected independent clauses that are not joined by a coordinating conjunction (*and, but, or, nor, for,* or *yet*):

> The expanding gases formed during burning drive the turbine; the gases are then exhausted through the nozzle.

When independent clauses joined by a coordinating conjunction have internal punctuation, then the comma before the coordinating conjunction may be changed to a semicolon:

> The front lawn has been planted with a Chinese Beauty Tree, a Bechtel Flowering Crab, a Mountain Ash, and assorted small shrubbery, including barberry and cameo roses; but so far nothing has been done to the rear beyond clearing and rough grading.

Series

When a series contains commas as internal punctuation within the parts, use semicolons between the parts:

> Included in the experiment were Peter Moody, a freshman; Jesse Gatlin, a sophomore; Burrel Gambel, a junior; and Ralph Leone, a senior.

SEXIST USAGE

sexist

Conventional usages often discriminate against both men and women, but particularly against women. For example, a problem often arises when someone is talking about some group in general but refers to members of the group in the singular, as in the following passage:

> The modern secretary has to be an expert with electronic equipment. She has to be able to use a computer and fix a fax machine. On the other hand, her boss still doodles letters on yellow pads. He has yet to come to grips with all the electronic gadgetry in today's office.

This paragraph makes two groundless assumptions: that all secretaries are female and all executives are male. Neither assumption, of course, is valid.

Similarly, in the past, letters began with "Dear Sir" or "Gentlemen." People who delivered mail were "mailmen" and those who protected our streets were "policemen." History books discussed "man's progress" and described how "man has conquered space."

Now we recognize the unfairness of such discriminatory usages. Most organizations make a real effort to avoid sexist usages in their documents. How can you avoid such usages once you understand the problem?

Titles of various kinds are fairly easy to deal with. *Mailmen* have become *mail carriers; policemen, police officers; chairmen, chairpersons* or simply *chairs;* and so forth. We no longer speak of "man's progress" but of "human progress."

The selection of pronouns when dealing with groups in general sometimes presents more of a problem. One way to deal with it is to move from the singular to the plural. You can speak of *secretaries/they* and *bosses/they*, avoiding the choice of either a male or a female pronoun.

You can also write around the problem. You can convert a sentence like the following one from a sexist to a nonsexist statement by replacing the *he* clause with a verbal phrase such as an infinitive or a participle:

The diver must close the mouthpiece shut-off valve before he runs the test.
The diver must close the mouthpiece shut-off valve before running the test.

If you write instructions in a combination of the second person (you) and the imperative mood, you avoid the problem altogether:

You must close the mouthpiece shut-off valve before you run the test.
Close the mouthpiece shut-off valve before running the test.

At times, using plural forms or second person or writing around the problem simply won't work. In an insurance contract, for example, you might have to refer to the policyholder. It would be unclear to use a plural form because that might indicate two policyholders when only one is intended. When such is the case, writers have little recourse except to use such phrases as *he or she* or *he/she*. Both are a bit awkward, but they have the advantage of being both precise and nonsexist.

You can use the search function in your word processing program to find sexist language in your own work. Search for male and female pronouns and *man* and *men*. When you find them, check to see if you have used them in a sexist or a nonsexist way. If you have used them in a sexist way, correct the problem, but be sure not to introduce inaccuracy or imprecision in doing so.

See also the entry for Pronoun–Antecedent Agreement.

sp SPELLING ERROR

The condition of English spelling is chaotic and likely to remain so. George Bernard Shaw once illustrated this chaos by spelling *fish* as *ghoti*—using the *gh* from *rough*, the *o* from *women*, and the *ti* from *condition*. If you have a spelling checker in your word processing program, it will help you avoid many spelling errors and typographical errors. Do remember, though, that a spelling checker will not catch the wrong word correctly spelled. That is, it won't warn you when you used *to* for *too*. You can obtain help from the spelling section in a collegiate dictionary, where the common rules of spelling are explained. You can also buy rather inexpensive books that explain the various spelling rules and provide exercises to fix the rules in your mind.

To assist you here, we provide a list of common words that sound alike, each used correctly in a sentence.

I **accept** your gift.
Everyone went **except** Jerry.

His attorney gave him good **advice.**
His attorney **advised** him well.

Her cold **affected** her voice.
The **effect** was rather froglike.

He was **already** home by 9 P.M.
When her bag was packed, she was **all ready** to go.

The senators stood **all together** on the issue.
Jim was **altogether** pleased with the result of the test.

He gave him **an** aardvark.
The aardvark **and** the anteater look somewhat alike.

The river **breached** the levee, letting the water through.
He loaded the cannon at the **breech.**

Springfield is the **capital** of Illinois.
Tourists were taking pictures of the **capitol** building.

Always **cite** your sources in a paper.
After the sun rose, we **sighted** the missing children.
She chose land near the river as the **site** for her house.

Burlap is a **coarse** cloth.
She was disappointed, of **course.**

His blue tie **complemented** his gray shirt.
I **complimented** him on his choice of ties.

Most cities have a governing body called a **council.**
The attorney's **counsel** was to remain quiet.

Being quiet, she said, was the **discreet** thing to do.
Each slice in a loaf of bread is **discrete** from the other slices.

"We must move **forward**," the president said.
Many books have **forewords.**

Am I speaking so that you can **hear** me?
He was **here** just a minute ago.

It's obvious why he was here.
The sousaphone and **its** sound are both big and round.

Lead (Pb) has a melting point of 327.5°C.
Joan of Arc **led** the French troops to victory.

Our **principal** goal is to cut the deficit.
Hold to high ethical **principles.**

A thing at rest is **stationary.**
Choose white paper for your **stationery.**

A **straight** line is the shortest distance between two points.
The **Strait** of Gibraltar separates Europe from Africa.

I wonder when **they're** coming.
Are they bringing **their** luggage with them?
Put your luggage **there,** in the corner.

He made a careful, **thorough** inspection.
He worked as **though** his life depended on it.
She **thought** until her head ached.

He **threw** the report on her desk.
His report cut **through** all the red tape.

Laurie moved **to** Trumansburg.
Gary moved to Trumansburg, **too.**
After one comes **two.**

We had two days of hot, sunny **weather.**
Whether he goes or not, I'm going.

Where **were** you on Monday?
The important thing is **we're** here today.
Where are you going tomorrow?

Whose house will you stay at?
Who's coming on the trip with us?

Is that **your** car you're driving?
You're right; it's my car.

vb VERB FORM

Improper verb form includes a wide variety of linguistic errors, ranging from such nonstandard usages as "He seen the show" for "He saw the show" to such esoteric errors as "He was hung by the neck until dead" for "He was hanged by the neck until dead." Normally, spending a few minutes with any collegiate dictionary will show you the correct verb form. College level dictionaries list the principal parts of the verb after the verb entry.

VERB–SUBJECT AGREEMENT

Most of the time, verb–subject agreement presents no difficulty to the writer. For example, to convey the thought "He speaks for us all," only a child or a foreigner learning English might say. "He speak for us all." However, various constructions exist in English that do present agreement problems, even for the adult, educated, native speaker of English. These troublesome constructions are examined in the following sections.

Words That Take Singular Verbs

The following words take singular verbs: *each, everyone, either, neither, anybody, somebody*. Writers rarely have trouble with a sentence such as "No one is going to the game." Problems arise when, as is often the case, a prepositional phrase with a plural object is interposed between the simple subject and the verb, as in this sentence: "Each *of these disposal systems* is a possible contaminant." In this sentence some writers are tempted to let the object of the preposition, *systems*, govern the verb and wrongly write, "Each of these disposal systems *are* a possible contaminant."

Compound Subject Joined by *Or* or *Nor*

When a compound subject is joined by *or* or *nor*, the verb agrees with the closer noun or pronoun:

Either the designer or the builders are in error.
Either the builders or the designer is in error.

In informal and general usage, one might commonly hear, or see, the second sentence as "Either the builders or the designer are in error." In writing you should hold to the more formal usage of the example.

Parenthetical Expressions

Parenthetical expressions introduced by such words as *accompanied by, with, together with,* and *as well as* do not govern the verb:

Mr. Roberts, as well as his two assistants, is working on the experiment.

Two or More Subjects Joined by *And*

Two or more subjects joined by *and* take a plural verb. Inverted word order does not affect this rule:

Close to the academy are Cathedral Rock and the Rampart Range.

Collective Nouns

Collective nouns such as *team*, *group*, *class*, *committee*, and many others take either plural or singular verbs, depending on the meaning of the sentence. The writer must be sure that any subsequent pronouns agree with the subject and verb:

> The team is going to receive its championship trophy tonight.
> The team are going to receive their football letters tonight.

Note: When the team was considered singular in the first example, the subsequent pronoun was *its*. In the second example the pronoun was *their*.

REFERENCES

Department of Energy, EMF in the Workplace. Washington, DC: National Institute for Occupational Health and Safety, 1996.

Levine, Minna. "Business Statistics." *MacUser,* April 1990.

Publication Manual of the American Psychological Association. 4th ed. Washington, DC: APA, 1994.

INDEX

ABOUT THE AUTHORS

Elizabeth Tebeaux is Director of Distance Education and Professor of English at Texas A&M University. She is a past president and fellow of the Association of Teachers of Technical Writing and Chair of the ATTW Fellows Committee. Dr. Tebeaux has published extensively in technical communication pedagogy, style, curriculum, and history. She is author of *Emergence of a Tradition: Technical Writing in the English Renaissance, 1475–1640* (1996) and coauthor of two other communication texts. She has received the NCTE Best Book Award and Best Article Award (teaching) and ATTW's Best Article Award.

Sam Dragga is Professor and Chair of English at Texas Tech University. He is a past president and fellow of the Association of Teachers of Technical Writing and manager of its e-mail discussion list. Dr. Dragga is coauthor of *A Reader's Repertoire: Purpose and Focus* (1996), *A Writer's Repertoire* (1995), and *Editing: The Design of Rhetoric* (1989). In addition, he serves as series editor of the Allyn & Bacon Series in Technical Communication. Dr. Dragga has received NCTE's Best Book Award and ATTW's Best Article Award.

Thomas E. Pearsall is Professor and Head Emeritus of the Department of Rhetoric at the University of Minnesota. He is coauthor of *How to Write for the World of Work*, Sixth Edition (2000). Dr. Pearsall is a past president of the Association of Teachers of Technical Writing and past president and founder of the Council for Programs in Technical and Scientific Writing. He is a fellow of the Society for Technical Communication and of the ATTW and has a Distinguished Service Award from the Council for Programs in Technical and Scientific Communication.

The late **Kenneth W. Houp** was Professor of English at Pennsylvania State University.

☑ Marking Symbols

This list of marking symbols refers you to sections and pages in Appendix A, Handbook, where you can find discussions of style and usage.

ab	Abbreviations	574
acro	Acronyms	577
apos	Apostrophe	578
brackets	Brackets	579
cap	Capitalization	580
colon	Colon	581
c	Comma	582
dm	Dangling Modifier	585
dash	Dash	586
d	Diction	586
ell	Ellipsis	586
exc	Exclamation Point	587
frag	Fragmentary Sentence	587
hyphen	Hyphen	588
ital	Italicization	589
mm	Misplaced Modifier	590
num	Numbers	591
paral	Parallelism	593
paren	Parentheses	593
per	Period	594
p/ag	Pronoun–Antecedent Agreement	594
pron	Pronoun Form	595
ques	Question Mark	596
quot	Quotation Marks	596
run-on	Run-On Sentence	597
semi	Semicolon	598
sexist	Sexist Usage	599
sp	Spelling Error	600
vb	Verb Form	602
v/ag	Verb–Subject Agreement	603